Welding Databook

Welding Databook

Edited by **Howard Currant**

NY RESEARCH PRESS
New York

Published by NY Research Press,
23 West, 55th Street, Suite 816,
New York, NY 10019, USA
www.nyresearchpress.com

Welding Databook
Edited by Howard Currant

International Standard Book Number: 978-1-63238-463-8 (Hardback)

Printed in China.

Contents

	Preface	IX
Section 1	**Laser Welding**	1
Chapter 1	**Hybrid Laser-Arc Welding** J. Zhou and H.L. Tsai	3
Chapter 2	**Laser Beam Welding of Austenitic Stainless Steels - Similar Butt and Dissimilar Lap Joints** Abdel-Monem El-Batahgy	33
Chapter 3	**Dissimilar Metal Joining of Zinc Coated Steel and Aluminum Alloy by Laser Roll Welding** Hitoshi Ozaki and Muneharu Kutsuna	57
Chapter 4	**Mitigating Zinc Vapor Induced Weld Defects in Laser Welding of Galvanized High-Strength Steel by Using Different Supplementary Means** Junjie Ma, Fanrong Kong, Blair Carlson and Radovan Kovacevic	79
Chapter 5	***In situ* Reaction During Pulsed Nd:YAG Laser Welding SiC_p/A356 with Ti as Filler Metal** Kelvii Wei Guo	101
Chapter 6	**Nd:YAG Laser Welding for Photonics Devices Packaging** Ikhwan Naim Md Nawi, Jalil Ali, Mohamed Fadhali and Preecha P. Yupapin	121

Section 2 Numerical Modeling of Welding Processes 139

Chapter 7 **Numerical Modelling to Understand Cracking Phenomena During Laser-GMA Hybrid Welding Nickel-Base Superalloys** 141
Zhiguo Gao

Chapter 8 **Development of a Comprehensive Process Model for Hybrid Laser-Arc Welding** 165
Fanrong Kong and Radovan Kovacevic

Chapter 9 **Numerical Simulation of Residual Stress and Strain Behavior After Temperature Modification** 191
Farag Soul and Nada Hamdy

Chapter 10 **Principles and Thermo-Mechanical Model of Friction Stir Welding** 221
Jauhari Tahir Khairuddin, Jamaluddin Abdullah, Zuhailawati Hussain and Indra Putra Almanar

Chapter 11 **Analytical Model for Estimating the Amount of Heat Generated During Friction Stir Welding: Application on Plates Made of Aluminium Alloy 2024 T351** 247
Miroslav Mijajlović and Dragan Milčić

Section 3 Sensing of Welding Processes 275

Chapter 12 **Visual Analysis of Welding Processes** 277
Yoji Ogawa

Chapter 13 **Monitoring of Arc Welding Process Based on Arc Light Emission** 305
Marek Stanisław Węglowski

Chapter 14 **Real-Time Measurement of Three Dimensional Weld Pool Surface in GTAW** 333
Wei Jie Zhang, Yu Kang Liu and Yu Ming Zhang

Section 4 General Topics in Welding 367

Chapter 15 **The Physics of Weld Bead Defects** 369
P. S. Wei

Chapter 16 **Welding Techniques in Dentistry** 389
Clébio Domingues da Silveira-Júnior, Morgana Guilherme de Castro, Letícia Resende Davi, Flávio Domingues das Neves, Verediana Resende Novais and Paulo Cézar Simamoto-Júnior

Chapter 17 **Optimized Stud Arc Welding Process Control Factors by Taguchi Experimental Design Technique** 413
Riyadh Mohammed Ali Hamza

Permissions

List of Contributors

Preface

The processes of welding are explained thoroughly in this elaborative book. Although there is a broad range of available literature on welding processes, a demand still exists to regularly update the engineering community on advancements in joining techniques of similar and dissimilar materials, in their numerical modelling, as well as in their sensing and control. This book has been compiled by a group of authors from different countries across five continents who joined hands to co-author this book on welding processes, giving detailed insight to the reader. Some of the main topics discussed here are sensing of welding procedures, welding of laser and numerical modelling.

Various studies have approached the subject by analyzing it with a single perspective, but the present book provides diverse methodologies and techniques to address this field. This book contains theories and applications needed for understanding the subject from different perspectives. The aim is to keep the readers informed about the progress in the field; therefore, the contributions were carefully examined to compile novel researches by specialists from across the globe.

Indeed, the job of the editor is the most crucial and challenging in compiling all chapters into a single book. In the end, I would extend my sincere thanks to the chapter authors for their profound work. I am also thankful for the support provided by my family and colleagues during the compilation of this book.

Editor

Section 1

Laser Welding

Chapter 1

Hybrid Laser-Arc Welding

J. Zhou and H.L. Tsai

Additional information is available at the end of the chapter

1. Introduction

Hybrid laser-arc welding has received increasing interest in both academia and industry in last decade[1,2]. As shown in Fig. 1, hybrid laser-arc welding is formed by combining laser beam welding and arc welding. Due to the synergic action of laser beam and welding arc, hybrid welding offers many advantages over laser welding and arc welding alone[3-6], such as high welding speed, deep penetration[7], improved weld quality with reduced susceptibility to pores and cracks[8-16], excellent gap bridging ability[17-22], as well as good process stability and efficiency, as shown in Fig. 2.

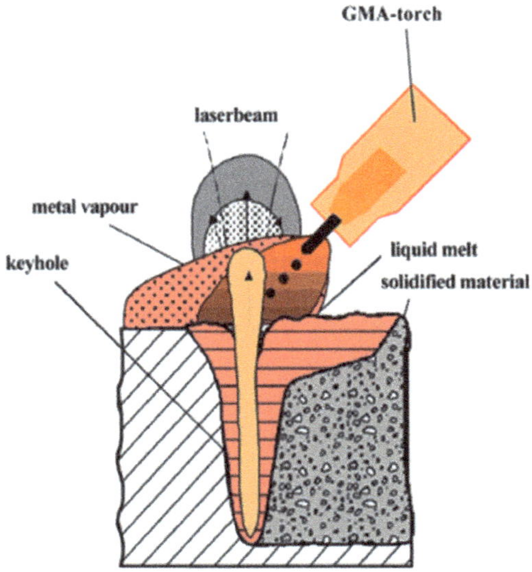

Figure 1. Schematic sketch of a hybrid laser-arc welding process.

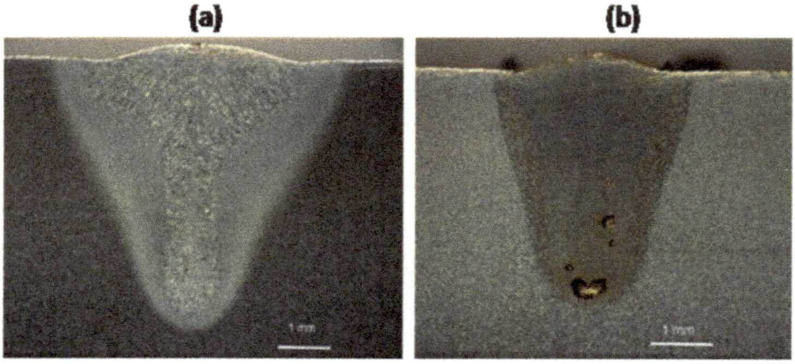

Figure 2. Comparison between (a) a laser weld and (b) a hybrid laser-arc weld in 250 grade mild steel.

Development of the hybrid laser-arc welding technique can be divided into three stages[1]. The concept of hybrid laser welding was first proposed by Steen *et al.*[3, 23, 24] in the late seventies. In their studies, a CO_2 laser was combined with a tungsten inert gas (TIG) arc for welding and cutting applications. Their tests showed clear benefits of combining an arc and a laser beam in the welding process, such as a stabilized arc behavior under the influence of laser radiation; a dramatic increase in the speed of welding of thin metal sheets; and an increase in penetration depth compared with laser welding. Japanese researchers continued Steen's effort and developed various methods and corresponding devices for laser-arc welding, cutting, and surface treatment. However, these efforts did not advance this joining technique into engineering applications particularly because laser welding itself was not yet an economic and viable joining technique at that time[25]. In the second stage of the development of the hybrid laser-arc welding technique, the observed influence of the arc column behavior by laser radiation was used to improve the efficiency of arc welding processes, which leads to the laser-enhanced arc welding technology[1]. A characteristic feature of this technology was that only a low-intensity laser beam was needed, i.e., the required laser power was small compared to the arc power. For TIG welding, Cui and Decker[26-28] demonstrated that a low-energy CO_2 laser beam with a power of merely 100 W could facilitate arc ignition; enhance arc stability; improve weld quality; and increase welding speed due to a reduced arc size and higher arc amperages. However, despite such reported improvements of the arc welding process through laser support, there were neither subsequent extensive investigations of this subject nor known industrial applications of the laser-enhanced arc welding technology. The third stage of hybrid welding technology started in the early 1990s with the development of combined welding processes using a high-power laser beam as the primary and an additional electric arc as the secondary heating source[29-37]. At that time, although the continuous wave CO_2 laser welding process was already well established in industry, it had some known disadvantages, e.g., high requirements of edge preparation and clamping; fast solidification leading to material-dependent pores and cracks; as well as the high investment and operating costs for the laser equipment. Additionally, some welding applications of highly practical interest could not be solved satisfactorily by the laser welding process alone, e.g., joining of tailored blanks in the automotive engineering; welding of heavy plates in shipbuilding industry; as well as high speed welding of crack-susceptible materials. In searching for suitable solutions, the hybrid laser welding was developed into a viable joining technique with significant industrial acceptance during the last decade.

According to the combination of various heating sources used, hybrid welding can be generally categorized as: (1) laser-gas tungsten arc (GTA) welding; (2) laser-gas metal arc (GMA) welding; and (3) laser-plasma welding[25]. Since laser welding offers deep penetration, primary heating sources commonly used in hybrid welding are CO_2, Nd:YAG, and fiber lasers. The first two types of lasers are well established in practice and used for various hybrid welding process developments. While the fiber laser is still in development for industrial applications, it seems to be a future primary heating source for hybrid welding due to its high beam quality. The secondary heating sources used in hybrid welding are mainly electric arcs. Dedicated processes can be divided into GMA welding with consumable electrodes and GTA welding with non-consumable tungsten electrodes. In GMA welding, the arc is burning between a mechanically supplied wire electrode and the workpiece. The shielding gas used in GMA welding was found to have significant effects on arc shape and metal transfer[38,39]. Hence, GMA welding can be subdivided into metal inert-gas (MIG) and metal active-gas (MAG) welding according to the type of shielding gas used. In GTA welding, a chemically inert gas, such as argon or helium, is often used. A special form of this is the plasma arc welding (PAW), which produces a squeezed arc due to a special torch design and results in a more concentrated arc spot.

In hybrid welding, laser and arc are arranged preferably in a way that they can compensate and benefit from each other during the welding process, which implies the creation of a common interaction zone with changed characteristics in comparison to the laser welding and the arc welding alone. In contrast to this is the arrangement in which laser and arc are serving as two separate heating sources during the welding process. Several configurations have been proposed. In a parallel arrangement, there is a distance in either the vertical or horizontal direction along the path between both heating sources. In a serial arrangement, the primary and secondary heating sources are moved along the same welding path with a certain working distance, and the secondary heating source can either lead or follow the primary heating source[1]. The first one enables a preheating of the region to be welded. It can increase the efficiency of the laser welding process because materials to be welded are locally preheated and energy losses through heat conduction are reduced. In comparison, the second one often acts like a short-time post-heat treatment of the weld that can change the weld microstructure favorably. There exists a key difference between parallel and serial process arrangement. In a serial arrangement, additional energy is dissipated within the weld seam region, whereas in the parallel arrangement, the heat flow is reduced only across the weld seam. The option to move the working area temporally enables flexibility in influencing the cooling rates in order to avoid defects.

In the hybrid laser-arc welding process, the workpiece is first heated up and melted due to the laser irradiation. The plasma arc between the consumable electrode and the workpiece continues to heat up and melt the base metal and the droplets generated at the electrode tip periodically detach and impinge onto the workpiece. Then a cavity with large depth-to-width ratio called keyhole was formed in the weld pool under the dynamical interaction of laser irradiation, plasma arc and filler droplets. An externally supplied shielding gas provides the protection of molten metal from exposing to the atmosphere. The successive weld pools create a weld bead and become a part of a welded joint when solidified. The numbers of process parameters are greatly increased in the hybrid welding, mainly including laser beam parameters, electric power parameters, laser-arc interval, electrode diameter, wire feed speed, welding speed and shielding gas. Bagger and Olsen[66] reviewed the fundamental phenomena occurring in laser-arc hybrid welding and the principles for choosing the process parameters. Ribic *et al.*[67] reviewed the recent advances in hybrid welding with emphases on the physical interactions between laser and arc, and the effects of the combined laser-arc heat source on the welding process.

Current understanding of hybrid laser-arc welding is primarily based on experimental observations. Hybrid laser-arc welding is restricted to specific applications, predominantly the joining of thick section plain carbon steels. In order to expand the applications of this joining technique and optimize the processes for its current applications, fundamental understanding of the transport phenomena and the role of each parameter becomes critical. Numerical investigations were often carried out for this purpose. Ribic *et al.*[67] developed a three-dimensional heat transfer and fluid flow model for laser-GTA hybrid welding to understand the temperature field, cooling rates and mixing in the weld pool. Kong and Kovacevic[68] developed a three-dimensional model to simulate the temperature field and thermally induced stress field in the workpiece during the hybrid laser-GTA process. Mathematical models have also been developed to simulate the weld pool formation and flow patterns in hybrid laser-GMA welding by incorporating free surfaces based on the VOF method. Generally, the typical phenomena in GMA welding such as droplets impingement into the weld pool, electromagnetic force in the weld pool and the typical phenomena in laser beaming welding such as keyhole dynamics, inverse Bremsstrahlung absorption and Fresnel absorption were considered in these models. Surface tension, buoyancy, droplet impact force and recoil pressure were considered to calculate the melt flow patterns. In the following, fundamental physics, especially transport phenomena involved in hybrid laser-arc welding will be elaborated.

2. Fundamentals of hybrid laser-arc welding

Since hybrid laser-arc welding involves laser welding, arc welding and their interactions as well, complicated physical processes like metal melting and solidification; melt flow; keyhole plasma formation; arc plasma formation and convection are typically involved, which results in very complex transport phenomena in this welding process[40]. As known, transport phenomena in welding, such as heat transfer; melt flow; and plasma flow, can strongly affect both metallurgical structures and mechanical properties of the weld[41-45]. In the following, transport phenomena in hybrid welding will be discussed and particular attentions are given to 1) arc plasma formation and its effect on metal transfer and weld pool dynamics; 2) laser-induced plasma formation and laser-plasma interaction; 3) recoil pressure and other possible mechanisms contributing to keyhole formation and dynamics; 4) the interplay among various process parameters; and 5) plasma - filler metal - weld pool interactions.

Due to the different natures of heat and mass transfer mechanisms in metal and plasma, different models are developed to study the fundamental physics in hybrid laser-arc welding. One is for the metal region containing base metal, electrode, droplets, and arc plasma. The other is for the keyhole region containing laser-induced plasma. There is a free surface (liquid/vapor interface) separating these two regions. For the metal region, continuum formation is used to calculate the energy and momentum transport[40]. For the keyhole plasma region, laser-plasma interaction and the laser energy absorption mechanism will be discussed. These two regions are coupled together and the VOF technique is used to track the interface between these two regions[40].

2.1. Transport phenomena in metal (electrode, droplets, and workpiece) and arc plasma

Differential equations governing the conservation of mass, momentum and energy based on continuum formulation are given below[46]:

Conservation of mass

$$\frac{\partial}{\partial t}(\rho)+\nabla\cdot(\rho V)=0 \tag{1}$$

where t is the time, ρ is the density, and V is the velocity vector.

Conservation of momentum

$$\begin{aligned}\frac{\partial}{\partial t}(\rho u)+\nabla\cdot(\rho \boldsymbol{V} u)=\nabla\cdot\left(\mu_l\frac{\rho}{\rho_l}\nabla u\right)-\frac{\partial p}{\partial x}-\frac{\mu_l}{K}\frac{\rho}{\rho_l}(u-u_s)-\frac{C\rho^2}{K^{0.5}\rho_l}\left|u-u_s\right|(u-u_s)\\ -\nabla\cdot(\rho f_s f_l \boldsymbol{V_r} u_r)+\nabla\cdot\left(\mu_s u\nabla\left(\frac{\rho}{\rho_l}\right)\right)+\boldsymbol{J}\times\boldsymbol{B}\Big|_x\end{aligned} \tag{2}$$

$$\begin{aligned}\frac{\partial}{\partial t}(\rho v)+\nabla\cdot(\rho \boldsymbol{V} v)=\nabla\cdot\left(\mu_l\frac{\rho}{\rho_l}\nabla v\right)-\frac{\partial p}{\partial y}-\frac{\mu_l}{K}\frac{\rho}{\rho_l}(v-v_s)-\frac{C\rho^2}{K^{0.5}\rho_l}\left|v-v_s\right|(v-v_s)\\ -\nabla\cdot(\rho f_s f_l \boldsymbol{V_r} v_r)+\nabla\cdot\left(\mu_s v\nabla\left(\frac{\rho}{\rho_l}\right)\right)+\boldsymbol{J}\times\boldsymbol{B}\Big|_y\end{aligned} \tag{3}$$

$$\frac{\partial}{\partial t}(\rho w)+\nabla\cdot(\rho \boldsymbol{V} w)=\rho g+\nabla\cdot\left(\mu_l \frac{\rho}{\rho_l}\nabla w\right)-\frac{\partial p}{\partial z}-\frac{u_l}{K}\frac{\rho}{\rho_l}(w-w_s)-\frac{C\rho^2}{K^{0.5}\rho_l}\left|w-w_s\right|(w-w_s)$$
$$-\nabla\cdot(\rho f_s f_l \boldsymbol{V_r} w_r)+\nabla\cdot\left(\mu_s w\nabla\left(\frac{\rho}{\rho_l}\right)\right)+\rho g\beta_T(T-T_0)+F_{drag}+\mathbf{J}\times\mathbf{B}\Big|_z \tag{4}$$

where u, v and w are the velocities in the x-, y- and z-directions, respectively, and $\mathbf{V}_r$ is the relative velocity vector between the liquid phase and the solid phase. **J** is the current field vector and **B** is the magnetic field vector. The subscripts s and l refer to the solid and liquid phases, respectively; Subscript 0 represents the reference conditions; p is the pressure; μ is the viscosity; f is the mass fraction; K, the permeability, is a measure of the ease with which fluid passes through the porous mushy zone; C is the inertial coefficient; β_T is the thermal expansion coefficient; g is the gravitational acceleration; and T is the temperature.

Conservation of energy

$$\frac{\partial}{\partial t}(\rho h)+\nabla\cdot(\rho\mathbf{V}h)=\nabla\cdot\left(\frac{k}{c_p}\nabla h\right)-\nabla\cdot\left(\frac{k}{c_p}\nabla(h_s-h)\right)-$$
$$-\nabla\cdot\left(\rho(\mathbf{V}-\mathbf{V}_s)(h_l-h)\right)+\frac{|\mathbf{J}|}{\sigma_e}-S_R+\frac{5k_b}{2e}\mathbf{J}\cdot\frac{\nabla h}{c_p} \tag{5}$$

where h is the enthalpy, k is the thermal conductivity, and c_p is the specific heat. The first two terms on the right-hand side of Eq. (5) represent the net Fourier diffusion flux. The third term represents the energy flux associated with the relative phase motion. σ_e is the electrical conductivity; S_R is the radiation heat loss; k_b is the Stefan-Boltzmann constant; and e is the electronic charge.

The third and fourth terms on the right-hand side of Eqs. (2)-(4) represent the first and second order drag forces of the flow in the mushy zone. The fifth term represents an interaction between the solid and the liquid phases due to the relative velocity. The second term on the right hand side of Eq. (5) represents the net Fourier diffusion flux. The third term represents the energy flux associated with the relative phase motion. All these aforementioned terms in this paragraph are zero except in the mushy zone. In addition, the solid phase is assumed to be stationary (V_S= 0).

Conservation of species

$$\frac{\partial}{\partial t}(\rho f^\alpha)+\nabla\cdot(\rho\mathbf{V}f^\alpha)=\nabla\cdot\left(\rho D\nabla f^\alpha\right)-\nabla\cdot\left(\rho D\nabla\left(f_l^\alpha-f^\alpha\right)\right)-\nabla\cdot\left(\rho(\mathbf{V}-\mathbf{V}_s)\left(f_l^\alpha-f^\alpha\right)\right) \tag{6}$$

where D is a mass diffusivity and f^α is a mass fraction of constitute. Subscript, l and s, represents liquid and solid phase respectively.

2.2. Transport phenomena in laser induced plasma

The vapor inside the keyhole is modeled as a compressible, inviscid ideal gas. No vapor flow is assumed in the keyhole and the energy equation is given in the following form[47]:

$$\frac{\partial}{\partial t}(\rho_v h_v) = \nabla \cdot \left(\frac{k_v}{c_v} \nabla h_v \right) + \nabla \cdot (-\mathrm{q}_r) + k_{pl} \cdot I_{laser} \cdot (1-\alpha_{iB,1}) + \\ + \sum_{mr=1}^{n} k_{pl} I_{laser} \cdot (1-\alpha_{iB,1}) \cdot (1-\alpha_{Fr}) \cdot (1-\alpha_{iB,mr}) \tag{7}$$

where h_v and ρ_v represent the enthalpy and density of the plasma; k_v and c_v represent the thermal conductivity and specific heat of the plasma. The first term on the right-hand side of Eq. (7) represents the heat conduction term. The second term represents the radiation heat term and q_r stands for the radiation heat flux vector. The fourth term represents energy input from the original laser beam. The last term represents the energy input from multiple reflections of the laser beam inside the keyhole.

2.3. Electrical potential and magnetic field

Arc plasma from GMA welding will not only provide heat to the base metal, but will also exert magnetic force on the weld pool. The electromagnetic force can be calculated as follows[48]:

Conservation of current

$$\nabla \cdot (\sigma_e \nabla \phi) = 0 \tag{8}$$

$$J = -\sigma_e \nabla \phi \tag{9}$$

where ϕ is the electrical potential. According to Ohm's law, the self-induced magnetic field B_θ is calculated by the following Ampere's law:

$$B_\theta = \frac{\mu_0}{r} \int_0^r j_z r dr \tag{10}$$

where $\mu_0 = 4\pi \times 10^{-7}$ H m^{-1} is the magnetic permeability of free space. Finally, three components of the electromagnetic force in Eqs. (2)-(4) are calculated via

$$\mathbf{J} \times \mathbf{B}\big|_x = -B_\theta j_z \frac{x - x_a}{r} \tag{11}$$

$$\mathbf{J} \times \mathbf{B}\big|_y = -B_\theta j_z \frac{y}{r} \tag{12}$$

$$\mathbf{J} \times \mathbf{B}\big|_z = -B_\theta j_r \tag{13}$$

2.4. Arc plasma and its interaction with metal zone (electrode, droplets, and weld pool)

In welding, shielding gas is ionized and forms a plasma arc between the electrode and workpiece. In the arc region, the plasma is assumed to be in local thermodynamic equilibrium (LTE)[49], implying the electron and the heavy particle temperatures are equal. On this basis, the plasma properties, including enthalpy, specific heat, density, viscosity, thermal conductivity and electrical conductivity, are determined from an equilibrium composition calculation[49]. It is noted that the metal vaporized from the metal surface may influence plasma material properties, but this effect is omitted in the present study. It is also assumed that the plasma is optically thin, thus the radiation may be modeled in an approximate manner by defining a radiation heat loss per unit volume[49]. The transport phenomena in the arc plasma and the metal are calculated separately in the corresponding arc domain and metal domain, and the two domains are coupled through interfacial boundary conditions in each time step.

Heat transfer

At the plasma-electrode interface, there exists an anode sheath region[49]. In this region, the mixture of plasma and metal vapor departs from LTE, thus it no longer complies with the model presented above. Since the sheath region is very thin, it is treated as a special interface to take into account the thermal effects on the electrode. The energy balance equation at the surface of the anode is modified to include an additional source term, S_a,[50,51] for the metal region.

$$S_a = \frac{k_{eff}(T_{arc}-T_a)}{\delta} + J_a\phi_w - q_{rad} - q_{evap} \tag{14}$$

The first term on the right-hand side of Eq. (14) is the contribution due to heat conduction from the plasma to the anode. The symbol k_{eff} represents the thermal conductivity taken as the harmonic mean of the thermal conductivities of the arc plasma and the anode material. δ is the length of the anode sheath region. T_{arc} is the arc temperature and T_a is the temperature of the anode. The second term represents the electron heating associated with the work function of the anode material. J_a is the current density at the anode and ϕ_w is the work function of the anode material. The third term q_{rad} is the black body radiation loss from the anode surface. The final term q_{evap} is the heat loss due to the evaporation of electrode materials.

Similar to the anode region, there exists a cathode sheath region between the plasma and the cathode. However, the physics of the cathode sheath and the energy balance at the nonthermionic cathode for GMA welding are not well understood[50-56]. The thermal effect due to the cathode sheath has been omitted in many models and reasonable results were obtained[50-54]. Thus, the energy balance equation at the cathode surface will only have the conduction, radiation, and evaporation terms.

$$S_a = \frac{k_{eff}(T_{arc}-T_c)}{\delta} - q_{rad} - q_{evap} \tag{15}$$

where k_{eff} is the effective thermal conductivity at the arc-cathode surface taken as the harmonic mean of the thermal conductivities of the arc plasma and the cathode material. δ is the length of the cathode sheath. T_c is the cathode surface temperature.

Force balance

The molten part of the metal is subjected to body forces, such as gravity and electromagnetic force. It is also subjected to surface forces, such as surface tension due to surface curvature, Marangoni shear stress due to temperature difference, and arc plasma shear stress and arc pressure at the interface of arc plasma and metal. For cells containing a free surface, surface tension pressure normal to the free surface can be expressed as[57]

$$p_s = \gamma \kappa \tag{16}$$

where γ is the surface tension coefficient and κ is the free surface curvature.

The temperature-dependent Marangoni shear stress at the free surface in a direction tangential to the local free surface is given by[58]

$$\tau_{Ms} = \frac{\partial \gamma}{\partial T}\frac{\partial T}{\partial s} \tag{17}$$

where s is a vector tangential to the local free surface.

The arc plasma shear stress is calculated at the free surface from the velocities of arc plasma cells immediately adjacent the metal cells

$$\tau_{ps} = \mu \frac{\partial V}{\partial T} \tag{18}$$

where μ is the viscosity of arc plasma.

The arc pressure at the metal surface is obtained from the computational result in the arc region. The surface forces are included by adding source terms to the momentum equations according to the CSF (continuum surface force) model[57]. Using F of the VOF function as the characteristic function, surface tension pressure, Marangoni shear stress, arc plasma shear stress, and arc pressure are all transformed to the localized body forces and added to the momentum transport equations as source terms for the boundary cells. Based on these assumptions, Hu *et al.* has successfully simulated the arc and droplet formation and effects of current density and the type of shielding gas on arc formation in a GMA welding process, as shown in Fig. 3.

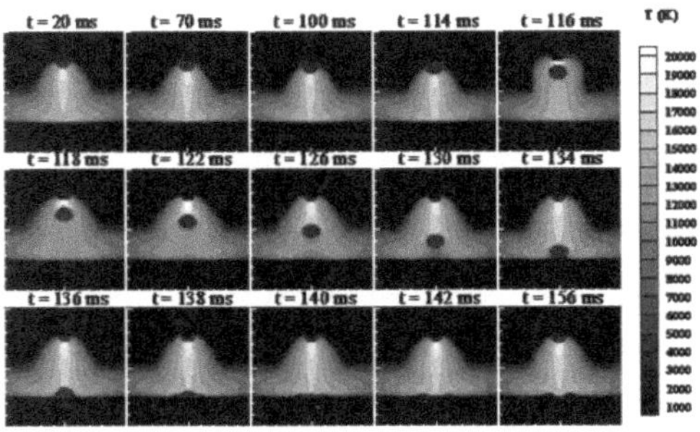

Figure 3. Arc formation in a GMA welding process.

2.5. Laser-induced recoil pressure and keyhole dynamics

In the laser welding process, the laser beam is directed to the metal surface, which first melts the material and produces a small molten pool in the workpiece. The liquid metal is then heated to high temperatures resulting in large evaporation rates. The rapid evaporation creates a large recoil pressure on the surface of the molten layer depressing it downwards. Thus, a cavity with large depth-to-width ration called keyhole is formed. Many investigators believe that the balance between the recoil pressure and surface tension force determines the shape of the keyhole. So, understanding the formation and behavior of the recoil pressure becomes very important for studying the laser welding process. The recoil pressure results from the rapid evaporation of the liquid metal surface. When the liquid metal on the surface is heated to its boiling point, evaporation begins to occur. There is a very thin layer called Knudsen layer adjacent to the liquid surface where the vapor escaping from the liquid surface is in a state of thermodynamic non-equilibrium, i.e., the vapor molecules do not have a Maxwellian velocity distribution. This occurs when the equilibrium vapor pressure (i.e., the saturation pressure) corresponding to the surface temperature is large compared to the ambient partial pressure of the vapor. Under these conditions the vapor adjacent to the surface is dominated by recently evaporated material that has not yet experienced the molecular collisions necessary to establish a Maxiwellian velocity distribution. The Knudsen layer is estimated to be a few molecular mean free paths thick in order to allow for the molecular collisions to occur that bring the molecules into a state of translational equilibrium at the outer edge of the Knudsen layer. The flow field around the Knudsen layer is shown in Fig. 4.

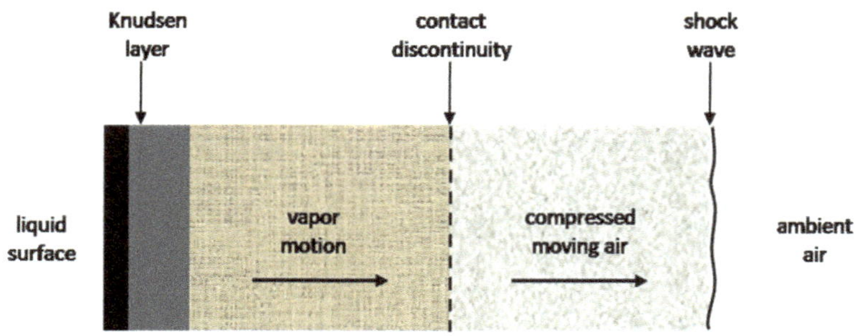

Figure 4. A schematic of the gas dynamic of vapor and air away from a liquid surface at elevated temperature.

Anisimov[59] and Knight[60] did the early investigations on the Knudsen layer. Here a kinetic theory approach[61] is used in the present study. The analysis proceeds by constructing an approximate molecular velocity distribution adjacent to the liquid surface. Equations describing the conservation of mass, momentum and energy across the Knudsen layer are developed in terms of this velocity distribution. This gives Eqs. (19) and (20), as given below, for gas temperature, T_K, and density, ρ_K, outside of the Knudsen layer as functions of the liquid surface temperature and the corresponding saturation density, ρ_{sat}.

$$\frac{T_K}{T_L}=\left[\sqrt{1+\pi\left(\frac{\gamma-1}{\gamma+1}\frac{m}{2}\right)^2}-\sqrt{\pi}\frac{\gamma-1}{\gamma+1}\frac{m}{2}\right]^2 \tag{19}$$

$$\frac{\rho_K}{\rho_{sat}}=\sqrt{\frac{T_L}{T_K}}\left[\left(m^2+\frac{1}{2}\right)e^{m^2}erfc(m)-\frac{m}{\sqrt{\pi}}\right]+\frac{1}{2}\frac{T_L}{T_K}\left[1-\sqrt{\pi}me^{m^2}erfc(m)\right] \tag{20}$$

The quantity, m, is closely related to the Mach number at the outer edge of the Knudsen layer, M_K, and is defined as, $m=u_K/\sqrt{2R_VT_K}=M_K\sqrt{2/\gamma_V}$, where γ_V and R_V are the ratio of specific heats and the gas constant for the vapor, respectively. The value of m depends on the gas dynamics of the vapor flow away from the surface. The gas temperature, pressure and density throughout the vapor region (outside of the Knudsen layer) are uniform. The contact discontinuity, that is, the boundary between vapor and air, is an idealization that results due to the neglect of mass diffusion and heat conduction. The velocity and pressure are equal in these regions, $u_K=u_S$ and $P_K=P_S$, where the subscript, S, denotes properties behind the shock wave. Note that, in general, $T_K\neq T_S$ and $\rho_K\neq\rho_S$.

The thermodynamic state and velocity of the air on each side of the shock wave are related by the Rankine-Hugoniot relations, where the most convenient forms to this application are given by Eqs. (21) and (22). M_K is the Mach number in the vapor, $M_K=u_K/\sqrt{2\gamma_VR_VT_K}$.

$$\frac{P_S}{P_\infty}=1+\gamma M_K\sqrt{\frac{\gamma_VR_VT_K}{\gamma_\infty R_\infty T_\infty}}\left[\frac{\gamma_\infty+1}{4}M_K\sqrt{\frac{\gamma_VR_VT_K}{\gamma_\infty R_\infty T_\infty}}+\sqrt{1+\left(\frac{\gamma_\infty+1}{4}M_K\sqrt{\frac{\gamma_VR_VT_K}{\gamma_\infty R_\infty T_\infty}}\right)^2}\right] \tag{21}$$

$$\frac{T_S}{T_\infty}=\frac{P_S}{P_\infty}\left(1+\frac{\gamma+1}{\gamma-1}\frac{P_S}{P_\infty}\right)/\left(\frac{\gamma+1}{\gamma-1}+\frac{P_S}{P_\infty}\right) \tag{22}$$

The saturation pressure, P_{sat}, is obtained from Eq. (23), where A, B and C are constants which depend on the material. This is used to obtain the saturation density, $\rho_{sat}=P_{sat}/(R_VT_L)$, assuming an ideal gas.

$$\log(P_{sat})=-\frac{A}{T_L}-B\log(T_L)+C \tag{23}$$

Eqs. (20)-(23) are solved as a function of T_L using an iterative solution method. The vapor was assumed to be iron in the form of a monatomic gas with a molecular weight of 56, and $\gamma_V=1.67$. Quantities of particular interest are the recoil pressure, P_r, and rate of energy loss due to evaporation, q_e, and they are given below.

$$P_r = P_K + \rho_K u_K^2, \qquad q_e = H_V \rho_K u_K \tag{24}$$

2.6. Laser-plasma interaction and multiple reflections of laser beam in keyhole

In the keyhole, the laser beam is reflected and absorbed multiple times on the keyhole wall. Each time when the laser beam travels inside the keyhole, it will interact with the keyhole plasma. Multiple reflections of the laser beam and its absorption mechanism are critical in determining the energy transfer in laser welding, which are discussed below.

Inverse Bremsstrahlung (IB) absorption

With the continuous heating of the laser beam, the temperature of the metal vapor inside the keyhole can reach much higher than the metal evaporation temperature, resulting in strong ionization, which produces keyhole plasma. The resulting plasma absorbs laser power by the effect of Inverse Bremsstrahlung (IB) absorption. Eqs. (25) and (26) define the IB absorption fraction of laser beam energy in plasma by considering multiple reflection effects[62]:

$$\alpha_{iB,1} = 1 - \exp\left(-\int_0^{s_0} k_{pl} ds \right) \tag{25}$$

$$\alpha_{iB,mr} = 1 - \exp\left(-\int_0^{s_m} k_{pl} ds \right) \tag{26}$$

here, $\alpha_{iB,1}$ is the absorption fraction in plasma due to the original laser beam; $\alpha_{iB,mr}$ is the absorption fraction due to the reflected laser beam. $\int_0^{s_0} k_{pl} ds$ and $\int_0^{s_m} k_{pl} ds$ are, respectively, the optical thickness of the laser transportation path for the first incident and multiple reflections, and k_{pl} is the plasma absorption coefficient due to inverse Bremsstrahlung absorption[63]:

$$k_{pl} = \frac{n_e n_i Z^2 e^6 2\pi}{6\sqrt{3m}\varepsilon_0^3 c h \omega^3 m_e^2} \left(\frac{m_e}{2\pi k T_e} \right)^{0.5} \left[1 - \exp\left(-\frac{\omega}{kT_e} \right) \right] \bar{g} \tag{27}$$

where Z is the average ionic charge in the plasma, ω is the angular frequency of the laser radiation, ε_0 is the dielectric constant, k is the Boltzmann's constant, n_e and n_i are particle densities of electrons and ions, h is Planck's constant, m_e is the electron mass, T_e is the excitation temperature, c is the speed of light, and $\bar{g}$ is the quantum mechanical Gaunt factor. For the weakly ionized plasma in the keyhole, the Saha equation[63] can be used to calculate the densities of plasma species:

$$\frac{n_e n_i}{n_0} = \frac{g_e g_i}{g_0} \frac{(2\pi m_e k T_e)^{1.5}}{h^3} \exp\left(-\frac{E_i}{kT_e}\right) \tag{28}$$

Fresnel absorption

As discussed before, part of the laser energy will be absorbed by keyhole plasma and part of the laser energy can reach the keyhole wall directly. So, the energy input (q_{laser}) for the keyhole wall consists of two parts: 1) Fresnel absorption of the incident intensity directly from the laser beam ($I_{\alpha,Fr}$) and 2) Fresnel absorption due to multiple reflections of the beam inside the keyhole ($I_{\alpha,mr}$).

$$q_{laser} = I_{\alpha,Fr} + I_{\alpha,mr} \tag{29}$$

$$I_{\alpha,Fr} = I_{laser} \cdot (1-\alpha_{iB,1}) \cdot \alpha_{Fr}(\varphi_1) \tag{30}$$

$$I_{\alpha,mr} = \sum_{mr=1}^{n} I_{laser} \cdot (1-\alpha_{iB,1}) \cdot (1-\alpha_{Fr}) \cdot (1-\alpha_{iB,mr}) \cdot \alpha_{Fr}(\varphi_{mr}) \tag{31}$$

where I_{laser} is the incoming laser intensity. We assume the laser beam has in the simplest case a Gaussiam-like distribution:

$$I_{laser}(x,y,z) = I_0 \left(\frac{r_f}{r_{f0}}\right)^2 \exp\left(-\frac{2r^2}{r_f^2}\right) \tag{32}$$

where r_f is the beam radius and r_{f0} is the beam radius at the focal position; I_0 is the peak intensity. α_{Fr} is the Fresnel absorption coefficient and can be defined it in the following formula[64]:

$$\alpha_{Fr}(\varphi) = 1 - \frac{1}{2}\left(\frac{1+(1-\varepsilon\cos\varphi)^2}{1+(1+\varepsilon\cos\varphi)^2} + \frac{\varepsilon^2 - 2\varepsilon\cos\varphi + 2\cos^2\varphi}{\varepsilon^2 + 2\varepsilon\cos\varphi + 2\cos^2\varphi}\right) \tag{33}$$

where φ is the angle of incident light with the normal of keyhole surface, n is the total number incident light from multiple reflections, $\vec{I}$ is the unit vector along the laser beam radiation direction and $\vec{n}$ is unit vector normal to the free surface. ε is a material-dependent coefficient.

2.7. Radiative heat transfer in laser-induced plasma

When an intense laser beam interacts with metal vapor, a significant amount of the laser radiation is absorbed by the ionized particles. The radiation absorption and emission by the vapor plume may strongly couple with the plume hydrodynamics. This coupling, shown on the right-hand side of Eq. (7), will affect the plasma laser light absorption and radiation cooling terms. The radiation source term $\nabla \cdot (-\mathbf{q_r})$ is defined via

$$\nabla \cdot \mathbf{q_r} = k_a (4\pi I_b - \int_{4\pi} I d\Omega) \tag{34}$$

where k_a, I_b and Ω denote the Planck mean absorption coefficient, blackbody emission intensity and solid angle respectively. For the laser-induced plasma inside the keyhole, the scattering effect is not significant compared with the absorbing and emitting effect. So it will not lead to large errors to assume the plasma is an absorbing-emitting medium. The radiation transport equation (RTE) has to be solved for the total directional radiative intensity I[65]:

$$(\mathbf{s} \cdot \nabla) I(\mathbf{r}, \mathbf{s}) = k_a (I_b - I(\mathbf{r}, \mathbf{s})) \tag{35}$$

where **s** and **r** denote a unit vector along the direction of the radiation intensity and the local position vector. The Planck mean absorption coefficient is defined in the following[65]:

$$k_a = \left(\frac{128}{27} k \right)^{0.5} \left(\frac{\pi}{m_e} \right)^{1.5} \frac{Z^2 e^6 \bar{g}}{h \sigma c^3} \frac{n_e n_i}{T_v^{3.5}} \tag{36}$$

where n_i and n_e represent the particle density of ions and electrons, T_v is the temperature of the plasma, Z stands for the charge of ions, e is the proton charge and m_e is the mass of electrons.

2.8. Tracking of free surfaces

The algorithm of volume-of-fluid (VOF) is used to track the moving free surface[48]. The fluid configuration is defined by a volume of fluid function, $F(x,y,z,t)$, which is used to track the location of the free surface. This function represents the volume of fluid per unit volume and satisfies the following conservation equation:

$$\frac{dF}{dt} = \frac{\partial F}{\partial t} + (\mathbf{V} \cdot \nabla) F = 0 \tag{37}$$

When averaged over the cells of a computing mesh, the average value of F in a cell is equal to the fractional volume of the cell occupied by the fluid. A unit value of F means a cell full of fluid and a zero value indicates a cell containing no fluid. Cells with F values between zero and one are partially filled with fluid and identified as surface cells.

3. Results and discussions

Based on the aforementioned scientific principles governing the hybrid laser-arc welding process, Zhou *et al.*[40,69,70] have successfully developed mathematical models to simulate the transport phenomena like heat and mass transfer, melt flow; energy transport in keyhole plasma, etc. in both pulsed and three-dimensional moving hybrid laser-MIG welding. Detailed discussions are given in the following sections.

3.1. Two-dimensional hybrid laser-MIG welding

In this study, the base metal is assumed to be a stainless steel 304 containing 100 ppm of sulfur. The laser energy is assumed to be in the Gaussian distribution and the divergence of the laser beam is negligible because the focus length of the laser beam is less than 3mm. The laser power and beam radius at the focus are 1800 W and 0.2 mm respectively. The laser power is turned on at t = 0 s and shut down at t = 15 ms. To simulate the MIG process, droplet is assumed to be spherical and is generated in a steady manner. The diameter of filler droplet is assumed to be 0.35 mm, its initial speed is 0.5 m/s right above the weld surface, its initial temperature is 2400 K, and its generation frequency is 1000 Hz. The droplet is assumed to be fed into the keyhole from the top. Droplet generation and formation are actually related to wire size and wire feed speed. Further information can be found in Ref. [13].

Fig. 5 shows the comparison of the cross-sectional view of a hybrid laser-MIG weld and a laser weld. As shown, there is a "pore" in the laser weld, which is due to the rapid solidification in laser welding. Detailed discussion on the formation of porosity in the weld can be found in Ref. [14]. It is also noticed that there are some "undercuts" near the top edge of the laser weld which is one of the major disadvantages of laser welding. In hybrid laser-MIG welding, the reason why there is no pore found in the final weld was believed to be mainly due to the addition of filler metal in the process. The momentum and energy carried by the filler droplets greatly impact the fluid flow and heat transfer in the weld pool and the shape of the solidified weld pool as well. The overall effect depends on the droplet size, droplet generation frequency and droplet generation duration as well. With an optimal operation window, a weld with desired shape and quality can be achieved in hybrid laser-MIG welding. In addition, it is found that the additional heat input from the arc in hybrid laser welding is transferred to the weld pool mainly in the region near the top of the weld, which makes the top portion of the weld wider than that in laser welding. It is further found the undercuts frequently observed in laser welding are eliminated and the shape of the final weld can be modified by the extra filler metal coming from the MIG process. However, the penetration depth in hybrid welding is noticed to be almost the same as that in laser welding, which means the penetration depth in hybrid laser-arc welding mainly depends on the laser power used, but not the arc power.

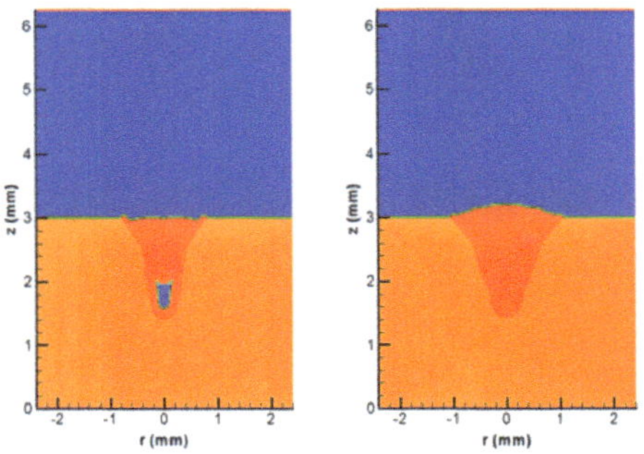

Figure 5. Comparison of weld bead shape between laser welding and hybrid laser-MIG welding.

3.2. Interaction between filler droplets and weld pool

Fig. 6 shows typical interactions between droplets and weld pool in hybrid laser-arc welding. The corresponding distributions of temperature, sulfur concentration, and melt flow velocity are given in Figs. 7, 8 and 9, respectively. Since only the interaction between filler droplets and weld pool is concerned in this discussion, the keyhole formation process is ignored which can be found in Ref. [14]. As shown in Fig. 6, after the laser is shut off at t = 15.0 ms, the laser-induced recoil pressure decreases quickly. Under the action of surface tension and hydrostatic pressure, the molten metal near the keyhole shoulder has tendency to "fill up" the keyhole. At about t = 17.5 ms, the first droplet impinges onto the liquid metal at the bottom of the keyhole. The downward momentum carried by the droplet causes the droplet liquid to flow downward and outward along the keyhole wall, which can be seen clearly by the sulfur composition shown in Fig. 8. Under the action of hydrostatic force and surface tension, the liquid along the keyhole wall has a tendency to flow downward along the keyhole wall. So the upward flow caused by the filler droplet impingement will be weakened. So when the subsequent droplets falls into the keyhole, the liquid level in the center of the keyhole rises, as shown in Fig. 6 at t = 21.5 ms. For the first several droplets, the filler metal mainly diffuses along the longitude direction. Only the first droplet can spread out along the solid-liquid interface driven by the downward momentum. However, as more and more droplets impinge into the weld pool, a vortex is created, which helps the filler metal to diffuse outwards in the latitude direction, as shown in Fig. 8.

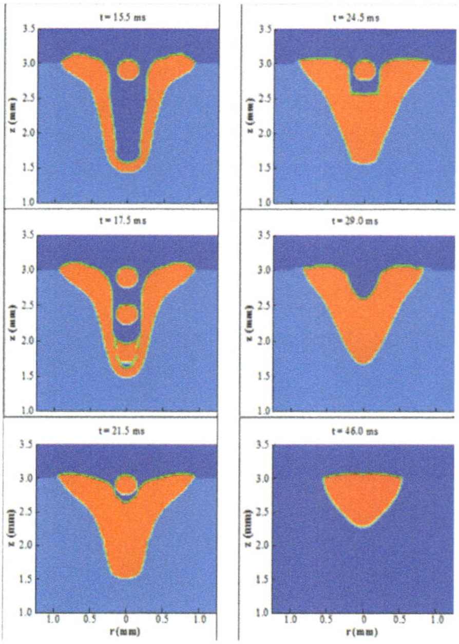

Figure 6. Droplet and weld pool interaction in hybrid laser welding.

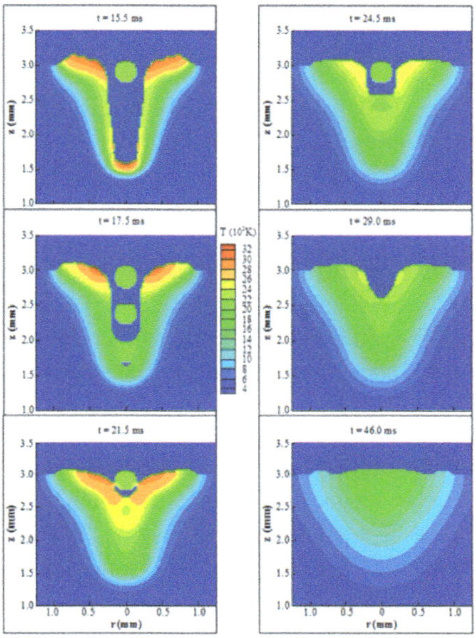

Figure 7. The corresponding temperature distributions as shown in Fig. 6.

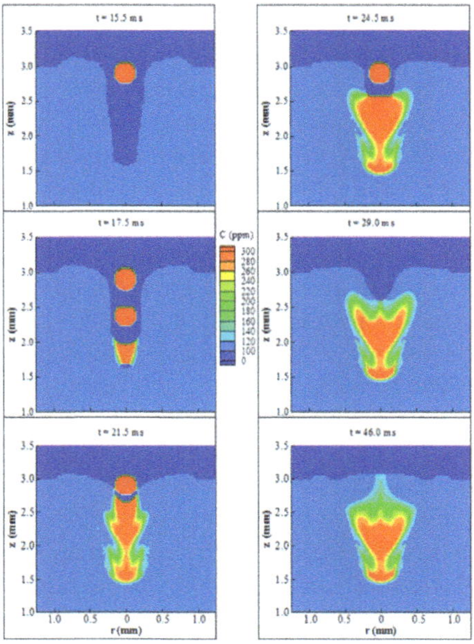

Figure 8. The corresponding sulfur concentration distributions as shown in Fig. 6

As shown in Fig. 9, there is an anticlockwise vortex in the middle waist of the keyhole. As mentioned before, the liquid on the shoulder of the keyhole has a tendency to fill back along the keyhole wall. When the droplets impinge into the keyhole, the outer liquid layer along the keyhole wall has the same flow direction as the filler metal. So the flow direction of the liquid metal here remains downward. Since the liquid is incompressible, the downward flow will push up some amount of liquid upward. The kinetic energy of the fluid flow in the center will be transferred into the potential and kinetic energy of the outward flow. So the downward momentum becomes smaller and smaller and finally it changes its direction. As shown in Fig. 9 at t = 21.5 ms, the flow direction changes from downward to outward at the bottom of the vortex and then bounds upward on the solid keyhole wall. During the upward flow process along the solid-liquid interface of the keyhole wall, the kinetic energy is transferred into the potential energy and the velocity becomes smaller and smaller. Finally the flow direction is changed to be inward by the back-filling momentum from the liquid on the shoulder of the keyhole. As droplets continue to drip into the keyhole, more and more downward momentum is added into the center of the keyhole, the vortex affected zone is enlarged and the strength of the vortex is enhanced, which helps the filler metal to distribute outward along with the vortex flow, as shown in Fig. 8. At t = 24.5 ms, the diffusion zone of filler metal is much larger compared with that at t = 21.5 ms. Since the latitude diffusion of filler metal has a close relationship with the vortex, the evolution of the vortex can be deduced from the shape of the diffusion zone of the filler metal in the final fusion zone. Moreover, at t = 24.5 ms, the downward velocity of the liquid in the center is quite large, the mass from droplets is not enough to compensate the downward mass flow in the center of the keyhole, which leaves the liquid surface decrease here.

After t = 25.0 ms, no droplet will be added into the keyhole. The fluid near the center of the keyhole is bounced back under the action of hydrostatic force and surface tension force. As shown at t = 29.0 ms in Fig. 9, the liquid in the keyhole starts to flow inward and downward, which causes the size of the keyhole to become smaller and smaller. Finally the keyhole will be filled, as shown at t = 49.0 ms in Fig. 6. During the backfill process, the vortex becomes weaker and weaker. So the diffusion of filler metal is not improved much in the latitude direction, which can be found by comparing those figures at t = 29.0 ms and at t = 46.0 ms in Fig. 5.9. Moreover, from the distribution of filler metal at t = 46.0 ms as shown in Fig. 8, it can be concluded that during the backfill process, majority of the filling metal comes from the upper shoulder of the keyhole because only a little of the filler metal is located near the center of the keyhole, which is brought here by the bouncing flow. As shown in Fig. 7, the filler droplet also brings some heat into the weld pool, which will delay the solidification process. Since the diffusion of filler in the fusion zone is greatly limited by the solidification, the delayed solidification will give more time for the filler to diffuse. After the termination of droplets, the heat input carried by droplets also decreases. Due to heat loss to the base metal through conduction and to the surroundings through radiation and convection, the size of the molten pool becomes smaller and smaller as a result of solidification. At t = 46.0 ms, the melt flow in the weld pool is almost diminished and the temperature distribution is more uniform than before, as shown in Figs. 9 and 7, respectively. The shape and composition of the weld will not change much comparing with the completely solidified one.

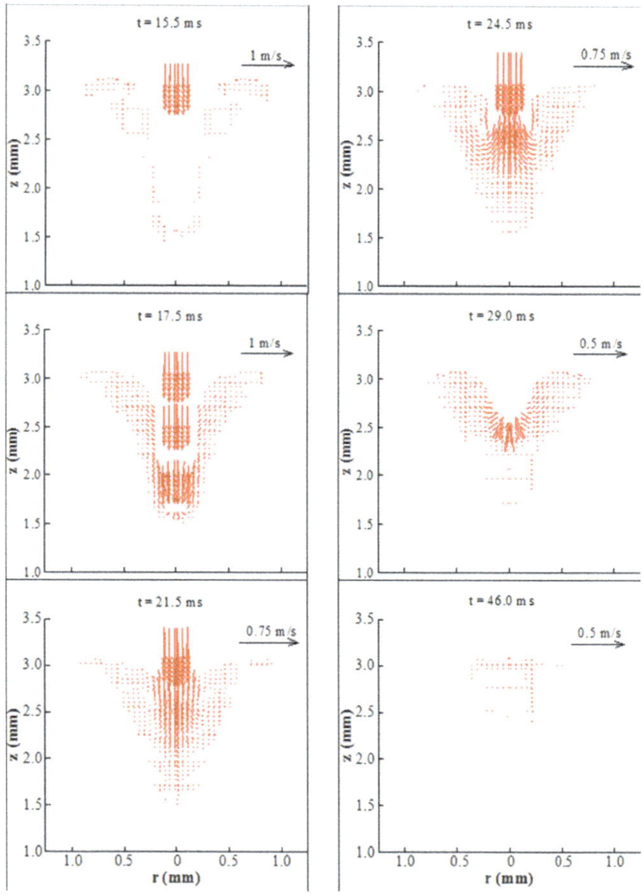

Figure 9. The corresponding velocity distributions as shown in Fig. 6.

3.3. Modification of composition by adding filler metal

Since crack sensitivity of the weld is believed to be strongly related with the composition of the weld pool, adding filler metal with anti-crack elements into the weld pool in hybrid laser welding can thus improve the weld bead quality. However, the effect depends greatly on the diffusion process in the weld pool. In the following, the effects of factors such as droplet size, droplet generation frequency, impingement velocity of the droplet and its lasting duration on the diffusion process are discussed by changing the condition of one specific parameter, while keeping the rest of the parameters unchanged. If not specially mentioned, the welding condition is defined as follows: the droplet diameter is 0.35 mm, its initial velocity is 0.5 m/s, the generation frequency is 1000 HZ and the duration of droplet feeding is 10.0 ms which starts at t = 15.0 ms and ends at t = 25.0 ms.

3.3.1. *Effect of droplet size on the diffusion process in hybrid laser welding*

Three studies are carried out with a droplet size of 0.3 mm, 0.35 mm and 0.4 mm respectively. As shown in Fig. 10, with the increase of droplet size, the latitude diffusion of filler metal is enlarged. From the previous discussions on the diffusion process, the latitude diffusion of the filler metal is found to be closely related to the vortex in the weld pool. The strength and the affected zone of the vortex depend on the downward momentum carried by droplets, which is the product of droplet mass and velocity. As the droplet size increases, the downward momentum increases, which will lead to a stronger vortex. So the diffusion zone is enlarged outward, especially in the middle depth of the keyhole where the vortex is located. This is clear shown by comparing those figures for d = 0.30 mm and d = 0.35 mm in Fig. 10. Meanwhile, larger downward momentum from larger droplet also leads to a strong bouncing flow near the center of keyhole after termination of droplet feeding, which helps filler metal to diffuse into the upper layer in the final weld, as shown in Fig. 10 for d = 0.40 mm. Moreover, larger droplet size brings more filler metal into the keyhole. The heat input carried by droplets also increases, which helps delay the solidification of the fusion zone. Thus, the filler metal has more time to diffuse into the weld pool before its solidification. More filler metal also helps to increase the filler concentration in the final weld. However, larger droplets also lead to some negative effects on the diffusion of filler metal near the center of the weld zone. After the termination of droplet feeding, the melt surface near the center of the keyhole will continue to go down due to the larger hydrodynamic pressure caused by the downward momentum. This will lead to a deep hole there. During the backfill process of this hole, some metal from the upper part of the keyhole may flow into the bottom part of this hole. Since the concentration of filler metal in the upper part of keyhole is very low, it leaves a low diffusion zone of filler metal in the center of the final weld, as shown in Fig. 10 for d = 0.40 mm.

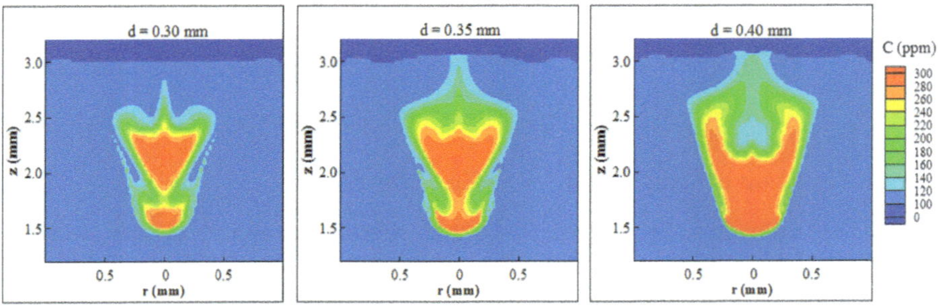

Figure 10. Effect of droplet size on diffusion process in hybrid laser welding.

3.4. Effect of droplet generation frequency on diffusion of filler metal in fusion zone

In the hybrid laser welding process, the droplet is generated at a specific frequency that is controlled by the wire feed rate. The effect of droplet generation frequency on diffusion of filler metal in the fusion zone is shown in Fig. 11. In the study, the droplet generation frequency is 500 HZ, 667 HZ and 1000 HZ, which corresponds to the generation of one droplet every 2.0 ms, 1.5 ms and 1.0 ms. As shown, the diffusion of filler metal in the weld pool is improved with the increase of generation frequency. This can be interpreted through the above analysis on the interaction between the droplets and weld pool. As mentioned before, the latitude diffusion of filler is mainly through the vortex flow induced by the impingement of droplets. With the increase of generation frequency, more droplets fall into the weld pool per unit time, which results in higher downward momentum per unit time. So the vortex in the weld pool is enhanced, which helps the filler metal diffuse in the latitude direction. Meanwhile, the total amount of filler metal added into the weld pool also increases with higher generation frequency, which also helps increase the concentration of filler metal in the final weld and increase the diffusion time, as mentioned before. Furthermore, the longitude distribution of filler metal is found to be improved with higher generation frequency. As shown, in the case f = 500 HZ, there exists a low filler metal concentration zone in the lower part of the keyhole due to the weak strength of the vortex in the weld pool and a long delay time between the droplet generation. When frequency increases to 1000 HZ, the size of this zone is greatly reduced.

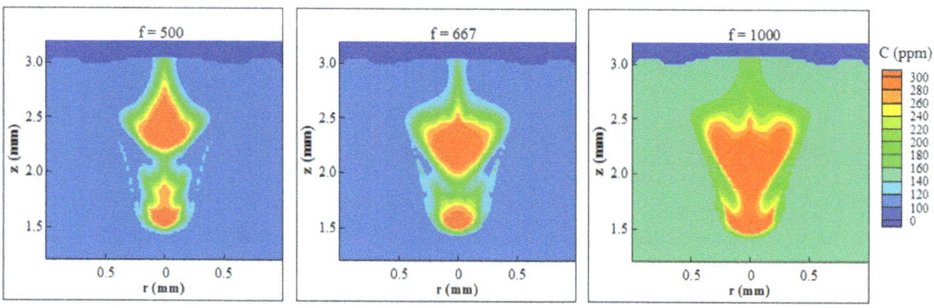

Figure 11. Effect of droplet generation frequency on diffusion process in hybrid laser welding.

3.5. Effect of droplet generation duration on diffusion of filler metal in fusion zone

In hybrid laser welding, the termination of droplet generation can be achieved through the control of removal of the filler wire. The effect of controlling the droplet generation duration on metal diffusion in the weld pool is investigated. As shown in Fig. 12, three cases are carried out with droplet generation duration at 5.0 ms, 10.0 ms and 15.0 ms, respectively. For short duration of 5.0 ms, the vortex induced by the downward momentum of the droplet is not completely developed because of lower downward momentum, which leads to poor latitude distribution of filler metal. In this case, most of the filler metal is located in the lower part of the keyhole. During the backfill process, the bounced flow is not strong enough to push the filler metal upward to the upper part of the keyhole. The keyhole is filled with the base metal liquid where no filler metal exists. So the longitude filler diffusion is also poor with a short duration of droplet generation.

With the increase of the duration length to 10.0 ms, more filler metal will fall into the keyhole. The vortex in the weld pool is enhanced with the increased downward momentum which improves the latitude diffusion. Meanwhile, the droplets are distributed along the entire depth of the keyhole, which leads to better longitude distribution of filler metal. Moreover, the total amount of filler metal also increases with the increase of duration, which also helps the diffusion of filler metal in the fusion zone, as mentioned before. So both the longitude and latitude diffusion of filler metal are improved, as shown. However, with a further longer droplet generation to 15.0 ms, the downward momentum is accumulated due to the continuous impingement of the droplets into the weld pool, which leads to a deep hole in the weld pool. During the backfill process of this hole, the filler metal is mainly located in the bottom, which cannot bounce back in time before the base metal fluid from the upper shoulder arrives at the bottom of this hole, which leaves a low diffusion zone of filler metal in the center of final weld, as shown in Fig. 12.

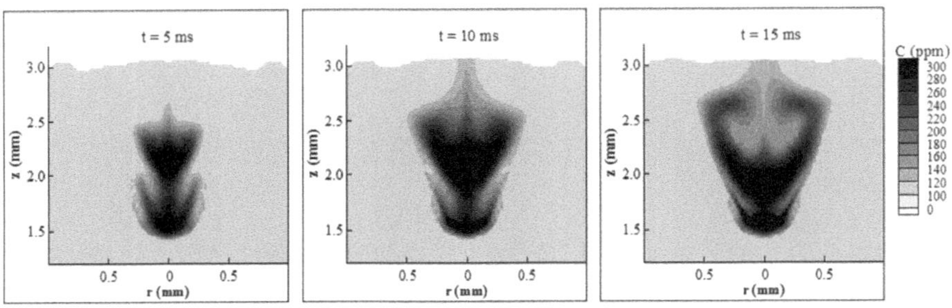

Figure 12. Effect of droplet impingement duration on diffusion process in hybrid laser welding.

3.6. Three-dimensional hybrid laser-MIG welding

Fig. 13 shows a schematic sketch of a three-dimensional hybrid laser-MIG welding. In this study, the laser power is 2.0 kW and the laser beam radius is 0.2 mm and the focal plane is on the top surface of the base metal. The laser beam is started at x = 3.75 mm. The laser beam begins to move after being held for 20.0 ms for the keyhole to reach a certain depth. The welding speed is 2.5 cm/s and the arc power is 1 kW. Droplet begins to fall onto the base metal at t = 20.0 ms and the radius of the droplet is 0.25 mm. The droplet feeding frequency is 86 Hz and its initial speed is 30 cm/s. The distance between arc center and laser beam center is 1 mm. Fig. 14 is the side-view (at Y = 0) of the hybrid laser welding process showing a sequence of a droplet impinging onto the weld pool at different times. Fig. 15 shows the corresponding sulfur concentration distribution during the hybrid welding process, indicating the mixing process in the welding. Fig. 16 shows the corresponding velocity distributions in the weld pool. As shown in Fig. 15, the filler droplet did not mix well with the base metal in this case. Most of the droplet is just stacking on the top of the weld coupon and only small amount of the filler metal is diffused into the base metal near the solid-liquid interface. The poor mixing may have been caused by the relative long distance between the laser beam and MIG arc center. The filler droplet is impinging into the weld pool where only a small amount of liquid metal exists. Since the temperature of this part of liquid metal is low, due to the quick solidification process there, the liquid metal there solidifies very quickly. The droplet flowing downward does not have enough time to flow around and exchange the momentum and mix with the base metal before it solidifies, as shown in Fig. 16. Therefore, most of the filler metals are just stacking on top surface of the base metal.

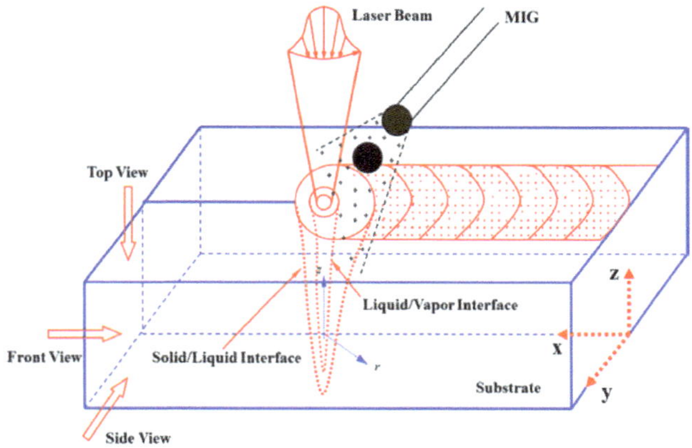

Figure 13. Schematic sketch of 3-D hybrid laser keyhole welding.

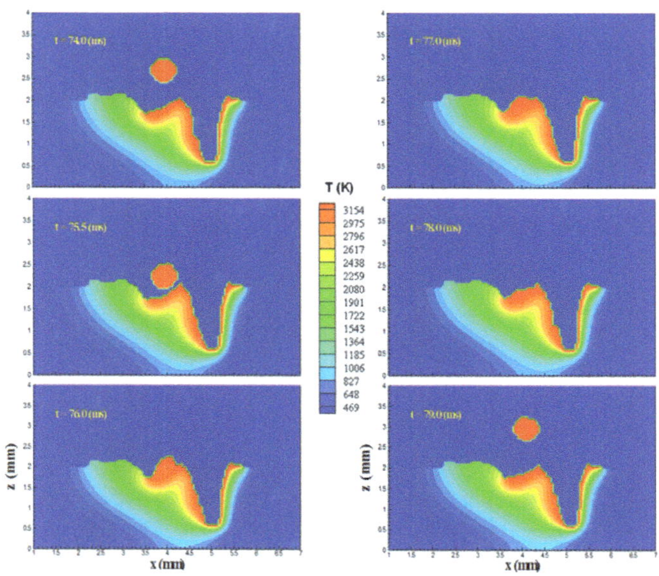

Figure 14. A typical sequence showing the impinging process and temperature distributions in 3-D moving hybrid laser keyhole welding.

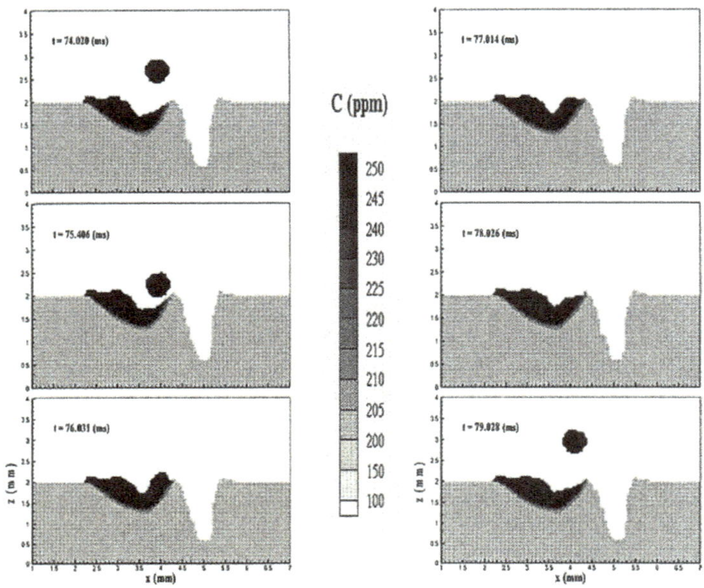

Figure 15. The corresponding sulfur concentration distributions as shown in Fig. 14.

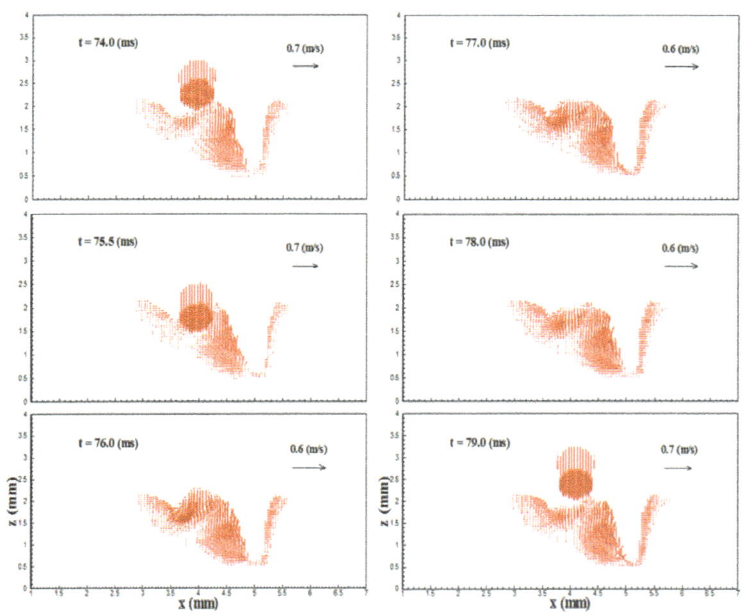

Figure 16. The corresponding velocity distributions as shown in Fig. 14.

There are a lot of process parameters which can affect the mixing of filler droplets into the weld pool in three-dimensional hybrid laser-arc welding. These include laser-arc distance, laser and arc powers, welding speed, wire feed speed and filler droplet size, etc. In the following study, the effect of laser-arc distance on diffusion is conducted by decreasing the laser-arc distance to 0.6 mm. Fig. 17 shows the mixing process during the welding. As shown, the droplet is now mixing with the base metal much better than in the previous case. In this case, since the laser-arc distance is decreased, the filler droplet can impinge into a region in the weld pool where there exists a lot of liquid metal with strong velocity and high temperature. This strong velocity liquid metal flow will interact with the impinging droplets, creating a strong momentum exchange between the droplets and weld pool, which can force the droplet to flow in all directions. Hence, a better mixing can be achieved. Also, in this case, there are more hot liquid metals in the droplet-weld pool interaction zone, thus creating relatively longer time for the droplet to mix and diffuse into the base metal. Hence, a better mixing of droplets into the weld pool is achieved.

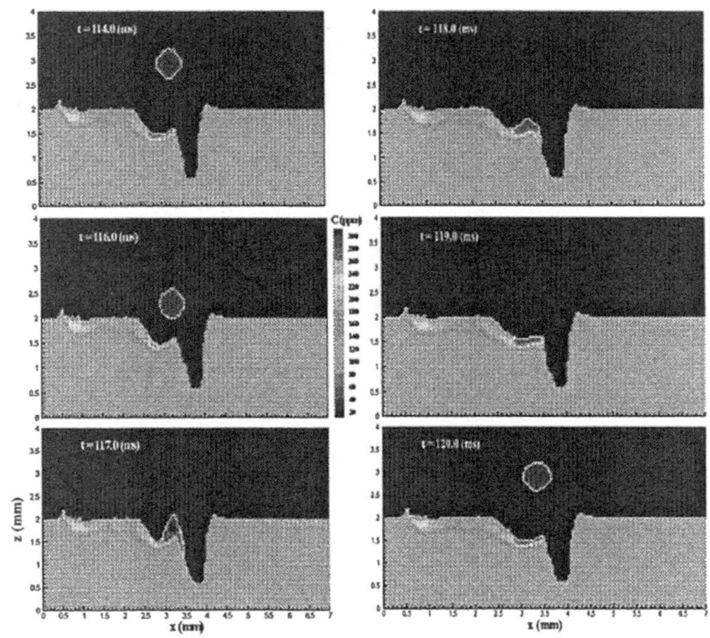

Figure 17. Diffusion process in 3-D hybrid laser keyhole welding with shorter laser-arc distance.

4. Future trends

Although hybrid laser-arc welding has been under investigation and development and gaining increasing acceptance in recent years, good understanding of the underlying physics remains a challenge. For example, the interaction between the laser and the arc has been observed to enhance arc stability and push the arc towards the laser keyhole, resulting in a deeper penetration. However, the origin of this synergistic interaction between the arc and laser plasma is not well understood. Measuring the distributions of electron temperatures and densities in the plasma can provide a better understanding of the laser-arc interaction[9]. Porosity formation is believed to be strongly related to the keyhole collapse process. Hence, better understanding of keyhole stability and dynamics through experimental and theoretical studies would be beneficial. Hybrid welding is known to produce welds with desirable widths and depths, but the maximum gap tolerance and weld penetration for various welding conditions have not been quantified. In the future, advanced mathematical modeling of the heat transfer and fluid flow will enable accurate predictions of weld profile and cooling rates in the welding process, which is critical in understanding the evolution of weld microstructures and residual stress formation in welds. Thus, the hybrid welding process can be optimized to obtain quality welds with no cracking, no brittle phase and less thermal distortion. Better sensing and process control of the hybrid welding process would also be helpful in expanding its applications[67].

Author details

J. Zhou
Department of Mechanical Engineering, Pennsylvania State University, The Behrend College, USA

H.L. Tsai
Department of Mechanical and Aerospace Engineering,
Missouri University of Science and Technology, USA

5. References

[1] Tusek, J., and Suban, M., 'Hybrid welding with arc and laser beam', *Science and Technology of Welding and Joining*, 4(5), 308-311, 1999.

[2] Mahrle, A., and Beyer, E., 'Hybrid laser beam welding – Classification, characteristics, and applications', *Journal of Laser Applications*, 18(3), 169-180, 2006.

[3] Steen, W.M., 'Arc-augmented laser processing of materials' *J. Appl. Phys.*, 51, 5636–5641, 1980.

[4] Campana, G., Fortunato, A., Ascari, A., Tani, G., and Tomesani, L., 'The influence of arc transfer mode in hybrid laser-MIG welding,' *J. Mater. Process. Technol.*, 191, 111–113, 2007.

[5] Casalino, G., 'Statistical analysis of MIG-laser CO2 hybrid welding of Al–Mg alloy', *J. Mater. Process. Technol.*, 191, 106–110, 2007.

[6] Hu, B., and Richardson, I.M., 'Microstructure and mechanical properties of AA7075(T6) hybrid laser/GMA welds', *Material Science Engineering A*, 459, 94-100, 2007.

[7] Feng, Z.L., *Processes and mechanisms of welding residual stress and distortion*, Woodhead Publishing Ltd, Cambridge, England, 2005.

[8] Metzbower, E.A., Denney, P.E., Moon, D.W., Feng, C.R., and Lambrakos, S.G., 'Thermal analysis and microharness maping in hybrid laser welds in a structural steel', *Materials Science Forum*, 426-432, 4147-4152, 2003.

[9] Ribic, B., Palmer, T.A., and Debroy, T., 'Problems and issues in laser-arc hybrid welding', *International Materials Reviews*, 54(4), 223-244, 2009.
[10] G. Song, L. M. Liu and P. C. Wang: *Material Science Engineering A*, 429, 312–319, 2006.
[11] Nonn, A., Dahl, W., and Bleck, W., 'Numerical modelling of damage behaviour of laser-hybrid welds', *Engineering Fracture Mechanics*, 75, 3251–3263, 2008.
[12] Pinto, H., Pyzalla, A., Hackl, H., and Bruckner, J., 'A Comparative Study of Microstructure and Residual Stresses of CMT-, MIG- and Laser-Hybrid Welds', *Materials Science Forum*, 524–525, 627–632, 2006.
[13] Zhang, J. X., Xue, Y., and Gong, S. L., 'Residual welding stresses in laser beam and tungsten inert gas weldments of titanium alloy', *Science and Technology of Welding & Joining*, 10, 643–646, 2005.
[14] Liu, L.M., Song, G., and Wang, J.F.,'', *Materials Science Forum*, 488–489, 361–364, 2005.
[15] Liu, L.M., Song, G., Liang, G.L., and Wang, J.F, 'Pore formation during hybrid laser tungsten inert gas arc welding of magnesium alloy AZ31B – mechanism and remedy', Material *Science and Engineering A*, 390, 76-80, 2005.
[16] Katayama, S., Uchiumi, S., Mizutani, M., Wang, J., and Fujii, K., 'Penetration and porosity prevention mechanism in YAG laser-MIG hybrid welding', *Welding International*, 21, 25–31, 2007.
[17] Fellman, A., and Kujanpaa, V., 'The effect of shielding gas composition on welding performance and weld properties in hybrid CO2 laser-gas metal arc welding of carbon manganese steel', *Journal of Laser Applications*, 18, 12–20, 2006.
[18] Kim, T., Suga, Y., and Koike, T., 'Welding of thin steel plates by hybrid welding process combined TIG arc with YAG laser', *JSME International Journal Series A*, 46A, 202–207, 2003.
[19] Kim, Y. P., Alam, N., and Bang, H. S., 'Observation of hybrid (cw Nd:YAG laser + MIG) welding phenomenon in AA5083 butt joints with different gap conditions', *Science and Technology of Welding & Joining*, 11(3), 295–307, 2006.
[20] Murakami, K., Mitooka, Y., Hino, M., Iogawa, H., Ono, H., and Katayama, S., '', *J. Japn. Inst. Met.*, 70, 134–137, 2006.
[21] Qin, G. L., Lei, Z., and Lin, S. Y., 'Effects of Nd:YAG laser + pulsed MAG arc hybrid welding parameters on its weld shape', *Science and Technology of Welding & Joining*, 12(2), 79–86, 2007.
[22] Swanson, P. T., Page, C. J., Read, E., and Wu, H. Z., 'Plasma augmented laser welding of 6 mm steel plate ', *Science and Technology of Welding & Joining*, 12(2), 153–160, 2007.
[23] Eboo, M., Steen, W.M., and Clarke, J., 'Arc-augmented laser welding', *Advances in Welding Processes*, Proc. of the 4th Int. Conf., Harrogate, UK, 257–265, 1978.
[24] Steen, W. M., and Eboo, M., 'Arc augmented laser welding', *Metal Construction*, 11, 332–335, 1979.
[25] Bagger, C., and Olson, F.O., 'Review of laser hybrid welding', *Journal of Laser Applications*, 17(1), 2-14, 2005.
[26] Cui, H., Decker, I., and Ruge, J., 'Wechselwirkungen zwischen WIGSchweißlichtbogen und fokussiertem Laserstrahl', *Proc. of the Conference Laser'89* , Springer, Berlin, 577–581, 1989.

[27] Finke, B. R., Stern, F., and Decker, I., 'Auswirkungen eines unterstützenden Laserstrahls auf den WIG-Schweißprozess," *DVS-Ber.*, 135, 149–152, 1991.
[28] Cui, H., Decker, I., Pursch, H., Ruge, J., Wendelstorf, J., and Wohlfahrt, H., 'Laserinduziertes Fokussieren des WIG-Lichtbogens', *DVS-Ber.*, 146, 139–143, 1992.
[29] Dilthey, U., Lüder, F., and Wieschemann, A., 'Expanded capabilities in the welding of aluminium alloys with the laser-MIG hybrid process', *Aluminium*, 75, 64–75, 1999.
[30] Graf, T., and Staufer, H., 'Laser hybrid process at Volkswagen', *International Institute of Welding*, IIW-Doc. XII-1730-02, 1999.
[31] Jokinen, T., Vihervä, T., Riikonen, H., and Kujanpä, V., 'Welding of ship structural steel A36 using a Nd:YAG laser and gas metal arc welding', *Journal of Laser Applications*, 12, 185–188, 2000.
[32] Höfemann, M., Fersini, M., Szinyur, J., Haferkamp, H., and Cordini, P., 'Laser-GMA-hybrid welding of zinc-coated steel for hydroforming applications', *Proc. of the 1st Int. WLT-Conf. on Lasers in Manufacturing*, Munich, Germany, 466–472, 2001.
[33] Dilthey, U., and Keller, H., 'Prospects in laser GMA hybrid welding of steel', *Proc. of the 1st Int. WLT-Conf. on Lasers in Manufacturing*, Munich, Germany, 453–465, 2001.
[34] Waltz, C., Seefeld, T., and Sepold, G., 'Laser-GMA welding and its influence on bead geometry and process stability', *Proc. of the 1st Int. WLT-Conf. on Lasers in Manufacturing*, Munich, Germany, 444–452, 2001.
[35] Shibata, K., Sakamoto, H., and Iwase, T., 'Laser-MIG hybrid welding of aluminium alloys', *Proc. of the 1st Int. WLT-Conf. on Lasers in Manufacturing*, Munich, Germany, 436–443, 2001.
[36] Jokinen, T., Jernström, P., Karhu, M., Vanttaja, I., and Kujanpää, V., 'Optimisation of parameters in hybrid welding of aluminium alloy', *Proc. Of the 1st Int. Symp. on High Power Laser Macroprocessing*, Proc. SPIE 4831, 307–312, 2003.
[37] Ishide, T., Tsubota, S., and Watanabe, M., 'Latest MIG, TIG Arc-YAG laser hybrid welding systems for various welding products', *Proc. of the 1st Int. Symp. on High Power Laser Macroprocessing*, 2003 Proc. SPIE 4831, 347–352,2003.
[38] Hyatt, C. V., Magee, K. H., Porter, J. F., Merchant, V. E., and Matthews, J. R., 'Laser-assisted gas metal arc welding of 25-mm-thick HY-80 plate', *Welding Journal Symposium*, Miami, FL, U. S.A, 163s–172s, 2001.
[39] Fellman, A., Jernstrøm, P., and Kujanpää, V., 'The effect of shielding gas composition in hybrid welding of carbon steel', *Proceedings 9th NOLAMP, Conference on Laser Materials Processing in the Nordic Countries*, Trondheim, Norway, 103–112, 2003.
[40] Zhou, J., and Tsai, H.L., 'Modeling of transport phenomena in hybrid laser-MIG keyhole welding', *International Journal of Heat and Mass Transfer*, 51, 4353-4366, 2008.
[41] Page, C. J., Devermann, T., Biffin, J., and Blundell, N., 'Plasma augmented laser welding and its applications ', *Science and Technology of Welding & Joining*, 7(1), 1–10, 2002.
[42] S. M. Joo, Y. P. Kim, C. S. Ro, H. S. Bang and J. U. Park, '', *Adv. Nondestr. Eval.*, 270–273, 2383–2388, 2004.
[43] Graf, T., and Staufer, H., 'Laser-hybrid welding drives VW improvements', *Welding Journal*, 82, 42–48, 2003.

[44] Ono, M., Shinbo, Y., Yoshitake, A., and Ohmura, M., 'Welding properties of thin steel sheets by laser-arc hybrid welding: laser-focused arc welding', *Proc. 1st Int. Symp. on 'High-power laser macroprocessing'*, Osaka, Japan, 369–374, 2002.
[45] Song, G., Liu, L. M., Chi, M. S., and Wang, J. F., 'Investigations of Laser-TIG Hybrid Welding of Magnesium Alloys', *Materials Science Forum*, 488–489, 371–375, 2005.
[46] Chiang, K.C., Tsai, H.L., 'Shrinkage-induced fluid flow and domain change in two-dimensional alloy solidification', *Int. J. Heat and Mass Transfer*, 35, 1763-1769, 1992.
[47] Zhou, J., Tsai, H.L., and Wang, P.C., 'Transport Phenomena and Keyhole Dynamics during Pulsed Laser Welding', *ASME Journal of Heat Transfer*, 128(7), 680-690, 2006.
[48] Wang, Y., 'Modeling of three-dimensional gas metal arc welding process', *Ph.D. Thesis*, University of Missouri-Rolla, 1999.
[49] Lowke, J.J., Kovitya, P., and Schmidt, H.P., 'Theory of free-burning arc columns including the influence of the cathode', *J. Phys. D: Appl. Phys.*, 25, 1600-1606, 1992.
[50] Haidar, J., 'A theoretical model for gas metal arc welding and gas tungsten arc welding', *J. Appl. Phys.*, 84, 3518-3529, 1998.
[51] Haidar, J., Lowke, J.J., 'Predictions of metal droplet formation in arc welding', *J. Appl. Phys. D: Appl. Phys.*, 29, 2951-2960, 1996.
[52] Haidar, J., 'An analysis of the formation of metal droplets in arc welding', *J. Phys. D: Appl. Phys.*, 1233-1244, 1998.
[53] Haidar, J., 'Prediction of metal droplet formation in gas metal arc welding. II.', *J. Appl. Phys.*, 84, 3530-3540, 1998.
[54] Haidar, J., 'An analysis of heat transfer and fume production in gas metal arc welding. III.', *J. Appl. Phys.*, 85, 3448-3459, 1998.
[55] Lowke, J.J., Kovitya, P., and Schmidt, H.P., 'Theory of free-burning arc columns including the influence of the cathode', *J. Phys. D: Appl. Phys.*, 25, 1600-1606, 1995.
[56] Zhu, P., Lowke, J.J., Morrow, R., 'A unified theory of free burning arcs, cathode sheaths and cathodes', *J. Phys. D: Appl. Phys.*, 25, 1221-1230, 1992.
[57] Brackbill, J.U., Kothe, D.B., and Zemach, C., 'A continuum method for modeling surface tension', *J. of Computational Physics*, 100, 335-354, 1992.
[58] Fan, H.G., and Kovacevic, R., 'Droplet formation, detachment, and impingement on the molten pool in gas metal arc welding', *Metall. Trans.*, 30B, 791–801, 1999.
[59] Anisimov, S.I., 'Vaporization of metal absorbing laser radiation', *Soviet Physics*, JETP 27 (1), 182-183, 1968.
[60] Knight, C.J., 'Theoretical modeling of rapid surface vaporization with back pressure', *AIAA J.*, 17(5) 519-523, 1979.
[61] Kanouff, M.P., Kassinos, A., Noble, D.R., and Schunk, P.R., 'Laser spot weld modeling using an ALE finite element method', Sandia National Laboratories Report.
[62] Kaplan, A., 'A model of deep penetration laser welding based in calculation of the keyhole profile', *J. Phys. D: Appl. Phys.*, 27, 1805-1814, 1994.
[63] Duley, W., *Laser welding*, John Wiley & Sons, 1999.
[64] Ducharme, R., Williams, K., Kapadia, P., Dowden, J., Steen, B., and Glowacki, M., 'The laser welding of thin metal sheets: an integrated keyhole and weld pool model with supporting experiments', *J. Phys. D: Appl. Phys.*, 27, 1619-1627, 1994.

[65] Ho, J.R., Grigoropoulos, C.P., Humphrey, J.A.C., 'Gas dynamics and radiation heat transfer in the vapor plume produced by pulsed laser irradiation of aluminum', *J. Appl. Phys.*, 79, 7205-7215, 1996.
[66] Bagger, C., Olsen, F.O., 'Review of hybrid laser welding', *J. Laser Appl.*, 17, 2-14, 2005.
[67] Ribic, B., Palmer, T.A., DebRoy, T., 'Problems and issues in laser-arc hybrid welding', *Int. Mater. Rev.*, 54 (4), 223-244, 2009.
[68] Kong, F. and Kovacevic, R., '3D Finite Element Modeling of the Thermally Induced Residual Stress in the Hybrid Laser/Arc Welding of Lap Joint', *Journal of Materials Processing Technology*, 941-950, 2010.
[69] Zhou, J., Zhang, W. H., Tsai, H. L., Marin, S. P., Wang, P. C., and Menassa, R., 'Modeling the transport phenomena during hybrid laser-MIG welding process', *Proceedings IMECE'03, ASME International Mechanical Engineering Congress & Exposition*, Washington, DC, IMECE2003-41763, 1-8, 2003.
[70] Zhou, J., and Tsai, H. L., 'Investigation of mixing and diffusion processes in hybrid spot laser – MIG keyhole welding', *Journal of Physics D: Applied Physics*, 42, 1-15, 2009.

Chapter 2

Laser Beam Welding of Austenitic Stainless Steels – Similar Butt and Dissimilar Lap Joints

Abdel-Monem El-Batahgy

Additional information is available at the end of the chapter

1. Introduction

1.1. Laser beam welding of similar butt joints of austenitic stainless steels

Because of its inherent corrosion resistance, austenitic stainless steels, known as 300 series, have become cost-effective, staple materials for long-term applications in many industrial sectors including gas, petroleum, petrochemicals, fertilizers, food processing, and pulp industries as well as power generating plants. They have found also widespread use for manufacturing of chemical installations including stationary pressure tanks and tanks for transport of liquid and compressed gases, pipelines of high diameter in water power plants, for manufacturing of ships for transport of chemicals and installations of drilling rigs, etc. Thick-section stainless steels are widely used in the components and structures for nuclear power plants.

For all applications of austenitic stainless steels, welding is of considerable importance since it is widely used in components' manufacturing. In comparison with ferritic steels, lower thermal conductivity and higher thermal expansion coefficient of austenitic stainless steels results in larger thermal distortions and internal stresses of the welded parts, which increase susceptibility of the weld to hot cracks.

Another possible welding problem of austenitic stainless steels is sensitization that occurs at 900-1400ºF (482-760ºC) during cooling after welding where chromium carbides form along the austenite grains and causes depletion of chromium from the grains resulting in decreasing the corrosion protective passive film.

In this concern, austenitic stainless steels poses distinct challenges when it is joined with gas tungsten arc welding (GTAW) due higher possibility of carbide precipitation and distortion in comparison with laser welding. In other words, joining austenitic stainless steels with

GTAW can be tricky, but with a laser, it can be done successfully. Previous studies of the weldability of stainless steels indicate that the basic condition for ensuring high quality of welded joints and reducing thermal distortions to minimum is reducing the heat input of welding that is ensured only by laser welding.

CO_2 laser beam welding with a continuous wave, which is widely used for stainless steels components, is a high energy density and low heat input process. The result of this is a small heat-affected zone (HAZ), which cools very rapidly with very little distortion, and a high depth-to-width ratio for the fusion zone.

The heat flow and the fluid flow in the weld pool can significantly influence the temperature gradients, the cooling rates and the solidification structure. In addition, the fluid flow and the convective heat transfer in the weld pool are known to control the penetration and shape of the fusion zone [1].

Generally, laser beam welding involves many variables; laser power, welding speed, defocusing distance and type of shielding gas, any of which may have an important effect on heat flow and fluid flow in the weld pool. This in turn will affect penetration depth, shape and final solidification structure of the fusion zone. Both the shape and microstructure of the fusion zone will considerably influence the properties of the weldment.

There are many reports [2-4] that deal with the shape and solidification structure of the fusion zone of laser beam welds in relation to different laser parameters. However, the effect of all influencing factors of laser welding has up to now not been extensively researched. More work is required for understanding the combined effect of laser parameters on the shape and microstructure of the fusion zone.

The present investigation is concerned with laser power, welding speed, defocusing distance and type of shielding gas and their effects on the fusion zone shape and final solidification structure of some austenitic stainless steels.

1.2. Experimental procedure

Three types of commercial austenitic stainless steels, 304L, 316L and 347, were used. Their chemical composition and mechanical properties are given in Table 1. The thickness of both 304L and 316L steels was 3 mm while that of 347 steel was 5 mm.

Base metal										TS (MPa)	YS (MPa)	Elong. (%)
	Cr	Ni	Mn	C	Si	P	S	Mo	Cb			
304L	18.2	8.5	1.7	0.025	0.35	0.03	0.01	0.20	0.01	235	309	32
316L	16.5	10.5	1.9	0.02	0.38	0.02	0.01	2.27	0.01	556	317	31
347	17.7	9.7	1.8	0.04	0.40	0.03	0.01	0.25	0.35	577	345	39

Table 1. Chemical composition (wt%) and mechanical properties of the used base metals

Both bead-on-plate and autogenous butt weld joints were made using a carbon dioxide laser capable of producing a maximum output of 5 kW in the continuous wave mode. Bead-on plate was made on plates with 3 mm thickness while autogenous butt weld joints were made on plates with 3 and 5 mm thickness. Specimens with machined surfaces were prepared as square butt joints with dimensions of 125x150 mm and were held firmly using fixture to prevent distortion

The laser beam welding parameters investigated are summarized in Table 2. Combinations of laser power (P) of 2-5 kW and speed (S) of 0.5-3 m/min resulted in nominal heat in nominal heat inputs (HI) ranging from 0.04 to 0.48 kJ/mm. The defocusing distance (D_d) was in the range of -5 to 3 mm. Shielding was made using either argon or helium gas.

Weld joint/ Thickness (mm)	P (kW)	S (m/min)	HI (kJ/mm)	D_d (mm)	Shielding gas/ Flow rate(1/min)
BOP / 3	2	3	0.04	0.0, -1.0, -2.0,-3.0, -4.0,-5.0, +1.0,+2.0, +3.0	argon / 15
Butt / 3	3	0.5, 1, 2, 3	0.36, 0.18, 0.09, 0.06	0.0	argon / 15
Butt / 3	4	0.5, 1, 2, 3	0.48, 0.24, 0.12, 0.08	0.0	argon / 15
Butt / 3	4	3	0.08	0.0, -0.2, -0.4, -0.6	argon/15
Butt / 3	4	3	0.08	0.0, -0.2	helium / 15
Butt / 5	5	1, 2, 3	0.3, 0.15, 0.1	0.0. -0.2, -0.4	argon/15

Table 2. Welding parameters used

After welding, the specimens were visually inspected then, sectioned transverse to the welding direction. The shape and microstructure of the fusion zone were examined using optical microscopy. Micro-compositional analysis of welds was performed using an electron probe micro-analyzer (EPMA) at an accelerating voltage of 25 kV.

Mechanical tests including tensile, bending and hardness measurements of butt welds having complete penetration were performed according to relevant standards. The data reported are the average of three individual results.

1.3. Results and discussion

1.3.1. Macrostructure of laser beam welds

1.3.1.1. Effect of laser power

The effect of heat input as a function of laser power, HI = P/S, was clarified using type 304L and type 316L steels. Both welding speed and defocusing distance were kept constant at 3 m/min and zero respectively.

The penetration depth increased sharply with increasing laser power from 2 to 3 kW as shown in Figure 1. Complete penetration for the 3mm base metal was obtained at laser power equal to or greater than 4 kW. Figure 9 shows an example of a cross section of type 304L steel butt weld made using laser power of 4 kW. The weld bead showed a characteristic of laser welding with dept / width ratio close to 3. No welding cracks or porosity were found in any of the welds, this may be partly due to the good crack resistance of the base metal and the welding conditions provided.

The results indicated also that the development of the weld pool is essentially symmetrical about the axis of the laser beam. Yet, lack of symmetry at the root side was observed particularly at higher welding speed (Figure 2) suggesting an unsteady fluid flow in the weld pool. This is due to the presence of two strong and opposing forces, namely, the electromagnetic and the surface tension gradient forces. At these locations, the electromagnetic force may have overcome the surface tension force, thereby, influencing convective heat transfer. As a result, any local perturbation in the weld pool can cause the flow field to change dramatically, resulting in the observed lack of local symmetry.

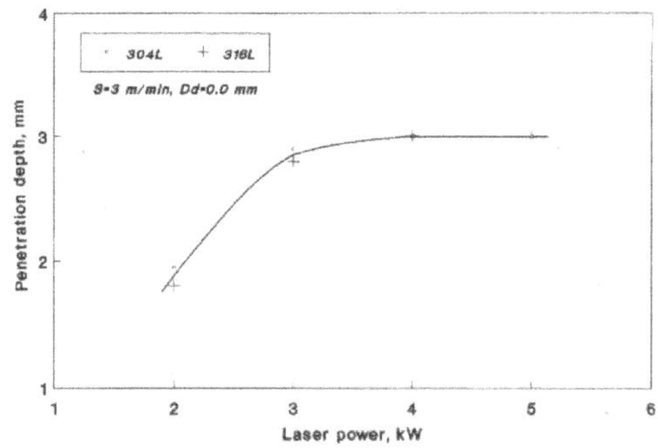

Figure 1. Effect of laser power on penetration depth of types 304L and 316L stainless steel welds.

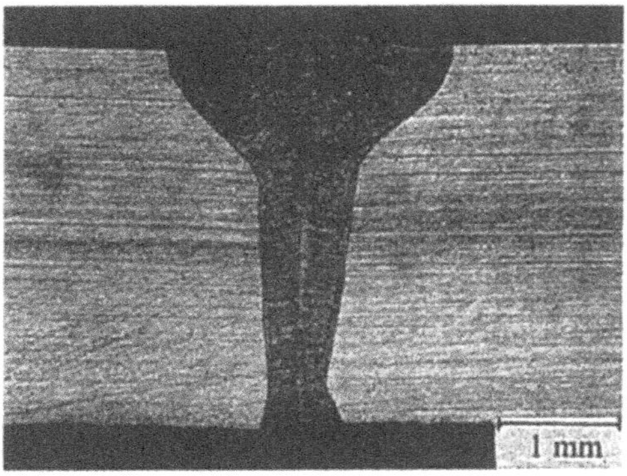

Figure 2. A cross section of type 304L stainless steel weld made using P=4kW, S=3m/min, D_d=0.0mm.

Laser power has a less influence on both weld profile and HAZ width in comparison with its effect on penetration depth. This is in agreement with other researchers work where they pointed out that changing laser power between 3 and 5 kW [5] did not result in any significant change in the size or shape of the weld.

It is expected that similar results concerning the dependence of penetration depth on laser power could be obtained in the case of type 347 steel due to similarity in both physical and mechanical properties. The optimum power for complete penetration with acceptable weld profile for the 5mm base metal thickness was 5 kW at a welding speed of 2 m/min as shown in Figure 3.

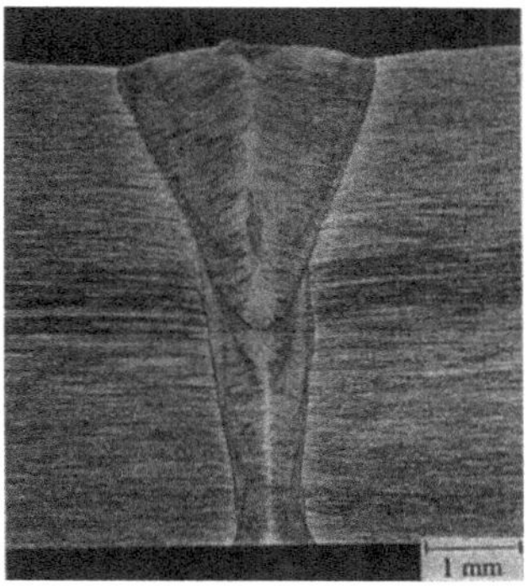

Figure 3. Optimum weld profile of type 347 stainless steel made using P=5kW, S=2m/min, D_d=-0.4mm.

1.3.1.2. Effect of welding speed

The effect of welding speed was investigated at the optimum laser power (4 kW) and zero defocusing distance. Figure 4 shows the relationship between welding speed and fusion zone depth/width ratio for both 304L and 316L base metals. The depth/width ratio increased sharply from 2.1 to 4.1 with the increase in welding speed from 0.5 to 3 m/min.

The dependence of depth/width ratio on welding speed was confirmed at a different laser power (3 kW). A lower welding speed resulted in a considerable increase in the fusion zone size and consequently a decrease in depth/width ratio leading to unacceptable weld profile. Complete penetration with relatively acceptable fusion zone size for the 3mm base metal thickness was obtained at welding speed of 2 m/min as shown in Figure 5. The fusion zone is symmetrical about the axis of the laser beam.

The above results have shown that the laser power and welding speed should be optimized in order to minimize heat input, then a satisfactory weld with reliable quality could be obtained. This reflects one of the most notable features of laser welding compared with other welding processes, which is small heat input.

Turning to the macrographs shown in Figures 5 and 8, complete penetration with relatively acceptable fusion zone profile could be obtained using either 4 kW, 3 m/min (Figure 5) or 3 kW, 2 m/min (Figure 8). However, 4 kW, 3 m/min resulted in a smaller fusion zone size with less inflection at its interface in addition to the high welding speed in this case.

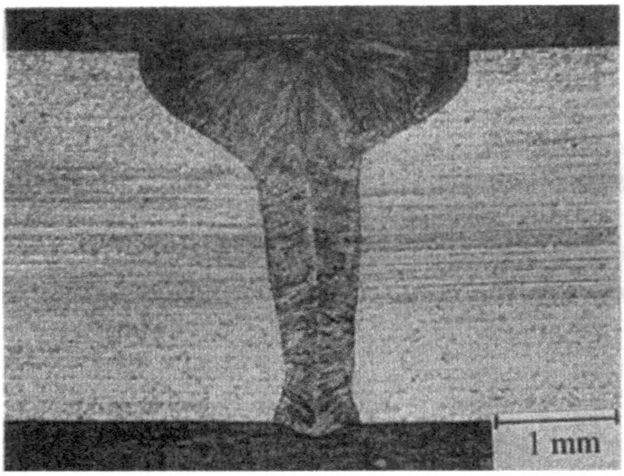

Figure 4. Influence of welding speed on weld depth/width ratio of types 304L and 316L stainless steels.

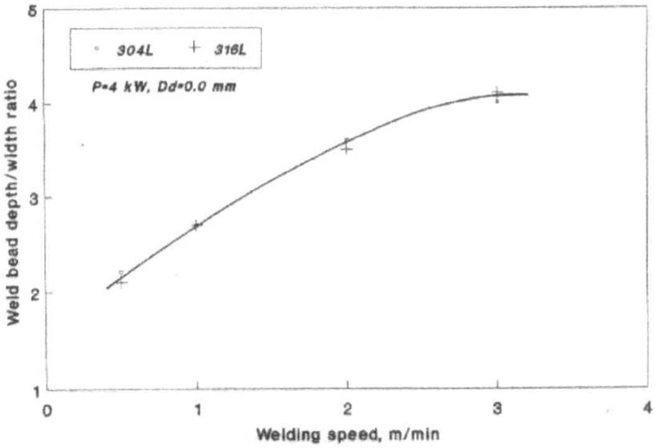

Figure 5. A cross section of type 304L stainless steel weld made using P=3kW, S=2m/min, D_d=0.0mm.

At high welding speed, attenuation of beam energy by plasma is less significant. This results in relatively more exposure of the laser beam on the sample surface. Consequently, the depth/width ratio would be increased and the fusion zone size would be minimized.

1.3.1.3. Effect of defocusing distance

Defocusing distance, focus position, is the distance between specimen surface and the optical focal point. In order to study its effect on both penetration depth and weld profile, bead-on-plate was made with changing defocusing distance between -5 and 3 mm. Low laser power (2 kW) and high welding speed (3 m/min) were selected to obtain incomplete penetration.

Examples of weld cross sections of type 304L steel made using different defocusing distances are shown in Figure 6. No cracking or porosity was observed in all welds. The penetration depth is considerably decreased with changing defocusing distance from zero (Figure 6-b) to either minus (Figure 6-a) or plus (Figure 6-c) values as a result of decreasing laser beam density.

The relationship between defocusing distance and penetration depth of both 304L and 316L steels is summarized in Figure 7. The penetration depth decreased from 1.9 to 1.6 mm on changing the defocusing distance from zero to either -1 or 1mm. Then, the penetration depth decreased sharply to about 0.2 mm on changing the defocusing distance to more negative (-5 mm) or positive (4 mm) values.

These results indicated that the most effective range of defocusing distance to get maximum penetration with acceptable weld profile lies between zero and -1 mm. In order to obtain the optimum value, complete penetration butt welds were made using previously obtained optimum laser power (4 kW) and optimum welding speed (3 m/min). The most acceptable weld profile was obtained at defocusing distance of - 0.2 mm for 3 mm thickness where weld bead depth/width ratio is maximum and fusion zone size is minimum with a slight taper configuration as shown in Figure 8. However, the optimum defocusing distance to attain acceptable weld profile for 5 mm thickness was -0.4 mm (Figure 3). The smooth curved and symmetrical fusion zone interface shown in Figures 3 and 8 suggests that the driving forces for fluid flow in the weld pool, buoyancy and surface tension gradient, augment each other, resulting in a coherent flow field.

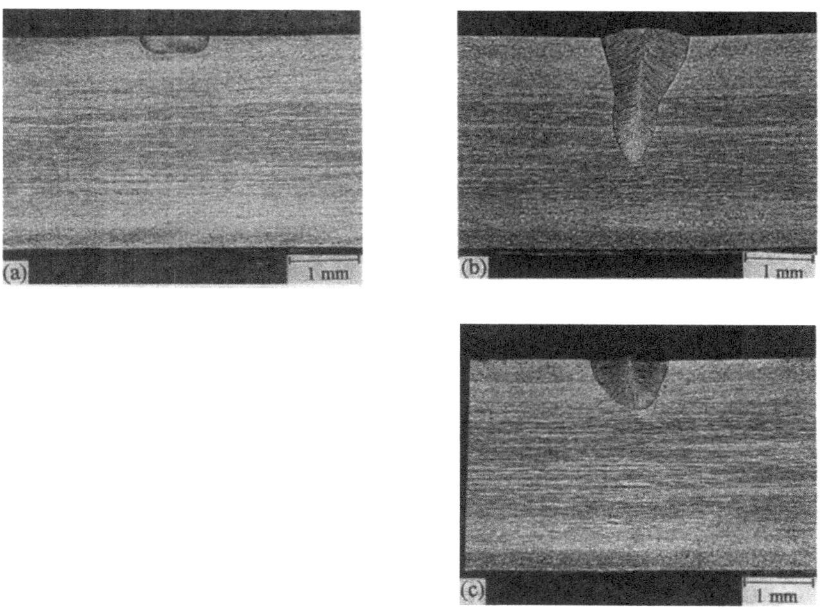

Figure 6. Cross section of type 304L stainless steel welds made using P=2kW, S=3m/min with different defocusing distances. (a) D_d=-3.0mm, (b) D_d=0.0mm, (c) D_d=+2.0mm.

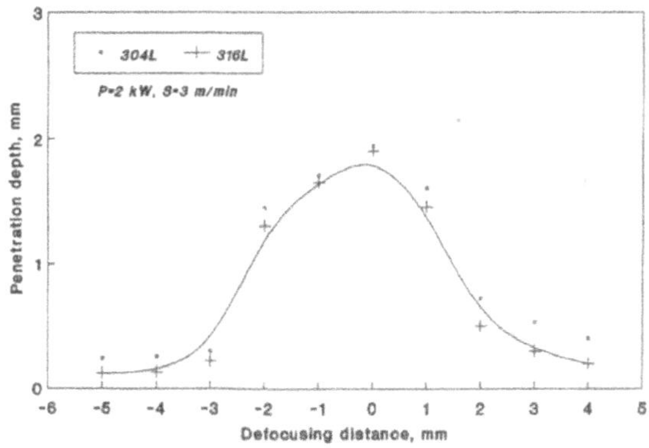

Figure 7. Relationship between defocusing distance and penetration depth of types 304L and 316L stainless steels.

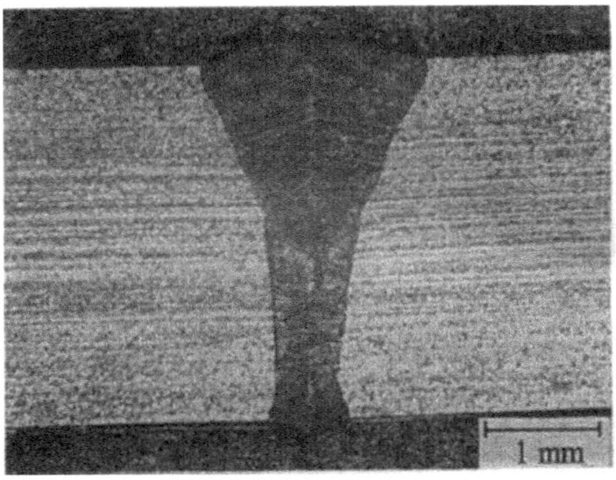

Figure 8. A cross section of type 304L stainless steel weld made using P=4kW, S=3m/min, D_d=-0.2mm.

1.3.1.4. Effect of type of shielding gas

In all previous experiments, argon was used as a shielding gas. For comparison, argon was replaced by helium while other laser parameters were kept constant. Weld profile is remarkably improved where fusion zone interfaces are almost parallel to each other as shown in Figure 9.

In general, when the laser beam interacts with the workpiece, a hole is drilled through the thickness of the material. This hole or cavity is filled with a plasma and surrounded by molten metal, thus, the high energy density of the focused beam could be lost easily. This plasma effect was reduced as a result of the higher ionization potential of helium then, the weld profile was improved.

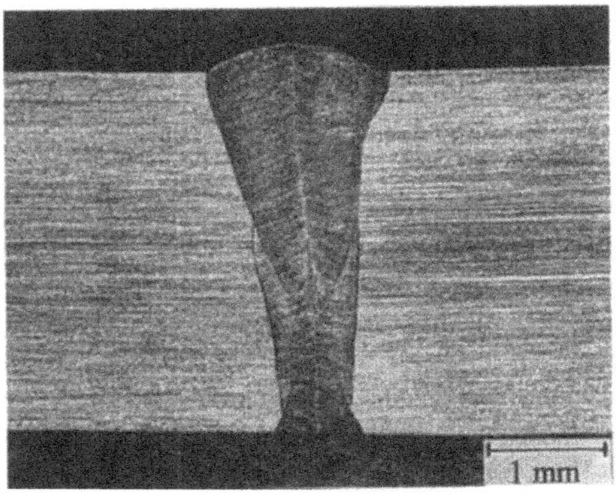

Figure 9. Optimum weld profile of type 304L stainless steel obtained using helium as a shielding gas instead of argon with the same laser parameters as of Figure 15.

1.3.2. Microstructure of laser beam welds

Microstructures of type 304L steel weld metals made using two different welding speeds, 1 and 3 m/min, with same laser power, 4 kW, are shown in Figure 10. The noticeable feature is the highly directional nature of the microstructure around the axis of the laser beam. This is due to solidification of the weld metal at high cooling rate compared to that of conventional GTA welding [6]. It can also be noticed that the higher the welding speed, the finer the dendritic structure (Figure 10-b). This is attributed to an increase in both solidification and cooling rates due to low heat input resulted from high welding speed.

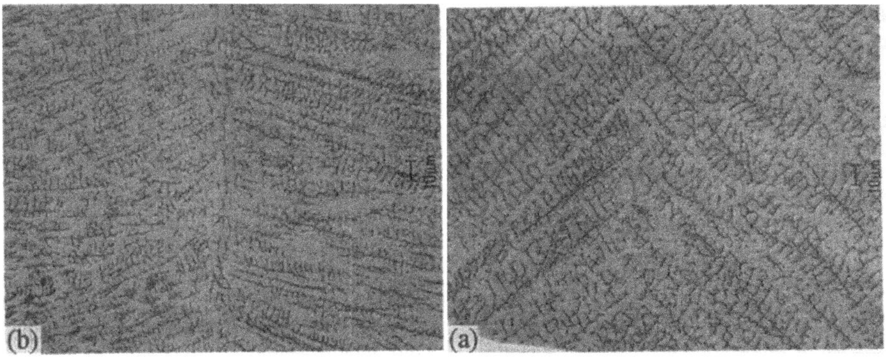

Figure 10. Optical micrographs of weld metals of type 304L stainless steel made using P=4kW with different welding speeds. (a) S=1m/min, (b) S=3m/min.

Concerning the effect of laser power, the higher the laser power, the coarser is the dendritic structure due to decreasing cooling rate. However, the effect of laser power was relatively less than that of welding speed.

The microstructures of all laser beam welds were always austenite with a few percent of delta-ferrite at the dendritic boundaries. The existence of delta-ferrite was confirmed by both colour etching and electron probe microanalysis (EPMA). The amount of delta-ferrite was estimated using Cr and Ni equivalents [7] of weld metal chemical composition. Based on the Schaeffler diagram, about 2 or 3 vol% ferrite was expected to exist in the austenitic matrix. It should be reported that no weld solidification cracking was observed in any of the welds evaluated.

Under normal weld solidification conditions, the solidification mode in austenitic stainless steels is primarily a function of composition, with a shift from primary ferrite to primary austenite accomplished by reducing the Cr_{eq}/Ni_{eq} ratio below 1.5 [8].

However, Suutala [2], Vitek and David [9], and Lippold [10] have illustrated that the boundary between primary austenite and primary ferrite solidification is not just a function of weld metal composition, but is a function of the growth rate.

The subject results are consistent with the modified Suutala diagram [11] for predicting microstructure and cracking of austenitic stainless steels under rapid weld solidification conditions encountered during laser welding. All the steels tested exhibited a Cr_{eq}/Ni_{eq} ratio greater than 1.7, the value suggested by the modified Suutala diagram as the demarcation of crack susceptibility. Consequently, it can be deduced that all welds concerned in this investigation were solidified as mixed mode of primary and massive austenite. The transition in primary solidification from ferrite to austenite could be attributed to weld pool undercooling as a result of extremely high solidification growth rate [12-15].

The results are also in agreement with Lippold [12] work where he pointed out that cracking in austenitic stainless steel welds is avoided as the proportion of primary ferrite in the mixed mode solidification increases.

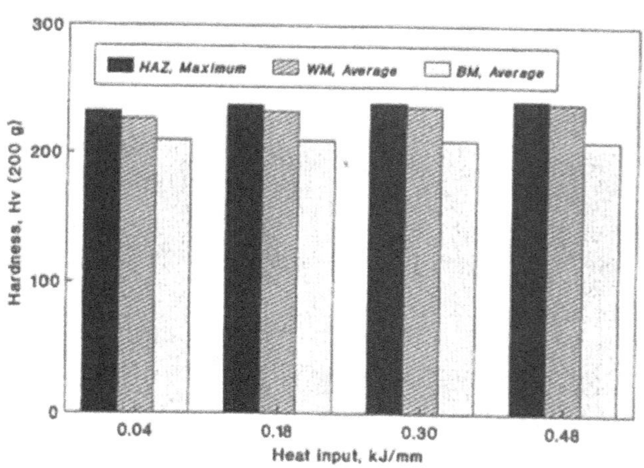

Figure 11. Hardness profiles of base metal, HAZ and weld metal of type 304L stainless steel as a function of heat input.

1.3.3. Mechanical properties

Tensile test results of all laser butt welds with complete penetration showed that failure has taken place in the base metal. The bending test at room temperature showed no cracks in all joints as a result of high ductility.

Typical hardness profiles of the base metal, HAZ and weld metal of type 304L steel as a function of heat input are shown in Figure 11. No significant difference between hardness of the base metal and that of weld metal or HAZ was obtained. Hardness of both weld metal and HAZ was slightly higher than that of the base metal regardless of heat input. These results were also valid for the other two base metals. This is expected because mechanical properties of steel, in general, are based on its microstructure.

2. Laser beam welding of austenitic stainless steel in lap joints with zn-coated carbon steel

2.1. Introduction

Because of its excellent corrosion resistance, austenitic stainless steel has found widespread use in the paper making equipment includes pressure vessels, storage tanks, piping, hopper, bins, chutes and structural components. For all of these applications, attachments such as access platforms, catwalks, stiffeners, column supports, stairways, washers and pipe hangers are welded to the outside surfaces of the equipment. Zn-coated carbon steel is often specified for these attachments due to its good corrosion resistance and lower cost. Lap weld joints of Zn-coated steel to austenitic stainless steel are used also in other fields such as plate-tube joints, radiators, washing machines as well as some components in the aeronautic field [16].

Although welding of Zn-coated steel to austenitic stainless steel is a common practice, it presents serious problems concerning with weld zone porosity and LME cracking of austenitic stainless steel base metal due to zinc vaporization. These welding problems have been studied in the case of conventional metal arc welding processes [17]. It is reported that joining gaps between the sheets to be lap welded are adjusted in order to enable the degasification of zinc vapour.

On the other hand, laser butt and lap weld joints of both similar and dissimilar materials are being used in many industrial applications. The fraction of laser welding in all industrial applications is about 15-25% which varies from country to country [18-20]. The most notable features of laser welding compared with other conventional welding processes are the high weld quality and high welding speed. This together with its low heat input makes laser a most hopeful candidate for thin sheet metal welding.

However, similar problems as in conventional metal-arc-welding of Zn-coated steel to austenitic stainless steel are expected also in case of laser beam welding. Therefore, more work is required for understanding these problems and the factors affecting them.

This investigation has been concerned with CO_2 laser welding of austenitic stainless steel in lap joints with Zn-coated carbon steel. The focus was made on weld joint quality in terms of weld profile, porosity in the weld zone and liquid metal emrittlement (LME) cracking of the austenitic stainless steel base metal. The influence of type and flow rate of shielding gas, gap between the sheets and zinc removal prior to welding was clarified. Quality of weld joints was evaluated as a function of weld zone shape, porosity and LME cracking of austenitic stainless steel base metal.

2.2. Experimental procedure

Commercial types of ASTM A36, 0.7 mm thick carbon steel sheet coated with 10 μm zinc on both sides and ASTM A240 Type 304L, 1mm thick stainless steel sheet were used for dissimilar lap joints. Table 3 shows their chemical composition and mechanical properties.

Base Metal	C	Mn	Si	S	P	Cr	Ni	YS (N/mm^2)	UTS (N/mm^2)	Elong (%)
Zn-coated	0.04	0.35	0.26	0.01	0.02	-	-	245	377	27
304L	0.04	1.70	0.35	0.01	0.03	18.2	8.5	359	558	32

Table 3. Chemical composition (wt%) and mechanical properties of used base metals

Pairs of these dissimilar steel sheets of 150x150 mm were welded with an overlap of 50 mm and with weld bead at the middle of the overlap. Zn-coated steel sheet was upper-most and the joint was clamped 15 mm on both sides of the weld line along its entire length. Configurations of laser lap weld specimen are shown in Figure 12. All specimens were ultrasonically cleaned to remove dirt and oil prior to welding.

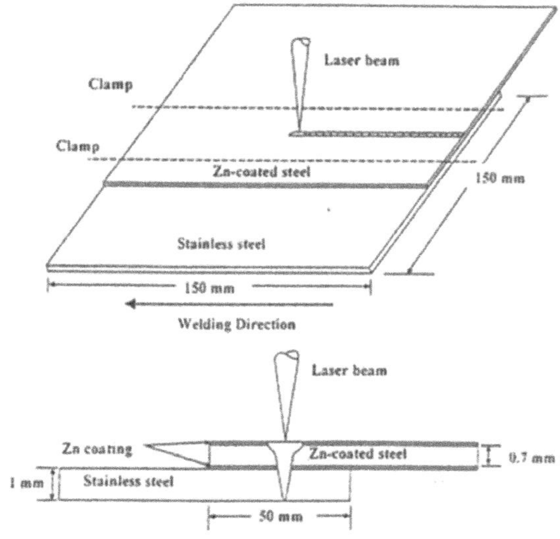

Figure 12. Configuration of the used laser lap weld specimen.

Welding was performed using CO_2 laser with a maximum output of 3 kW operated in multi-mode. The beam was focused using a parabolic mirror with 150 mm focal length. Laser beam welding parameters used are summarized in Table 4. Optimizing laser power, welding speed and focal point position is of considerable importance for the weld quality in terms of fusion zone size and profile. In order to clarify the influence of shielding gas, gap between the sheets and pre-weld zinc removal on weld quality, laser power, welding speed, and defocusing distance (focal point position) were optimized and kept constant at 2.5 kW, 3 m/min, and 0.1 mm below specimen surface respectively. Shielding was done using argon or helium with flow rate of 15~30 l/min. Prescribed gap ranging from 0.025 to 0.3 mm was introduced between the sheets along the clamped areas of the welding fixture. Pre-weld zinc removal from the weld area was done by grinding.

P (KW)	S (m/min)	Dd (mm)	Shielding gas Type Flow rate (1/min)	Gap between Sheets (mm)	Zinc removal prior to Welding
2.5	3	-0.1	Argon 15 ~ 30	No	No
2.5	3	-0.1	Helium 15 ~ 30	No	No
2.5	3	-0.1	Argon 15	0.025 ~ 0.3	No
2.5	3	-0.1	Helium 15	0.025 ~ 0.3	No
2.5	3	-0.1	Argon 15	No	Yes
2.5	3	-0.1	Helium 15	No	Yes

Table 4. Laser welding parameters
P: Laser power, S: Welding speed, Dd: Defocusing distance
Working distance: 10 mm at Dd = 0.0 mm, Nozzle diameter: 4mm

After welding, the specimens were subjected to non-destructive testing including visual and dye penetrant test methods then, sectioned transverse to the welding direction. Three sections of each seam weld were prepared for metallographic examinations using standard technique. Quality of the dissimilar lap joints was evaluated as a function of weld profile, porosity level in fusion zone, LME cracking in austenitic stainless steel base metal. Tensile shear test was carried out for all laser lap welded joints and the data reported are the average of three individual results.

2.3. Results and discussion

2.3.1. Effect of shielding gas

Examples of macrographs laser welds produced using argon and helium as a shielding gas with a flow rate of 15 l/min are shown in Figure 13-a and b respectively. It is clear that non-uniform weld beads with large pores were obtained in both cases. However, wide seam width combined with an increase of the frequency of pores was obtained with argon (Figure 13-a). In other words, the number of pores was much less with using helium as a shielding gas (Figure 13-b). This means that shielding gas type is an essential factor to improve weld quality since it is used to protect the molten metal against oxidation and blow the plasma away from the beam path.

Generally, when welding zinc coated steel with stainless steel there is a very strong plasma formation due to the low boiling point of zinc (906 °C) and it's high vapour pressure, which is about eight orders of magnitude greater than that of Fe [21]. The high vapourization of zinc increases the pressure of the vapour, which is transformed to plasma in the laser beam, to expand further into the free space above the metal surface. This will affect the absorption and fluctuation of the plasma and in practice this is shown as increased splattering and porosity in the weld. This plasma effect was reduced as a result of the higher ionization potential of helium then, weld quality was improved.

With this relatively low flow rate, the process seemed to change swiftly between deep penetration welds and vapour assisted welds. This may be explained by the presence of zinc, which makes the process unstable due to plasma fluctuation as has reported in prior investigation [22]. Consequently, optimized shielding gas flow rate makes the difference between a good or poor weld.

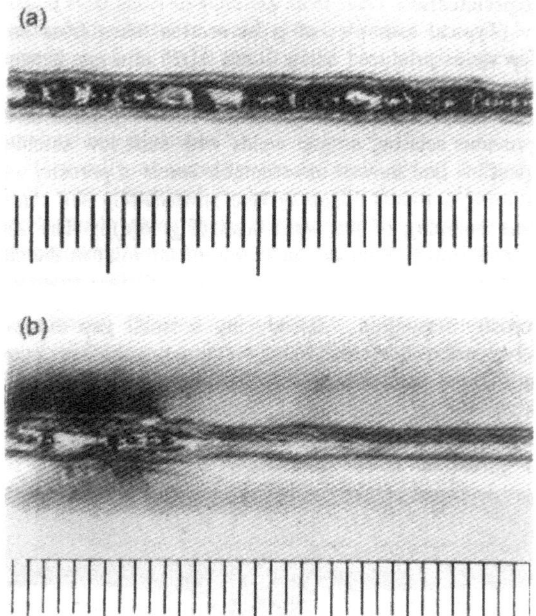

Figure 13. Macrographs of laser welds produced using (a) argon and (b) helium as a shielding gas with 15 l/min flow rate.

With increasing shielding gas flow rate above 15 l/min, the number of pores in seem welds was decreased and weld pool penetration was increased. Examples of macrographs of laser welds produced using higher flow rate; 30 l/min of argon and helium as a shielding gas are shown in Figure 14. It is obvious that smooth and homogeneous seam welds free from pores were obtained in both cases. The increase in penetration depth obtained in this case is consistent with the expected effect of increased plasma suppression with increased flow rate, i.e. more of the beam was allowed to reach the work-piece. However, there appeared to be a trade-off between plasma suppression effects and weld pool stability with increase shielding gas flow rate.

Turning to shielding gas type, it remarkably affected weld zone profile. Low magnification stereoscopic photographs of cross sections taken from laser welds of Figure 14-a and b are shown in Figure 15-a and b respectively. The use of helium has resulted in complete penetration with higher depth/width ratio and a slight taper configuration which means minimum fusion zone size (Figure 15-b) in comparison with that obtained using argon (Figure 15-a). In other words, helium has a more favorable effect on the molten metal than argon at optimized flow rates, which make the welds more homogeneous and free from pores. A flow rate of 22 l/min for helium was found to be satisfactory in comparison with 30 l/min for argon. These results do conform with prior results of other investigators where they have shown that the weld defects, due to the vapourization of zinc, can be reduced by optimizing shielding gas parameters [23-25].

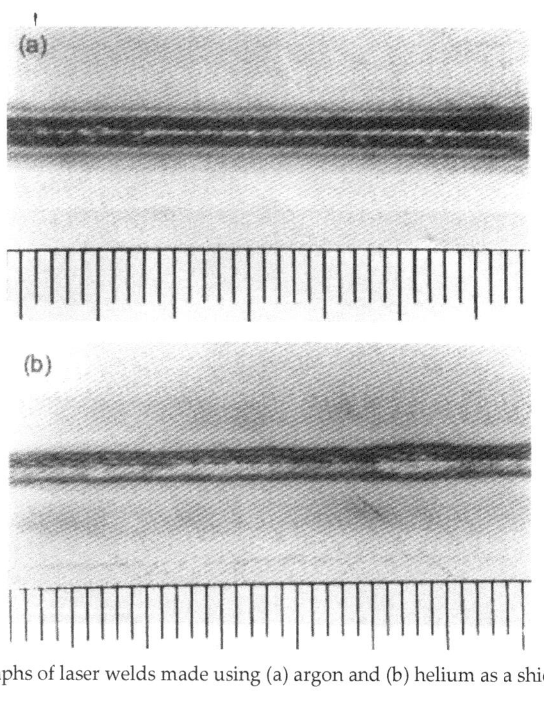

Figure 14. Macrographs of laser welds made using (a) argon and (b) helium as a shielding gas with 30 l/min flow rate.

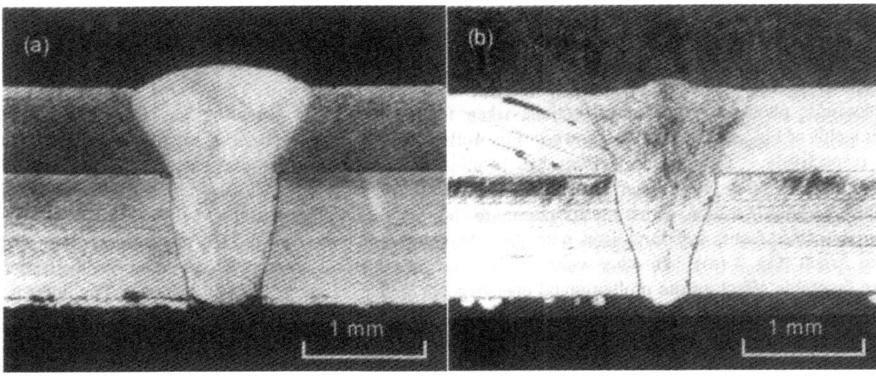

Figure 15. Low magnification stereoscopic photographs of cross sections taken from laser welds of Figure 14

On the other hand, it is found that both type and flow rate of shielding gas have no effect on LME cracking of austenitic stainless steel. Figure 16 shows a typical example of optical micrograph of a cross section taken from lap weld produced using helium with its optimum flow rate; 30 l/min. In spite of obtaining homogeneous, sound, complete penetration and acceptable weld profile with such high flow rate, the noticeable feature is the formation of severe cracking at the stainless steel base metal. These cracking were extended for a distance of about 0.7 mm around both sides of lap weld joints and propagated on grain boundaries. This type of cracking is typical LME cracking of the austenitic stainless steel which occurs above 750°C when it is exposed to molten zinc and tensile stresses. Molten zinc can be produced by the heat of welding and tensile stresses can be generated from the heating and cooling cycles during welding. This cracking type is characterized by extremely rapid crack propagation perpendicular to the applied stress [26]. It should be mentioned also that similar results were obtained for lap weld joints produced using argon gas shielding.

Figure 16. Typical example of optical micrograph of a cross section taken from laser lap weld produced using helium as a shielding gas with its optimum flow rate; 30 l/min.

2.3.2. *Effect of gap between the sheets*

In the above section, lap welds were made with a good contact, i.e. without a gap between the sheets. In order to clarify its effect on weld joint quality, prescript gap was introduced between the sheets and shielding was done using either argon or helium with its lower flow rate (15 1/min) which resulted in weld porosity in the previous experiments.

Typical examples of cross sections taken from laser lap welds produced using 0.0 and 0.025 mm gap between the sheets in case of helium shielding are shown in Figure 17-a and b respectively. As has been explained in the previous section, no-gap welds with such low shielding gas flow rate showed unacceptable levels of porosity with varying amounts of top surface undercutting and center-line humping of the weld bead (Figure 17-a). The zinc vapourization from the underside of the joint in the case of complete weld penetration is not sufficient to prevent porosity formation. Introducing a small gap distance between the sheets with same welding conditions resulted in a sound weld where porosity was not observed (Figure 17-b). Also, weld profile was remarkably improved where a smooth curved and symmetrical fusion zone interface was obtained.

In other words, acceptable quality for laser lap welds concerning soundness and profile could not be obtained with low shielding gas flow rate and without a gap between the sheets. Once the heat input was sufficient to permit melting through the top sheet, there was explosive ejection of molten weld metal due to vapourization of the zinc layers at the Zn-coated steel sheet-to-stainless steel sheet interface. This resulted in extensive weld metal porosity or complete expulsion of the weld metal in the case of no-gap weld leading to undercutting the top steel sheet, as shown in Figure 17-a. This is in a good agreement with results of previous investigations [27, 28].

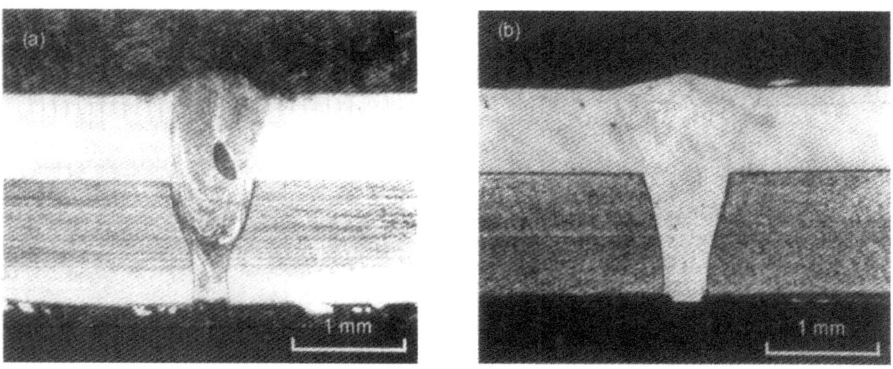

Figure 17. Typical examples of cross sections taken from laser lap welds produced using 15 l/min helium and different gap distances. (a) Gap: 0.0mm, (b) Gap: 0.025mm.

Generally, no porosity was found in any of the welds made with introducing gap between the sheets. However, welds made by gap larger than 0.05 mm showed unacceptable weld profile where weld depth /width ratio decreases sharply and the weld geometry begins to deteriorate. Photographs of laser weld cross sections produced using 0.1 and 0.3 mm gaps are shown in Figure 18-a and b respectively. A gap distance of 0.1 mm gave a concave top surface, with a relatively low depth/ width ratio (Figure 18-a). The tendency of the molten pool to collapse increased significantly with increasing the gap to higher value. This resulted in remarkable undercut in the welds and excessive drop-through of the weld metal into the gap (Figure 18-b).

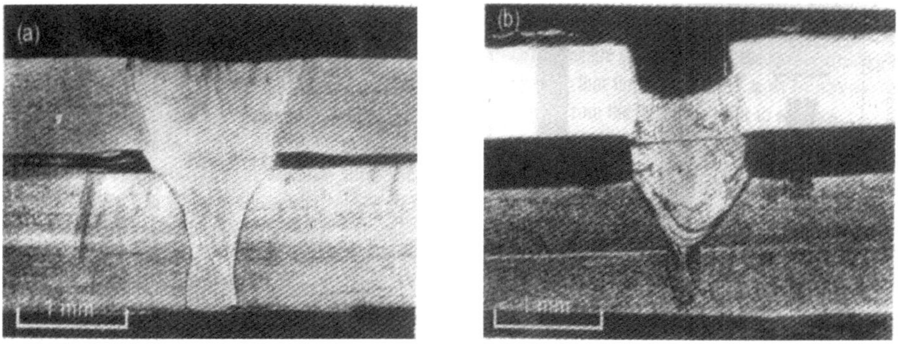

Figure 18. Photographs of cross sections taken from laser lap welds produced using (a) 0.1mm and (b) 0.3mm gap between the sheets.

Generally, there are two mechanisms in laser welding, one is heat conduction under low energy density and the other is deep penetration (keyhole effect) under high energy density. In these experiments, the laser power density when laser beam touched the surface of the top sheet was high enough to melt the metal rapidly and formed the deep penetration. The power density greatly decreased when approached the bottom plate particularly in case of large air gap between both plates that obstructed heat transfer. At this time, the heat transfer was mainly by means of conduction, which could be proved by the weld shape and penetration depth. Based on the weld shape, the fusion lines at both sides of welds were approximately parallel under deep penetration welding conditions, while the fusion line was half circle under heat conduction welding. Consequently, under such experimental conditions, the welding mechanism of lap joints was combination of deep penetration and heat conduction.

Although introducing a small gap between these dissimilar material sheets has resulted in avoiding porosity in weld zone, it has no effect on zinc induced LME cracking in austenitic stainless steel base metal. Figure 19 shows typical example of an optical micrograph of a cross section taken from lap welded joint produced using helium with 15 1/min flow rate and 0.05 mm gap. It is noticed that a sound and uniform weld seam was obtained. However, the most important notice is the formation of LME cracking on grain boundaries of stainless steel base metal as has been explained in the previous section. Cracking were extended for a distance of about 0.5 mm around both sides of lap weld joints. These results of laser welding do conform to other research work concerned with arc welding processes [17].

Generally, the influence of gap between the sheets on weld geometry and quality can be explained using schematic illustrations shown in Figure 20. Since the zinc vapour has no enough access escape route with zero gap, both pores weld with unacceptable profile and LME cracking in stainless steel will be obtained (Figure 20-a). Introducing a small gap between the sheets give the zinc vapour an alternative escape route during welding then, sound and acceptable weld joint will be produced. It should be mentioned that this gap should be limited to an optimum value as has been reported by other researchers [22]. However, the problem is still concerned with LME cracking in austenitic stainless steel base metal around both sides of the joint since it can not be prevented by these measures (Figure 20-b).

2.3.3. *Effect of zinc removal prior to welding*

The results of the previous two sections confirmed that LME cracking of austenitic stainless steel in laser lap joint with Zn-coated steel is attributed mainly to molten zinc resulted from welding heat. Consequently, this section is concerned with studying this type of cracking as a function of zinc removal prior to welding. In this respect, the effect of both one and two sides grinding of the weld area of Zn-coated sheets was clarified using welding parameters previously resulted in weld zone porosity.

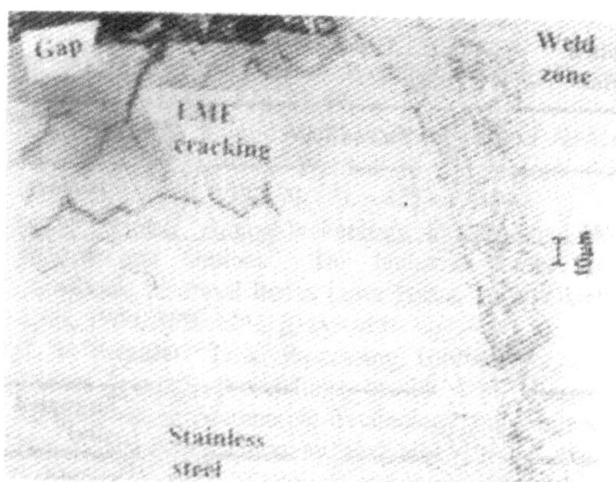

Figure 19. Typical example of optical micrograph of a cross section of lap welded joint produced using 0.05mm gap and 15 l/min helium.

Photographs of laser welds produced without gap between the sheets and with 15 1/min of helium shielding after zinc removal by grinding of weld area from one and two sides are shown in Figure 21-a and b respectively. It can be noticed that removing zinc coating from only one side of Zn-coated sheet was not effective to obtain sound welds since porosity were observed in weld zone (Figure 21-a). In the case of two sides grinding before welding, molten zinc was avoided due to removing of zinc coating then, molten metal was not ejected and this in turn could result in a sound and uniform weld seam (Figure 21-b).

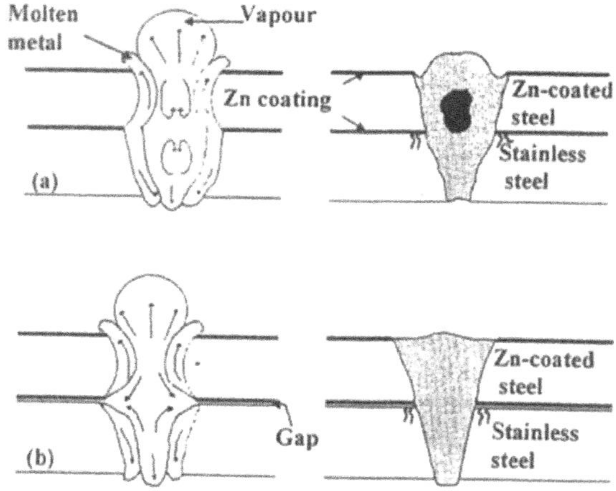

Figure 20. Schematic illustrations showing the effect of introducing a gap between the sheets being welded on weld zone profile, porosity and LME cracking.

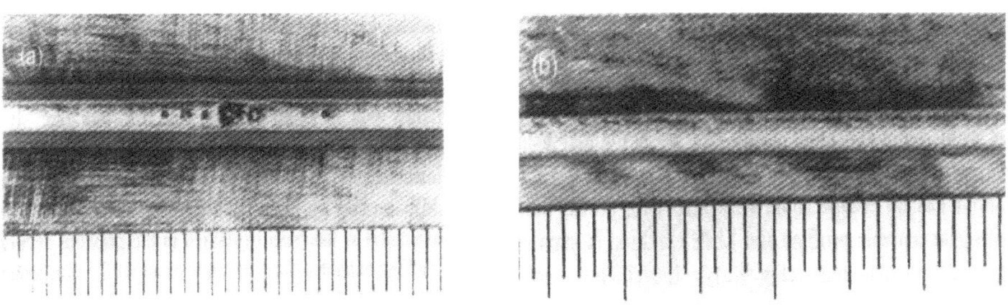

Figure 21. Photographs of laser welds produced using 15 l/min helium shielding and zero gap between the sheets after zinc removal from (a) one side and (b) two sides of the weld area of Zn-coated sheet.

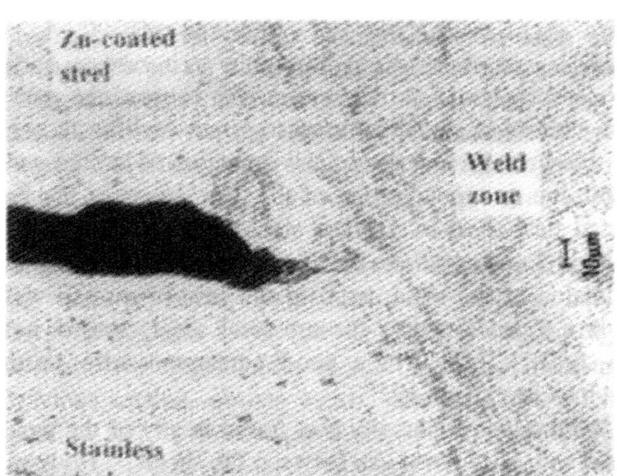

Figure 22. Effect of two sides grinding of weld area of Zinc-coated sheet on LME cracking susceptibility.

In addition, the most important finding in the case of two sides grinding is the disappearance of LME cracking in austenitic stainless steel as shown in Figure 22 that could not be attained in the above two sections. This is due to the complete removal of zinc coatings from the weld area prior to welding.

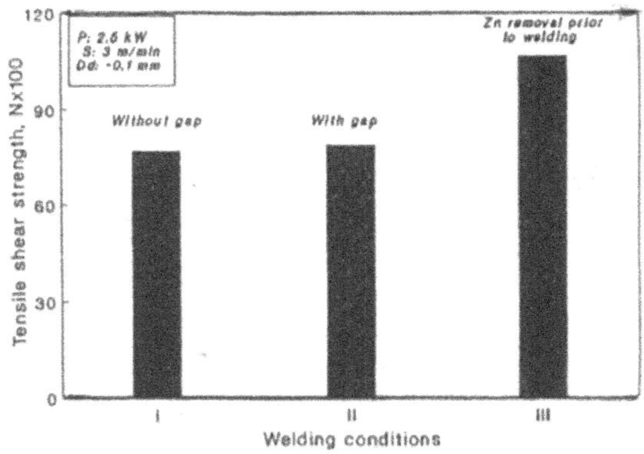

Figure 23. Tensile shear strength of laser lap welded joints as a function of welding conditions used. Welding conditions I, II and III are the optimum conditions used in section 3.3.1, 3.3.2 and 3.3.3, respectively.

Results of tensile shear test of laser lap joints as a function of welding conditions used are shown in Figure 23. Tensile shear strength of joints produced with zinc removal from both sides prior to welding was considerably higher than that of all other joints produced with and without gap regardless of shielding gas type and flow rate. This is attributed to the absence of LME cracking in case of two sides grinding.

Recently, serious industrial incidents of zinc-induced LME cracking in austenitic stainless steel have been reported [29, 30]. The potential for cracking during field welding is certainly greater than the cracking potential in these test specimens. This is due to higher tensile stresses in the case of field welding. Therefore, the removal of galvanized zinc coating prior to welding should be done properly to avoid contamination of austenitic stainless steel with any molten zinc during welding.

3. Conclusion

For CO_2 laser butt welding of similar butt joints of austenitic stainless steels, the following conclusions can be drawn.

- The penetration depth increased with the increase in laser power. However, laser power has a less effect on weld profile.
- Unlike laser power, welding speed has a pronounced effect on size and shape of the fusion zone. Increase in welding speed resulted in an increase in weld depth/width ratio and hence a decrease in the fusion zone size.

- Minimizing heat input and optimizing energy density through optimizing laser power, welding speed, and defocusing distance is of considerable importance for the weld quality in terms of fusion zone size and profile. Helium is more effective than argon as a shielding gas to obtain acceptable weld profile.
- Fusion zone composition was insensitive to change in heat input. However, increase in welding speed and/or decrease in laser power resulted in a finer solidification structure due to low heat input. A dominant austenitic structure with no solidification cracking was obtained for all welds. This could be associated with primary ferrite or mixed mode solidification based on Suutala and Lippold diagrams.
- Mechanical properties, tensile, hardness and bending at room temperature, were not significantly affected by heat input.

For CO_2 laser welding of austenitic stainless steel in lap joints with Zn-coated carbon steel, the following conclusions can be drawn.

- One way to produce sound and uniform laser lap welds of Zn-coated steel with austenitic stainless steel without gap between the sheets is the optimizing shielding conditions. This is of considerable importance for avoiding plasma or preventing porosity and obtaining full penetration without deteriorating the surface quality of the weld. Helium shielding produced noticeably deeper welds while argon exhibited the smoothest top surface. A flow rate of 22 1/min was found to be satisfactory in the case of helium in comparison with 30 1/min for argon.
- The other way to produce sound and homogeneous laser lap welds of these dissimilar materials is the introducing of a small gap (0.025~0.05 mm) between the sheets. Maintaining such gap between the sheets give the zinc vapour an alternative escape route. Smaller gap resulted in pores weld and random instabilities in the weld bead surface while larger gap showed unacceptable levels of drop-through of the weld metal between the sheets.
- To preclude both weld porosity and cracking of the stainless steel by molten zinc in making attachments of Zn-coated steel to 300 series austenitic stainless equipment which in turn will improve tensile shear strength, the choice seems to be very clear that is the zinc coating must be scrupulously removed from the joint area prior to welding.

Recently, new generation of lasers, such as fiber and disk lasers is receiving great attention due to its high efficiency, high power and high beam quality, which can produce an ultra-high peak power density of MW/mm2 levels corresponding to a focused electron beam. These features and advantages of fiber and disc lasers are of considerable importance for deep penetration and high speed welding of austenitic stainless steels with thick sections, which are used in some critical applications such as nuclear power plants. It has been reported that such new generation of lasers is promising to be among the desirable heat sources for deep-penetration high speed welding of thick-section austenitic stainless steels [31-34].

In this regard, multi-passes narrow-gap welding of 50mm thick 316L plates has been investigated using 8 kW disk laser where the effect of welding conditions on the weld bead geometry and welding defects was studied. It is reported that butt joint of 50 mm thick plates with narrow gap could be performed with eight-layers welding at laser power of 6 kW and welding speed of 0.4 m/min. In order to reduce the weld passes further, gas jet assisted laser welding was tried to weld thick 316L plates with a 10 kW fiber laser. The result showed that butt-joint welding of 40 mm plates without filler wire could be carried out at 0.3 m/min welding speed with no porosity or other welding defects. As for 50 mm thick plate, a good weld bead could be obtained with bead-on-plate welding from both sides at 0.2 m/min welding speed [35].

Author details

Abdel-Monem El-Batahgy
Manufacturing Technology Department, Central Metallurgical R & D Institute, Cairo, Egypt

Acknowledgement

The author would like to thank Laser-X Company Ltd., Japan for conducting CO_2 laser beam welding experiments.

4. References

[1] Zacharia, T.; David, S. A.; Vitek, J. M. & Debroy, T. (1989). *Metall. Trans.* 20 A, 1125.
[2] Suutala, N. (1983). *Metall. Trans.* 14 A, 191.
[3] Klimpel, A. & Lisiecki, A. (2007). Laser Welding of Butt Joints of Austenitic Stainless Steel AISI 321. *Journal of Achievements in Materials and Manufacturing Engineering*, 25, 1.
[4] Curcio, F.; et al. (2006). Welding of Different Materials by Diode Laser. *Journal of Materials Processing and Technology*, 175, 83-89.
[5] Zacharia, T.; David, S. A.; Vitek, J. M. & Debroy, T. (1989). *Welding Journal*, 68, 12.
[6] David, S. A.; Vitek, J. M. & Hebble, T. L. (1987). *Welding Journal*. 66, 289.
[7] Schaeffler, A. L. (1949). *Metal Progr.* 56, 680.
[8] Kujanpaa, V.; Suutala, N.; Takalo, T. & Moisio, T. (1979). *Welding Research Int.* 9, 55.
[9] Vitek, J. M. & David, S.A. (1982). *ASM Conference Proceedings*, Trends in Welding Research in the United States (ASM, Metals Park, Ohio,) pp.243-258.
[10] Lippold, J. C. (1985). *Welding Journal*, 64, 127.
[11] Kotecki, D. & Siewert, T. A. (1992). *Welding Journal* 71, 171.
[12] Lippold, J. C. (1994), *Welding Journal*. 73, 129.
[13] Dilthey, U. & Risch, A. (2001). Laser Welding of Stainless Steels and Stainless Low-Alloy Material Combinations, *Welding in the World*, 36, 67-71.
[14] Elmer, J. W.; Alien, S. M. & Eagar, T. W. (1990). *Recent Trends in Science and Technology*, eds. David S. A. & Vitek J. M. (ASM International, Materials Park, Ohio, pp. 165-170.
[15] Brooks, J. A. & Thompson, A. W. (1991). *Inst. Met. Reviews*, 36, 16.
[16] Metzbower, E. A. (1991). Laser Welding, *Naval Engineers Journal*, Vol. 8, p. 41-49

[17] Bruscato, R. M. (1992). Liquid Metal Embrittlement of Austenitic Stainless Steel When Welded to Galvanized Steel. *Welding Journal*, 71(12), p.455s-459s.
[18] Belforte, D. A. (1990). Annual technology, Industry, and Market Review and Outlook. *The Industrial Laser Annual Handbook*, Pennwell Books Laser Focus, Tulsa, Oklahoma USA, SPIE 1241, p. xvi-xxxi.
[19] Roessler, D. M. (1989). Laser Processing-Global Overview and Future Trends. *Proceedings of the 21st International Symposium on Automotive Technology and Automation*, Vol. 1, Wiesbaden, Germany, November 1989, p. 494-516.
[20] Marinono, G.; et al. (1989). Technical and Economic Comparison of Laser Technology with The Conventional Technologies For Welding. *Proceedings of the 6th Int. Conf. on Laser in Manufacturing*, IFS, May 1989, ISBN 1-85423-047-6, p.105.
[21] Beyer, E. & Gasser, E. (1987). Plasma Fluctuations in Laser Welding With CW CO2-Lasers. *Proc. of The 6th Int. Cong. on Appl. Lasers and Electro-Optics, ICALEO '87*, San Diego, California, November 1987, p. 17-23.
[22] Bagger, C.; Miyamoto, I.; Olsen, F. & Maruo, H. (1992). Process Behaviour During High Power CO2 Laser Welding of Zinc Coated Steel. *Proceedings of LAMP*, Nagaoka, Japan, June 1992.
[23] Akhter, R.; Steen, W. M. & Cruciani, D. (1988). Laser Welding of Zinc Coated Steel. *Proc. of The 5th Int, Cong. on Lasers in Manufacturing, LIM 5*, Stuttgart, West Germany, September 1988, p. 195-206.
[24] Heyden, J.; Nilsson, K. & Magnusson, C. (1989). Laser Welding of Zinc Coated Steel, *Proc. of The 6th Int, Conf. on Lasers in Manufacturing, LIM 6*, May 1989, p. 93-104.
[25] Graham, M. P.; Hirak, D. M.; Kerr, H. W. & Weckman, D. C. (1994). Nd-YAG Laser Welding of Coated Sheet Steel. *Journal of Laser Applications*, Vol. 6, p. 212-222.
[26] Stoloff, N. S. (1977). Recent Development in Liquid Metal Embrittlement. *Proceedings Conference on Environment Sensitive Mechanical Behavior*, AIME, Chicago, 1977.
[27] Nicholas, M. G. & Old, C. F. (1979). Review of Liquid Metal Embrittlement. *Journal of Material Science*, Vol. 14, p. 1-18.
[28] Tzeng, Y. (2006). Gap-free lap welding of zinc-coated steel using pulsed CO2 laser. *International Journal of Advanced Manufacturing and Technology*, 29, 287-295.
[29] Johnson, J. M.; Berry, M. R. & Gutzeit, J. (1982). Zinc Embrittlement of Stainless Steel welds. *Proceedings of AIME Meeting on Embrittlement by Liquid and Solid Metals*, St. Louis, MO, 1982.
[30] Shinohara, T. & Matsumoto, K. (1982). Welding Cracks of Zn-contaminated Stainless Steel Pipe. *Corrosion Science*, 22(8), p.723-737.
[31] Thomy, C.; Seefeld, T. & Vollertsen, F. (2005). Proceedings of the Third International WLT-Conference on Lasers in Manufacturing, Munich, Germany, pp.27-32.
[32] Verhaeghe, G. & Hilton, P. (2005). Proceedings of ICALEO, Miami, USA, pp.264-271.
[33] Liu, S.; Kutsuna, M. & Xu, G. (2006). Proceedings of ICALEO, Scottsdale, USA, pp.562-568.
[34] Reem, S. (2006). Proceedings of ICALEO, Scottsdale, USA, pp.586-594.
[35] Zhang, X.; Ashida, E.; Tarasawa, S.; Anma, Y.; Okada, M.; Katayama, S. & Mizutani, M. (2011). Welding of thick stainless steel plates up to 50 mm with high brightness lasers. Journal of Laser Applications, **23**, 022002 (2011); http://dx.doi.org/10.2351/1.3567961

Chapter 3

Dissimilar Metal Joining of Zinc Coated Steel and Aluminum Alloy by Laser Roll Welding

Hitoshi Ozaki and Muneharu Kutsuna

Additional information is available at the end of the chapter

1. Introduction

Nowadays, the car industry has targets to improve fuel consumption due to the problems of the global environment. For example, in the Corporate Average Fuel Economy, CAFE, which was approved in the U. S. Congress in July 2011, attaining the average fuel efficiency of 54.5 mpg (23.2 km/l) by 2025 is called for (Yamamoto, 2012). In addition, it's obliged to improve the average fuel efficiency to 35.5 mpg (15.1 km/l) by 2017. This is a significant increase compared with the present 27.5 mpg (11.7 km/l) which have been fixed for the past 11 years. Hence, the further weight reduction of cars has become imperative.

On the other hand, the weight of vehicle which influences the fuel consumption directly is increasing owing to a rise of safety awareness of automobile user's, tightening of safety standards, and diversification and sophistication of needs. Therefore, the car industry has conflicting targets of the low fuel consumption by "lightening car body" and "safety improvement". Additionally these targets should be achieved in together. Then, a "hybrid body structure" concept has been proposed.

In this concept, high strength steel and light alloy are arranged in the right places of the car body. For instance, by using high strength steel for pillars, a lightening with improving strength is achieved. In addition, by using aluminum alloys for the bonnet, the door panel, and the trunk lid, a further lightening is achieved. In order to make this "hybrid body structure", the welding technology for joining of steel to aluminum with high reliability and productivity is required. However, it's difficult to join steel to aluminum by conventional fusion welding.

Generally, in respect to dissimilar metal joining by the fusion welding, because of melting and alloying largely both metals, intermetallic compound layer formed at joint interface grows thick; oxide firm is formed; hot cracking is generated. Consequently, high joint strength can not be gotten. Moreover, when melting point of both metals has large difference, burn-through occurs in the metal of lower melting point (Nishimoto et al., 2005).

In order to solve these problems, many studies of the dissimilar metal joining have been conducted such as resistance welding (Okita, 2004), laser welding (Katayama, 2004), explosive welding (Satou, 2004), friction stir welding (Okamura & Aota, 2004; Katoh & Tokisue, 2004), and diffusion bonding (Ohashi, 2004). As a result, solid-phase bonding is mainly put to practical use. However, higher productivity, joint strength and flexibility are needed to expand the coverage of application of the dissimilar metal joints. Since laser welding has advantages such as local heat input, short process time and high flexibility compared with other welding processes, there has been considerable research on the welding of steel and aluminum. In the recent, the studies of keyhole welding (Torkamany, et al., 2010), laser welding-brazing (Dharmendra, et al., 2011), dual-beam YAG laser welding (Yan, et al., 2010), and thermal cycle during laser welding (Fan, et al., 2011) are carried out.

Then, the welding process of steel and aluminum which have been regarded as difficult, Laser Roll Welding has been developed for joining of dissimilar metals by M. Kutsuna, M. Rathod and A. Tsuboi in 2002. This welding is a hybrid welding process, as shown in Fig.1, combined high-temperature heating by a 2.4 kW CO_2 laser with pressurizing by a pressure roller. It's registered as Japanese Patents No. 3535152 and No. 3692135.

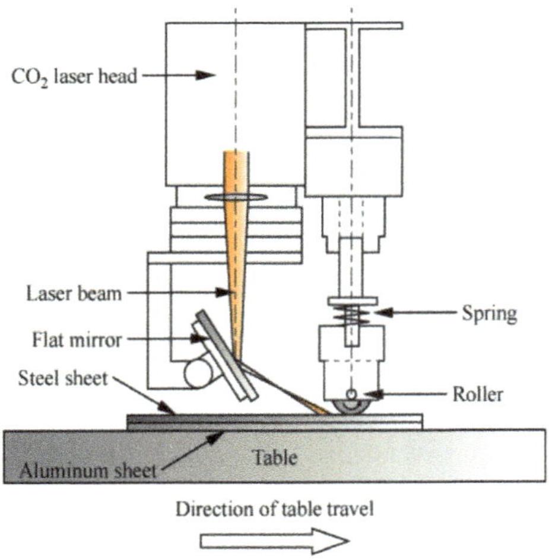

Figure 1. Schematic diagram of Laser Roll Welding process

Fig.2 shows Fe-Al phase diagram (Massalski, et al., 1986). In this figure, there are varioius intermetallic compounds, hereafter, called as IMC's, and they are grouped as Fe-rich compounds, FeAl and Fe_3Al, and Al-rich compounds, $FeAl_2$, Fe_2Al_5 and $FeAl_3$. M. Yasuyama et al. (1996) have shown the mechanical properties of these IMC's. Vickers hardness of cast IMC's is shown in Table 1, and mechanical properties by compressive test are shown in Fig.3.

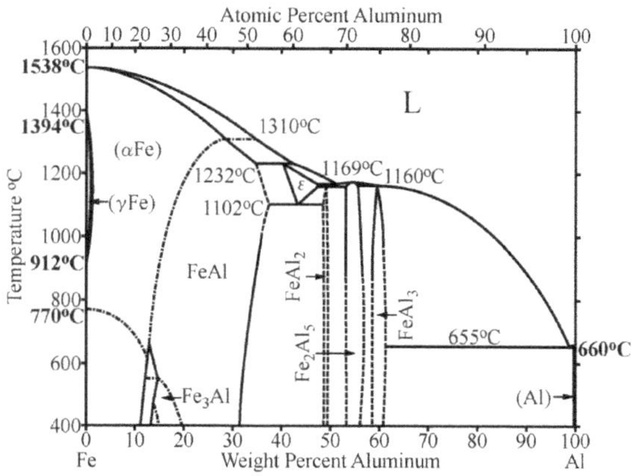

Figure 2. Fe-Al binary equilibrium diagram

	Vickers hardness
$FeAl_3$	892
Fe_2Al_5	1013
FeAl	470
Fe_3Al	330

Table 1. Vickers hardness of Fe-Al intermetallic compounds

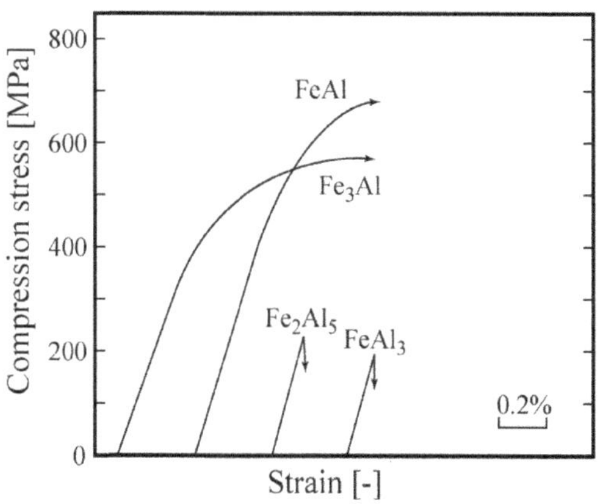

Figure 3. Stress-strain curves in compressive test of Fe-Al intermetallic compounds

While Al-rich IMC's are hard and brittle, Fe-rich IMC's show slight ductility and high strength. These brittle IMC's constitute barriers to the dissimilar metal welding of steel to aluminum. Kutsuna et al. thought if the thickness of brittle IMC layer is minimized and the formation of more ductile IMC's is promoted, high reliable joints are able to be obtained. Therefore, they developed the Laser Roll Welding process. In this process, since the thermal cycle for joining can be shortened by laser heating, the formation of the brittle IMC's can be easily controlled. Furthermore, good contact of a steel sheet and an aluminum sheet and rapid heat transfer from the steel sheet to the aluminum sheet are conducted by the pressure roller.

Kutsuna et al. produced Laser Roll Welidng equipment experimentally, and conducted basic studies of the welding of low carbon steel to aluminum alloy (Rathod &Kutsuna, 2003, 2004). As a result, when the IMC layer thickness was from 4 to 8 μm, failure of base metal sheet was obtained after tensile shear test of specimen.

Until now, it's found that Laser Roll Welding has two types of welding mechanisms. The combination of iron and aluminum is that of metals with large difference in their melting points (Ozaki & Kutsuna, 2007). The combination of titanium and aluminum also corresponds to this category (Ozaki et al., 2008). In this case, only the metal with a higher melting point is heated up to an elevated temperature for avoid the formation of brittle IMC's. For example, in the case of steel to aluminum joint, steel is heated up to 1200 °C, because the max formation temperature of Fe_2Al_5 is 1169 °C. The interlayer is formed and cooled rapidly to minimize the thickness of brittle IMC's.

The other is a combination of metals with eutectic reaction at a lower temperature than melting point. For instance, the combination of titanium and iron falls into this type (Ozaki et al., 2007). In this case, the eutectic reaction occurs at a lower temperature, about 1085 °C, than melting points of both metals. The interlayer of the eutectic phase, β-Ti and TiFe, is formed at the interface.

In this way, uncoated materials have been mainly used in the previous study of Laser Roll Welding. However, coated materials such as zinc coated steel have not investigated enough yet. In this chapter, Laser Roll Welding of zinc coated steel and 6000 series aluminum alloy was conducted, and the weldability was investigated. Two types of zinc coated steel were used. One is hot-dip galvanized steel called as GI, the other is hot-dip galvannealed steel called as GA. The former is used mainly in Europe, the latter is in Japan. Then, the influences of process parameters, such as welding speed and roll pressure, on the formation of intermetallic compound layer and the change of zinc coated layer have been investigated to get sound joints with these galvanized steel. Furthermore, the effects of process parameters on joint performance have been also discussed.

2. Experimental procedure

2.1. Materials used

As materials, two types of zinc coated steel and A6000 series aluminum alloy, Al-0.5Mg-1.0Si, were used for joining. The zinc coated steels were hot-dip galvanized steel, hereafter, called as GI, and hot-dip galvannealed steel, hereafter, called as GA, respectively. The dimension of the zinc coated steel sheets were 125 x 180 x 0.55 mm and that of the A6000 series aluminum alloy sheet, hereafter, called as A6000, was 125 x 180 x 1 mm. The zinc coating weight of the GI sheet was 60 g/m^2 and the GA was 45 g/m^2.

The surface of the steel sheets was coated with graphite by using a graphite spray to increase the absorption rate of laser beam. The thickness of the coating layer was approximately 10 μm. The faying surface of the steel sheets was only degreased with ethanol alcohol. The faying surface of the A6000 sheet was wire-brushed, polished by emery paper #600 and cleaned by ethanol before welding. Then, to remove the oxide firm on aluminum alloy sheet, the surface of the aluminum alloy was spread with flux, $KAlF_4$:$K_2AlF_5 \cdot H_2O$, 17-25 wt%, with particle size of 15 to 21 μm.

The setup of specimens is shown in Fig.4. A zinc coated steel sheet is placed as a top plate; an A6000 sheet is placed as a bottom with 3 mm width overlapping.

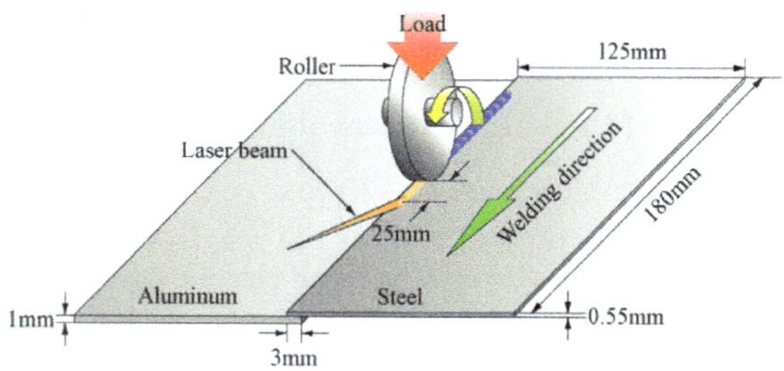

Figure 4. Setup of specimens in Laser Roll Welding

Laser type	CO_2 laser (Pulse mode)
Laser peak power	2.0 kW
Duty cycles	75%
Frequency	150 Hz
Beam spot size	Minor axis : 2.5 mm Major axis : 3.5 mm
D_{LR}	25 mm
Welding speed	3.3 ~ 15.0 mm/s
Overlapped width	3 mm
Roll pressure	100 ~ 175 MPa
Shielding gas	Ar : 25 l/min

D_{LR}: Distance between Laser beam spot and Roller

Table 2. Process parameters for Laser Roll Welding

2.2. Process parameters

In this study, to investigate the effect of the welding condition on the joint properties of two types of zinc coated steel and aluminum alloy by Laser Roll Welding, welding speed and roll pressure were varied. Process parameters are shown in Table 2.

Pulsed laser was used by controlling a 2.4 kW CO_2 laser with continues wave. Laser peak power, duty cycles and frequency were constant 2.0 kW, 75 % and 150 Hz respectively. These parameters have been optimized in the previous study. Beam spot shape was an quasi-elliptical with major axis of 3.5 mm and minor axis of 2.5 mm. Distance between the center of laser beam spot and the pressurizing axis of roller was 25 mm. Welding speed was varied from 3.3 to 15.0 mm/s. Overlapped width was 3 mm, and laser beam was irradiated the center of overlap. The roller is mounted with a calibrated compression spring for applying predetermined roll pressure. The roll pressure was calculated by assuming that the contact area between the roller and the steel sheet surface was a rectangle of 15 mm^2. Roll pressure was varied from 100 to 175 MPa. For the sake of the protection of condenser lens and the oxidation prevention of Laser Roll Welded joints, argon gas with flow rate of 25 l/min was used.

2.3. Evaluation method of Laser Roll Welded joints

After welding, Laser Roll Welded specimens were cut across the lap joint seam for macrostructure and microstructure observation. Etching with 3% nital was made for observation of the interface layer. Hardness test and electron-probe microanalysis (EPMA) were conducted to analyze the interface layer and to identify the IMC present.

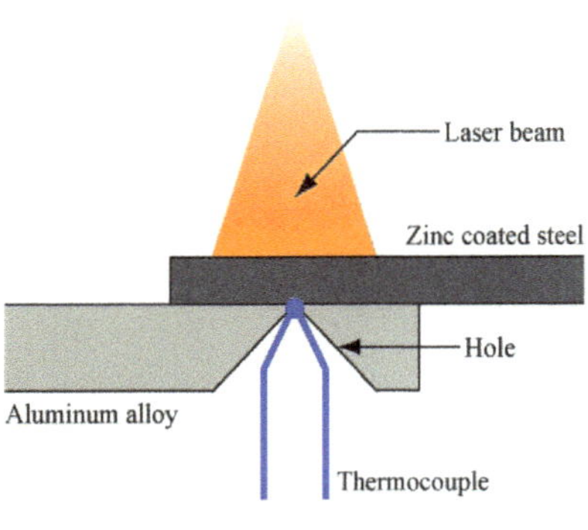

Figure 5. Schematic diagram of measurement method for thermal cycle at weld interface

Thermal cycle was measured using a Pt-PtRh thermocouple with a diameter of 0.3 mm. As shown in Fig.5, a cone-shaped hole of 2 mm in diameter was drilled at the center of the overlap width in the aluminum sheet. The hole was used for placing thermocouple at the lower surface of the steel sheet. In addition, tensile shear test and Erichsen cupping test were carried out to investigate the mechanical properties of welded joint. The specimens with 20 mm width were used for tensile shear test and with 77 mm square for Erichsen cupping test.

3. Experimental results and discussions

3.1. Bead appearance and cross-section of Laser Roll Welded joints

Bead appearance of Laser Roll Welded specimen with the welding speed of 8.3 mm/s and roll pressure of 150 MPa are shown in Figs.6-7. Top bead is shown in Fig.6, and bottom bead is in Fig.7. Furthermore, cross-section of the weld bead is shown in Fig.8. In these figures, (a) shows GI/A6000 joint and (b) shows GA/A6000, respectively.

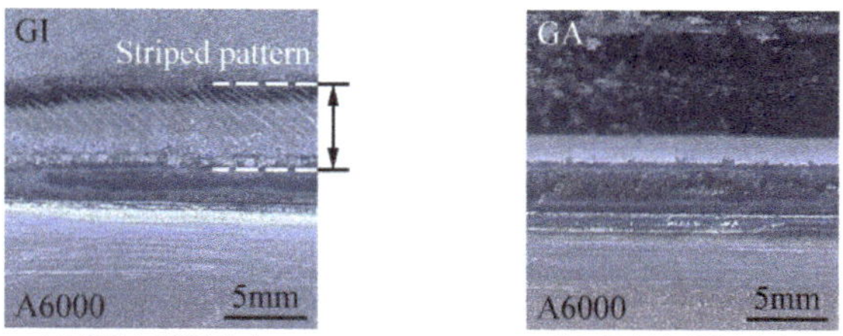

(a) GI/A6000 joint (b) GA/A6000 joint

Figure 6. Top bead appearance of GI/A6000 and GA/A6000 Laser Roll Welded joints

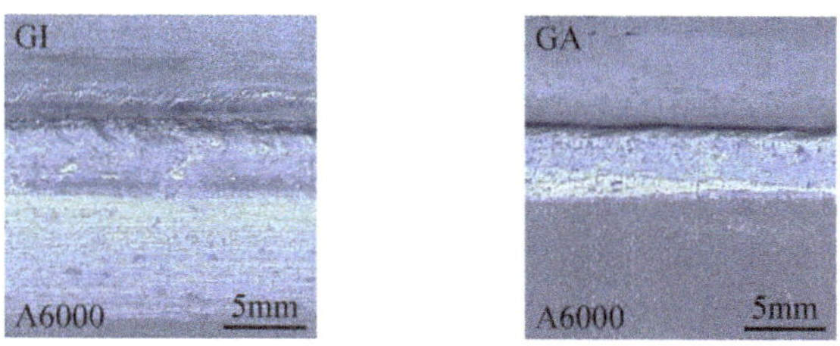

(a) GI/A6000 joint (b) GA/A6000 joint

Figure 7. Bottom bead appearance of GI/A6000 and GA/A6000 Laser Roll Welded joints

(a) GI/A6000 joint (b) GA/A6000 joint

Figure 8. Macro cross-section of GI/A6000 and GA/A6000 Laser Roll Welded joints

As shown in Fig.6, both of the bead turned black due to the effect of the graphite spray and the laser heating. In Fig.6 (a), striped pattern is observed on the GI sheet near the weld bead. This is the first time this striped pattern has been observed in the past study of Laser Roll Welding, the pattern seems to be unique to this welding of galvanized steel sheet. A GI sheet has galvanized layer, which exists as a thin layer at the surface of steel sheet. Thus it's thought that the pattern was caused by the laser heating and the roller pressure. On the other hand, the striped pattern doesn't exist on the GA sheet as shown in Fig.6 (b). This is because the coated layer of a GA sheet is alloyed zinc with iron.

As shown in Figs.7-8 (a), partial melting and spreading of molten aluminum alloy on the GI sheet occurs in the A6000 sheet. The GI sheet was heated by laser, and the heat transferred to A6000 sheet by the contact of GI and A6000 sheets. Hence A6000 sheet was supposed to be melted. In contrast, molten A6000 alloy doesn't spread on the GA sheet.

In order to compare the wettability of the GI sheet with that of the GA sheet, bonding width was measured from the cross-section of the weld bead as shown in Fig.9 (a). Effect of the welding speed and the roll pressure on the bonding width of GI/A6000 and GA/A6000 joints are shown in Fig.9 (b) and (c), respectively.

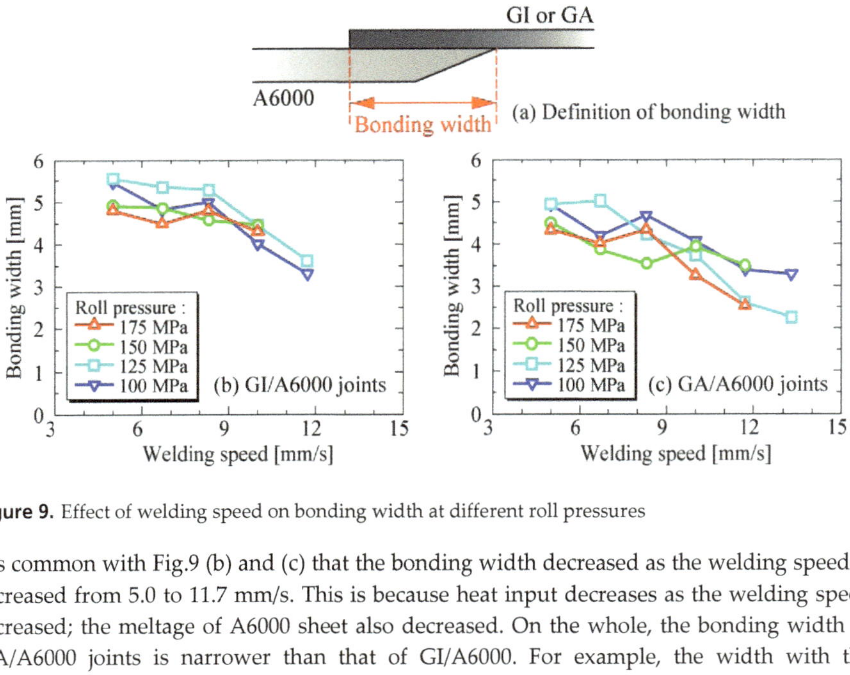

Figure 9. Effect of welding speed on bonding width at different roll pressures

It's common with Fig.9 (b) and (c) that the bonding width decreased as the welding speed is increased from 5.0 to 11.7 mm/s. This is because heat input decreases as the welding speed increased; the meltage of A6000 sheet also decreased. On the whole, the bonding width of GA/A6000 joints is narrower than that of GI/A6000. For example, the width with the welding speed of 8.3 mm/s and roll pressure of 150 MPa of GI/A6000 joint is 4.6 mm, and that of GA/A6000 is 3.5 mm. This is considered to be produced from the difference in the wettability of the surface of the GI and GA sheet.

3.2. Microstructures at weld interface

Interface microstructures of GI/A6000 joint with welding speed of 6.7 mm/s and roll pressure of 150 MPa are shown in Fig.10.

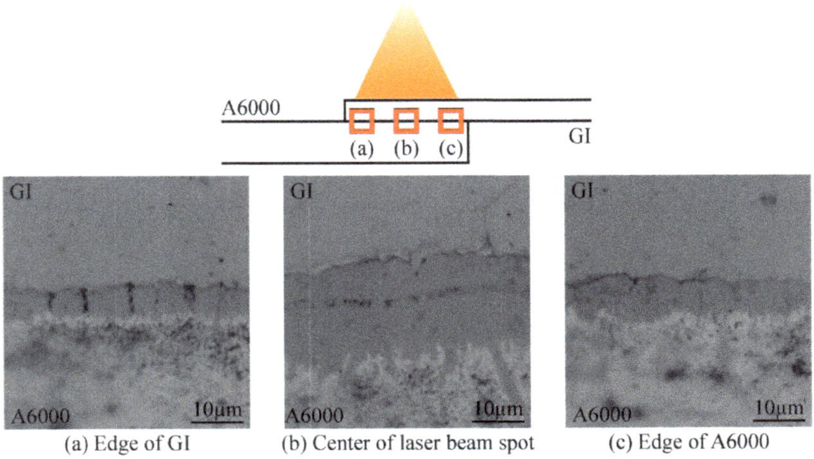

Figure 10. Microstructures at joint interface of GI and A6000 sheet

IMC layer was confirmed at the joint interface between the GI and the A6000 sheet. The IMC layer at the edge of the GI sheet is shown in Fig.10 (a), at the center of the laser beam spot is shown in (b) and at the edge of the A6000 sheet is show in (c). The IMC layer thicknesses in each place were 4.1 μm, 13.0 μm, and 4.9 μm respectively. Therefore, the thickness right under the center of laser beam spot was the thickest. Due to the high energy density near the center of the beam spot of the CO_2 laser used, it's thought that the area under the center of the beam spot is heated the highest temperature and cooled the slowest speed.

Interface microstructures of GA/A6000 joint with welding speed of 8.3 mm/s and roll pressure of 150 MPa are shown in Fig.11. IMC layer was also observed at the interface between the GA and the A6000 sheet. In addition, the thickness right under the center of laser beam spot was the thickest, and this was the same as that of the case of GI/A6000 joint.

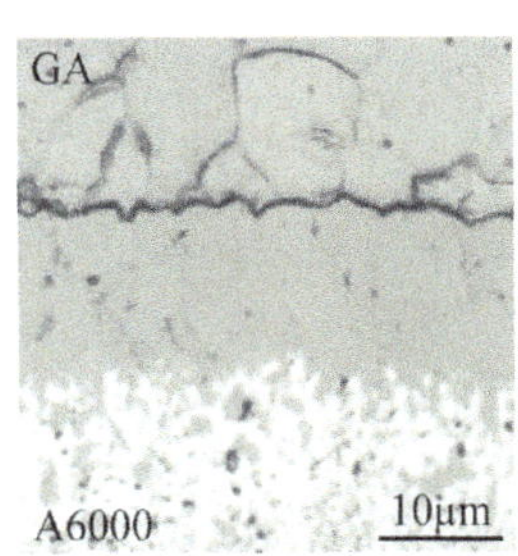

Figure 11. Microstructure at joint interface of GA and A6000 sheet

3.3. Electron-probe microanalysis (EPMA) of interlayer

As the optical microscope doesn't reveal the details of the IMC layers, EPMA of iron, aluminum and zinc across joint interface were made to identify IMC's and existence of zinc. The results for GI/A6000 and GA/A6000 joints at different welding speeds with roll pressure of 150 MPa are shown in Figs.12-13. Bottom pictures show the SEM images of the IMC layer. In these figures, the EPMA results with the welding speed of 8.3 mm/s are shown in (a), with the welding speed of 10.0 mm/s are shown in (b).

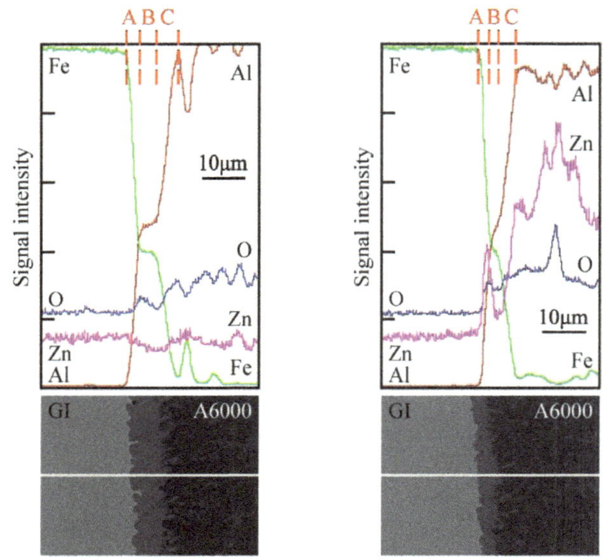

(a) Welding speed = 8.3 mm/s (b) Welding speed = 10.0 mm/s

Figure 12. Results of EPMA at interlayer of GI/A6000 joints

From EPMA results of iron, aluminum and zinc across interface, IMC's and existence of zinc were identified. At the center of these EPMA results which are shown by the layer A to C in Fig.12 as an example, the signal intensity of Fe and Al has changed similarly. At the layer A, the intensity of Fe decreases rapidly, and Al rises. At the layer B, the IMC's are observed as stepped lines. At the layer C, the intensity of Fe decreases further and Al rises. From the estimation of composition from stepped lines at the layer B, it's suggested that main IMC's were brittle $FeAl_3$ and Fe_2Al_5.

When the welding speed is increased to 10.0 mm/s, zinc can be seen in aluminum alloy. From this result, it's thought that zinc tends to diffuse into aluminum when the welding speed becomes more than 10.0 mm/s. This reason is considered as follows.

Most of the zinc layer is vapored by the laser heating from the faying surface of the GI sheet at slow welding speeds. On the other hand, at fast welding speeds, the diffusion amount of the zinc into aluminum alloy by the roll pressure is larger than the evaporation amount of the zinc by the laser heating.

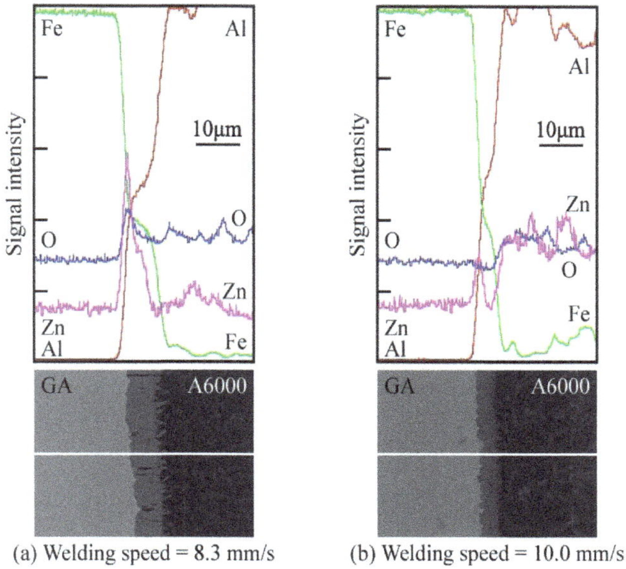

Figure 13. Results of EPMA at interlayer of GA/A6000 joints

As shown in Fig.13 (a), the zinc line has higher peak than any other zinc line at the interface of the GA sheet side. This is because the zinc on GA sheet surface exists as Fe-Zn alloy. Then it's suggested that zinc was hard to be evaporated by the laser heating, and a lot of zinc was remained at the interface of the GA sheet side.

3.4. Vickers hardness measurements

From the EPMA results, it was presumed that main IMC's were brittle $FeAl_3$ and Fe_2Al_5. However, in order to obtain further evidence, Vickers hardness measurement was conducted. SEM image of GI/A6000 and GA/A6000 weld interface after the measurement with welding speed of 8.3 mm/s and roll pressure of 150 MPa are shown in Fig.14 (a) and (b), respectively.

Indentation size became small in order of the base metal of A6000, zinc coated steel, and the IMC layer. In particular, the indentation size of the IMC layer is much smaller than that of both base metals. As shown in Fig.14 (a), the Vickers hardness of the base metal of the GI, the A6000, and the IMC layer were 137 Hv, 94 Hv and 940 Hv on average respectively. The results show a large difference between IMC layer hardness and the surrounding base metals. The IMC hardness was about 10 times more than the A6000 base metal and 7 times more than the GI base metal. In the same way, as shown in Fig.14 (b), the Vickers hardness of the base metal of the GA, the A6000, and the IMC layer were 141 Hv, 97 Hv and 857 Hv on average respectively, the IMC layer was the hardest. Because the hardness of these IMC layers are between 892 Hv of $FeAl_3$ and 1013 Hv of Fe_2Al_5 according to Table 1, it's thought that $FeAl_3$ and Fe_2Al_5 are mainly formed at the interface as above.

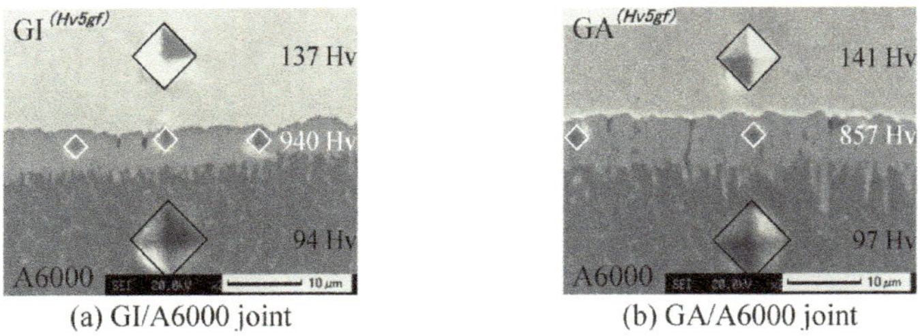

Figure 14. Results of Vickers hardness measurement at joint interface

3.5. Effect of welding speed on thickness of intermetallic compound layer

Effect of the welding speed on the IMC layer thickness of GI/A6000 and GA/A6000 joints at different roll pressure are shown in Fig.15 (a) and (b), respectively. The thicknesses were measured right under the center of the laser beam spot.

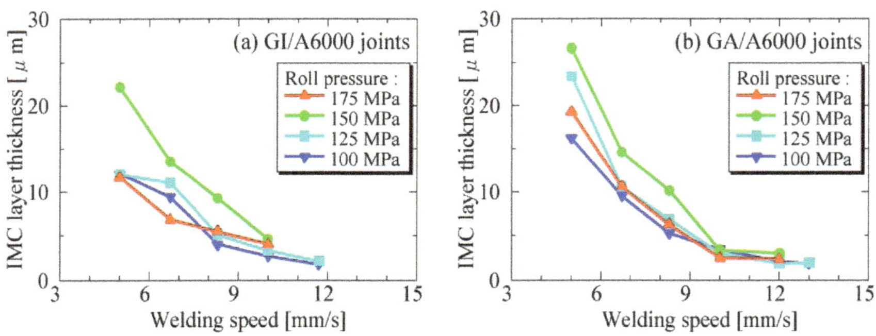

Figure 15. Effect of welding speed on IMC layer thickness of Laser Roll Welded joints interface

The IMC layer thickness decreases significantly as the welding speed is increased from 5.0 to 11.7 mm/s regardless of the roll pressure as shown in Fig.15 (a) and (b). This is because heat input decreases as the welding speed in increased. Therefore, this result indicates that the IMC layer thickness could be suppressed by heat input as was mentioned in previous study of Laser Roll Welding of steel to aluminum alloy.

3.6. Effect of roll pressure on thickness of intermetallic compound layer

Effect of the roll pressure on the IMC layer thickness of GI/A6000 and GA/A6000 joints at constant welding speed of 8.3 mm/s is shown in Fig.16.

The IMC layer thickness of both joints increases as the roll pressure is increased from 100 to 150 MPa. This is because increment of the roll pressure augments the contact between the zinc coated steel and A6000 sheet in this region. In contrast, the IMC layer thickness decreases as the roll pressure is increased from 150 to 175 MPa. This reason is considered as follows.

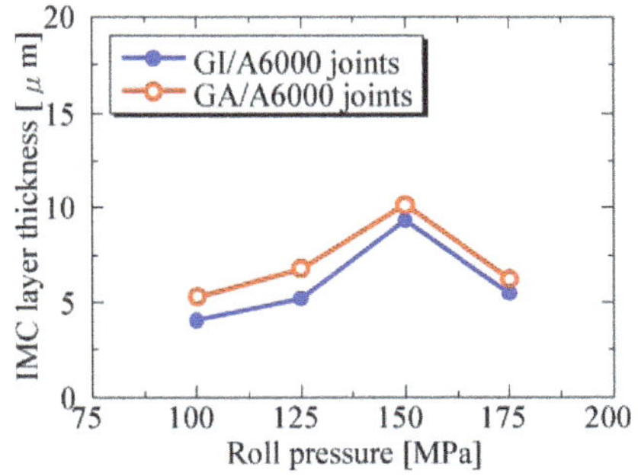

Figure 16. Effect of roll pressure on IMC layer thickness of Laser Roll Welded joints interface

The zinc coated steel and the A6000 sheet contact most widely at roll pressure of 150 MPa. However, when the roll pressure becomes more than 150 MPa, the steel sheet might be curved greatly by high roll pressure. Thus, it's thought that the contact area between the zinc coated steel and the A6000 sheet is narrowed in this region.

3.7. Effect of types of zinc coated steel on thickness of intermetallic compound layer

As shown in Fig.16, the IMC layer thickness of GA/A6000 joints is thicker than that of GI/A6000. This difference is attributed to the evaporation of zinc on the surface of each zinc coated steel sheets.

The boiling point of aluminum, iron and zinc are 2477 °C, 2887 °C and 906 °C respectively, that of zinc is far below those of aluminum and iron. Hence, the zinc on the surface of zinc coated steel sheet is melting, heat of fusion of 7.12 kJ/mol, and evaporating, heat of vaporization of 113.4 kJ/mol (The Japan Institute of Metals, 2004), in the process of Laser Roll Welding. However, because of the zinc on the surface of the GA sheet exists as Fe-Zn alloy, it's hard to be evaporated by the laser heating. Thus, laser energy is little-used for the zinc evaporation, and heat is conducted to A6000 sheet. On the other hand, the zinc on the surface of the GI sheet is easier to be evaporated than that of the GA. Therefore, it's thought that more laser energy for the zinc evaporation is used, and the heat conduction to A6000 sheet decreases.

3.8. Discussion of formation of intermetallic compound layer by thermal cycle at weld interface

Thermal cycle at the weld interface was measured to discuss about the effects of the welding speed and the roll pressure on the IMC layer thickness of Laser Roll Welded joints. The results of GI/A6000 and GA/A6000 joints at constant roll pressure of 150 MPa is shown in Fig.17 (a) and (b), respectively.

When the welding speed increases from 5.0 to 11.7 mm/s, different thermal cycles were obtained as shown in Fig.17 (a) and (b). Then, in order to quantitative the shape of these thermal cycles, peak temperature and holding time more than 500 °C were focused on.

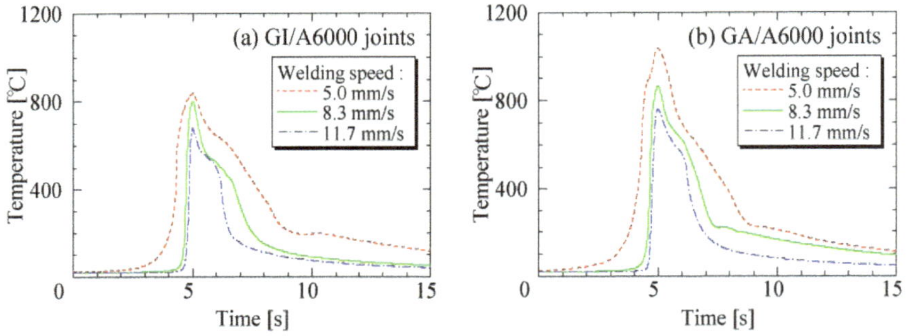

Figure 17. Effect of welding speed on interface thermal cycles of Laser Roll Welded joints

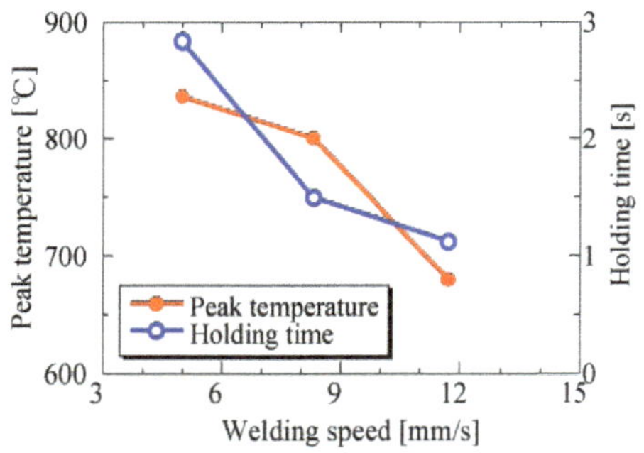

Figure 18. Effect of welding speed on peak temperature and holding time more than 500 °C at interface of GI/A6000 joints

The peak temperature and the holding time more than 500 °C of the thermal cycles in Laser Roll Welding of the GI and the A6000 at different welding speeds are shown in Fig.18. When the welding speed is increased from 5.0 to 11.7 mm/s, the peak temperature decreases from 850 to 680 °C, and the holding time more than 500 °C shortens from 2.8 to 1.1 sec at the weld interface. Hence the reduction of the IMC layer thickness by the increment of welding speed was attributed to the decline of the peak temperature and the shortening of the holding time. The similar results were obtained in Laser Roll Welding of the GA and the A6000.

When the welding speed is slow, there is excessive heat input and the cooling rate is slow. This provides surplus time for the formation of a thick interlayer containing a large amount of the Al-rich brittle IMC's. When the welding speed is fast, there is suitable heat and time for melting of the aluminum and the diffusion process to take place. Therefore, it's thought that the change of the thermal cycle at the interface affect the formation of the IMC layer when the welding speed is varied.

In addition, when Fig.17 (a) is compared with (b), the peak temperatures of GA/A6000 joints are higher than those of GI/A6000. This fact caused the formation of thicker IMC layer in GA/A6000 joints than in GI/A6000 as shown in Fig.16.

Effect of the roll pressure on the interface thermal cycles of GA/A6000 joints at constant welding speed of 8.3 mm/s is shown in Fig.19. Thermal cycles are shown in Fig.19 (a) and the peak temperature and the holding time more than 500 °C of them are in (b). The reason of the changing of IMC layer thickness by the increment of the roll pressure as shown in Fig.16 is considered by using Fig.19.

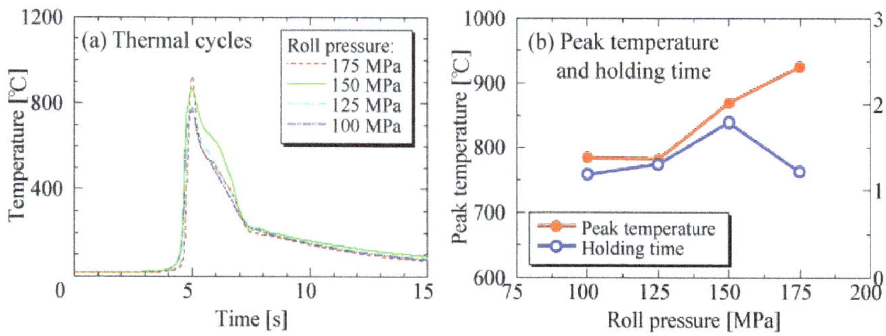

Figure 19. Effect of roll pressure on interface thermal cycles of GA/A6000 joints

From 100 to 150 MPa, the peak temperature and the holding time increase as the roll pressure is increased in Fig.19. Hence the IMC layer thickness increased as the roll pressure is increased in Fig.16. This is because increment of the roll pressure augments the contact between zinc coated steel and A6000 sheet in this region as above. In contrast, from 150 to 175 MPa, the peak temperature rises, but holding time shortens as the roll pressure is increased in Fig.19. It's supposed that this tendency caused the decrement of IMC layer thickness as the roll pressure is increased at high region in Fig.16.

3.9. Results of tensile shear test

Tensile shear test was conducted to investigate the influence of the welding conditions on the weldability. The tensile shear specimens were prepared by cutting the welded specimen with 20 mm width.

Fig.20 shows the tensile shear specimen after testing of GI/A6000 joints. Failure in base metal of the GI sheet is shown in Fig.20 (a), and failure in interface is shown in (b). It's found that the base-metal-failure specimen failed far from the weld bead as shown in Fig.20 (a).

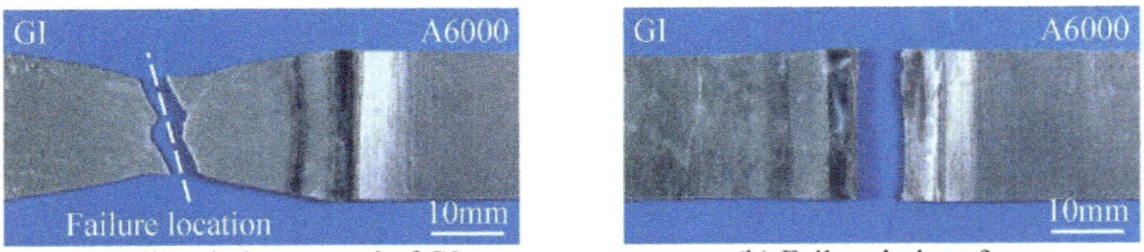

(a) Failure in base metal of GI (b) Failure in interface

Figure 20. Tensile shear specimen after testing of GI/A6000 Laser Roll Welded joints

Results of tensile shear test of GI/A6000 and GA/A6000 joints at various welding speed with the roll pressure of 150 MPa is shown in Fig.21 (a) and (b), respectively. Here the tensile shear strength was converted into the tensile shear load per millimeter of weld length, N/mm, as adopted by Peyre et al. (2007) and Sasabe et al. (2007). Failure in the base metal of the zinc coated steel sheet is shown as a circle in the figure, and failure in interface is shown as a solid mark.

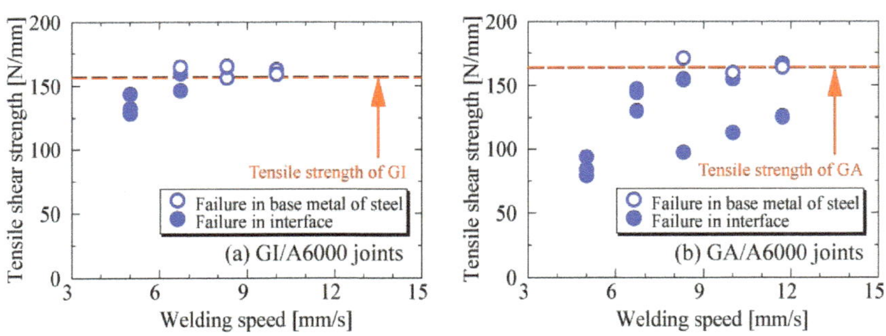

Figure 21. Effect of welding speed on tensile shear strength for Laser Roll Welded joints

When the welding speed is between 6.7 and 10.0 mm/s, the tensile shear strength shows high, and there are many specimens failed in the base metal. This result indicates that the IMC layer thickness is appropriate at these welding speeds. As shown in Fig.21 (a), specimens failed in the base metal could be obtained from 6.7 to 10.0 mm/s. In this region, the IMC layer thickness was less than 10 μm, and failure of specimen occurred in the base metal of the steel sheet. Therefore, it's thought that the IMC layer thickness should be less than 10 μm to get a good joint. As shown in Fig.21 (b), specimens failed in base metal were confirmed when the IMC layer thickness was less than 10 μm. This result corresponded to the articles by other researchers (Bruckner, 2005; Furukawa, 2005).

Additionally, when Fig.21 (a) is compared with (b), the tensile shear strength of GA/A6000 joints is lower than those of GI/A6000. There are three points which can be considered as this reason.

The first is the bonding width of GA/A6000 joints was narrower than that of GI/A6000 as shown in Fig.9. Therefore, when tensile shear load was applied to the joints with narrow bonding width, the load concentrated to the weld. The second is the IMC layer thickness of GA/A6000 joints was thicker than that of GI/A6000 as shown in Fig.16. Moreover, the IMC layers were mainly composed of Al-rich brittle IMC's from the results of the EPMA and the Vickers hardness. The third is the residual zinc at the interface of GA/A6000 joints influenced the tensile shear strength. From the results of the EPMA, a lot of zinc was remained as Fe-Zn alloy layers at the interface of GA sheet side. The zinc coated layer of the GA sheet is formed from the steel side in order of Γ-phase (Fe_3Zn_{10}), $Γ_1$- (Fe_5Zn_{21}), δ- ($FeZn_7$) and ζ- ($FeZn_{13}$) (The Iron and Steel Institute of Japan, 1982). Since each phase has a difference in generation speed, the Γ-phase is thin and alloy layers are composed mostly of δ- and ζ-. Their phases are hard and brittle. Therefore, the interface of the GA/A6000 joints is weaker than that of the GI/A6000.

The maximum tensile shear strength of GI/A6000 joint, 162 N/mm, was obtained at the welding speed of 8.3 mm/s and the roll pressure of 150 MPa. This strength is equal to the tensile strength of the GI sheet, 157 N/mm, and 63% of the A6000 sheet, 256 N/mm. On the other hand, the maximum strength of GA/A6000 joint, 160 N/mm, was obtained at the welding speed of 10.0 mm/s and the roll pressure of 100 MPa. This strength is equal to the tensile strength of the GA sheet, 164 N/mm, and 62% of the A6000 sheet, 256 N/mm.

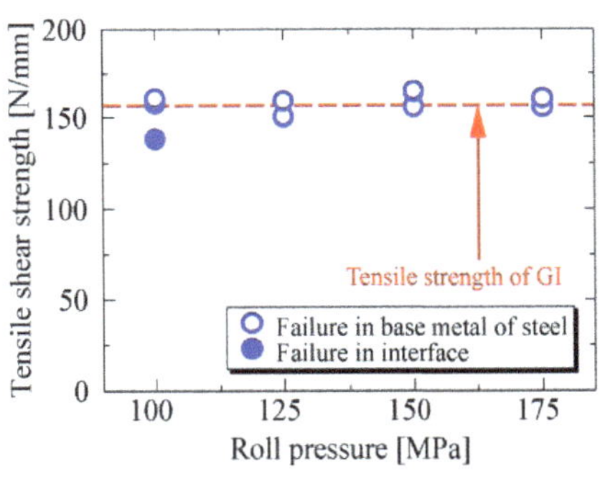

Figure 22. Effect of roll pressure on tensile shear strength for GI/A6000 joints

In addition, results of tensile shear test at various roll pressure with welding speed of 8.3 mm/s is shown in Fig.22. All specimens were failure in the base metal of the GI sheet except for the roll pressure 100 MPa. It's considered that the IMC layer thickness is less than 10 μm at these roll pressures as shown in Fig.16. Therefore, it seems that there is little influence of the roll pressure on the tensile shear strength.

3.10. Discussion of relation between intermetallic compound layer thickness and tensile shear strength

From the above results, it has became clear that there is a close relation to the IMC layer thickness to the tensile shear strength also in Laser Roll Welding of zinc coated steel and aluminum alloy. There, the relationship between the IMC thickness and the tensile strength was discussed about all joints in this experiment. Effect of the IMC layer thickness on the tensile shear strength of GI/A6000 and GA/A6000 joints with different welding speed and roll pressure is shown in Fig.23 (a) and (b), respectively.

As above, only when the IMC layer thickness was less than 10 μm, the base-metal-failure specimens were obtained. When the IMC layer thickness is from 4 to 6 μm, high tensile shear strength could be obtained. When the IMC layer is thicker than 6 μm, the tensile strength is decline. This is because the increment of the brittle IMC's at the joint interface might lead to weaken of the welded joint, and the joint isn't able to resist a heavy load.

On the other hand, when the IMC layer is thinner than 4 μm, the tensile strength also declines. In this case, due to low heat input might lead to incomplete welding at the joint interface. As shown in Fig.9, the bonding width is decreased by the decrement of the heat input.

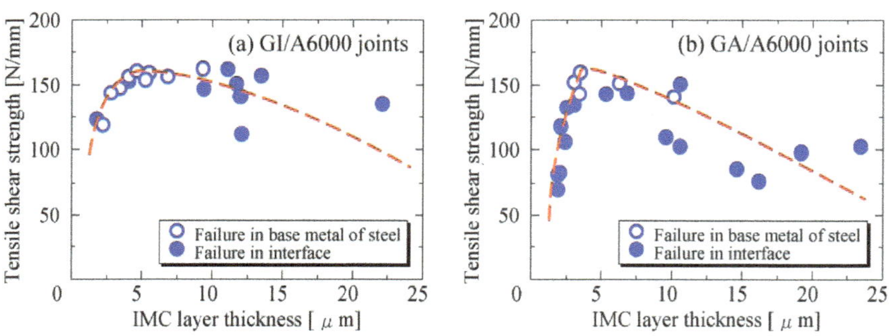

Figure 23. Effect of IMC layer thickness on tensile shear strength for Laser Roll Welded joints

3.11. Results of Erichsen cupping test

Finally, in order to investigate the formability of GI/A6000 and GA/A6000 joints, Erichsen cupping tests were carried out. The specimens were prepared by cutting the welded joints with the welding speed of 8.3 mm/s and roll pressure of 150 MPa into 77 mm square. A punch is pushed into the specimen. When the specimen is failed somewhere, the cupping height is evaluated as Erichsen value.

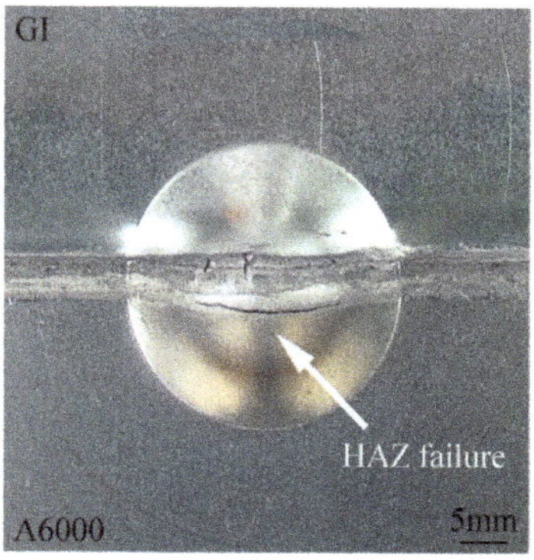

Figure 24. Erichsen cupping test specimen after testing of GI/A6000 joint

The Erichsen cupping test specimen after testing of GI/A6000 joint is shown in Fig.24. The specimen was failed at HAZ of the A6000 sheet side and the Erichsen value was 7.9 mm. With the GI base metal, the Erichsen value was 11.9 mm; with the A6000 base metal, the value was 8.6 mm. Therefore, this value was 92% of the base metal of A6000 sheet. In contrast, the specimen of GA/A6000 joint was failed at interface and the value was 3.6 mm. The same tendency was seen in the tensile shear test.

4. Conclusions

The present study is focused on the dissimilar metal joining of zinc coated steel and aluminum alloy by Laser Roll Welding. The following conclusions can be drawn.

1. The IMC layer was observed at the interface of all welded joins. It was suggested that most of the IMC's are brittle $FeAl_3$ and Fe_2Al_5 from the results of EPMA and Vickers hardness measurement. As the welding speed was faster than 10.0 mm/s, zinc was confirmed in aluminum alloy.
2. Increase in the welding speed led to decrease the bonding width and the IMC layer thickness at the joint interface. When the roll pressure was increased, the IMC thickness at the pressure of 150 MPa was the thickest. The IMC layer thickness of GA/A6000 joints was thicker than that of GI/A6000 on the whole.
3. Increase in the welding speed led to lowering of the peak temperature and shortening of the holding time more than 500 °C at the interface. The peak temperature at the roll pressure of 175 MPa was the highest, and the holding time at the pressure 150 MPa was the longest. The peak temperature of GA/A6000 joints was higher than that of GI/A6000 at the same welding condition.
4. When the IMC layer was less than 10 μm, failure of specimen occurred at the base metal of zinc coated steel in tensile shear test. The joint properties of GI/A6000 joints were better than those of GA/A6000 from the results of tensile shear test and Erichsen cupping test.

5. The welding speed influenced the joint performance such as the IMC layer thickness and the tensile shear strength to a greater degree than the roll pressure in Laser Roll Welding.

Author details

Hitoshi Ozaki
Graduate School of Engineering, Mie University, Japan

Muneharu Kutsuna
Advanced Laser Technology Research Center Co., Ltd., Japan

Acknowledgement

The authors would like to thank Mr. S. Nakagawa and Mr. K. Miyamoto from Research Center, Nissan Motor Co., Ltd. for their support in this research.

5. References

Bruckner, J. (2005). Cold Metal Transfer Has a Future Joining Steel to Aluminum. *Welding Journal*, Vol.84, No.6, pp.38-40, ISSN 0043-2296

Dharmendra, C., Rao, K. P., Wilden, J. & Reich, S. (2011). Study on Laser Welding-brazing of zinc Coated Steel to Aluminum Alloy with a Zinc Based Filler. *Material Science and Engineering A*, Vol.528, pp.1497-1503, ISSN 0921-5093

Fan, J., Thomy, C. & Vollertsen, F. (2011). Effect of Thermal Cycle of the Formation of Intermetallic Compounds in Laser Welding of Aluminum-Steel Overlap Joints. *Physics Procedia*, Vol.12, pp.134-141, ISSN 1875-3892

Furukawa, K. (2005). Welding Process of Iron-Aluminum. *Welding Technology*, Vol.53, No.8, pp.94-102, ISSN 0387-0197

Katayama, S. (2004). Laser Welding of Aluminum Alloys and Dissimilar Metals. *Journal of Light Metal Welding & Construction*, Vol.42, No.1, pp.16-25, ISSN 0368-5306

Katoh, K. & Tokisue, H. (2004). Dissimilar Friction Welding of Aluminum Alloys to Other Materials. *Journal of Light Metal Welding & Construction*, Vol.42, No.2, pp.11-18, ISSN 0368-5306

Massalski, T. B. (1986). *Binary Alloy Phase Diagrams Volume 1*, American Society for Metals, p.148, ISBN 0871702614, Ohio

Nishimoto, K., Atagi, K., Fujii, H. & Katayama., S. (2005). Laser Pressure Welding of Dissimilar Metals Welding. *Proceedings of the 63rd Laser Materials Processing Conference*, pp.133-138, ISBN 4947684585, May, 2005

Ohashi, O. (2004). Diffusion Bonding of Aluminum to Different Metals. *Journal of Light Metal Welding & Construction*, Vol.42, No.2, pp.19-23, ISSN 0368-5306

Okamura, H. & Aota, K. (2004). Joining of Dissimilar Materials with Friction Stir Welding, *Journal of Light Metal Welding & Construction*, Vol.42, No.2, pp.1-10, ISSN 0368-5306

Okita, T. (2004). Resistance Welding of Aluminum Alloys to Dissimilar Metals. *Journal of Light Metal Welding & Construction*, Vol.42, No.1, pp.2-15, ISSN 0368-5306

Ozaki, H., Hayashi S. & Kutsuna M. (2008). Laser Roll Welding of Dissimilar Metal Joint of Titanium to Aluminum Alloy. *Quarterly Journal of Japan Welding Society*, Vol.26, No.1, pp.24-30, ISSN 0288-4771

Ozaki, H., Ichioka R. & Kutsuna M. (2007). Laser Roll Welding of Dissimilar Metal Joint of Low Carbon Steel and Titanium. *Quarterly Journal of Japan Welding Society*, Vol.25, No.1, pp.173-178, ISSN 0288-4771

Ozaki, H. & Kutsuna M. (2007). Laser Roll Welding of Dissimilar Metal Joint of Low Carbon Steel to Aluminum Alloy Using 2kW Fiber Laser. *Quarterly Journal of Japan Welding Society*, Vol.25, No.4, pp.473-479, ISSN 0288-4771

Peyre, P., Sierra, G., Deschaux-Beaume, F., Stuart, D. & Fras, G. (2007). Generation of Aluminum-steel Joints with Laser-induced Reactive Wetting. *Material Science and Engineering A*, Vol.444, pp.327-338, ISSN 0921-5093

Rathod, M. & Kutsuna M. (2003). Laser Roll Bonding of A5052 Aluminum Alloy and SPCC Steel. *Quarterly Journal of the Japan Welding Society*, Vol.21, No.2, pp.282-294, ISSN 0288-4771

Rathod, M. J. & Kutsuna M. (2004). Joining of Aluminum Alloy 5052 and Low-Carbon Steel by Laser Roll Welding. *Welding Journal*, Vol.83, No.1, pp.16s-26s, ISSN 0043-2296

Sasabe, S., Iwase, T., Matsumoto, T., Hattori, Y. & Miono, T. (2007). Dissimilar Metal Joining of Aluminum Alloys to Steel by using Newly-developed Hot-dip Aluminized Steel Sheet. *Journal of Light Metal Welding & Construction*, Vol.45, No.2, pp.23-33, ISSN 0368-5306

Satou, D. (2004). Aluminum Clad of Explosive Welding. *Journal of Light Metal Welding & Construction*. Vol.42, No.1, pp.26-30, ISSN 0368-5306

The Iron and Steel Institute of Japan (1982). *Iron and Steel Handbook Vol. 6 (3rd Edition)*, Maruzen, pp.421-434, Tokyo

The Japan Institute of Metals (2004). *Metals Data Book (Revised 4th Edition)*, Maruzen, p.11, ISBN 4621073672, Tokyo

Torkamany, M., Tahamtan, S. & Sabbaghzadeh, J. (2010). Dissimilar Welding of Carbon Steel to 5754 Aluminum Alloy by Nd:YAG Pulsed Laser. *Materials and Design*, Vol.31, pp.458-465, ISSN 0261-3069

Yamamoto, S. (2012). Welding in the World –Picking up from Foreign Magazine. *Welding Technology*, Vol.60, No.3, (2012), p. 75, ISSN 0387-0197

Yan, S., Hong, Z., Watanabe, T. & Jingguo, T. (2010). CW/PW Dual-beam YAG Laser Welding of Steel/Aluminum Alloy Sheets. *Optics and Lasers in Engineering*, Vol.48, pp.732-736, ISSN 0143-8166

Yasuyama, M., Ogawa, K. & Taka, T. (1996). Spot welding of aluminum and steel sheet with insert of aluminum clad steel sheet – Part 1. *Quarterly Journal of Japan Welding Society*, Vol.14, No.2, pp.314-320, ISSN 0288-4771

Chapter 4

Mitigating Zinc Vapor Induced Weld Defects in Laser Welding of Galvanized High-Strength Steel by Using Different Supplementary Means

Junjie Ma, Fanrong Kong, Blair Carlson and Radovan Kovacevic

Additional information is available at the end of the chapter

1. Introduction

Laser beam welding is a process where a focused laser beam is used as a moving heat source to join pieces of metal. The focused laser beam has a high power density that allows high speed welding, a deep penetration and a narrow heat affect zone (HAZ). There are two distinct types of laser welding modes: a conduction welding mode and a keyhole welding mode. When the laser beam intensity reaches 10^9 W/m^2, the molten pool starts to evaporate. As the laser intensity increases above 10^{10} W/m^2, the recoil pressure of the metal vapor pushes the molten metal downward and aside and a deep capillary called the "keyhole" is generated (Dawes, 1992). In the keyhole mode welding process, the keyhole maintains open due to the dynamic balance between the liquid metal surface tension and the pressure of the metal vapor and laser-induced plasma (Bakowski et al., 1984). When the laser radiates on the wall of the keyhole, the laser reflects multiple times on the wall of the keyhole. The laser beam energy is absorbed by Fresnel absorption directly by the walls of the keyhole, and a fraction of the laser energy is absorbed during each reflection (Dowden, 2009). Due to the multiple reflections of the laser beam, the keyhole behaves like an optical black body, making the keyhole mode welding process a highly energy efficient one (Steen, 2003).

Lap joint is the most common type of joint in the automotive assembly application; the traditional car body assembly method in a lap joint configuration uses resistance spot welding. However, the heavy and big spot guns limit the flexibility and accessibility of the welding process (Park et al., 2010); moreover, the localized joints are not particularly strong compared to those acquired by laser welding. On the other hand, the laser welding provides several benefits including a high scanning speed, high strength and low distortion of the joints, and the flexible implementation of the system for the automotive industry. Because of these advantages, laser welding shows immense potential over the conventional resistance spot welding and has been widely used in the automotive industry in the fabrication of different auto bodies parts (Forrest et al., 2004).

In order to reduce the weight of the vehicles and improve fuel efficiency and safety, the development of lightweight, and high-strength vehicles has prompted an increased use of advanced high strength steels (AHSS) in the automotive industry. These new steel grades include dual phase (DP) steels, transformation-induced plasticity (TRIP) steels, high hole expansion (HHE) steels, complex-phase (CP) steels, martensitic steels (MS), and twining induced plasticity (TWIP) steels (UltraLight Steel Auto Body- Advanced Vehicle Concepts, 2001). Additionally, these steels are galvanized in order to improve the surface corrosion resistance for automotive parts. However, it is still a great challenge to laser weld of galvanized steels in a zero-gap lap joint configuration. When laser welding of galvanized steels in a zero-gap lap-joint configuration, the zinc coating at the contact interface will vaporize; due to the lower boiling point (906 ºC) of zinc as compared to the melting temperature of steel (above 1500 ºC), the highly pressured zinc vapor expels the liquid metal out of the weld pool, resulting in blowholes and pores which dramatically decrease the mechanical properties of the weld (Akhter et al., 1988) (Fig. 1).

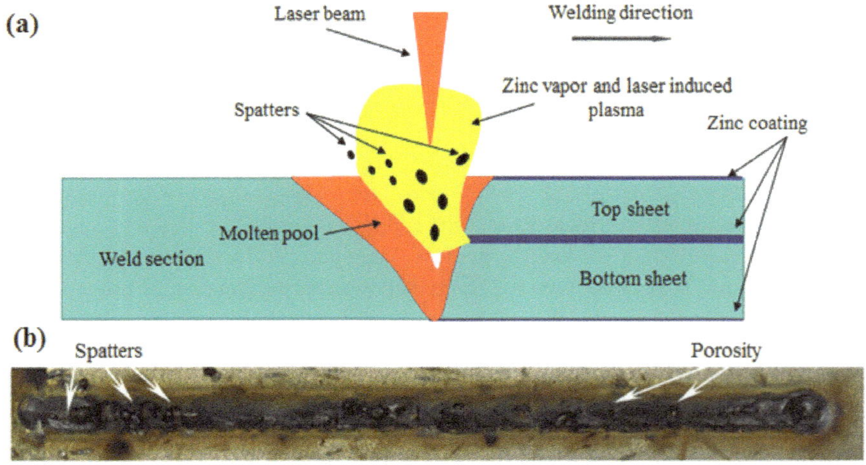

Figure 1. (a) The schematic view of the laser welding of galvanized steel in a zero gap lap-joint configuration and (b) the acquired weld bead with pores

2. Review of the laser welding of galvanized steels in a lap joint configuration

Over the past several decades, industry and academic researchers have been seeking new technologies that will successfully join galvanized steels in a lap joint configuration. Many techniques were proposed, and some are listed in Table 1.

Methods	Schematic	Technical details	Drawbacks
Elongating the laser beam (Fabbro et al., 2006) or using low power / low speed laser welding (Ma et al., 2012)	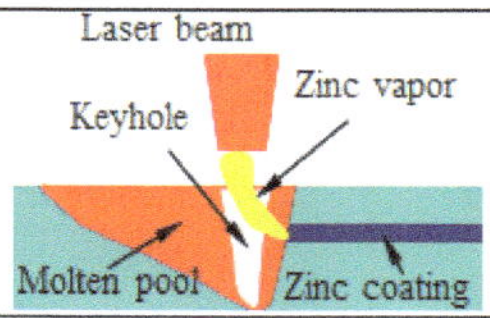	The zinc vapor was degassed through the keyhole or the enlarged molten pool during welding	Low processing speed or unstable weld qualities that limit the application in production
Using pulsed laser welding (Heydon et al., 1989; Kennedy et al., 1989; Norris et al., 1992; Tzeng, 1999; Tzeng, 2006)	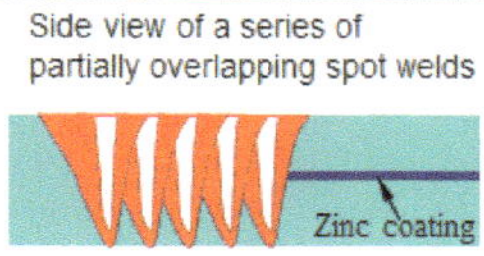	Zinc vapor was mitigated in the pulsed laser welding and effectively exhausted through a stable keyhole	
Using various shielding gas combinations (Berlinger, 1987; Akhter et al., 1990; Ream, 1991; Mitsubishi Co., 1993; Chung et al., 1999; Briand et al., 2008; Yang et al., 2011)	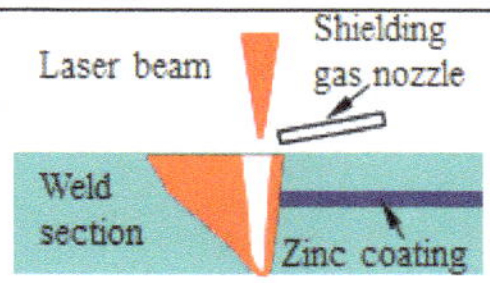	Suppressed the formation of the laser induced zinc plasma or interacted of the zinc vapor with the shielding gas during welding	
Pre-placing a thin metal sheet or powder along the centerline of the weld seam (Dasgupta et al., 2000; Li et al., 2007)	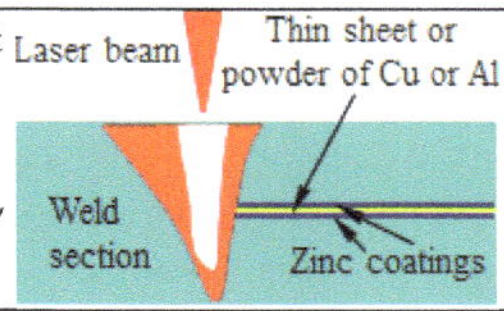	The zinc reacted chemically with the added metal before the steel started to melt	Difficulties will be implemented in production
Applying an appropriate spacer at the faying surfaces (Akhter et al., 1988; Imhoff et al., 1988)	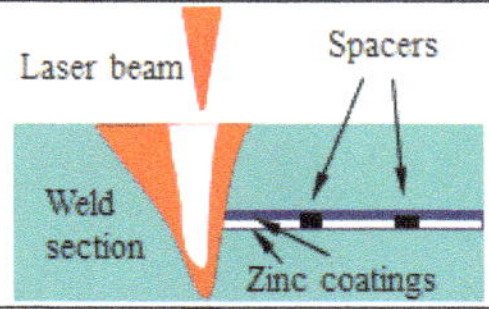	The generated zinc vapor vented out through the gap	The additional pre-welding procedure increases the production cost
Using a laser to create humps on the bottom sheet to create a gap at the faying surfaces (Gu et al., 2011)	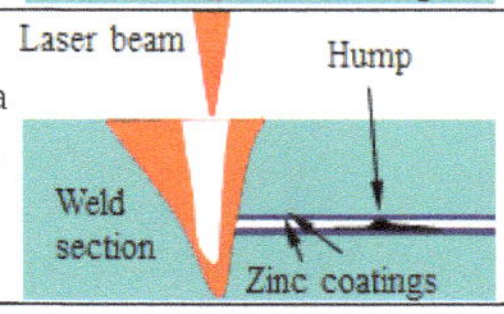	The generated zinc vapor vented out through the gap	
Creating vent holes on the bottom steel sheet (Chen et al., 2009)	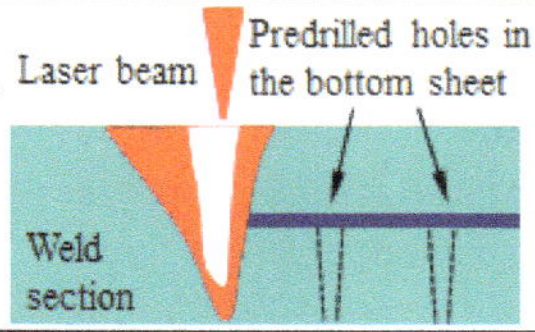	The generated zinc vapor vented out through the vent holes	
Adding a second laser heat source or splitting the laser beam into two laser beams in order to weld galvanized steel (Loredo et al., 2002; Xie et al., 2001)	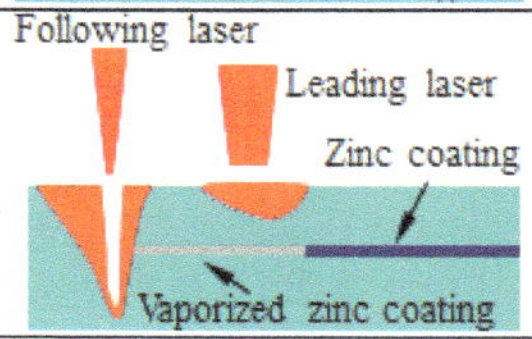	The leading laser melted the zinc coating at the interface	Complex equipment that would be difficult to implement in the production environment.

Using a leading beam to cut a slot along the joint line (Iqbal et al., 2010; Milberg et al., 2009)	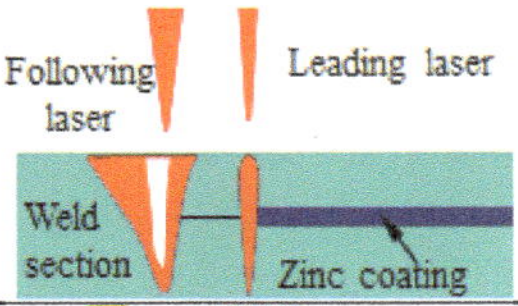	The leading laser cut the slot through which the zinc vapor was vented out	Moreover, a specific offset is needed between these two heat sources that could limit the application of this welding procedure
Applying gas tungsten arc welding (GTAW) as an auxiliary preheating heat source (Gu et al., 2001; Kim et al., 2008; Yang et al., 2009)	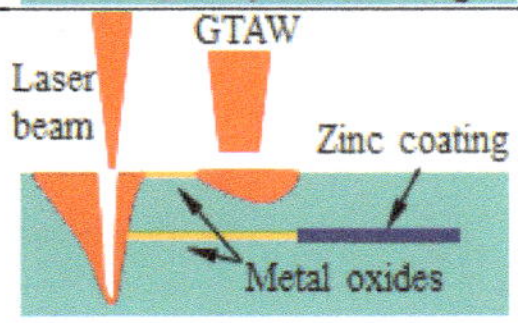	The preheating increased the absorption of the laser beam that contributes to a formation of a stable keyhole through which the zinc vapor is evacuated	

Table 1. The alternatives for the welding of galvanized steels in a lap joint configuration

As mentioned above, all of these techniques have some drawbacks that limit their application in the industry. So far, there is no report on a cost-effective, efficient, and easy-to-apply welding procedure that is capable of joining galvanized high-strength steels in a zero-gap lap joint configuration without the material addition or without the assistance of a second heat source. In order to avoid highly pressurized zinc vapor caused weld defects like spatters and blowholes, either the zinc coating should be removed before the steel melts or the vaporized zinc needs to be vented out properly during the welding process. The most direct way of accomplishing this is to mitigate the presence of high pressure zinc vapor during the welding process. In order to achieve this goal several techniques are proposed which will be described in the following sections of this chapter.

3. Low power / low speed laser welding of galvanized steels in a zero-gap lap joint configuration

3.1. Experimental procedure

Ribic et al. (2009) concluded that the generated zinc vapor escapes from the weld pool if the solidification time is longer. A lower welding speed will generate an enlarged weld pool that will require a longer solidification time. An experimental work is presented to show that the effect of zinc vapor on the quality of a weld in a zero-gap lap joint configuration may be successfully mitigated. A fiber laser of 4 kW in power with a focused spot of 0.6 mm in diameter was used as the welding heat source, and a 6-axis high precision robot was used to implement the welding procedure of galvanized steels (see Fig. 2). Pure argon with a flow rate of 30 standard cubic feet per hour (SCFH) was employed as side shielding gas to suppress the laser-induced plasma and to protect the molten material against corrosion. The base material used in this work was galvanized high strength dual phase (DP) steel DP980, whose nominal chemical composition is listed in Table 2 (Burns, 2009). The coupons of galvanized DP980 steel sheets were 1.2 mm and 1.6 mm in thickness, with a zinc coating weight of about 60 g/m^2.

Figure 2. Experimental setup for low power / low speed laser welding

C	Si	Mn	Cr	Mo	Al	Cu	Ni	Fe
0.15	0.31	1.5	0.02	0.01	0.05	0.02	0.01	Balance

Table 2. Chemical composition of galvanized DP980 steel, wt% (Burns, 2009)

3.2. Experimental results for low power / low speed laser welding

The top and bottom views of the weld obtained under a laser power of 2.0 kW and a welding speed of 5 mm/s show an acceptable weld surface quality (see in Fig. 3). A high-speed CCD camera with a frame rate of 4000 fps combined with a green laser with a band pass filter wavelength of 532 nm as the illumination source were used for real time monitoring of the dynamic behavior of the molten pool under different laser welding conditions. The weld pool formed under a relatively low welding speed was larger and relatively stable (see Fig.4a). On the other hand, the molten pool acquired under a higher welding speed shows sever fluctuation, and the high pressured zinc vapor generated at the faying surface jetted into the molten pool causing blowholes (see Fig. 4b). According to Ribic's work, an enlarged weld pool has a longer solidification time which obviously decreased the probability that the zinc vapor would be trapped in the molten pool under a relatively low welding speed (around 5 mm/s), and a visually acceptable weld quality could be acquired. However, if the welding speed is exceedingly low, the sagging may be generated, which also reduces joint strength. Fig. 5 shows the tensile shear test results of the joints acquired under different welding speeds. A higher failure load was acquired under a lower welding speed. The trapped zinc vapor may result in pores inside the joints which could decrease the failure load with respect to the joints acquired under the same welding conditions but without zinc at the faying surface (see Fig. 5). Although an acceptable quality of joints could be achieved by this low power / low welding speed procedure, this procedure is not accepted by the industry because of a low productivity.

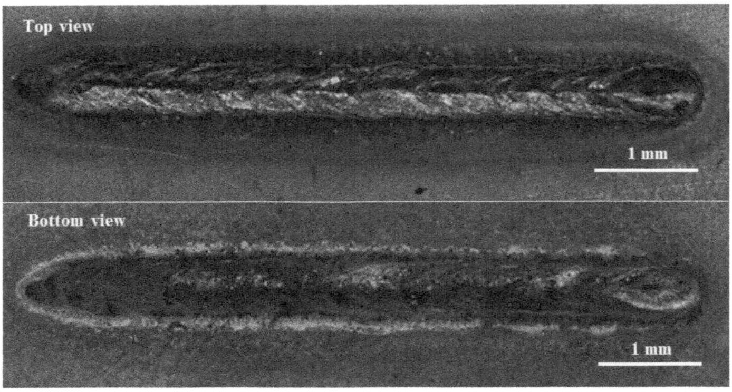

Figure 3. Top and bottom views of the weld obtained under a laser power of 2.0 kW and welding speed of 5 mm/s

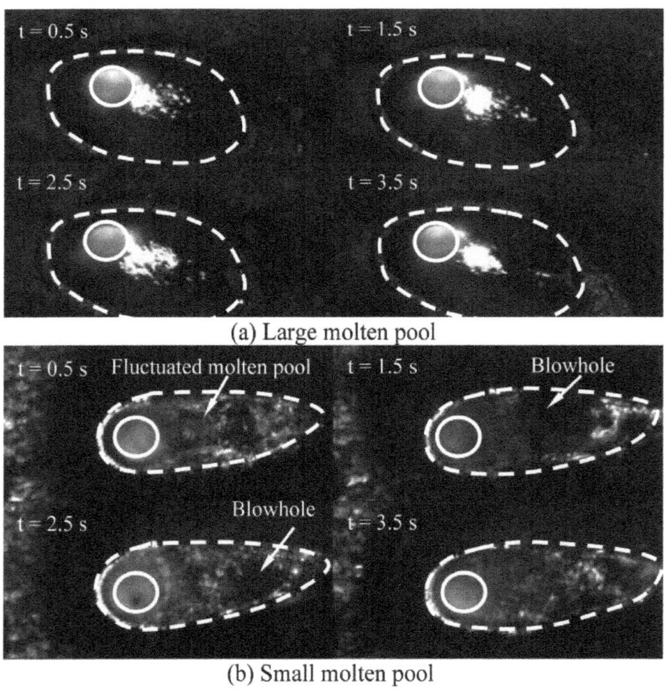

Figure 4. The dynamic behavior of the molten pool acquired under different welding parameters: (a) laser power of 1.5 kW, welding speed of 5 mm/s, and (b) laser power of 4.0 kW, welding speed of 30 mm/s at different time steps during the welding process

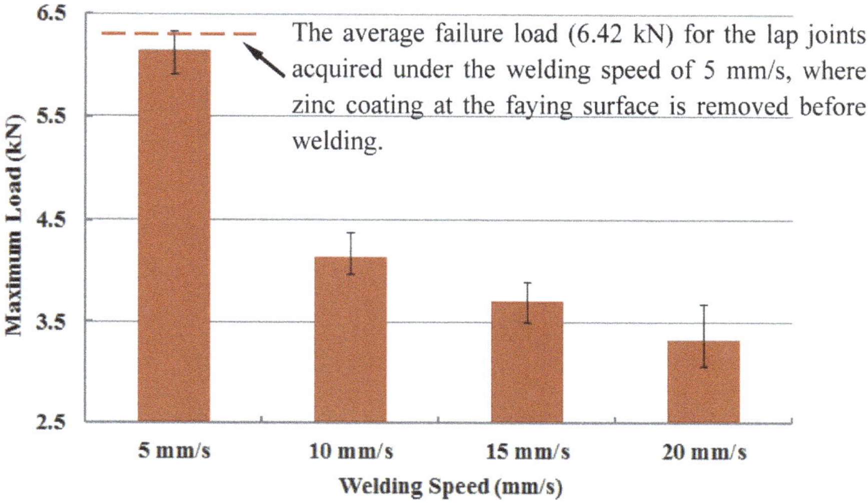

Figure 5. The failure load for tensile shear testing of the DP980 lap joints acquired under different welding speeds and with the laser power of 2.0 kW

4. Two-pass laser welding of galvanized steels in a zero-gap lap joint configuration

4.1. Experimental procedure

In order to improve the production efficiency, the automotive industry requires a welding technique that can join of overlapped galvanized high strength steels successfully under a higher working speed. As discussed previously, if the zinc coating is removed before the steel starts to melt, a much higher welding speed can be achieved. Therefore, a two-pass laser welding process that is capable of successfully joining galvanized steel sheets in a zero-gap lap joint configuration is presented. Fig. 6 shows the schematic view of the two-pass laser welding process. The defocused laser beam shown in Fig. 6a is used to preheat the two overlapped sheets during the preheating pass. Only when the width of the area where the zinc coating is vaporized by preheating is larger than the distance between the zinc boiling isotherms (906 °C), a sound weld could be acquired (see Fig. 6c). The laser power was set at its maximum value of 4.0 kW in order to allow a higher scanning speed. The laser welding speed was set at 60 mm/s in order to acquire a partial penetrated joint that was determined by the preliminary executed experimental trails. The preheating parameters, like the defocused position of the laser beam and the laser scanning speed, are critical in achieving a final weld quality. As shown in Table 3, three levels of defocused off-set distance (26 mm, 44 mm, and 62 mm which corresponds to the defocused diameters of laser beam of about 3 mm, 5 mm, and 7mm, respectively) and four levels of scanning speeds (20 mm/s, 30 mm/s, 40 mm/s, and 50 mm/s) were chosen to optimize the preheating procedure.

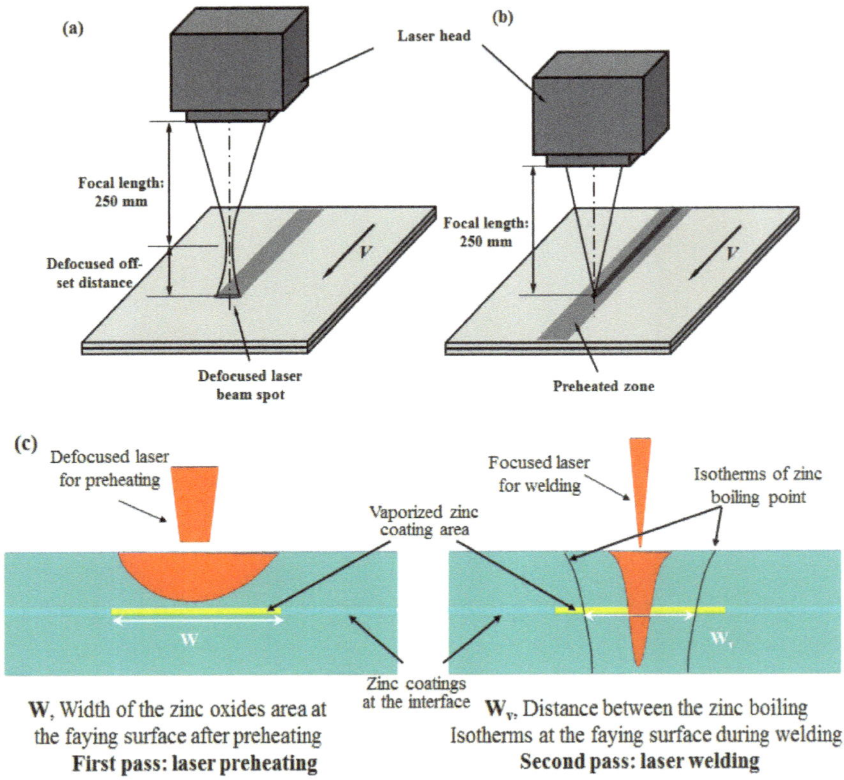

Figure 6. The schematic view of the two-pass welding process: (a) laser preheating, (b) laser welding, and (c) geometrically defined width of zinc coating treated during preheating and welding

Process parameters	Experiment series A				Experiment series B				Experiment series C			
Scanning speed (mm/s)	20	30	40	50	20	30	40	50	20	30	40	50
Defocused off-set distance (mm)	26				44				62			

(Laser preheating and welding power: 4.0 kW, laser welding speed: 60 mm/s)

Table 3. The preheating parameters

4.2. Experimental results for two-pass laser welding

As the defocused off-set distance increased, the defocused laser beam spot became larger, and the laser energy distribution was dispersed. A defocused off-set distance combined with a lower scanning speed generated too much energy that penetrated the top sheet resulting in spatters and permanent defects which could not be mitigated by the following laser welding pass (see Fig. 7a). A longer defocused off-set distance combined with a higher scanning speed could not vaporize a sufficient amount of zinc coating; the remaining zinc coating at the contact interface caused weld defects (see Fig. 7b). Thus, only for the optimized laser defocused off-set distance and the scanning speed, will a reasonable width of the zinc coating be vaporized (see Fig. 7c). Fig. 8 shows the experimental results for the selected preheating parameters (shown in Table. 3). The optimized preheating parameters that allowed a sound weld are shown in Fig. 8, area A.

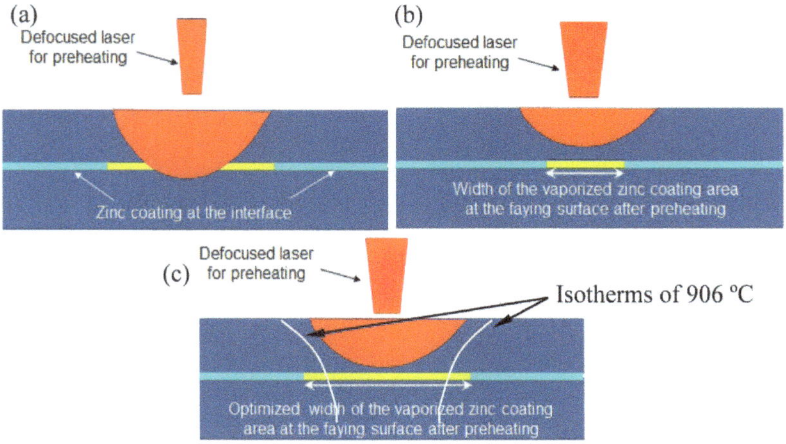

Figure 7. The schematic view of the preheating process: (a) molten pool penetrates the interface, (b) narrow vaporized zinc coating area, and (c) optimized width of the vaporized zinc coating area

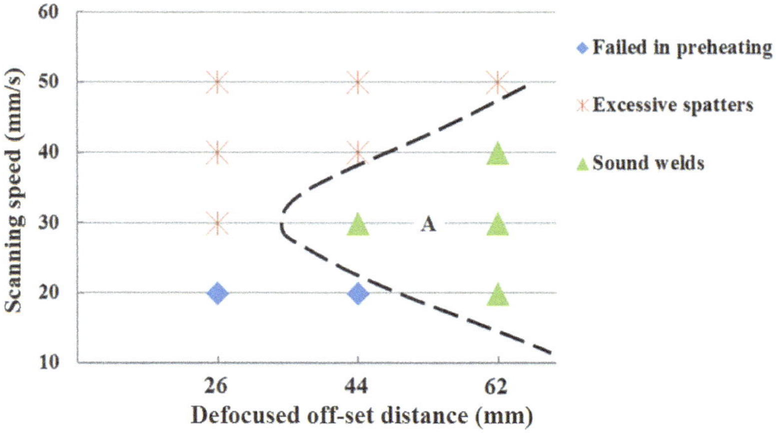

Figure 8. Experimentally determined combinations of defocused off-set distance and scanning speed that result in a good weld quality

Figs. 9 and 10 show the preheated interfaces of the coupons and the cross-sections corresponding to the different locations marked on the preheated interfaces. The zinc coatings far from the preheated zones at the top and bottom sheets are not affected (see Figs. 9b and 10b); the zinc coatings are melted and deformed at the edges of the preheated zone (see Figs. 9c and 10c); and the zinc coatings are vaporized at the center of the preheated zone (see Figs. 9d and 10d).

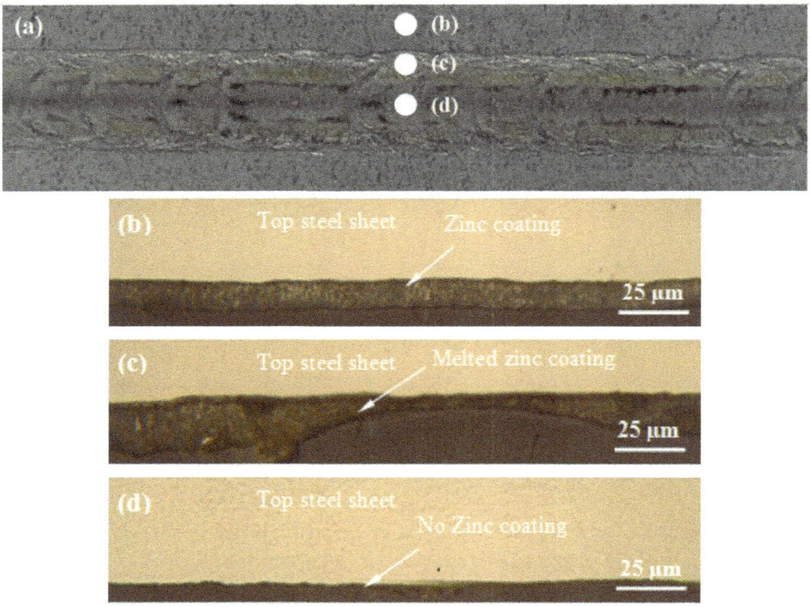

Figure 9. Preheated surface of the top steel sheet obtained with a laser power of 4.0 kW, a 30 mm/s scanning speed, and a 62 mm defocused laser beam off-set distance and the corresponding cross-section at different locations

During the laser preheating pass, the defocused laser beam burns the zinc at the top surface, and melts and partially vaporizes the zinc coatings at the interface of the two overlapped steel sheets, and improves the absorption of the laser beam which results in the formation of a stable keyhole through which any zinc vapor formed at the interface will be vented out (Yang et al., 2009). Fig. 11 shows the top and bottom views of the weld obtained under preheating and welding with a laser power of 4.0 kW, a defocused off-set distance of 62 mm, a scanning speed of 30 mm/s, and a welding speed of 60 mm/s. Fig. 12 shows the weld cross-section of the weld shown in Fig. 11.

The tensile shear test was carried out in order to determine the mechanical strength of the welded joints obtained by the two-pass laser welding procedure. The experimental results demonstrated that the two-pass welded joints were broken in the HAZ of the bottom steel sheet. One of the tensile shear results is shown in Fig. 13. The tensile shear test for the welded coupons without a zinc coating at the interface was also performed. In order to use the data as a reference, the welded coupons without a zinc coating at the interface had a average failure load value of 5295.88 N which was lower than that of the two-pass welded coupons (6127.58 N). The reason for this difference in results is explained by the fact that the preheating process increased the laser beam absorption of the coupons, which contributed to a stronger (wider) partially penetrated weld joint.

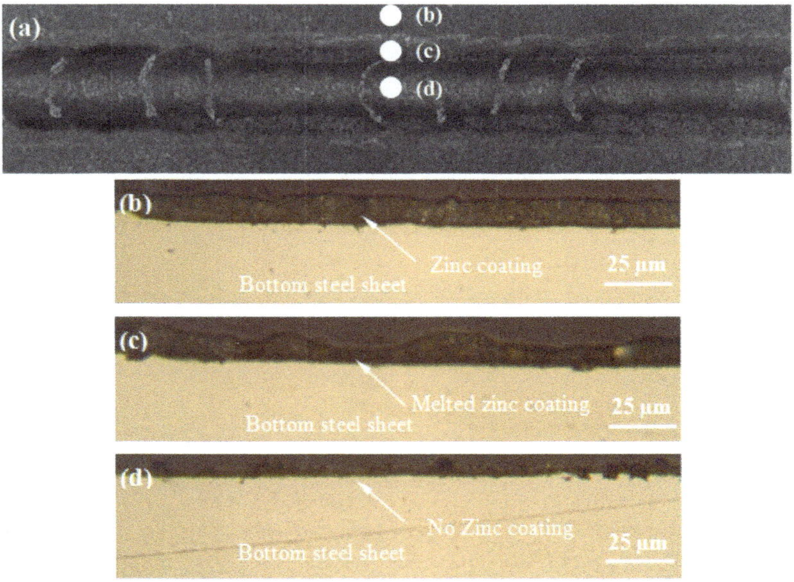

Figure 10. Preheated surface of the bottom steel sheet obtained with a laser power of 4.0 kW, a 30 mm/s scanning speed, and a 62 mm defocused laser beam off-set distance and the corresponding cross-section at different locations

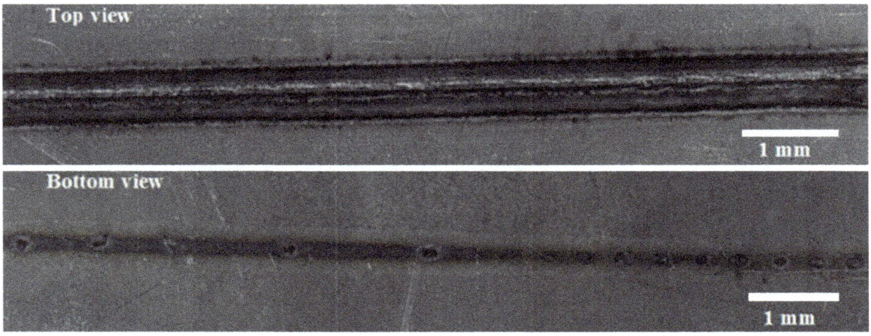

Figure 11. Top and bottom views of weld obtained under a scanning speed of 30 mm/s and a defocused off-set distance of 62 mm (the preheating and welding laser power is 4.0 kW; the welding speed is 60 mm/s)

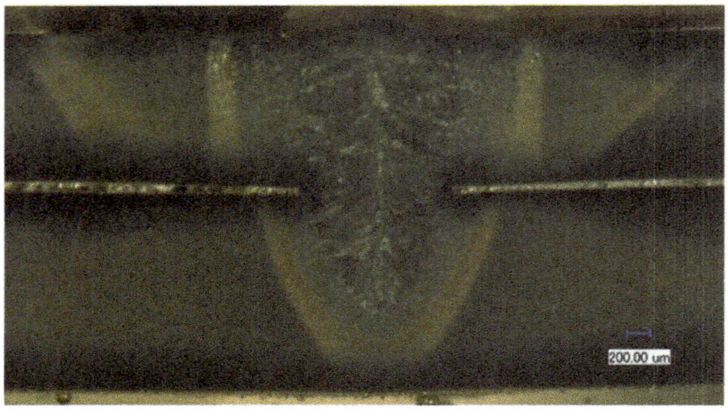

Figure 12. Cross-sectional view of the weld obtained under a scanning speed of 30 mm/s and a defocused off-set distance of 62 mm (the preheating and welding laser power is 4.0 kW; the welding speed is 60 mm/s)

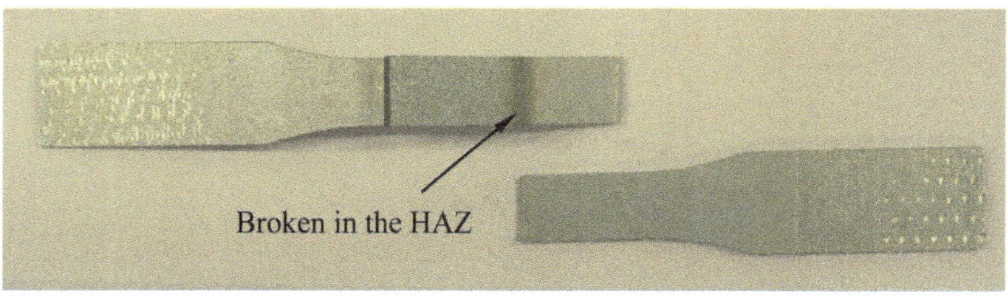

Figure 13. Tensile shear test results of the weld joint obtained under a scanning speed of 30 mm/s, and a defocused laser beam off-set distance of 44 mm (the preheating and welding laser power is 4.0 kW; the welding speed is 60 mm/s)

5. Laser welding of galvanized steels in a lap joint configuration with a pressure wheel

5.1. Experimental procedure

Based on the experimental study performed, it was found that the stability of the laser welding process was sensitive to the clamping conditions. A relatively loose clamp condition resulted in a better weld than a very tight clamp condition. The gap ahead of the weld pool is the key to performing the laser welding of galvanized steel in a lap joint configuration successfully. Moeckel et al. (2003) developed a device for controlling the gap at the faying surface of the overlapped galvanized steel sheets in order to degas the generated zinc vapor during the welding process. The Fraunhofer Institute developed a pressure wheel system which could control the roller clamping force that allows for the controlling of the gap at the faying surface (Fraunhofer Institute website). In order to achieve an over lapped galvanized steel joint with a single laser beam without a pre- and/or post-weld process, a force-controllable pressure wheel (ZM YW50 PW P300 II) is used to control the gap near the laser focused spot during the laser welding. The laser welding of galvanized steels for a lap joint configuration with a pressure wheel control system is shown in Fig. 14. Fig. 15a shows the close-up of the pressure wheel set-up. The laser head is set-up under a 30 °decline with respect to the pressure wheel, and Fig. 15b shows the pressure wheel controller.

Figure 14. Laser welding of galvanized steel for a lap joint configuration with a pressure wheel control system

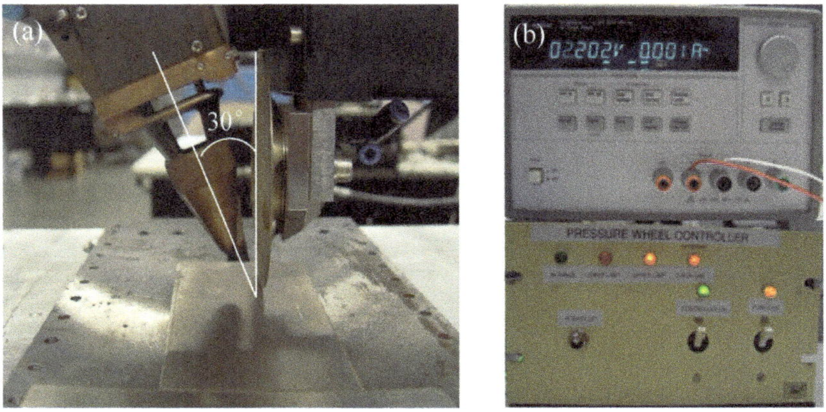

Figure 15. (a) The close-up of pressure wheel set-up and (b) the pressure wheel controller

5.2. Experimental results for laser welding with a pressure wheel

The feasibility of welding galvanized steel sheets in a lap joint configuration by controlling the pressure wheel force during the fiber laser welding process is discussed. Fig. 16 shows the welds obtained by various levels of pressure wheel force under a laser power of 4.0 kW and a welding speed of 50 mm/s. The corresponding weld cross-sections are shown in Fig. 17. The cross-sections of the welds show that the weld beads are under the angle because the laser head is set-up with an angle of 30° with respect to the pressure wheel (see Fig. 15a). A sound weld was obtained by using a single laser beam with a force-controllable pressure wheel under the optimized force. As shown in Figs. 16 and 17, with an increased pressure wheel force, the weld quality decreased, and lots of spatters and blowholes were generated (see Fig. 16d). An increased pressure wheel force larger than 12N resulted in a decreased gap between the overlapped sheets near the laser focused area; the gap became too narrow to evacuate the generated high pressured zinc vapor. The jet of high pressured zinc vapor generated spatters and blowholes during the welding process.

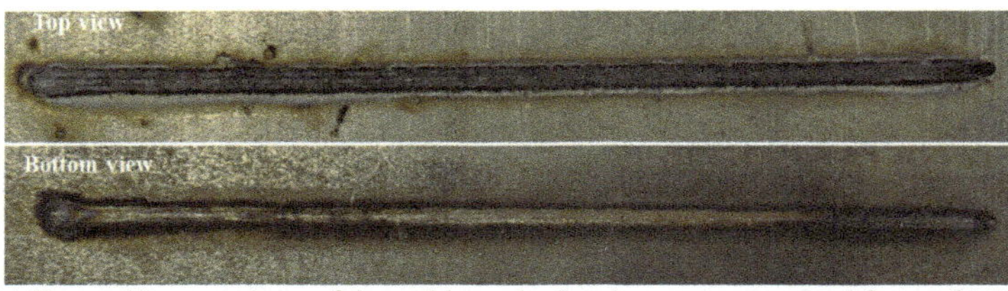

(a) Top and bottom views of the weld acquired under a pressure wheel force of 0.3 N

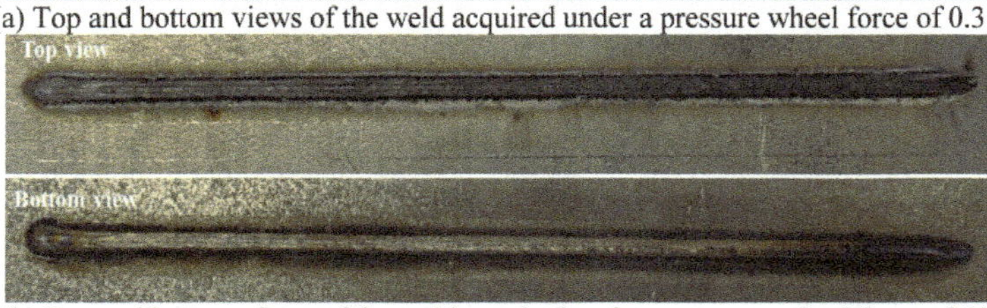

(b) Top and bottom views of the weld acquired under a pressure wheel force of 6 N

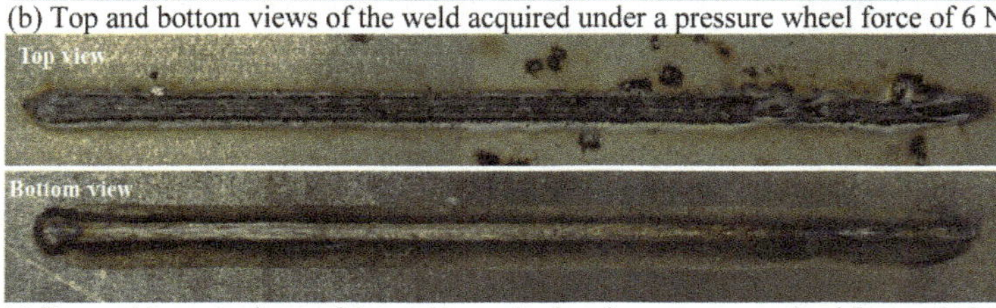

(c) Top and bottom views of the weld acquired under a pressure wheel force of 12 N

(d) Top and bottom views of the weld acquired under a pressure wheel force of 18 N

Figure 16. Top and bottom views of the welds acquired by various pressure wheel forces under a laser power of 4.0 kW and a welding speed of 50 mm/s

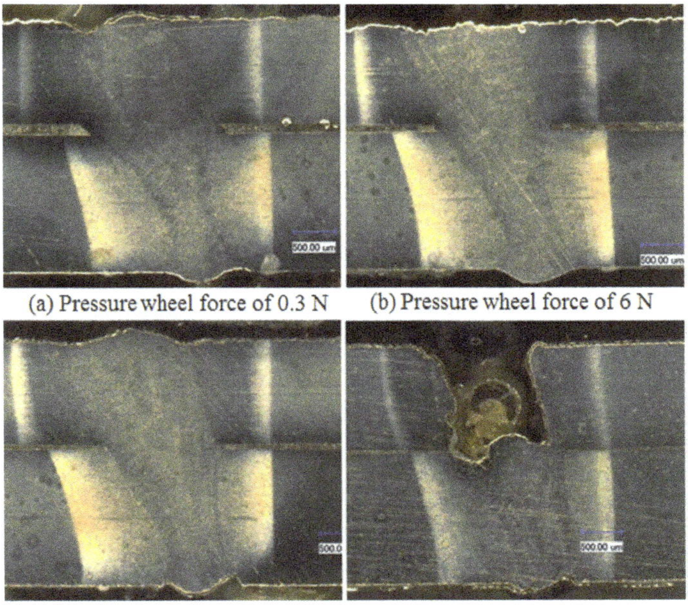

(a) Pressure wheel force of 0.3 N (b) Pressure wheel force of 6 N

(c) Pressure wheel force of 12 N (d) Pressure wheel force of 18 N

Figure 17. Cross-sectional views of the welds acquired by various pressure wheel forces under a laser power of 4.0 kW and a welding speed of 50 mm/s

Kong et al. (2012) reported that there is a correlation between the optical emission of the plasma and zinc vapor induced welding defects in the laser welding of galvanized steel for an overlapped joint configuration. Therefore, the spectroscopy was used to on-line monitor the laser welding of galvanized steel with a pressure wheel for an overlap joint configuration. The set-up to monitor the optical emission of the plasma in laser welding is shown in Fig. 18.

The emission line intensities detected from plasma during the laser welding of galvanized steel under various pressure wheel forces are shown in Fig. 19. The intensities of the emission lines above the weld pool were much lower when the pressure wheel force is larger than 12 N; while, the intensities of the emission lines were relatively higher when the pressure wheel force is set at 0.3, 6 or 12 N. The reason for this stems from the fact that when welding of galvanized steel under a higher pressure wheel force (18 N), the gap near the focal laser spot became too narrow to evacuate the high pressured zinc vapor; the zinc vapor caused spatters that disturbed the stability of the plasma which affected the intensity of the detected spectrum (Kong et al. 2012). The evolution of iron electron temperature within the laser-induced plasma along the weld bead length is shown in Fig. 20. The iron electron temperature was calculated by using the Boltzmann Plot method expressed by Equation (1) (Kong et al., 2012, Griem, 1997 and Marotta, 1994):

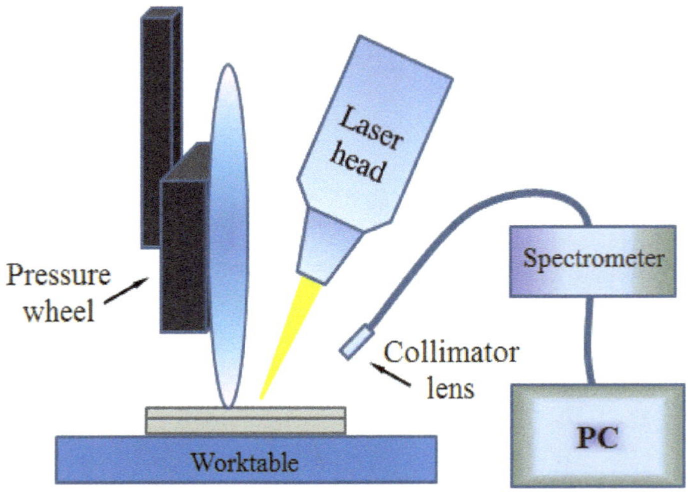

Figure 18. Schematic view of the setup for on-line monitoring the optical properties of plasmas during laser welding

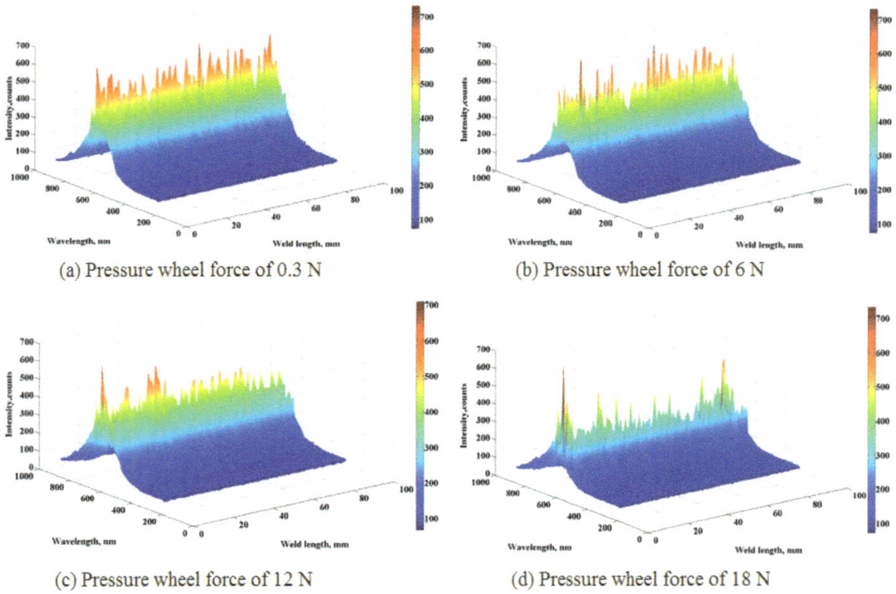

Figure 19. Spectrum of laser induced plasma captured by a spectrometer in the laser welding process by various pressure wheel forces under a laser power of 4.0 kW and a welding speed of 50 mm/s

$$T_e = \frac{E_m(2) - E_m(1)}{k \ln\left[\frac{E_m(1) I(1) A_m(2) g_m(2) \lambda(1)}{E_m(2) I(2) A_m(1) g_m(1) \lambda(2)}\right]} \quad (1)$$

where T_e is the plasma electron temperature, E_m is the energy of the upper state, k is the Boltzmann constant, I_m is the emission line relative intensity, A_m is the transition probability, g_m is the statistical weight, and λ_m is the wavelength.

As shown in Fig. 20, under lower pressure wheel forces (0.3, 6, and 12 N), the electron temperature showed lower intensity and less fluctuation compared to a higher pressure wheel force (18 N). The presence of the zinc vapor induced spatters in the plasma which increased the iron electron concentration which, in turn, increased the iron electron temperature value (Kong et al., 2012).

Thus, there is a correlation between the optical emission of the plasma and zinc vapor induced welding defects during the laser welding; and this optical signal could be further used as feedback for the closed-loop control of the laser welding of galvanized steel with a pressure wheel, which is shown in Fig. 21.

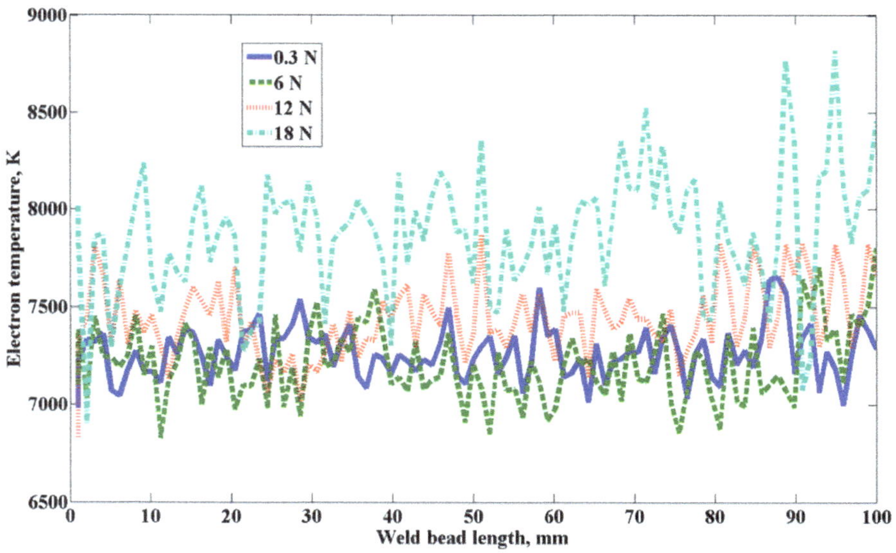

Figure 20. Electron temperatures of iron in laser-induced plasma captured by a spectrometer in the laser welding process by various pressure wheel forces under a laser power of 4.0 kW and a welding speed of 50 mm/s

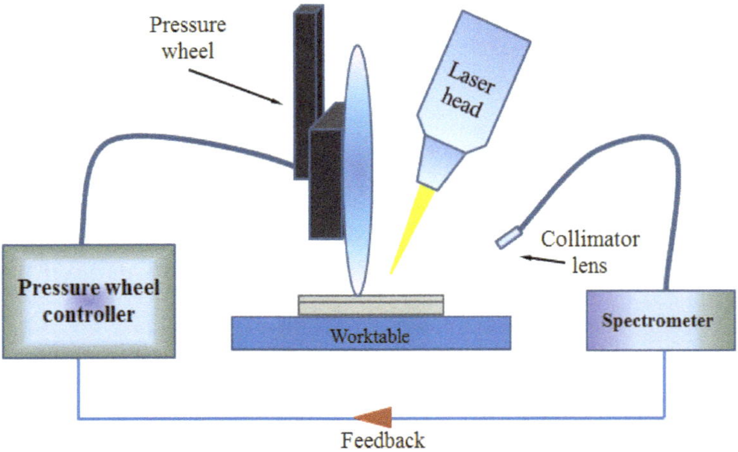

Figure 21. A schematic presentation of a closed-loop control system of the laser welding of galvanized steel by controlling the clamping force

6. Conclusions

In this chapter, issues related to the laser welding of galvanized steels in a zero-gap lap joint configuration are discussed. The authors' recent research results on the laser welding of galvanized steel in a lap joint configuration are reviewed. The different welding procedures, namely, low power / low speed welding, two-pass laser welding and laser welding with a pressure wheel are introduced for the laser welding of galvanized steels in a lap joint configuration. It was found that acceptable weld quality could be achieved by a low power / low welding speed procedure; however, the relatively low welding speed limits its application in the industrial environment. A high quality weld could be obtained by introducing a preheating pass with a defocused laser beam. By using the optical signal acquired during the laser welding process as a feedback signal in the pressure wheel force control, it is possible that a controllable clamping force could be a solution for achieving a good weld quality without using a pre- and post- welding procedure.

Author details

Junjie Ma, Fanrong Kong and Radovan Kovacevic*

Center for Laser-aided Manufacturing, Lyle School of Engineering, Southern Methodist University, Dallas, TX, USA

Blair Carlson

General Motors R&D Center, Warren, MI, USA

* Corresponding Author

Acknowledgement

This work was financially supported by the NSF's Grant No. IIP-1034652.

7. References

Akhter R., Steen W. M. and Cruciani D. (1988). Laser welding of zinc-coated steel. Proceedings of the 5th International Conference on Lasers in Manufacturing, Stuttgart, West Germany, September 13-14

Akhter R. and Steen W.M. (1990). The gap model for welding zinc-coated steel sheet. In: Proceedings of the International Conference on Lasers Systems Application in Industry, Torino, Italy, 219-236

AWS WZC (D19.0-72), Welding Zinc-Coated Steels (1972). American Welding Society.

Bakowski L., Beyer E., Herziger G. and Poprawe R. (1984). Development and optical absorption properties of a laser induced plasma during CO_2-laser processing. SPIE Proceedings Series 455: 75-80.

Berlinger G. and Speranza J.J. (1987). U.S. Patent 4,684,779

Briand F., Chouf K. and Lefebvre P. (2008). Method and installation for laser welding with a N_2/He gas mixture, the N2/He content being controlled according to the laser power, U.S. patent 7,385,158

Burns T. Weldability of a dual-phase sheet steel by the gas metal arc welding process. Master Thesis, the University of Waterloo, 2009.

Chen W., Ackerson P. and Molian P. (2009). CO2 laser welding of galvanized steel sheets using vent holes. Materials and Design (30): 245-251

Chung B.G., Rhee S. and Lee C.H. (1999). The effect of shielding gas types on CO2 laser tailored blank weldability of low carbon automotive galvanized steel, Materials Science and Engineering 272: 357-362

Dasgupta A., Mazumder J. and Bembenk M. (2000). Alloying based laser welding of galvanized steel, Proceedings of International Conference on Applications of Lasers and Electro Optics, Dearborn MI.

Dawes C. Laser welding: a practical guide, Abington publishing, 1992.

Dowden J. M. The theory of laser materials processing: Heat and mass transfer in modern technology, Springer, 2nd edition, 2009.

Fabbro R., Coste F., Geobels D. and Kielwasser M. (2006). Study of CW Nd-Yag laser welding of Zn-coated steel sheets. Journal of physics D: Applied physics 39:401-409

Forrest M. G. and Lu F. (2004). Advanced dual beam laser welding of zinc-coated steel sheets in lap joint configuration with zero gap at the interface. 23th International Congress on Applications of Lasers and Electro-Optics, ICALEO.

Graham M. P., Hirak D. M., Kerr H. W. and Weckman D. C. (1996). Nd: YAG laser beam welding of coated sheet steels using a modified lap joint geometry. Welding Journal 75(5): 162-170

Griem H. R. Principle of plasma spectroscopy. Cambridge Monographs on Plasma Physics. Cambridge University Press, Cambridge, 1997.

Gu H. and Mueller R. (2001). Hybrid welding of galvanized steel sheet. 20th International Congress on Applications of Lasers & Electro-optics, ICALEO, 130-139

Gu H. and Shulkin B. (2011). A practical use of humping effect in laser beam welding. Journal of Laser Applications 23(1): 1-6

Gualini M.M.S., Iqbal S. and Grassi F. (2006). Modified dual-beam method for welding galvanized steel sheets in lap configuration. Journal of Laser Applications 18(185): 185-191

Heydon J., Nilsson K. and Magnusson C. (1989). Laser welding of zinc coated steel. In: Proceedings of the 6th International Conference on Lasers in Manufacturing, 93-104

Homepage Fraunhofer - Gesellschaft: Http://www.fraunhofer.de/en.html

Imhoff R., Behler K., Gatzweiler W. and Beyer E. (1988). Laser beam welding in car body making. Proceedings of the 5th International Conference on Lasers in Manufacturing, Stuttgart, West Germany, September 13-14

Iqbal S., Gualini M.M.S. and Rehman A.U. (2010). Dual beam method for laser welding of galvanized steel: Experimentation and prospects. Optics & Laser Technology 42(1): 93-98

Kim C., Choi W., Kim J. and Rhee S. (2008). Relationship between the welding ability and the process parameters for laser-TIG hybrid welding of galvanized steel sheets. Materials Transactions 49(1): 179-186

Kennedy S.C. and Norris I.M. (1989). Nd-YAG laser welding of bare and galvanized steels. SAE Technical Paper Series 890887, International Congress and Exposition, Detroit, MI

Kong F., Ma J., Carlson B. and Kovacevic R. (2012). Real-time monitoring of laser welding of galvanized high strength steel in lap joint configuration. Optics & Laser Technology 44(7): 2186-2196

Lacroix D., Jeandel G. and Boudot C. (1997). Spectroscopic characterization of laser induced plasma created during welding with a pulsed Nd: YAG laser. Journal of Physics D: Applied Physics, 81(10): 6599-6606

Li X., Lawson S. and Zhou Y. (2007). Novel technique for laser lap welding of zinc coated sheet steels. Journal of Laser Applications 19(4): 259-264

Li X., Lawson S. and Zhou Y. (2008). Lap welding of steel articles having a corrosion resisting metallic coating. U.S. Patent No. 2008/0035615 Al.

Loredo A., Martin B., Andrzejewski H. and Grevey D. (2002). Numerical support for laser welding of zinc-coated sheets process development, Applied Surface Science 195(1-4): 297-303

Ma J., Kong F. and Kovacevic R. (2012). Finite-element thermal analysis of laser welding of galvanized high-strength steel in a zero-gap lap joint configuration and its experimental verification. Materials and Design (36): 348-358

Marotta A. (1994). Determination of axial thermal plasma temperatures without Abel inversion. Journal of Physics D: Applied Physics (27): 268-272

Mazumder J., Dasgupta A. and Bembenk M. (2002). Alloy based laser welding. U.S. patent 6,479,168.

Milberg J. and Trautmann A. (2009). Defect-free joining of zinc-coated steels by bifocal hybrid laser welding. Production Engineering. Research and Development 3(1): 9-15

Mitsubishi Co. (1993). U.S. Patent 5,618,452

Moeckel A., Stein H., Jucht H., (2003). Device for degassing laser welded seams Patent no DE 10160156

Norris I., Hoult T., Peters C., Wileman P., (1992). Material processing with a 3 kW Nd:YAG laser. Proceedings of Laser Advanced Materials Processing (LAMP), Nagaoka, Japan

Pan Y. and Richardson I. M. (2011). Keyhole behavior during laser welding of zinc-coated steel. Journal of physics D: Applied physics 44:045502

Park H.S. and Choi H.W. (2010). Digital Laser Welding System for Automobile Side Panel, Laser Welding, Na, X.D., Stone (Ed.), ISBN: 978-953-307-129-9, InTech.

Ream S. (1991). Laser welding of zinc-coated steel. In: Laser applications in materials processing and manufacturing. Society of Manufacturing Engineering (SME), Southfield, MI

Ribic B., Palmer T.A. and DebRoy T. (2009). Problems and issues in laser-arc hybrid welding. Int Mater Rev 54(4):223-44

Schmidt M., Otto A., Kägeler C. (2008). Analysis of YAG laser lap-welding of zinc coated steel sheets. CIRP Annals - Manufacturing Technology, 57(1): 213-216.

Steen W. M. Laser material processing, 3rd edition, Springer, 2003.

Tzeng Y.F. (1999). Pulsed Nd:YAG laser seam welding of zinc-coated steel. Welding J 78(7): 238s-244s

Tzeng Y.F. (2006). Gap-free lap welding of zinc-coated steel using pulsed CO2 laser. The International Journal of Advanced Manufacturing Technology, (29)3-4: 287-295

ULSAB-AVC Body Structure Materials, Technical Transfer Dispatch No. 6, May, 2001.

Wiese W.L. and Martin G.A. Wavelengths and transition probabilities for atoms and atomic ions. Part II. Transition Probability. 1st edition, 1980

Xie J. and Denney P. (2001). Galvanized steel welding with lasers. Welding Journal 80(6): 59-61

Yang S. and Kovacevic R. (2009). Welding of galvanized dual-phase 980 steel in a zero-gap lap joint configuration. Welding Journal 88(8): 168-178

Yang S., Carlson B. and Kovacevic R. (2011). Laser welding of high-strength galvanized steels in a zero-gap lap joint configuration under different shielding conditions. Welding Journal 90: 8s-18s

In situ Reaction During Pulsed Nd:YAG Laser Welding SiC_p/A356 with Ti as Filler Metal

Kelvii Wei Guo

Additional information is available at the end of the chapter

1. Introduction

As aluminum matrix composites are generally low-cost and exhibit higher specific strength, high wearability, and good design performance property and functionality. They are widely applied in aerospace-flight, aviation structure, and automobile and in the heat resistant-wearable parts of engine [1-4]. Hence, a great deal of contemporary research effort is focusing upon their development and applications, typically on the discontinuously reinforced aluminum matrix composites like matrix with particle, short fiber, whisker and so forth. Additionally, a great deal of attention has also been drawn into the investigation of their secondary processing technologies like machining, joining and plastic forging. Welding is an important process for joining these materials. There has extensive effort to be devoted to developing appropriate process for joining the similar or dissimilar composites in literatures [2-4]. These processes can be mainly categorized as: (i) fusion welding like arc welding, shielding gas welding, laser welding and electron beam welding, *etc.*; and (ii) solid-state welding like soldering, explosive welding, friction welding and diffusion welding, *etc.* However, there still exist many problems in joining of the discontinuously reinforced aluminum matrix composites using conventional arc-welding processes and those high energy density welding methods like: laser welding and electron beam (EB) welding. These problems include: (i) the formation of poor weldment and the unsatisfactory properties of welded joints – mainly due to the high viscosity and poor flowability of the liquid welding pool causing mixing difficulty of the composite base material with filler materials; (ii) the occurrence of micro-segregation or inhomogeneous distribution of the reinforcement phases of SiC_P, Al_2O_{3P}, AlN and *etc.*, and whiskers like SiC_w – typically owing to the rejection by their solidification front in the welding poor as cooling down, which subsequently prompts for many micro and macro defects in the weld and very poor properties of the welded joints; (iii) the formation of aluminum carbide – mainly as a result of harmful interfacial reaction between aluminum matrix and reinforcement phases; and (iv) *so on*. A typical example of the interfacial reaction likely to have pernicious effects on the mechanical and chemical behavior of the composite is $4Al + 3SiC \rightarrow Al_4C_3 + 3Si$. This is owing to: (a) the formation of brittle and weak aluminum carbide Al4C3 in the interfacial reaction sacrifices the reinforcement-materials in the composite; and (b) the unstable aluminum carbide in wet environments causes corrosion of the composite because of its rapid hydrolysis *etc*.

Aiming at developing or improving the conventional welding technique, this paper studies the technique of welding the stir-cast aluminum matrix composite SiC_p/A356 by Nd:YAG laser welding with pure titanium as filler. This study has been specifically concerned on the *in situ* reinforcement effect of Ti on the microstructures of laser welded joints, which have been analyzed by means of Scanning Electron Microscope (SEM), Transmission Electron Microscope (TEM) and Electron Diffusion X-ray analysis (EDX) *etc*. The study aims at providing some ground works for further studies in this field.

2. Experimental material and process

2.1. Experimental material

Stir-cast SiC_p/A356 aluminum matrix composite, reinforced with 20 % volume fraction SiC particle of 12 μm mean size, was used as the welding specimens. The tensile strength of such specimens was 240 MPa and their solid-liquid phase transformation temperature was in the range of 562.6～578.3 °C. Figure 1 showed their corresponding microstructure while Table 1 tabulated the chemical composition of the A356 matrix alloy. Pure titanium was used as the filler metal.

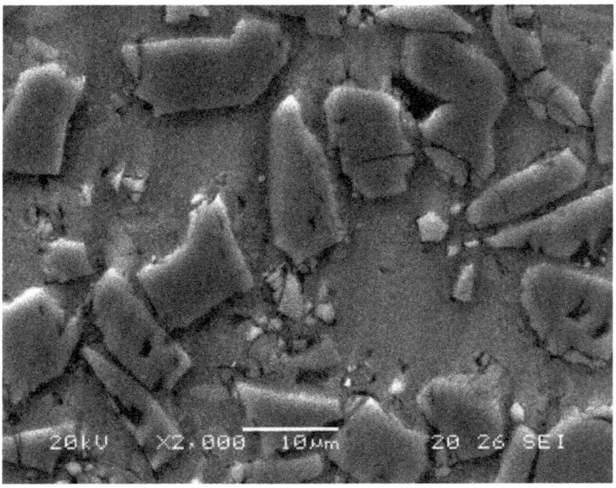

Figure 1. Microstructure of SiC_p/A356 aluminum matrix composite

Composition (Wt %)			
Si	Mg	Ti	Al
6.5~7.5	0.3~0.5	0.08~0.2	Bal.

Table 1. Composition of A356

2.2. Experimental process

The stir-cast aluminum matrix composite specimens were individually wire-cut to the size of 3 mm × 10 mm × 35 mm (Fig. 2). The quench-hardened layer induced by wire-cut and the oxide on the surfaces of specimens were polished away by 400 # emery cloth. The pure titanium filler was then machined to depth 3 mm × width 10 mm × thickness of 0.15 mm, 0.3 mm, 0.45 mm, 0.5 mm, 0.6 mm and 0.75 mm, respectively The specimens were then mounted into a clamping devise on the platform of a GSI Lumonics Model JK702H Nd:YAG TEM_{00} mode laser system, and their welding surfaces were properly cleaned by acetone and pure ethyl alcohol so as to remove any possible contaminant. The prepared pure titanium filler was also thoroughly cleaned and carefully sandwiched between the two composite specimens in the clamp. Thereafter, specimens were welded immediately by the Nd:YAG laser with wavelength of 1.06 μm, defocused distance of 10 mm so as to give a focus spot diameter of approximately 1.26 mm, laser fluence energy 2 J, frequency 25 Hz, and pulse duration 4 ms. In the welding, the relative moving speed of the laser and the welding pieces (i.e. feedrate) was set at 300 mm/min

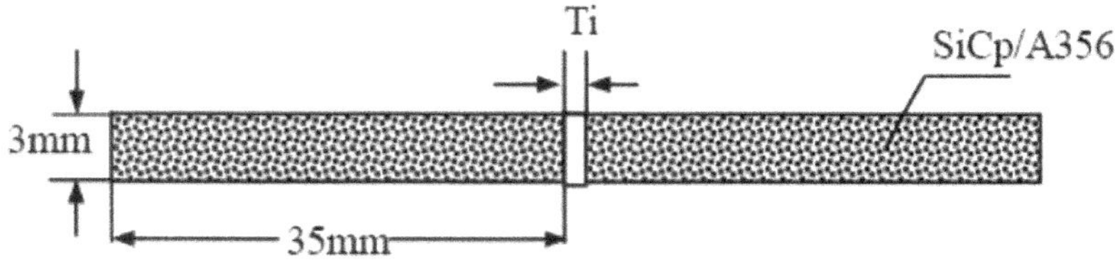

Figure 2. Schematic illustration of laser welding with Ti filler

Tensile strength of the joint was performed by a MTS Alliance RT/30 electron-mechanical material testing machine with a straining velocity of 0.5mm/min.The cross-section of welded joints was wire-cut for optical microscope investigation, and Scanning Electron Microscope (SEM) and Transmission Electron Microscope (TEM) analyses. SEM was used to analyze in detail the microstructure at the weld joints and the fractured tensile test-pieces of the joints. Optical microscope was used for observing the structure of a large area. TEM and Electron Diffusion X-ray analysis (EDX) were used to analyze the interface between the newly-formed phases and aluminum matrix, the distribution of chemical elements and spectra at the joints. The Nd:YAG laser with the similar setting conditions and feedrate was also used to weld the aluminum matrix composite specimens without any filler.

3. Results and discussion

3.1. Microstructures and properties of welded joints

The microstructure of the *in situ* reinforcement of Ti by Nd:YAG laser welding with 0.3 mm thick Ti filler was shown in Fig. 3. Appearance of *in situ* reinforcement particles distributed uniformly in the welded joint was seen. It also showed the disappearing of the drawbacks like incomplete fusion and pernicious phase Al4C3. These subsequently resulted in higher tensile strength (Table 2) of the joint. Comparatively, the reinforcement particles distributed more compactly than that of parent composite (cf. Fig. 1 and Fig. 3). The relatively more highly compacting reinforcement particles improved distinctly the properties of welded joints. The presence of Ti effectively improved the flowability of the liquid welding pool and the newly formed *in situ* reinforcement particles (Fig. 3) replaced those initial reinforcement particles (Fig. 1). Those dimples appeared in the SEM of the corresponding fractured surface (Fig. 4) suggested that: (i) the new-formed reinforcement particles had been perfectly wet; and (ii) the harmful composite structure of the initial welding viz. reinforcement/Ti/reinforcement had been changed to the state of reinforcement/matrix /reinforcement. XRD pattern of the fractured surface (Fig. 5) of the weld joint did not give sign of any harmful phase or brittle phase of Al4C3. This suggested the occurrence of effective interface transfers between reinforcement particles and matrix in the laser welded joint that subsequently provided favorable welding strength (Table 2). By the newly formed *in situ* reinforcement particles as detected by EDX (Table 3) and the intensity spectra shown in Fig. 5, the newly-formed reinforcement particle in the weld was identified as TiC.

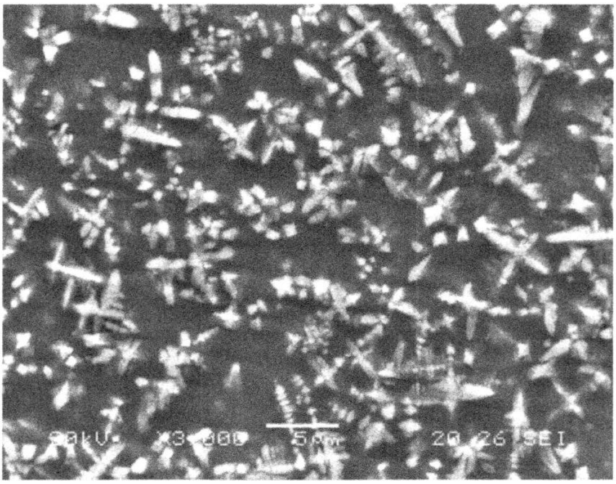

Figure 3. Microstructure of *in situ* reinforcement by laser welding with 0.3 mm thick Ti filler

Thickness of Ti filler (mm)	0.00	0.15	0.3	0.45	0.60	0.75
Mean strength (MPa)	76-91	126-135	168-180	143-160	107-113	79-96

Table 2. Mean tensile strength of laser welded specimens with various Ti filler thicknesses

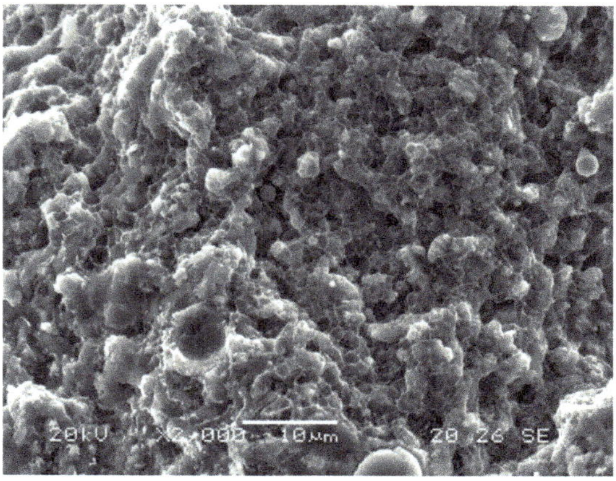

Figure 4. Fractograph of the laser welded joint with 0.3 mm thick Ti filler

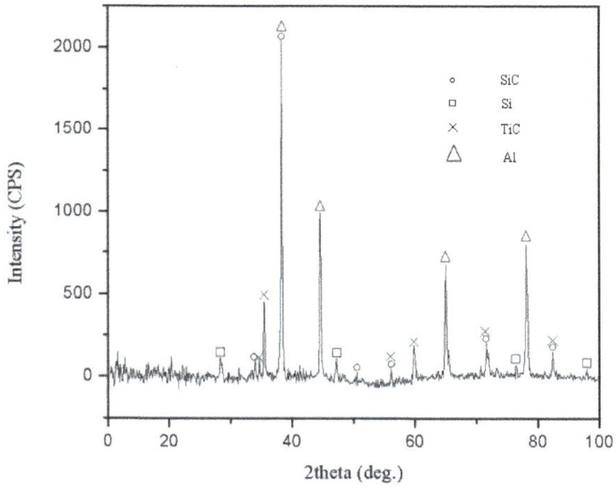

Figure 5. XRD pattern of the fracture surface for laser welding with 0.3 mm thick Ti filler

Element	Ti	Si	Al
Wt (%)	81.84	4.42	Bal.

Table 3. EDX analysis of newly-formed particle in the laser weld with 0.3 mm thick Ti filler

Figure 6. Macro-structure of the laser welded joint with 0.3 mm thick Ti filler

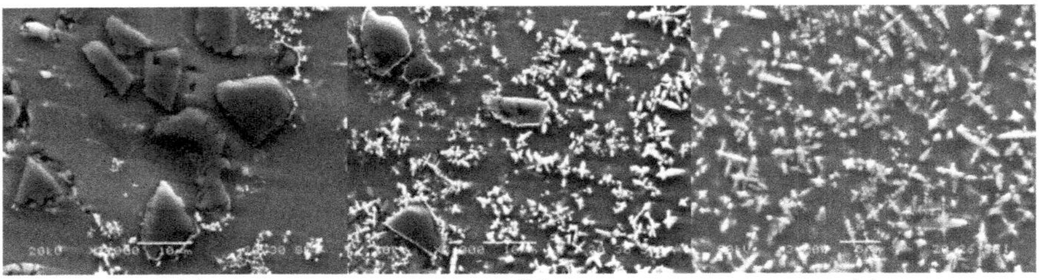

(a) (b) (c)

Figure 7. Microstructures of the different areas in the laser weld with 0.3 mm thick Ti filler
(a) Area A (b) Area B (c) Area C

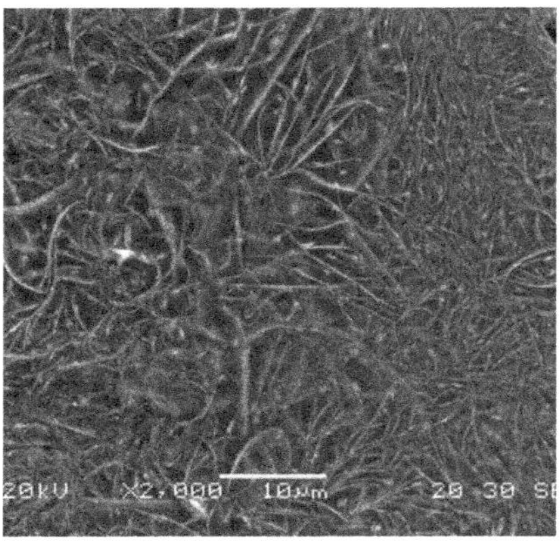

Figure 8. Microstructure of the weld with no-filler

Figure 6 illustrates the macro-structure of welded joint with Ti filler. Basically, the weld consisted of three main areas, namely: the *in situ* reinforcement area A, the two transitional areas B and C, and the no-*in situ* reinforcement area D. Their individual microstructures were shown in Fig. 7. The microstructures indicated the initial reinforcement SiC particles were completely replaced by the newly-formed *in situ* reinforcement TiC particles that mainly resulted in the formation of the *in situ* reaction area A (Fig. 7a). In the area B, the newly-formed *in situ* reinforcement TiC particles and the initial reinforcement SiC particles were coexistent (Fig. 7b). In the area C, very little newly-formed *in situ* reinforcement TiC particles were found (Fig. 7c). In the area D, there only existed the initial reinforcement particles SiC (Fig. 1). It was found that the typical pernicious acicular Al4C3 microstructure had been effectively alleviated in the welded area. Hence, it improved markedly the properties of the welded joints and their achievable tensile strength was up to 180 MPa (Table 2) that was about 75 % of the strength of SiCp/A356. The microstructure (Fig. 8) of the traditional Nd:YAG laser welding with no-filler showed that there were lots of pernicious acicular Al4C3 in the weld, which led to lower weld joint tensile strength (Table 2), i.e. only as high as 91 MPa (that was about 37.9 % of its parent SiCp/A356).

3.2. Element distribution in the transition area

Figure 9 illustrates the element distribution of the area B in the weld as shown in Fig. 6 and Fig. 7b. It showed the newly-formed *in situ* reinforcement particles were around the initial reinforcement SiC particles which offered a high density nucleus area for the nucleation of *in situ* TiC. The stirring effect in the welding pool by laser irradiation and the initial reinforcement SiC particles would be replaced by the newly-formed *in situ* TiC (cf. Figs 9b and 9c) following the chemical reaction process of:

$$Ti\,(l) + SiC\,(s) \longrightarrow TiC\,(s) + Si\,(s)$$

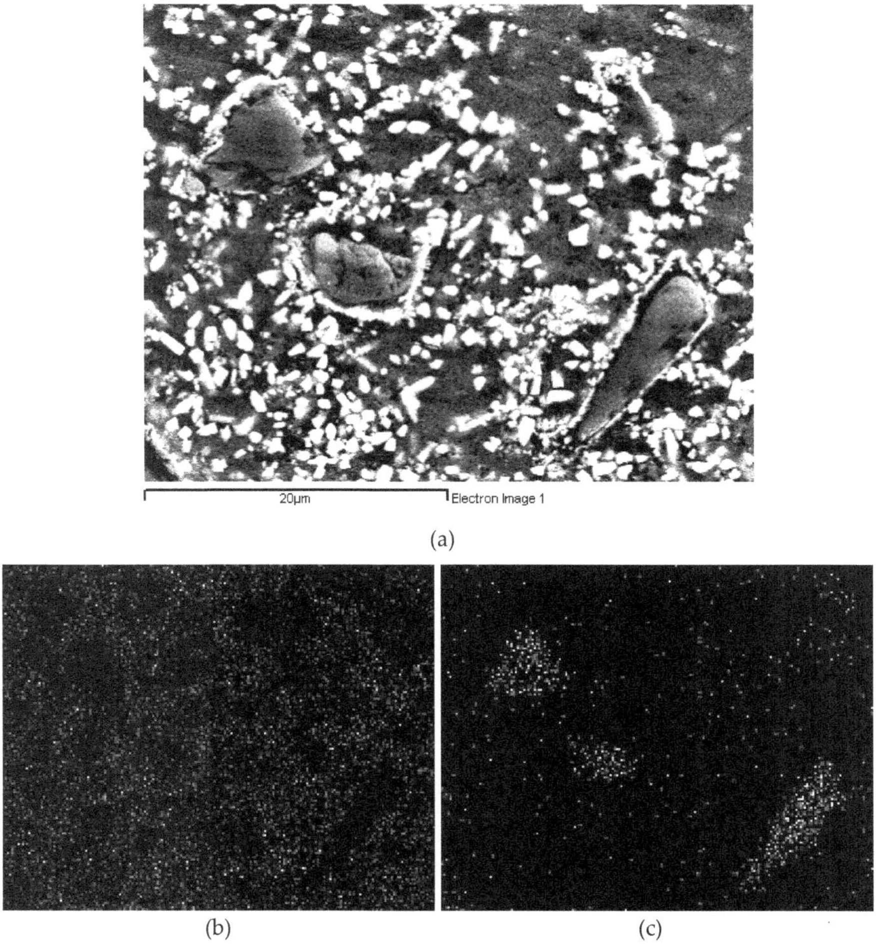

Figure 9. Element distribution of B area in the weld; (a) micrograph of the area B (b) Ti element face distribution (c) Si element surface distribution

According to the free energy for forming the SiC, TiC and Al_4C_3 carbides as elaborated in Fig. 10 of literatures [5, 6, 7], the free energy required to form TiC was much lower than that for Al_4C_3 when the reaction temperature was above 800 ºC. The affinity between Ti and C in the Nd:YAG laser welding was therefore greater than that of Al and C. The chemical reaction between Ti and SiC in the welding pool would subsequently take precedence over the reaction between Al and SiC and thus resulted in restraining the formation of the pernicious acicular Al_4C_3.

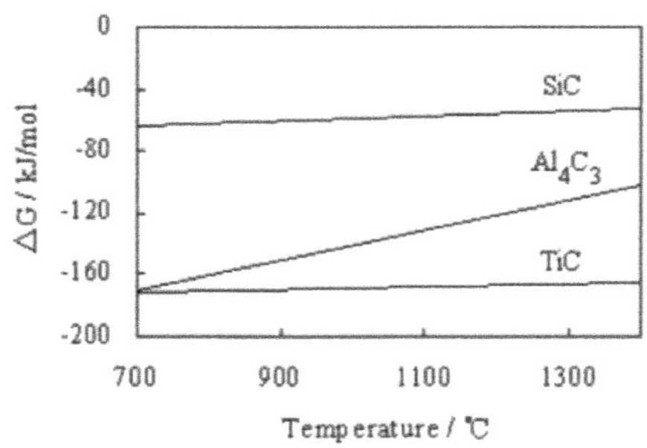

Figure 10. Free energy of formation of several metallic carbides [5, 6, 7]

3.3. Influence of Ti filler thickness

The microstructures of *in situ* reinforcement with various thicknesses of Ti filler were illustrated in Fig. 11 and its corresponding fractographs were shown in Fig. 12. It illustrated that the amount of the formed *in situ* TiC was distinctly increased with the increase in the thickness of Ti filler. Test indicated that maximum strength of welded joints (Table 2) was achieved at Ti filler thickness 0.3 mm (Figs. 3 and 4). This was because the newly-formed *in situ* reinforcement particles TiC were uniformly distributed in the weld and the initial irregular (mostly in hexagonal shape as shown in Fig. 1) reinforcement SiC particles in the weld were no longer observed (Fig. 3). Moreover, the pernicious acicular Al_4C_3 was successfully restrained (Figs 3 and 7a). At the thickness of Ti filler below 0.3 mm, there was little sign of the newly-formed *in situ* reinforcement TiC particles to be observed (Fig. 11a) and a number of pernicious acicular Al_4C_3 were formed in the weld. When the thickness of Ti filler was just beyond 0.3 mm, the properties of the joints tended to become poorer again (Fig. 12b). This was because the input energy was mainly used for melting the Ti filler and resulted in the coexistence of coarse columnar crystals and fine equiaxed crystals (Fig. 11b). When the thickness of Ti filler was further increased (Fig. 11c), the percentage of liquid Ti in the welding pool would also be increased. Subsequently, the weld zone would form coarser columnar crystals, as displayed in the SEM micrograph of Fig. 11c, after the resolidification of the melt. From the phase diagram (Fig. 13) of Ti-Al binary system [8], it can be anticipated that increasing the content of Ti would lead to the intermetallic compounds like TiAl and Ti_3A_l, *etc*., to be formed during the Nd:YAG laser welding. As illustrated by the XRD pattern of the fractured surface of a laser weld joint with the thicker Ti filler in Fig. 14, there were some distributing brittle intermetallic compounds like TiAl and Ti_3Al to be detected. Available literature [9] has demonstrated that TiAl and Ti_3Al are the harmful intermetallic compounds in the weld and tend to decrease obviously the strength of welded joints. Such harmful effect may follow the chemical reaction of: 5Ti[Al l] +3Al[l] + SiC[s]—→TiC[s] + Si[Al [l]] + Al[l] + (TiAl + Ti_3Al). Hence, too thick of the Ti filler led to: (i) the appearance of the large block of columnar crystals in the microstructure (Fig. 15); and (ii) the new-formed reinforcement TiC to be replaced by the melted/re-solidified Ti and subsequently only the melted/re-solidified Ti existing in the weld. Results (Figs 3, 7 and 11) indicated that there existed an optimal thickness of Ti filler in the individually set parameters in the Nd:YAG laser welding of SiCp/A356. With the optimal thickness of Ti filler, the initial reinforcement particles SiC distributed in the aluminum matrix composite SiCp/A356 would offer a highly dense nucleus area for the *in situ* TiC nucleation. This would effectively alleviate the forming of intermetallic compounds like TiAl and Ti_3Al in the weld. It was in this manner that ultimately created favorable condition to provide relatively superior strength of the welded joint to that of conventional laser welding.

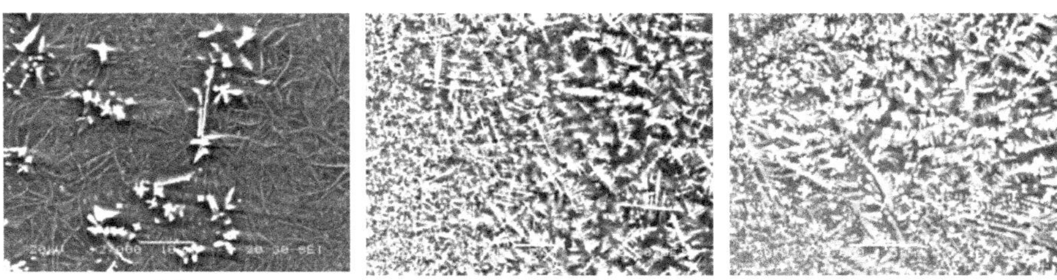

(a) (b) (c)

Figure 11. Microstructures in the A area with various thicknesses of Ti filler; (a) δ=0.15 mm (b) δ=0.45 mm (c) δ=0.60 mm

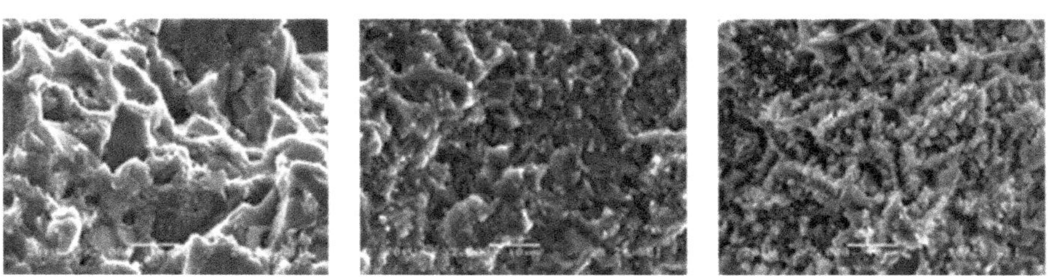

(a) (b) (c)

Figure 12. Fractographs of welded joints with various thicknesses of Ti filler; (a) δ=0.15 mm (b) δ=0.45 mm (c) δ=0.60 mm

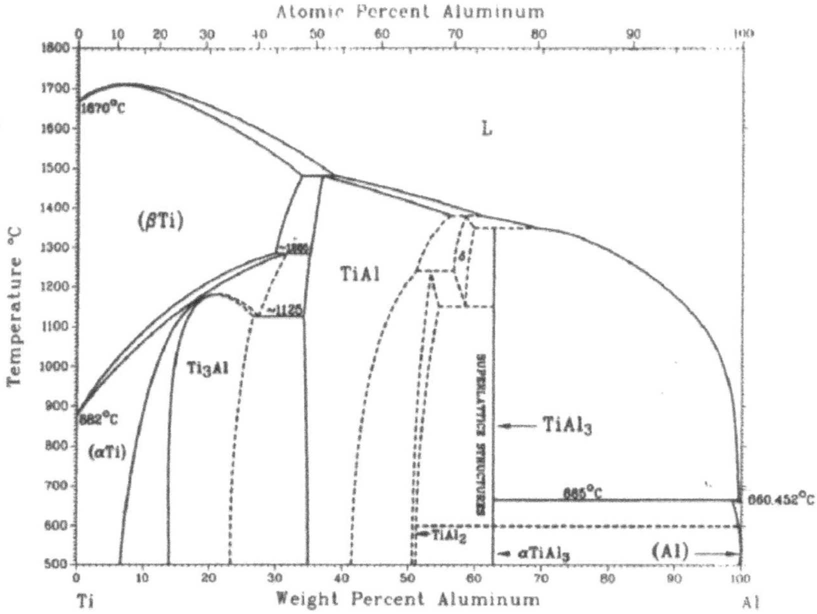

Figure 13. Binary phase diagram of Ti—Al [8]

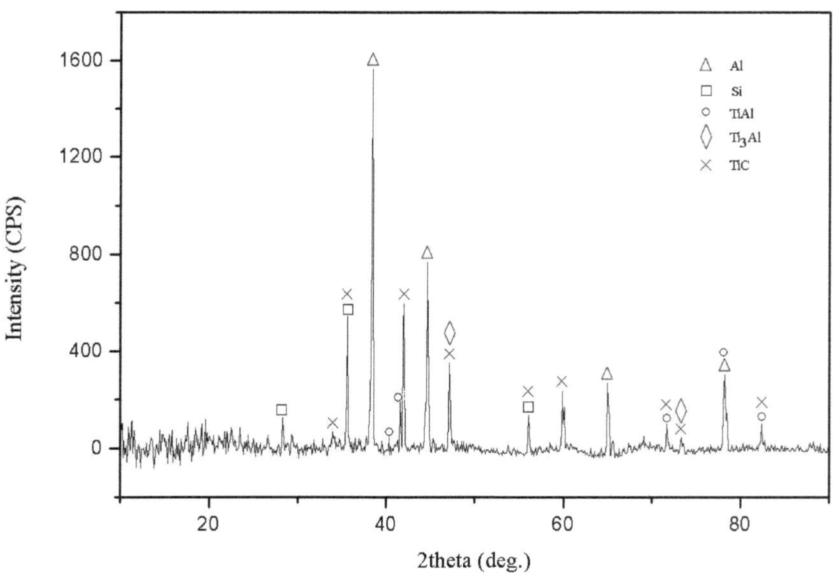

Figure 14. XRD pattern of fracture surface (δ=0.6 mm)

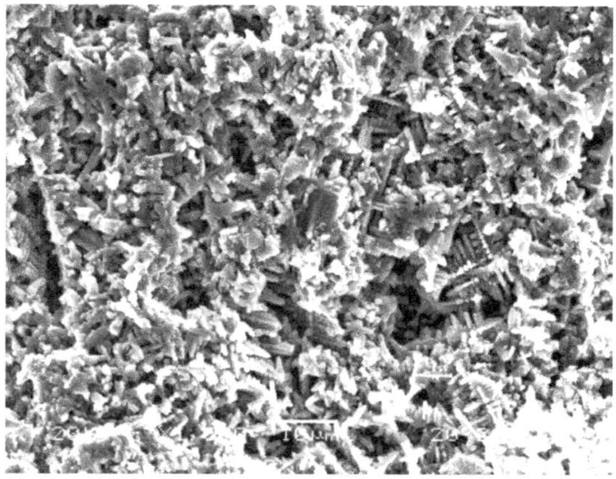

Figure 15. Columnar crystals in the laser weld with 0.6 mm thick Ti filler

3.4. TEM of the interface between *in situ* reinforcement particle TiC and matrix

The interface between *in situ* reinforcement particle TiC and the matrix was analyzed by the TEM micrograph displayed in Fig. 16. It demonstrated a clear interface between the newly-formed *in situ* reinforcement particle TiC and matrix. This clearly distinct interface (Fig. 16) suggested the occurrence of prominent *in situ* reaction to integrate the reinforcement particle with matrix (cf. Figs 4 and 16), and the high probability of successfully transferring load from the matrix to TiC and vice versa. It also gives indication that the aluminum matrix composite SiCp/A356 would be welded satisfactorily by Nd:YAG laser.

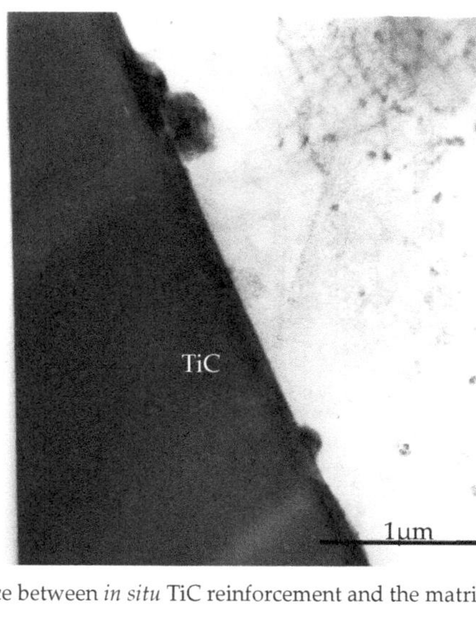

Figure 16. TEM of interface between *in situ* TiC reinforcement and the matrix for laser welding with 0.3 mm Ti filler

4. Microstructure evolution during the welding

4.1. Temperature field of laser welding

The heat is assumed to be released instantaneously at time t=0 on the surface of the substrate. This causes a temperature rise in the material as follows [10-12]:

$$T - T_0 = \frac{Q}{\rho c\left(4\pi\alpha t\right)^{\frac{3}{2}}}\exp\left(\frac{-R^2}{4\alpha t}\right) \tag{1}$$

where ρ is the material density, C is specific heat, α is thermal diffusivity, λ is thermal conductivity, Q is the input energy.

When temperature distribution is quasi-steady state:

$$T - T_o = \frac{q_o}{2\pi\rho c\alpha R}\exp\left(-\frac{v}{2\alpha}\left(R + x\right)\right) \tag{2}$$

During the Nd:YAG laser welding, q_0 can be expressed as

$$q_0 = \eta\left(N\left(f,v\right)\right) \times \frac{P}{\frac{\pi D^2}{4} \times PD} = \eta\left(N\left(f,v\right)\right) \times \frac{4P}{\pi D^2 \times PD} \tag{3}$$

Where

$\eta\left(N\left(f,v\right)\right)$ is the coefficient of laser welding input energy, which is direct proportional to number of overlaps or pulse frequency. With number of overlaps increasing, or pulse frequency increasing with constant the feed-rate, $\eta\left(N\left(f,v\right)\right)$ will be increased synchronously. With the velocity (feedrate) increasing, number of overlaps with the constant pulse frequency will be decreased correspondingly led to lower heat input, $\eta\left(N\left(f,v\right)\right)$ will be decreased accordingly.

Eq. (2) can be rewritten as:

$$T - T_o = \frac{2 \times \eta\left(N\left(f,v\right)\right) \times P}{\pi^2 D^2 \rho c\alpha R \times PD}\exp\left(-\frac{v}{2\alpha}\left(R + x\right)\right) \tag{4}$$

Define $\xi = \frac{1}{2\pi\rho c\alpha}$, then Eq. (4) can be written as:

$$T - T_o = \frac{\xi \times \eta\left(N\left(f,v\right)\right) \times P}{\frac{\pi D^2}{4} \times PD \times R}\exp\left(-\frac{v}{2\alpha}\left(R + x\right)\right) \tag{5}$$

$$\text{Define } q=\frac{P}{\frac{\pi D^2}{4}\times PD}, \tag{6}$$

then Eq. (6) can be written as

$$T-T_o=\frac{\xi\times\eta\left(N(f,v)\right)\times q}{R}\exp\left(-\frac{v}{2\alpha}(R+x)\right) \tag{7}$$

4.2. Simulation model

4.2.1. Equations for temperature distribution

Using energy balance, a differential equation can be obtained for the steady temperature distribution in a homogeneous isotopic medium, that is

$$\frac{\partial}{\partial x}\left(K_x\frac{\partial\theta}{\partial x}\right)+\frac{\partial}{\partial y}\left(K_y\frac{\partial\theta}{\partial y}\right)+\frac{\partial}{\partial z}\left(K_z\frac{\partial\theta}{\partial z}\right)=-q^B \tag{8}$$

Where the boundary conditions are $\theta\big|_{S_1}=\theta_e$, $K_S\frac{\partial\theta}{\partial x}\Big|_{S_2}=q^s$.

For

$$\pi=\int_V\left\{K_x\left(\frac{\partial\theta}{\partial x}\right)^2+K_y\left(\frac{\partial\theta}{\partial y}\right)^2+K_z\left(\frac{\partial\theta}{\partial z}\right)^2\right\}d_V-\int_V\theta q^s d_V-\int_{S_2}\theta^s q^s d_s \tag{9}$$

After Eq. 9 is discrete for the element, and according to $\delta\pi=\sum_{e=1}^{n}\delta\pi_e=0$, it will be obtained

$$\overline{K}\,\overline{\overline{K}\theta}=\overline{\theta_S}+\overline{\theta_B}-\overline{C\theta}+\overline{K^C}(\theta_C-\overline{\theta^S})+\overline{K^r}(\theta_r-\overline{\theta^S}) \tag{10}$$

where S: isothermal boundary, B: the heat-input, c: the conductive and r: the irradiative.

4.2.2. Hypothesis and mesh

Based on the situations during the laser welding and mainly focused on the temperature distribution, it is supposed that the laser resource is considered as a Gaussian distribution. Also, on the basis of specimen size wire-cut, the calculating size is set as 25 mm (x) × 20 mm (y) × 3 mm (z), the schematic of its finite element (FE) mesh is shown in Fig. 17. Moreover, Ti filler is considered as a section of the substrate with the different properties to ignore the effect of gap between the Ti filler and the substrate.

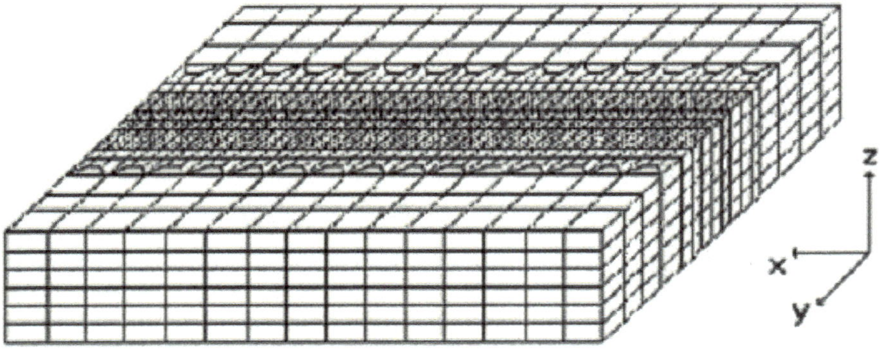

Figure 17. FE mesh for 3D numerical analysis

4.2.3. Temperature distribution

The simulated results are shown in Fig. 18 to Fig. 23. It shows that the temperature without Ti filler is same as the traditional laser welding. Simultaneously, due to the heat input into the substrates directly, without the additional heat resource for melting Ti filler, the peak of temperature (heat input) is relatively higher to form the weld. As a result, increasing the heat input into the substrate will decrease the tensile strength of the welded joint and wide the heat affected zone (HAZ) resulted in lower properties in the succedent practical applications (Table 2 and Fig. 19). Furthermore, a large amount of coarser acicular Al_4C_3 distributed in the fracture surface as shown in Fig. 19, which decreased the tensile strength of the welded joints seriously.

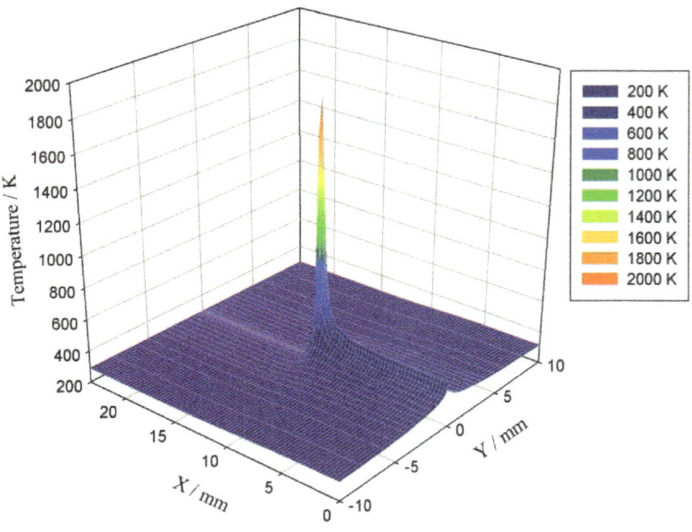

Figure 18. Temperature distribution without Ti filler

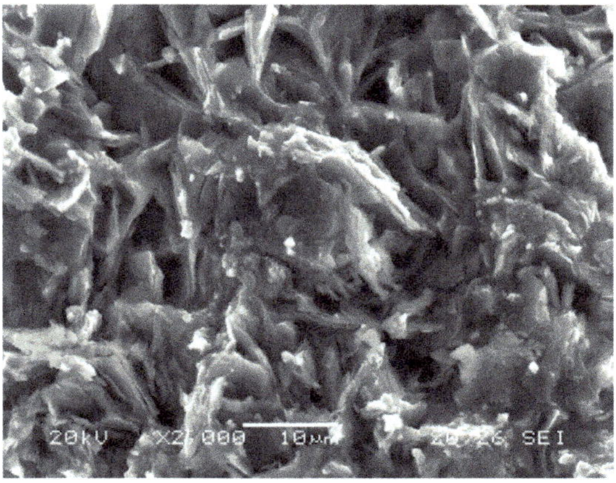

Figure 19. Fractograph of the laser welded joint without Ti filler

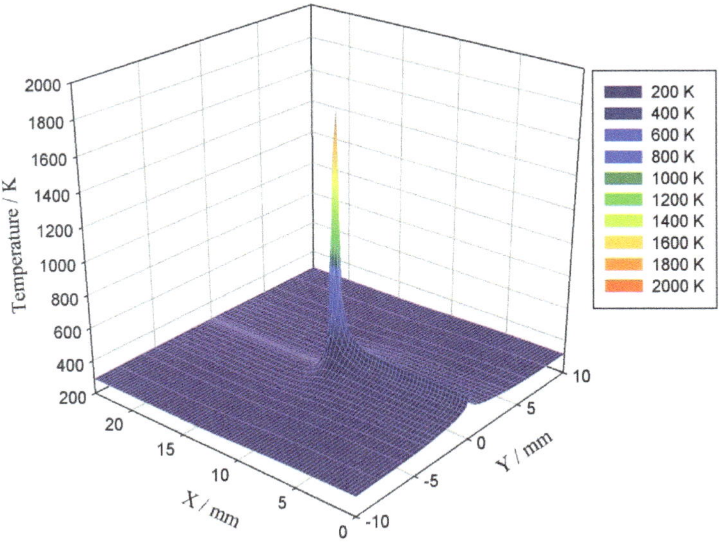

Figure 20. Temperature distribution with Ti filler

Figure 20 shows the temperature field of laser welding SiCp/A356 with Ti filler. Considering the Ti melting and *in situ* reaction in the welding pool as an endothermic reaction, the welding temperature decreases and will be lower than that of laser welding directly (cf. Figs. 18 and 20), and its temperature field is distributed more smoothly with *in situ* reaction than that of laser welding without Ti filler as shown in Fig. 21. Also, the width of HAZ is decreased to some extent (Fig. 21b). Furthermore, it shows that according to the real effect of laser beam diameter, the thickness of Ti filler is about 0.3 mm will be optimal for *in situ* welding which conformed to the experimental results as shown in Table 2.

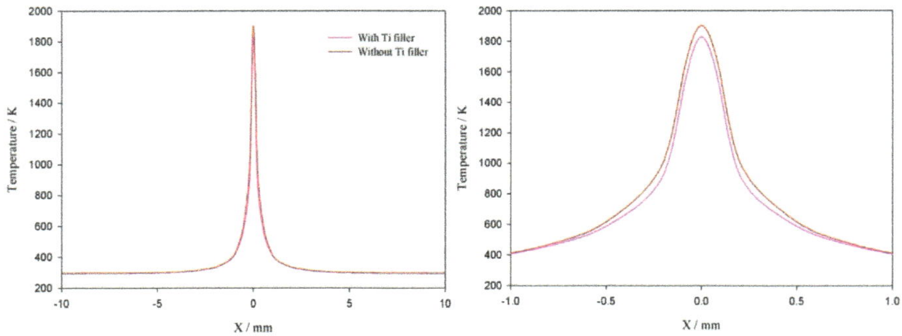

(a) (b)

Figure 21. Temperature distribution of central heating on XOZ plane
(a) Temperature distribution on XOZ plane (b) Magnification of (a)

In addition, the effect of Ti on the temperature distribution on the central line is shown in Fig. 22. It illustrates that the peak of the temperature is changed distinctly. Because of the sandwiched Ti between the substrates and *in situ* endothermic reaction, the temperature of substrate ahead of laser resource is lower than that of without Ti filler. Moreover, the temperature at the succedent distance is increased or accumulated a little bit due to the different conductive coefficient between Ti and substrate. On the other side, its corresponding trend of the temperature behind the laser resource (resolidification) is same as that of without Ti filler except for a peak appearance induced by more serious exothermic potential during the crystallization.

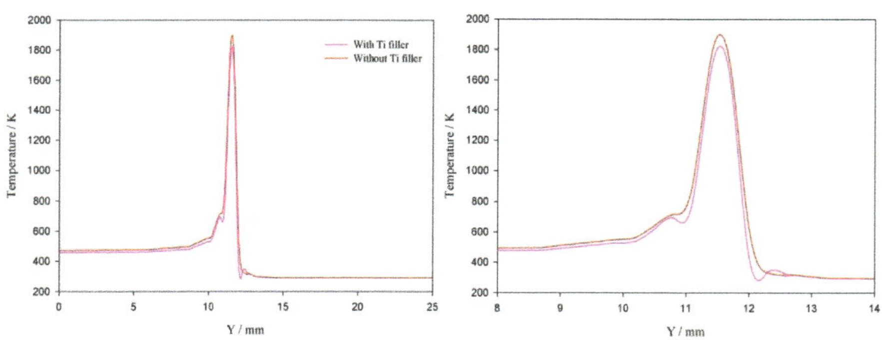

(a) (b)

Figure 22. Temperature distribution of central line on YOZ plane
(a) Temperature distribution on YOZ plane (b) Magnification of (a)

Figure 23 shows the temperature distribution when Ti filler is thick. The peak of temperature is decreased obviously and leads to the welding failure.

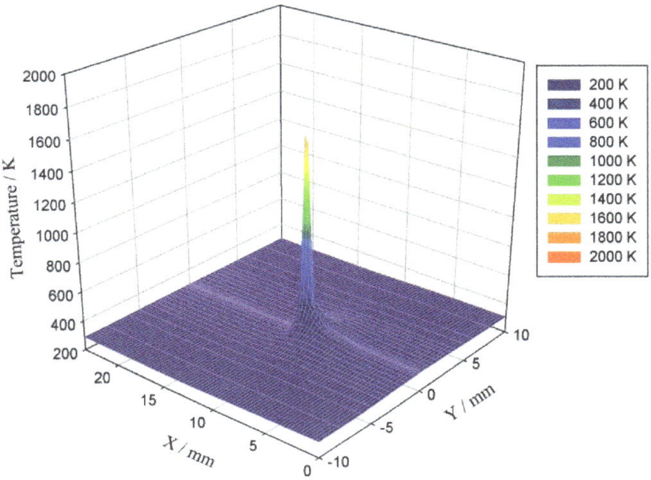

Figure 23. Temperature distribution with thick Ti filler

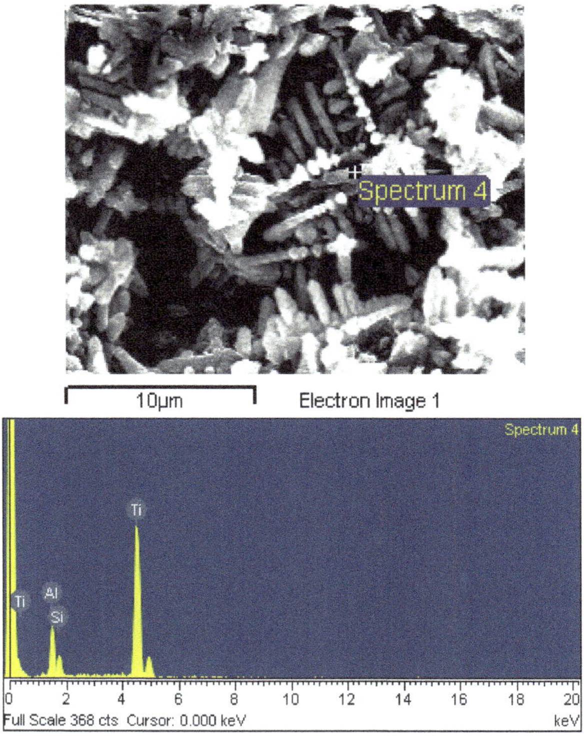

Figure 24. Microstructure and EDX of laser weld with thick Ti filler

Figure 24 shows the microstructure of laser welded joint with thick Ti filler and its corresponding energy dispersive X-ray spectroscopy (EDX) results. It can be observed that a large number of columnar Ti crystallization is distributed in the weld. From Figs. 23 and 24, it elucidates that with the increase of Ti thickness, the heat input into the substrate is decreased and most of energy is used for melting Ti led to the insufficient *in situ* reaction and stirring in the welding pool resulted in lower properties of welded joints.

Furthermore, in order to verify the temperature field, noncontact thermometer (model AZ9881) was used to measure the spot temperature on-line. The measured temperature results are shown in Fig. 25. It shows that the measured results agree well with the simulated results.

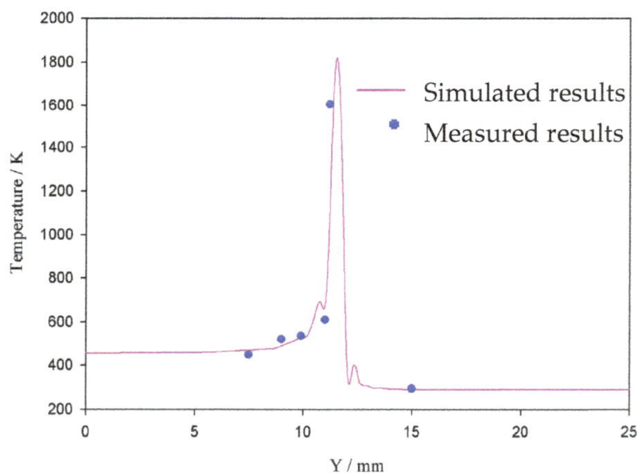

Figure 25. Surface temperature distribution in the processing center

4.3. Microstructure evolution simulation

According to the temperature calculation, the simulation of the evolution of the microstructure based on thermodynamic equilibria, diffusion [5, 6, 7] was shown in Fig. 26. It showed that during the welding pool solidification, the *in situ* reinforcement particles TiC would be formed around the initial reinforcement SiC particles. With the increase of cooling time, the initial reinforcement SiC particles would be replaced by the newly-formed *in situ* reinforcement particles TiC. It was well matched with the results shown in Figs. 7 and 9.

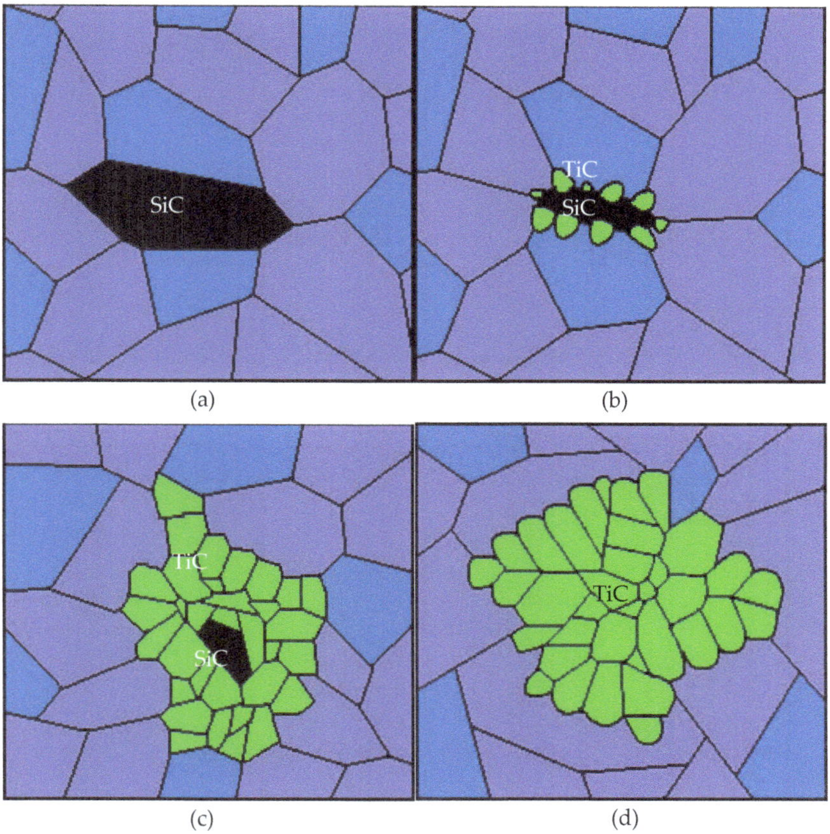

Figure 26. Simulation on microstructure evolution
(a) Initial (b) Start of solidification (c) Middle of solidification (d) End of solidification

5. Conclusions

The use of titanium as a filler metal in Nd:YAG laser welding of SiC_p/A356 provided beneficial *in situ* reinforcement effect. The effect of *in situ* reinforcement of the Ti filler allowed the newly-formed reinforcement TiC particles to distribute uniformly in the weld that subsequently resulted in successfully welding the SiC_p/A356 composite. Moreover, the typical pernicious interfacial reaction microstructure such as Al_4C_3 was effectively restrained from the interface between aluminum matrix and reinforcement particles in the Nd:YAG laser welding of SiC_p/A356 with Ti filler. Furthermore, according to the temperature calculation, the evolution of the microstructure was simulated based on thermodynamic equilibria, diffusion. Results were well matched with the corresponding experiments.

Author details

Kelvii Wei Guo
MBE, City University of Hong Kong, Hong Kong, China

Acknowledgement

The work is supported by a RGC general research fund (GRF) (Grant No.:9041503.) and a Strategic Research Grant (SRG) from City University of Hong Kong (Grant No.: 7002287.)

6. References

[1] J. M. Go'mez de Salazar, M. I. Barrena: Dissimilar Fusion Welding of AA7020/MMC Reinforced with Al_2O_3 Particles. Microstructure and mechanical properties. Materials Science and Engineering A 2003; 352(1-2): 162-168.
[2] W. Guo, M. Hua, H. W. Law and J. K. L. Ho: Liquid-Phase Impact Diffusion Welding of SiCp/6061Al and Its Mechanism. Materials Science and Engineering: A, 2008, 490, (1-2), 427-437.
[3] W. Guo, M. Hua, and J. K. L. Ho: Study on Liquid-Phase-Impact Diffusion Welding SiCp/ZL101. Compos. Sci. Technol., 2007, 67, (6), 1041-1046.
[4] L. M. Marzoli, A. V. Strombeck, J. F. Dos Santos, C. Gambaro, L. M. Volpone: Friction Stir Welding of an AA6061/Al2O3/20p Reinforced Alloy. Composites Science and Technology 2006; 66(2): 363-371.
[5] D. A. Porter, K. E. Easterling: Phase Transformations in Metals and Alloys, 2nd. Cheltenham: Nelson Thornes, 2001.
[6] R. Riedel: Handbook of Ceramic Hard Materials. New York: Wiley-VCH, Weinheim, 2000.
[7] R. Boyer, G. Welsch, E. W. Collings: Materials Properties Handbook: Titanium Alloys. Materials Park, Ohio: ASM International, 1994.
[8] J. R. Davis: ASM Specialty Handbook – Aluminum and Aluminum Alloys. Materials Park, Ohio: ASM International, 1993. p.557.
[9] S. Mall, T. Nicholas: Titanium Matrix Composites – Mechanical Behavior. Lancaster, Pa.: Technomic Pub. Co. Inc., 1998.
[10] W. M. Steen: Laser Material Processing, 3rd ed. London: Springer-Verlag; 2003.
[11] K. W. Guo: Influence of In Situ Reaction on the Microstructure of SiCp/AlSi7Mg Welded by Nd:YAG Laser with Ti Filler. J. Materials Engineering and Performance, 2010, 19, 52-58.
[12] A.F. Mills: Heat and Mass Transfer. P. R. Donnelly & Sons Company; 1995.

Chapter 6

Nd:YAG Laser Welding for Photonics Devices Packaging

Ikhwan Naim Md Nawi, Jalil Ali, Mohamed Fadhali and Preecha P. Yupapin

Additional information is available at the end of the chapter

1. Introduction

The state-of-the art of the pulsed Nd:YAG laser spot welding for photonics device packaging has been introduced by Marley (Marley, 2002), which utilizes the laser for high precision joining and alignment. The advantages of laser welding over conventional fusion welding processes include precise welds with a high aspect ratio, narrow heat affected zone (HAZ), very little thermal distortion, ease of automation, high welding speed, enhanced design flexibility, clean, high energy density, low heat input and an efficient process (Zhou & Tsai, 2008; Kazemi & Goldak, 2009). One of the key features of laser welding is the ability to weld without filler materials and it offers distinct advantages (Pang et al., 2008). Laser welding is a liquid-phase fusion process. It joins metals by melting the interfaces and resulting the mixing of liquid molten metal. Then, it solidifies on the removal of the laser beam irradiation (Ready, 1997). Photonics devices used for telecommunications of military applications are usually required to operate for a long life of operation in fields with potentially humid, corrosive and mechanically turbulent environments. Therefore, long term reliability in such hostile operating conditions requires strong fixing of the aligned components and hermetic sealing of the photonics devices inside metal hybrid housings (Fadhali et al., 2007a). It is worth to mention that 60% to 80% of the photonics devices modules cost are due to the coupling and packaging processes. Therefore, understanding the effective packaging technique is very important to produce efficient and reliably packaged photonics devices. Moreover, care must be taken to assemble functional packages because only packages that can be manufactured reliably at competitive costs will survive in the business world. For photonics packaging applications, most of the welds are of butt or lap joints which require the weld penetration depth to be larger than the bead width. Moreover, for miniature packages that contain some sensitive coupling components, the penetration depth should be large enough to achieve a strong attachment (Fadhali et al.,

2007a, 2007b, 2007c, 2007d, 2008). At the same time, the bead width should be small to minimize the HAZ and hence prevent the damage to the sensitive optical components. The desired material for this application requires a low thermal conductivity or a higher electrical resistivity (Dawes, 1992). The lower the thermal conductivity of a material the more likely it is to absorb laser energy. For this reason, several weldable grades of steel and stainless steel are ideal for laser welding. The low carbon austenitic stainless steel (300 series steel) which has carbon level less than 0.1% produces good quality welds and reliable weld performance (Fadhali et al., 2007a; Naim et al., 2009). Hence, in this chapter, a stainless steel 304 is utilized as a base material for laser welding. It has been reported that other types of stainless steel were studied by other researchers. For instance, Mousavi and Sufizadeh investigated stainless steel 321 and 630 (Mousavi & Sufizadeh, 2008), while Beretta et al. studied stainless steel 420 (Beretta et al., 2007), for the application of pulsed Nd:YAG laser welding. Whilst a great deal of effort has been focused on developing processing systems there is an urgent need to understand the strength of the weldment. It is the aim of this chapter to examine the strength of a stainless steel 304 welded joint. Despite the pulsed Nd:YAG laser welding has been widely used in microelectronic and photonics packaging industry, a full understanding of various phenomena involved is still a matter of trials and speculations.

2. Mathematical model of penetration depth of a spot weld

Penetration depth of laser spot welding is one of the vital parameters that contributes to a good laser spot micro welding outcome. Practical penetration depth measurements are time consuming and laborious like cutting the samples in order to obtain the weld cross section. Thus, this section shows a theoretical model to estimate penetration depth inside the welded samples by controlling laser parameters or to predict the laser parameters at the required penetration depth. Figure 1 illustrates the penetration depth on a welding specimen. The 1-Dimensional model is developed base on the heat conduction and energy balance equations (Naim et al., 2010).

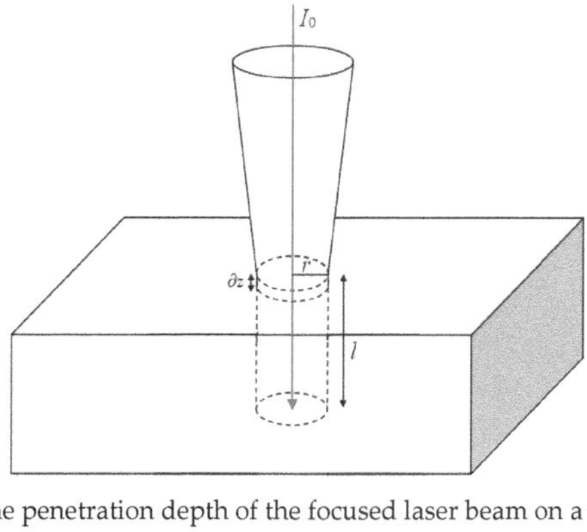

Figure 1. Illustration of the penetration depth of the focused laser beam on a welding specimen.

Applying the energy balance equation at the laser spot gives (Naim et al., 2010),

$$(1-R)I_0 = \rho L_m \frac{\partial l}{\partial t} - K\left(\frac{\partial T}{\partial z}\right)_{z=0} \tag{1}$$

Here I_0 is laser power density, L_m is latent heat of fusion, R is material reflectivity, ρ is material density, K is thermal conductivity, $\partial T / \partial z$ is temperature gradient at the welding front, l is the penetration depth, t is irradiation time and $\partial l / \partial t$ is penetration velocity. To determine the penetration velocity which is the change of penetration depth with time, the temperature gradient at the welding front should first be determined. The temperature distribution inside the target material is governed by the 1-Dimension heat conduction equation given as (Naim et al., 2010),

$$\frac{1}{\alpha}\frac{\partial T}{\partial t} = \frac{\partial^2 T}{\partial z^2} \tag{2}$$

where, α is thermal diffusivity. Equation (2) can be rewritten as

$$-\frac{1}{\alpha}\frac{\partial l_p}{\partial t}\frac{dT}{dz} = \frac{d^2 T}{dz^2} \tag{3}$$

The boundary conditions are taken as,

$z=0$, $T=T_m$ and at $z \to \infty$, $T=T_0$.

The maximum attainable temperature is the melting temperature of the material, T_m. Substituting these boundary conditions into Equation (3) and the solution for temperature distribution inside the material as

$$\frac{T-T_0}{T_m-T_0} = \exp\left[-\frac{1}{\alpha}\left(\frac{dl}{dt}\right)z\right] \tag{4}$$

The temperature gradient at the welding front can be obtained using Equation (4),

$$\left(\frac{dT}{dz}\right)_{z=0} = -\frac{1}{\alpha}\left(\frac{dl}{dt}\right)\left(T_m - T_0\right) \tag{5}$$

Substituting Equation (5) into the energy balance equation,

$$\left(1-R\right)I_0 = \rho L_m\left(\frac{dl}{dt}\right) + \rho c\frac{dl}{dt}\left(T_m - T_0\right) \tag{6}$$

From Equation (6), the welding velocity can be rewritten as,

$$\frac{dl}{dt} = \frac{\left(1-R\right)I_0}{\rho\left[L_m + c\left(T_m - T_0\right)\right]} \tag{7}$$

Integrating Equation (7) gives,

$$l = \frac{(1-R)Pt}{\pi r_o^2 \rho \left[L_m + c(T_m - T_0) \right]} \tag{8}$$

Here P is the laser peak power, t is the irradiation time or pulse duration and r is the laser beam radius. From Equation (8), one can observe that penetration depth is proportional to laser peak power and pulse duration as well as the material reflectivity.

Figure 2 illustrates the penetration depth versus peak power and pulse duration. Any changes of either peak power or pulse duration gives the same influence to the penetration depth. This is in agreement with the fact that the penetration depth increases linearly with both the peak powers and pulse durations. The penetration depth increases linearly with an increase of laser peak power. The higher laser peak power means more laser energy is absorbed into the welding material. When a higher laser energy is absorbed, it will produce a deeper melting pool. After solidification, the melting pool produces deeper weld or penetration depth. The trend is similar for variation of pulse duration. Higher pulse duration means longer heating duration. This will provide more time for laser beam to penetrate into the welding material. Thus, a deeper penetration depth will be produced. From Figure 3, it can be summarized that the penetration depth decreases significantly with laser spot radius. This is because the laser power density, I is proportional to $1/r^2$ and the power density will drop significantly as the laser spot radius increases. Laser power density determines the quantity of energy applied on welding material at a specific time. Higher power density will produce a deeper penetration depth.

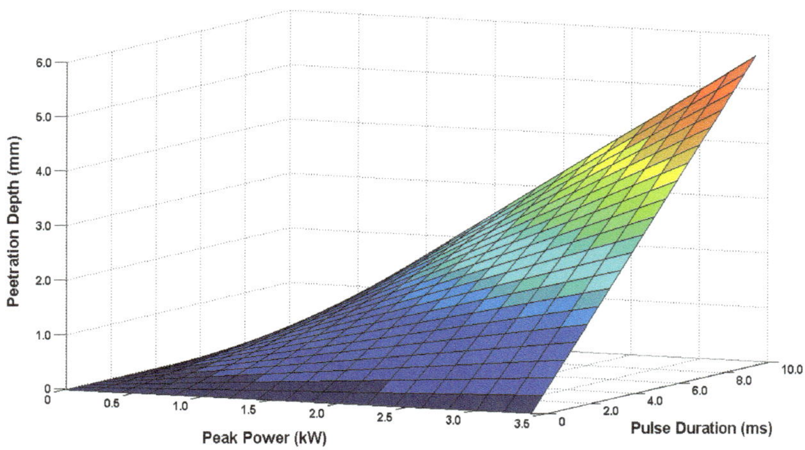

Figure 2. Penetration depth versus peak power and pulse duration.

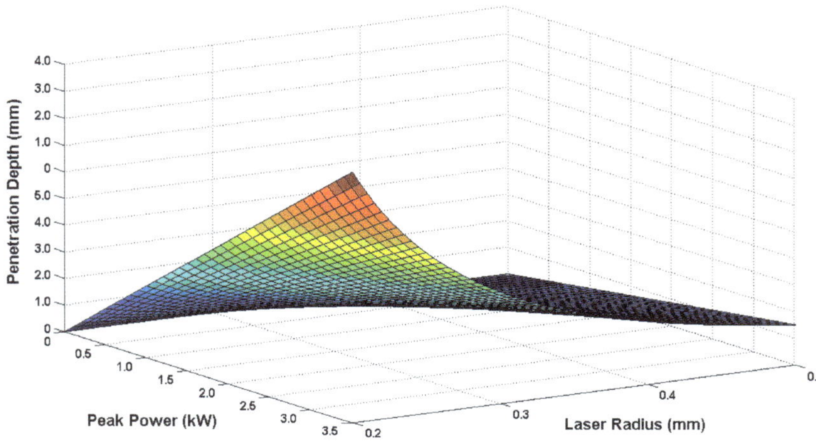

Figure 3. Penetration depth versus peak power and laser beam radius.

3. Mathematical model of laser beam power penetration

In order to produce a spot weld, a laser beam is incident on a specimen surface. The incident laser beam will be absorbed by the specimen depending on the material absorption coefficient. Thus, the absorbed laser beam heats up the specimen surface, raising its temperature. It will melt the specimen and produce a weld when it solidifies. The laser beam penetration in welding material has been derived using the continuity equation (Naim et al., 2010). It is assumed that laser beam penetrates the material based on a cylindrical coordinate. Figure 4 illustrates the laser beam power penetration in stainless steel specimen.

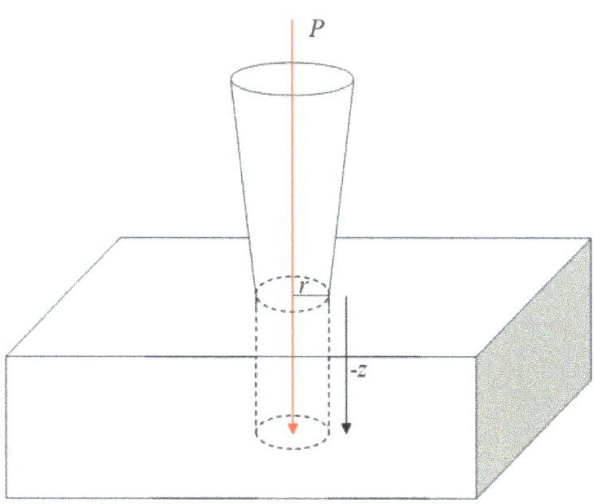

Figure 4. Laser beam power penetration in stainless steel specimen.

The laser beam power penetration is derived from a continuity equation given as (Naim, 2010),

$$\frac{\partial P}{\partial t} - k\nabla P = 0 \tag{9}$$

In cylindrical coordinate, Equation (9) then becomes,

$$\frac{\partial P}{\partial t} = k\left(\frac{\partial^2 P}{\partial r^2} + \frac{1}{r}\frac{\partial P}{\partial r} + \frac{\partial^2 P}{\partial z^2}\right) \tag{10}$$

Here $P=P(r,z,t)$ is the laser beam power penetration, r is the laser spot radius, z is the depth, k is the thermal conductivity and t is time. To solve this equation, let $P = RZT = R(r)Z(z)T(t)$ and the solutions are given by,

$$T = c_1 e^{-\mu^2 kt}, \tag{11}$$

$$R = c_2 J_0(\mu r) + c_3 Y_0(\mu r), \tag{12}$$

$$\text{and } \; Z = c_4 e^{\mu z} + c_5 e^{-\mu z} \tag{13}$$

where c_1, c_2, c_3, c_4, c_5 and μ are the unknown constants and determined by the boundary conditions, J_0 is first order of Bessel function and Y_0 is second order of Bessel function. Equation (10) can be solved by using Equations (11), (12) and (13) giving,

$$P(r,z,t) = [c_1 e^{-\mu^2 kt}][c_2 J_0(\mu r) + c_3 Y_0(\mu r)][c_4 e^{\mu z} + c_5 e^{-\mu z}] \tag{14}$$

Let $A= c_1= c_2= c_3= c_4= c_5$ as the incident laser beam peak power and μ is the material absorption coefficient. In this calculation, only first order of Bessel function is considered. It is also assumed that the laser beam propagates only directing into the material which is $-z$ direction. Equation (14) can be written as,

$$P(r,z,t) = Ae^{-k\mu^2 t}[J_0(\mu r)][e^{-\mu z}] \tag{15}$$

For stainless steel material, the absorption coefficient, μ is 0.3 (Kazemi & Goldak, 2009) and the laser beam peak power used in this consideration is 3.5kW. Equation (15) then specified as,

$$P(r,z,t) = 3500e^{-0.3^2 kt}[J_0(0.3r)][e^{-0.3z}] \tag{16}$$

Equation (16) provides the time dependent laser beam penetration. The laser beam penetration is computed for its radius and depth. Figure 5 shows the profile of applied peak power versus the depth of specimen and laser spot radius. The peak power 3.500kW decreases exponentially with depth of stainless steel material and decreases according to the first order of Bessel function in terms of the spot radius. The peak power incident on the surface is 3.500kW at the centre of laser spot target, but decreases to 3.0248 kW at the spot radius of 0.25mm. Peak power reduces to 0.1743kW at the centre and 0.1506kW at spot radius of 0.25mm for a depth of 1.00mm. At the laser spot radius of 2.00mm, the laser beam is almost fully penetrated with only peak power of 0.0087kW at the centre of the penetration.

Figure 6 illustrates the time dependent laser beam penetration for stainless steel material. According to the illustration, the peak power reduces exponentially for both depth and time. From observation, laser beam peak power decreases faster through the depth rather than through the time, relatively. This is explained by Equation (16) which states penetrated laser beam is based on absorption coefficient through the depth is -0.03 m but only -0.0009 m through time. At a depth of 2.00 mm and time of 4.0 ms, the peak power is 0.0002 kW. At this point, laser beam is almost fully penetrated. On the interface between stainless steel material and laser beam spot, laser beam penetration can be illustrated by Figure 7 at certain time. As discussed before, laser beam penetration decreases according to the first order of Bessel function with the spot radius and exponentially with time. At the central of the interface, the peak power is 3.5 kW and it reduces to 0.0026 kW after 8.0 ms. At laser spot radius of 0.25 mm, the laser beam penetration is 3.0248 kW and it reduces to 0.0023 kW after 8.0 ms where the laser beam is about to fully absorbed by the stainless steel material.

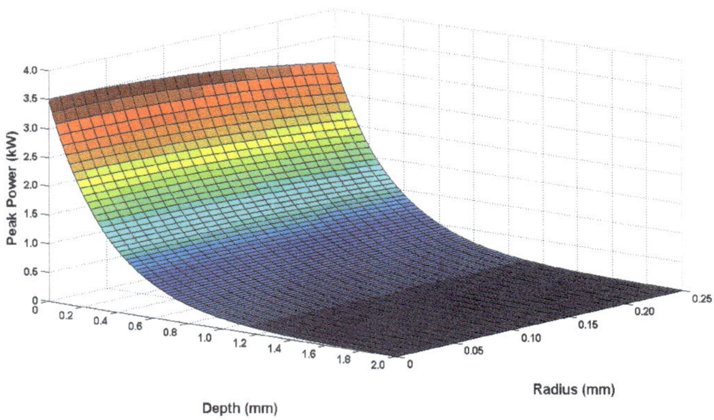

Figure 5. Peak power versus depth and radius for stainless steel with absorption coefficient, μ=0.3 and t=1 ms.

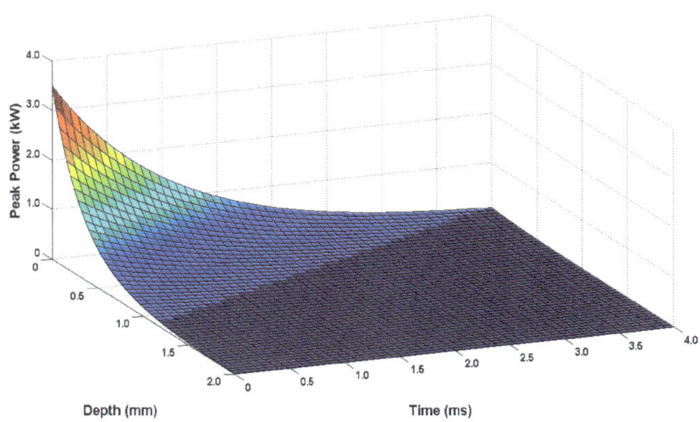

Figure 6. Peak power versus depth and time for stainless steel with absorption coefficient, μ=0.3 at the centre of penetration.

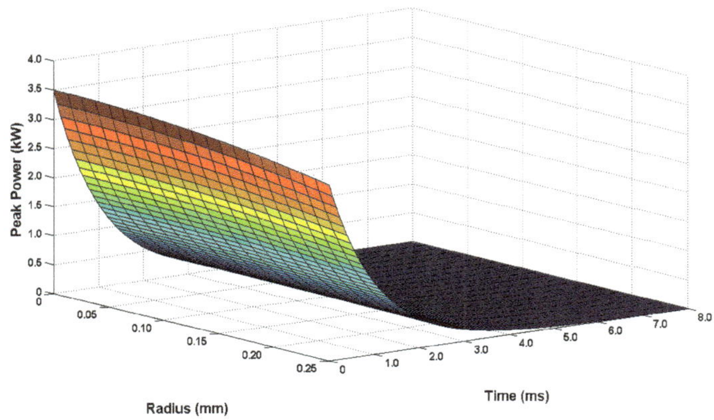

Figure 7. Peak power versus radius and time for stainless steel with absorption coefficient, μ=0.3 at the welding front.

4. Analysis of the pulsed Nd:YAG laser spot weld

The profiles for the penetration depth and bead width produced by the pulsed Nd:YAG laser beam are depicted in Figure 8 and Figure 9. The weld dimensions (weld penetration depth and bead width) for different laser beam peak powers are illustrated in Figure 10. The results display an increase in penetration depth and bead width with an increase in the laser beam peak power. The deepest penetration depth produced is 1.31 mm and the largest bead width is 0.57 mm when laser beam peak power is set at 3.5 kW. When laser beam peak power is reduced to 0.5 kW, the reading of penetration depth and the bead width are only 0.36 mm and 0.24 mm. The linear gradients of penetration depth and bead width are 0.301 and 0.10, respectively. These values show the laser beam peak power is almost 3 times more effective on the penetration depth rather than the bead width. This suggests that the laser beam peak power is a reliable parameter to control the desired penetration depth. It is observed in Figure 10 that the penetration depth and bead width increases when the pulse duration is increased. From the Figure 11, linear gradient of penetration depth and bead width are 0.0039 and 0.035, respectively. Only slight difference is noted and this means that the pulse duration has no significant effect either on penetration depth or bead width. As compared with the effect of laser peak power, the pulse duration is a much more better parameter to control the desired bead width.

Laser beam can be defocused by moving the focus point position forward or backward from the specimen surface. Figure 12 depicts that the penetration depth decreases significanly when the laser beam focus point position is moved from 0 to 4.0 mm with respect to the specimen surface. After 4.0 mm, the penetration depth decreases gradually untill 6.0 mm and there is no penetration depth that can be traced after 6.0 mm. Otherwise, when the focus point position is moved forward or backward, the bead width increases. But after 6.00 mm, there is no more bead width and only the sign of burning can be observed on the welding surface. As the focus point position is moved forward or backward from the specimen surface, the laser spot size increases. This will reduce the intensity of the laser beam which is given by, $I_0=P/\pi r^2$. Here, P is the laser beam peak power and r is the laser spot radius. It indicates that the intensity of laser beam is negatively proportional with r^2. Hence, the laser beam does not have sufficient intensity to penetrate the material. As illustrated in Figure 13, it is observed that the penetration depth and bead width changes with the laser beam incident angle. As the laser beam incident angle increases, the laser spot becomes elliptical and wider. Hence the bead width also becomes elliptical and wider. The wider laser spot will reduce the intensity of the laser beam. This will result in a decrease of the penetration depth. However, when the laser beam incident angle is increased, the reflectivity of material surface drops due to the influence of light polarization. When the reflectivity drops, it will transport more laser energy into the welding material. Hence, it will increase the penetration depth. It can be seen that when weld penetration depth at angle of incidence of 45 degrees is much higher than that at 25 degrees. But, even though the reflectivity of the material surface drops greatly at 65 degrees, the penetration depth is only 0.82 mm because laser beam is less intense due to the significant increases in the laser spot area for a slight increase of the incident angle. In Figure 14, it can be seen that the penetration depth increases slightly with the number of shots. The penetration depth for the first and seventh pulse shots are 0.91 mm and 1.11 mm. When the first shot is applied it produces a shallow concave hole on the specimen surface. This is due to the material ablation produced by laser pulse pressure when it strikes the material surface. Then, when the second shot is given, it is able to go towards the bottom of the concave hole. The second shot weld penetration depth is similar to the first shot but with an increase in the depth of the concave hole. The first and seventh shot produce 0.52 mm and 0.57 mm of bead width. The difference is relatively small because the same laser beam parameters are used for each shot.

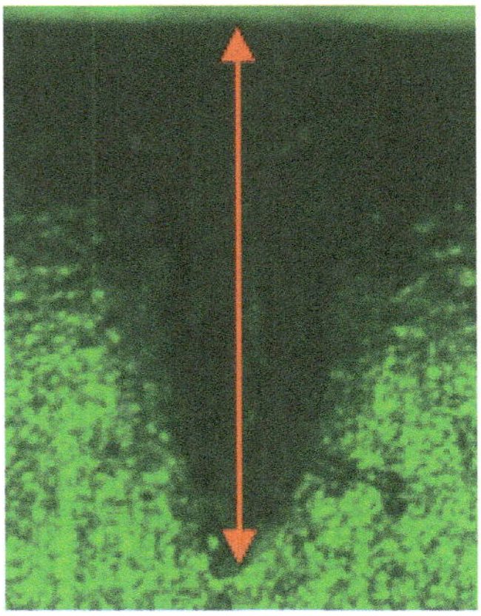

Figure 8. Cross-section of a spot weld produced by 2.5kW laser beam peak power and 2.5ms pulse duration. The penetration depth is labelled by the vertical line.

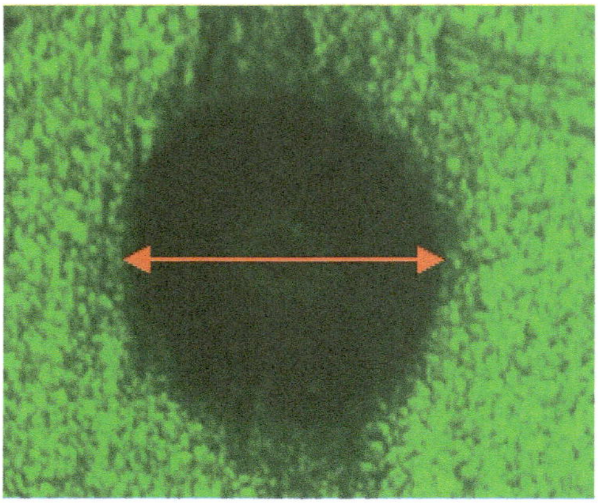

Figure 9. Figure 9. Top view of a spot weld produced by 2.5kW laser peak power and 2.5ms pulse duration. The bead width is labelled by the horizontal line.

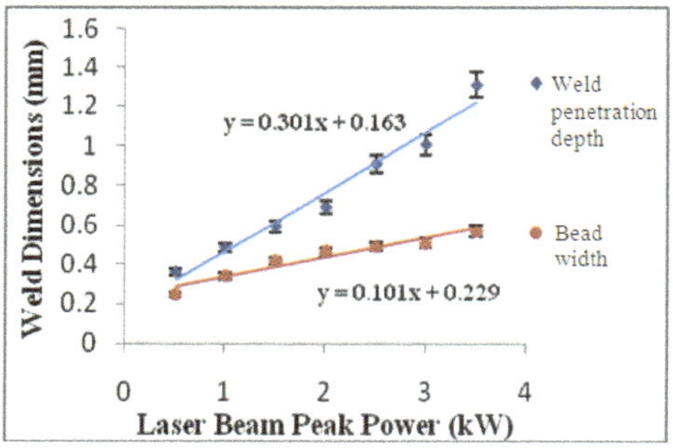

Figure 10. Characteristics of a spot weld dimensions as a function of laser peak powers conducted by 2.5 ms pulse duration with laser beam incidence is vertical in respect to the surface normal.

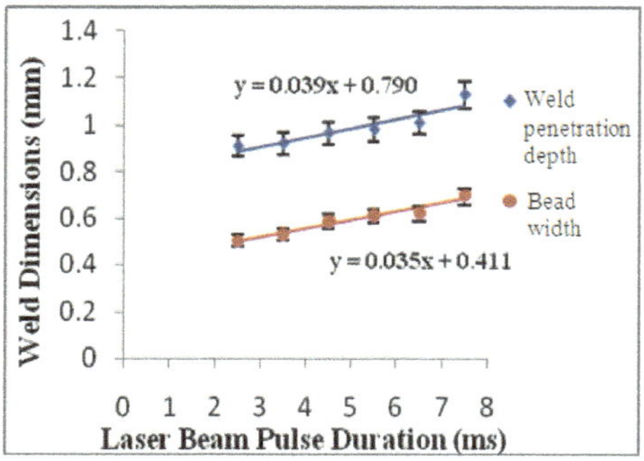

Figure 11. Characteristics of a spot weld dimensions as a function of laser pulse durations conducted by 2.5 kW laser peak power with wavelength of 1064 nm.

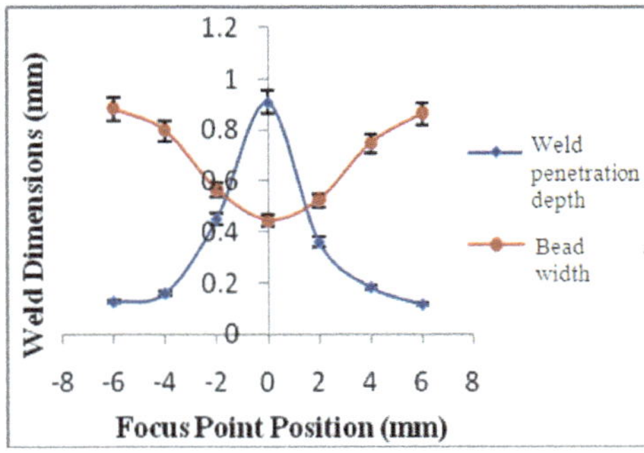

Figure 12. Characterictics of a spot weld dimensions when Nd:YAG laser spot is positioned forward and backward from the stainless steel 304 surface with 2.5 kW laser peak power and 2.5 ms pulse duration.

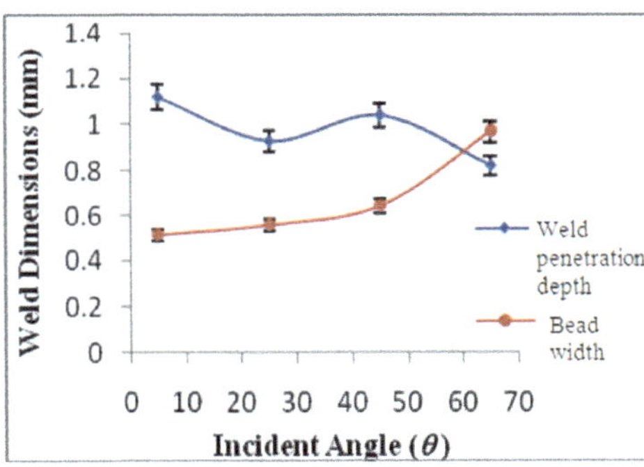

Figure 13. Characterictics of a spot weld dimensions when Nd:YAG laser spot incident angle is varied with respect to the stainless steel 304 surface normal by employing 2.5 kW laser peak power and 2.5 ms pulse duration.

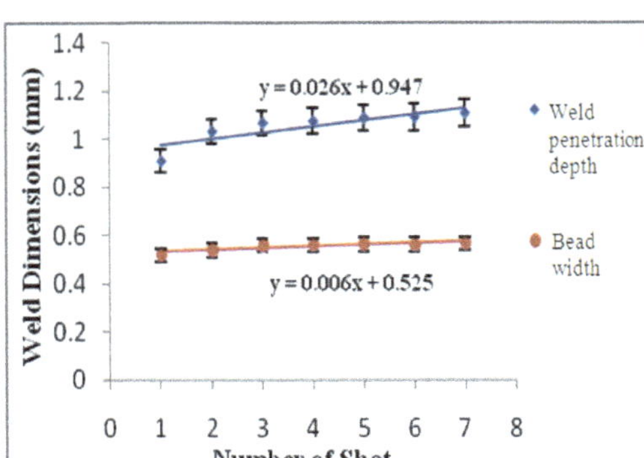

Figure 14. Spot weld produced by 2.5 kW laser peak power, 2.5 ms pulse duration with laser beam incidence is vertical in respect to the surface normal and laser focus point is on the stainless steel 304 surface, while the laser beam shot numbers are varied.

5. The strength analysis of a welding joint

In this section, the strength test of a welding joint of a single spot welding is discussed for stainless steel 304. A test is also executed for a seam welding which is produced by shooting the laser pulses continuously along the 10.0 ± 0.1 mm stainless steel 304 interfaces. The strength for seam welding of stainless steel 304 is also compared to an Invar™, which is a commercial welding material for photonics device packaging. A Unitek Miyachi LW10E ultra compact pulsed Nd:YAG laser with wavelength of 1.064 μm is employed to produce a weld joint. The energy per pulse output is in the range of 1 to 20 J, laser beam peak power is up to 3.5 kW and pulse duration ranging from 0.3 ms to 10.0 ms. However, in this investigation, only 3.5 kW laser beam peak power and 6.5 ms pulse duration is employed. This parameter is chosen because it produces a deepest penetration depth on stainless steel 304 which is 1.2 mm (Nawi et al., 2011). The penetration depth is important and should be large enough to achieve a strong attachment. The welding base material used is a stainless steel 304 sheet with a thickness of 1.0±0.1mm. The experimental setup to produce a weld consists of a laser source, fiber optics delivery system, and focusing lenses with focal length of 100.00 mm as illustrated in Figure 14.

The optical fiber cables transmit the laser beam to the welding focusing lenses inside the lens housing. The laser welding system is also equipped with an aiming diode laser beam, which simplifies positioning of the laser spot on the stainless steel sheet. A spot weld and a seam weld are produced with two types of joint; butt joint and lap joint, as illustrated in Figures 15(a) and 15(b). The strength of the weld joints is conducted by pulling the joint materials using INSTRON Series IX/s Automated Materials Tester System. The crosshead speed utilized for pulling the joint is set to 0.2 mm/min. The strength for a weld joint is determined by measuring the applied load during the test.

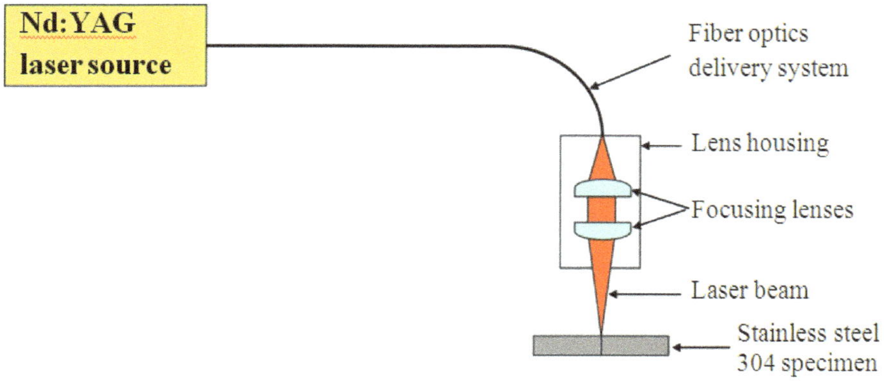

Figure 15. Nd:YAG laser experimental setup to produce a spot weld on stainless steel 304 specimen.

The element composition of stainless steel 304 and welded stainless steel 304 are examined by using Energy Dispersive X-ray (EDX) technique. The stainless steel 304 consists of iron (Fe) 69.59wt%, chromium (Cr) 18.33wt% and nickel (Ni) 12.08wt%. Figure 16 shows the element composition for stainless steel 304 changes when it is welded by Nd:YAG laser beam and it is now comprising of iron (Fe) 68.83wt%, chromium (Cr) 17.10wt% and nickel (Ni) 14.08wt%. It is observed that there are no changes in the elemental composition before and after welding of stainless steel 304. The applied load for the test and the pulling displacement for spot and seam welding are shown in Figure 17 and Figure 18, respectively. Maximum applied load is considered as the strength of the weld joint. Figure 16 illustrates maximum load for butt and lap joint which are 0.2296 kN and 0.0691 kN with a pulling displacement of 0.43 mm and 0.40 mm, respectively.

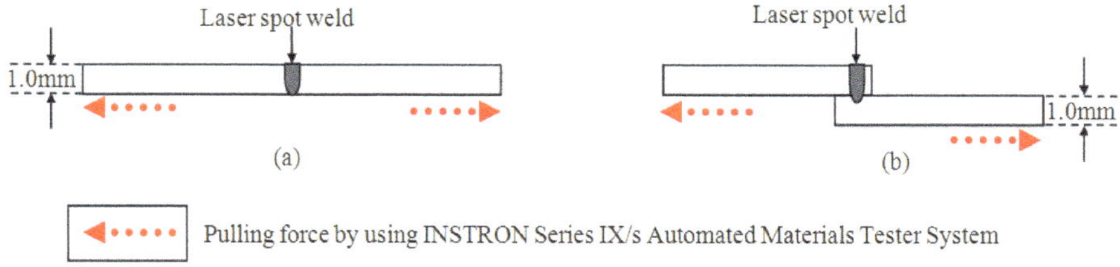

Figure 16. Configuration of the (a) butt joint and (b) lap joint, for stainless steel 304 with a thickness 1.0±0.1mm.

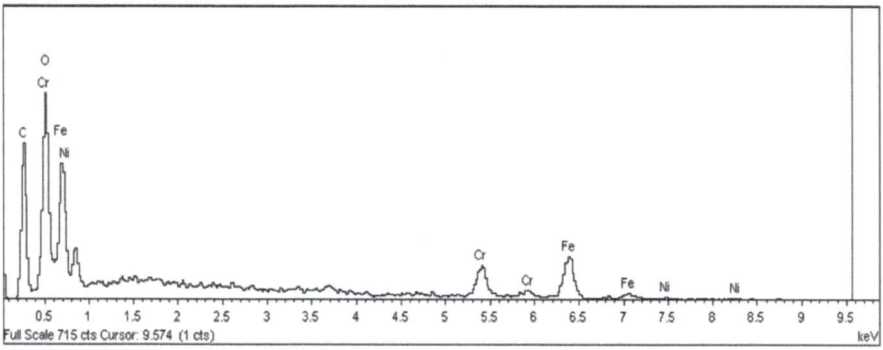

Figure 17. EDX analysis for the elemental composition of welded stainless steel 304.

The Invar^{TM} maximum load is recorded as 0.3760 kN with 0.43 mm pulling displacement. The outcome indicates that the butt joint produces much stronger attachment than the lap joint. The stronger attachment is contributed by a larger volume of welded stainless steel at the joint interfaces as shown in Figure 19(a). For lap joint, the thickness of the stainless steel 304 becomes double as the upper and lower stainless steel sheets are overlapped. If the laser beam produces insufficient penetration depth, thus only a small volume of welded stainless steel 304 is produced at the interfaces of the joint as shown in Figure 19(b). This will lead to a weaker joint. But, the penetration depth can be increased by controlling the laser beam parameters such as laser beam peak power and pulse duration. Hence, it will increase the strength of a lap joint. In terms of welding materials comparison, the commercial welding material, Invar™, provides more robust joint than stainless steel 304. Figure 18 indicates the seam welding maximum load for butt joint is 2.6339 kN with 0.62 mm pulling displacement. Lap joint recorded the maximum load of 1.0466 kN at 0.49 mm. This implies that the butt joint of a seam welding provides a much stronger attachment than the lap joint similar with the case of a single spot weld. Figure 18 also indicates between the butt joint pulling displacement of 0.48 mm and 0.72 mm, there are no significant changes in the joint strength because each point along the seam welding has the similar maximum strengths but it is not achieved simultaneously depending on its geometry.

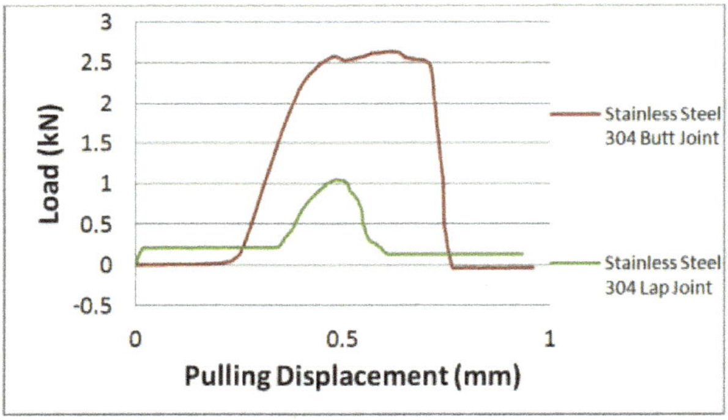

Figure 18. Strength profile of a single spot weld produced on stainless steel 304 and Invar™ by using 3.5 kW Nd:YAG laser beam peak power and 6.5 ms pulse duration.

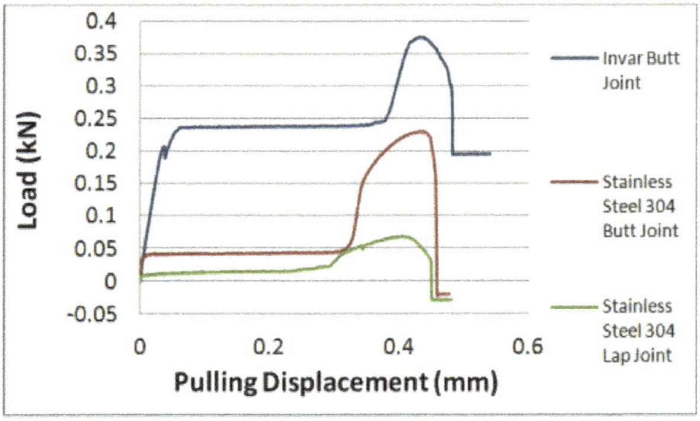

Figure 19. Strength profile of a seam weld produced on a stainless steel 304 along 10.0 ± 0.1 mm joint interfaces.

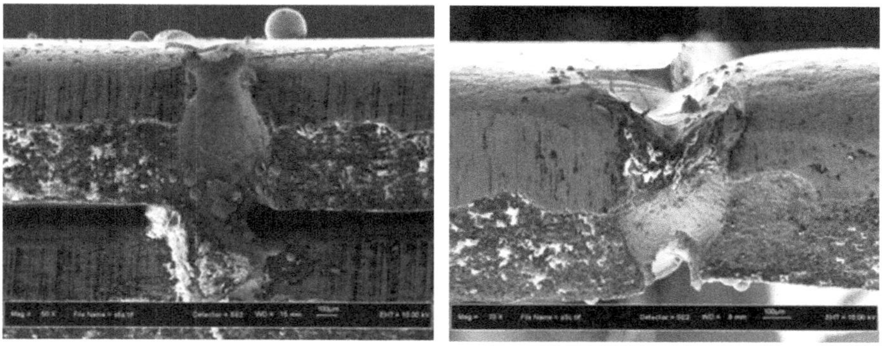

a) b)

Figure 20. FESEM micrographs of a stainless steel 304 spot weld joint; (a) butt joint and (b) lap joint, produced by 3.5 kW laser beam peak power and 6.5 ms laser pulse duration.

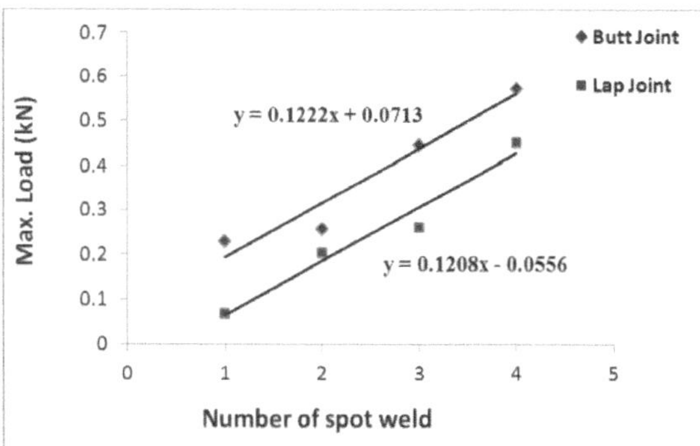

Figure 21. The strength comparison between butt joint and lap joint for variation number of spot weld.

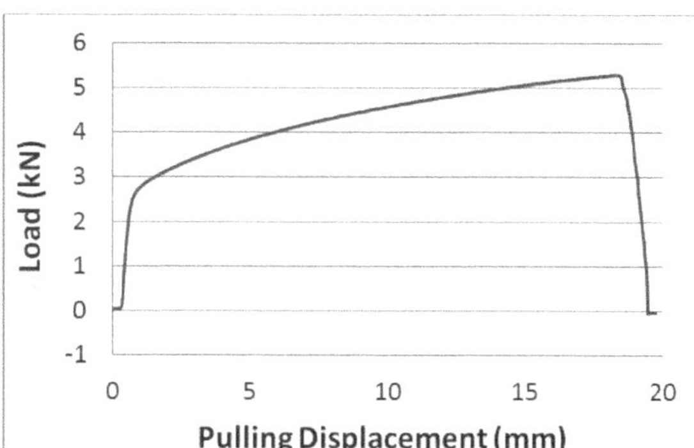

Figure 22. The strength of a stainless steel 304 with 1.0 ± 0.1 mm thickness and 10.0 ± 0.1 mm width.

In miniature assemblies, only a small number of laser spot weld is used for an attachment because of the relative small parts (Fadhali et al., 2007a, 2007b, 2007c, 2007d, 2008). The strength on the number of laser spot welding up to four spots is investigated for butt joint and lap joint. Then, the maximum strength for both joint types is compared. Figure 20 illustrates that the strength increases when the number of spots increases for butt and lap joint. A linear relationship is obtained from both joints strength against the number of spots and it has equal gradient which are 0.122 and 0.120, respectively. This suggests there is no significant difference in the influence of number of spots on the strength for both types of joint. The strength of stainless steel 304 is shown in Figure 21. It produces a strength of 5.2866 kN with 18.23 mm pulling displacement. The strength of stainless steel 304 is almost twice stronger than the seam welded stainless steel 304. Thus the welded material cannot attain the strength of the base material itself. The pulling displacement indicates that the elastic modulus of stainless steel 304 is much more greater than the welded stainless steel 304.

6. Conclusion

In conclusion, it is shown that the penetration depth and bead width increase when the laser beam peak power, pulse duration and number of shot are increased. It is found that the laser peak power is more effective to produce the deeper penetration depth rather than the bead width. Otherwise, the laser beam pulse duration is an accurate parameter to control if the desired bead width is required rather than the penetration depth. As the focus point positions are placed forward and backward from the stainless steel 304 specimen surface, the laser spot size increases. This leads to the reduction of the laser beam intensity. Hence, the penetration depth decreases. The increase of the laser spot size will increase the bead width. When the laser beam incident angle is varied, the bead width also changes due to the widening of the laser spot size. The penetration depth depends on the both widening laser spot size and material surface reflectivity that drops due to the influence of light polarization. The penetration depth increases slightly with the number of shots due to the material ablation produced by laser pulse pressure when it strikes the specimen surface. For welding joint strength, a butt joint produces much stronger attachment than a lap joint. The stronger attachment is contributed by a larger volume of welded stainless steel 304 at the joint interfaces. It is observed that each point along the seam welding has a similar maximum strengths but it is not achieved simultaneously. The similar gradient obtained in a linear relationship suggests that there is no significant difference in the influence of number of spots on the butt and lap joint strength. The commercial welding material, Invar™, provides more robust joint rather than stainless steel 304. The welded stainless steel 304 cannot attain the strength and elastic modulus of the stainless steel 304 itself. From the strength test results, it is found that stainless steel 304 is a good candidate to be used as a welding material for photonics device packaging by using pulsed Nd:YAG laser welding technique.

Author details

Ikhwan Naim Md Nawi
Centre of Foundation Studies & Faculty of Applied Sciences, Universiti Teknologi MARA, Malaysia

Jalil Ali
Universiti Teknologi, Malaysia

Mohamed Fadhali
Physics Department, Faculty of Science, Ibb University, Yemen

Preecha P. Yupapin
King Mongkut's Institute of Technology Ladkrabang, Bangkok, Thailand

7. References

Beretta, J.R.; Rossi, W.; Neves, M.D.M.; Almeida, L.A. & Vieira, N.D. (2007). Pulsed Nd:YAG laser welding of AISI 304 and 420 Stainless Steel. *Journal of Optical Lasers Engineering*, Vol. 45, pp. 960-966.

C. Dawes, C. (1992). *Laser welding: A practical guide*. Woodhead Publication Ltd., England.

Fadhali, M.; Zainal, J.; Munajat, Y.; Ali, J. & Rahman, R. (2007a). Reliable pigtailing of photonic devices employing laser microwelding. *Journal Engineering and Applied Science*, Vol. 2, pp. 1724-1728.

Fadhali, M.; Zainal, J.; Munajat, Y.; Ali, J. & Rahman, R. (2007b). Investigation of the application of Nd:YAG laser welding to couple photonic devices and packaging. *Lasers Engineering*, Vol. 17, pp. 273-286.

Fadhali, M.; Zainal, J.; Munajat, Y.; Ali, J. & Rahman, R. (2007c). Laser diode pigtailing and packaging using Nd:YAG laser welding technique. *Jurnal Komunikasi Fisika Indonesia*, Vol. 5, pp. 203-208.

Fadhali, M.; Zainal, J.; Munajat, Y.; Ali, J. & Rahman, R. (2007d). Analysis of laser microwelding applied for photonic devices packaging. Journal Solid State Science and Technology Letter, Vol. 14, pp. 18-19.

Fadhali, M. (2008). Efficient coupling and reliable packaging of photonic devices using laser welding technique. *Ph.D Thesis: Universiti Teknologi Malaysia*, Skudai, Johor, Malaysia.

Kazemi, K & Goldak, J.A. (2009). Numerical simulation of laser full penetration welding. *Computer Material Science*, Vol. 44, pp. 841–849.

Marley, C. (2002). Laser welding photonics devices: Proper design guidelines enable users to achieve high-precision alignment requirements. *Industrial Laser Solutions for Manufacturing*, 2002.

Mousavi, S.A.A. & Sufizadeh, A.R. (2008). Metallurgical investigations of pulsed Nd:YAG laser welding of AISI 321 and AISI 630 stainless steels. *Journal of Material Design*, Vol. 15, pp. 44-49.

Naim, I.; Saktioto, T.; Fadhali, M.; Ali, J.; & Yupapin, P.P. (2009). An investigation of a pulsed Nd:YAG laser welding technique. *Proceedings of IEEE-Regional Symposium on Micro and Nanoelectronics*, pp. 274-280, Kelantan, Malaysia, August 10-12, 2009.

Naim, I.; Saktioto, T.; Hamdi, M.; Fadhali, M.; Ali, J.; & Yupapin, P.P. (2010). Penetration depth estimation for stainless steel 304L using Nd:YAG laser spot micro welding. Proceedings of 2nd Topical Meeting on Lasers and Optoelectronics, Vol. 1, pp. 278-288, Terengganu, Malaysia, March 13-15, 2010.

Nawi, I.N.; Saktioto; Fadhali, M.; Hussain, M.S.; Ali, J. & Yupapin, P.P. (2011). A Reliable Nd:YAG Laser Welded Stainless Steel 304 for Photonics Device Packaging. *Procedia Engineering*, Vol. 8, pp. 380-385.

Pang, M.; Yu.G.; Wang, H.H. & Zheng, C.Y. (2008). Microstructure study of laser welding cast nickel-based superalloy K418. *Journal of Material Processing Technology*, Vol. 207, pp. 271–275.

Ready, J.F. (1997). *Industrial applications of lasers*. New York: Academic Press.

Zhou, J. & Tsai, H.L. (2008). Modeling transport phenomena in hybrid laser-MIG keyhole welding. *Internattonal Journal of Heat and Mass Transfer*, Vol. 51, pp. 4353-4366.

Section 2

Numerical Modeling of Welding Processes

Numerical Modelling to Understand Cracking Phenomena During Laser-GMA Hybrid Welding Nickel-Base Superalloys

Zhiguo Gao

Additional information is available at the end of the chapter

1. Introduction

Laser welding is a highly efficient and precise welding method, which is greatly desired in the automotive, aero and ship building industries. It elucidates some excellence, such as higher depth to width ratio, concentrated heat input, minimal thermal distortion, minimized weld fusion and heat affected zone, reduced post weld rework and possibly joining widely dissimilar materials. But laser welding has also clearly demonstrated some drawbacks. First of all, the cost of laser equipment and maintenance are high. Second, highly reflective materials, such as aluminum, copper and gold, are more difficult to process with some laser sources. Finally, the high welding speed leads to high solidification rate, which in turn induces metallurgical problems, such as cracking, pores and brittle phase structures.

Laser-gas metal arc (GMA) hybrid welding is one of the state-of-the-art technologies and is designed to overcome problems commonly encountered during either laser or arc welding. Hybrid welding offers some additional advantages over mere laser and arc welding. Firstly, the combination of a laser beam and gas metal arc in welding will reduce the propensity for cracking, brittle phase formation and porosity. Secondly, the feeding wire modifies the metallurgy of the weld. Thirdly, the effect of the coupled process exceeds the effects of the laser and arc heat sources taken separately, and the laser beam located in front of the GMA weld pool could suppress hump formation during high travel speed (Claus et al., 2005; Gao et al., 2009; Ribic et al., 2009).

Nickel-base superalloys are extensively used as aeronautical material due to their high temperature properties. However, commonly used types of commercial welding for nickel-base superalloys have three major hurdles that need to overcome in order to make refurbishment and repair feasible. First, the single crystal nature of nickel-base superalloys is easily lost during welding due to stray grain formation. This phenomenon is called a columnar-to-equiaxed transition (CET). Secondly, polycrystalline nickel-base superalloys are very prone to cracking during welding, and these cracks are mostly thermally induced cracks which originate in the grain boundary in the heat affected zone (HAZ). The combination of welding stress and increased concentration of gamma-prime are thought to be the cause of hot cracks in the HAZ. Finally, on-equilibrium solidification, elemental partitioning and subsequent solid state transformation can yield non-ideal microstructures and will not produce material with the desired properties.

It has been found that the mechanical strengths at elevated temperatures of columnar grains are better than those of gas turbine blades with equiaxed grains. Columnar grains are desirable, and equiaxed grains are treated as casting defects in directionally solidified or single crystal castings. Some researchers have implemented theoretical analyses on the solidification behaviours of nickel-base superalloys. Hunt (1984) and Gäumann et al. (2001) theoretically analyzed stray grain formation to consider the parameter Φ for describing the extent of nucleation and growth for stray grain formation ahead of the advancing solidification front. Φ represents the volume fraction of equiaxed grains and varies from 0 to 1. The value of Φ can be calculated by the following equation:

$$\frac{G^n}{V} = a\left\{\sqrt[3]{\frac{-4\pi N_0}{3\ln(1-\Phi)}}\frac{1}{n+1}\right\}^n \tag{1}$$

where G is the thermal gradient, V is the growth velocity, a and n are material constants and N_0 is the nucleation density. $a=1.25\times10^6 K^{3.4}m^{-1}s^{-1}$, $n=3.4$ and $N_0=2\times10^{15}m^{-3}$ are used for a similar nickel-base superalloy. When constitutional supercooling is minimal, $\Phi=0$, the weld microstructure is free of stray grains and maintains its single crystal nature. When $\Phi=1$, solidification is equiaxed. A critical value of Φ that corresponds to a stray grain free microstructure is hypothesized to be 0.0066.

Gäumann et al. (2001) derived a criterion based on the G^n/V ratio which states that the microstructure is columnar when the following condition is satisfied everywhere in the weld melt pool.

$$\frac{G^{3.4}}{V} > 2.7\times10^{24}\left(\mathrm{K}^{3.4}/\mathrm{m}^{4.4}\mathrm{s}\right) \tag{2}$$

Park et al. (2003) used the degree of constitutional supercooling (CS) at the solidification front as a metric to assess the effects. The basic criterion to avoid CS, which leads to the growth of new grains at the solidification front, is given by:

$$\frac{G}{V} > \left(\frac{G}{V}\right)_{cr} \tag{3}$$

where $(G/V)_{cr}$ is a critical value of G/V, which is proportional to $\Delta T/D_L$ (ΔT is the solidification temperature range, and D_L is the diffusion coefficient of solute in the liquid). In cases where the planar solid/liquid interface breaks down into cellular or dendritic solidification, the theory of CS predicts fairly closely. However, no theory can predict the $(G/V)_{cr}$ for the CET.

Liu and DuPont (2004, 2005) studied the effects of melt-pool geometrical parameters on crystal growth and microstructure development during laser surface melting of single-crystal alloys.

Thanks to the development of welding techniques, much progress has been made to discuss cracking phenomena in the laser join nickel-base superalloys (Egbewande et al., 2010; Hong et al., 2008; Liu et al., 2011). Recently, feasible finite element models were established to provide substantial insights into cracking and how thermal gradient, transient stress and strain were developed in the laser welding process (Dai&Shaw,2001; Hee et al.,2010; Hu&Richardson,2006;Long et al.,2008;Luo et al.,2002; Nishimoto et al., 2002; Wang et al., 2004; Yann et al.,2010 ;Yilbas et al.,2010).

As mentioned previously, albeit laser welding is achieved by rapid heating and rapid cooling, solidification cracking and liquation cracking are prone to be generated in the weld and HAZ. The effect of weld pool shape on the solidification cracking and HAZ cracking susceptibility still remain obscure in many respects for the laser welding. Some experimental conclusions from previous literature have proved that the number of micro cracks at the fusion line in the weld cross-section is reduced with increasing heat input for laser-arc hybrid welding, and cracking might be prevented by adjustment of filler material and heat input for the laser-arc hybrid welding. However, limited information and rationale are available in the literature about how the weld solidification cracking and HAZ crack susceptibility are minimized during laser-arc hybrid welding.

In this chapter, in order to properly explore the capability of the new welding process, i.e. laser-arc hybrid welding, it is necessary to know and understand how laser-arc hybrid welding affects solidification cracking and liquation cracking. A major factor control susceptibility to solidification cracking and liquation cracking are stray grain formation and stress-strain generation during weld pool solidification, respectively. It is important to study and understand stray grain formation and stress-strain generation mechanism during laser-arc hybrid welding, whereas these information are currently scarce in the literature.

2. Solidification cracking and liquation cracking modeling

In this section two kinds of theoretical analysis models are developed to understand cracking phenomena, one is modelling analysis of hybrid laser-arc welding of single-crystal nickel-base superalloys and the other is modelling analysis HAZ liquation cracking of laser hybrid welding polycrystalline nickel-base alloy and schematic diagram is shown in Fig.1.

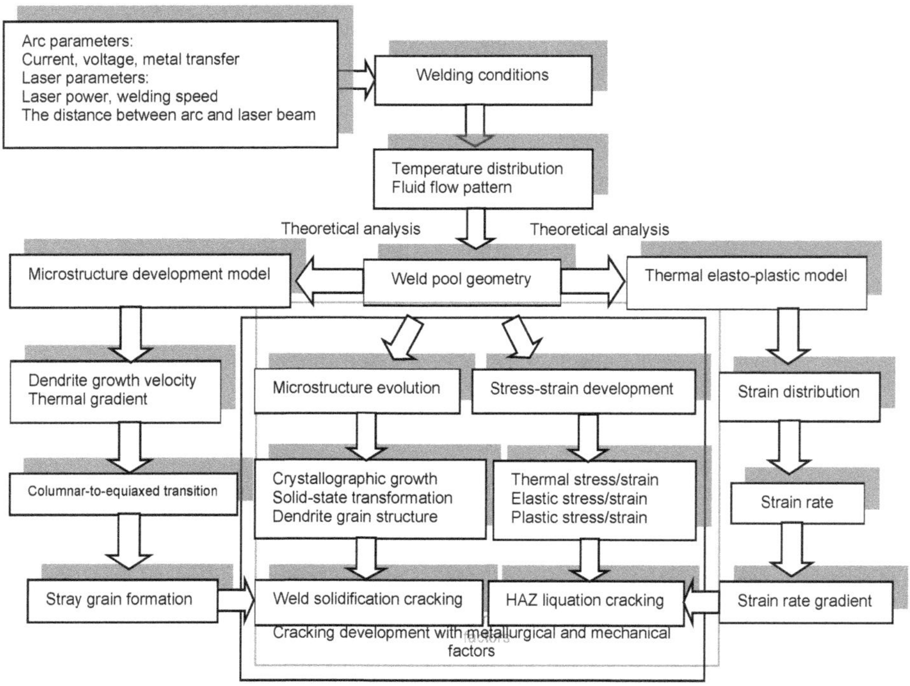

Figure 1. Schematic diagram of cracking development and theoretical analysis for the laser-GMA hybrid welding nickel-base superalloys

2.1. Solidification cracking modeling for single-crystal superalloys welding

2.1.1. Mathematical model description

To execute the analysis on the dynamic behavior and solidification of molten pool, three-dimensional essential governing equations, such as continuity, Navier-Stokes, energy and VOF method equations are iteratively solved with the assumption that the liquid flow is Newtonian, laminar and incompressible.

Continuity equation

$$\nabla \cdot V\rho_l = \dot{m}_s \tag{4}$$

Momentum equation

$$\rho_l\left(\frac{\partial V}{\partial t}+V\cdot\nabla V\right)=\mu\nabla^2 V-\nabla P+\dot{m}_s\cdot V+F-KV \tag{5}$$

Energy equation

$$\rho_l\left(\frac{\partial(U)}{\partial t}+V\cdot\nabla(U)\right)=\nabla\cdot(k_l\nabla T)+\dot{U}_s \tag{6}$$

$$U=\int_T c(T)dT+(1-f_s)h_{st} \tag{7}$$

where V is the molten metal velocity, $\dot{m}_s$ is a mass source term, P is the hydrodynamic pressure, μ is the dynamic viscosity, F is the body force, ϱ_l is the fluid density, U is internal energy per unit mass, k_l is thermal conductivity, T is a local temperature, $\dot{U}_s$ is an energy source term due to mass source term, K is the drag coefficient for the porous media model, f_s is the solid fraction, and c (T) is specific heat.

According to the kinematic motion equation, the VOF function moves according to the velocity field in the fluid, as shown by Equation (8).

$$\frac{\partial F}{\partial t}+\nabla\cdot(VF)=\dot{F}_s \tag{8}$$

For a single fluid, incompressible problem F represents the volume fraction occupied by the fluid. Thus, fluid exists where F=1, and void regions correspond to locations where F=0. $\dot{F}_s$ represents the change of volume fraction of fluid associated with the mass source .

The local temperature gradient G_{hkl} and dendrite growth velocity V_{hkl} can be calculated to predict the microstructure development across the entire solidification interface. The solidification front produced at the rear part of the weld pool is identified by a solidus curve. The effect of keyhole penetration on weld pool geometry is considered in the analysis of the microstructure development. Fig. 2 illustrates a schematic diagram of the weld pool and associated geometry variables, which illustrate the relationships among the welding velocity $\vec{V}_b$, the solidification front unit vector $\vec{n}$, and the dendrite growth velocity $\vec{V}_{hkl}$.Ψ is the angle between the solidification front normal $\vec{n}$ and a particular dendrite growth direction [hkl]. θ is the angle between the surface normal $\vec{n}$ and the welding direction, and φ is the angle between the y-axis and the projection of $\vec{n}$ on the y-z plane. L_{al} is the distance between the laser beam and GMA, c_l and c_a are the penetration of laser and arc welds, respectively.

Based on previous analytical methods(Liu&DuPont,2004,2005; Rappaz et al.,1990), a mathematical model is extended for the calculation of stray grain fraction during solidification microstructure development in the hybrid weld pool. The 3-D melt-pool geometry on the rear part of the weld can be described in an ellipsoid formula.

$$\frac{x^2}{a^2}+\frac{y^2}{b^2}+\frac{z^2}{c^2}=1 \tag{9}$$

where a, b and c are the half-axes of the ellipsoid in the x, y and z directions, respectively. As well, a is the length between the location of the maximum width and trailing point, b is the maximum width of the arc pool, c is the maximum depth and in consideration of the solidification history of weld pool penetration, c_l is used in this model.

The relationship of the velocity of the dendrite tip, $\vec{V}_{hkl}$, along a specific [hkl] direction is given by:

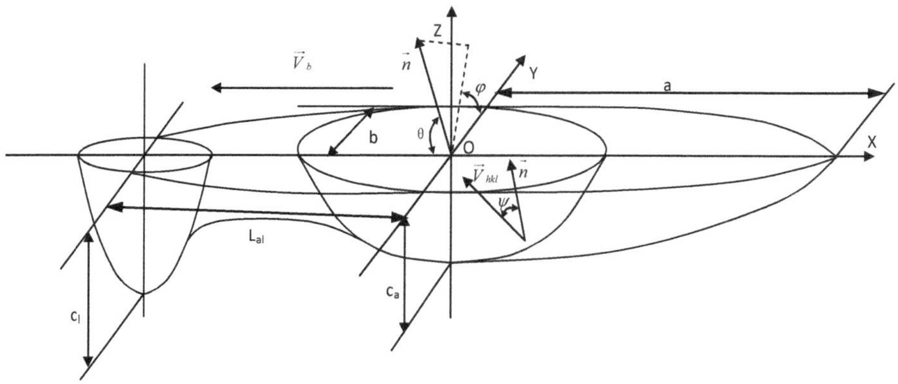

Figure 2. Schematic diagram of weld pool geometry mathematical model and associated variables (Gao&Ojo,2012a)

$$\left|\vec{V}_{hkl}\right|=\frac{\left|\vec{V}_n\right|}{\cos\psi}=\left|\vec{V}_b\right|\frac{\cos\theta}{\cos\psi} \tag{10}$$

where $\vec{V}_n$ is the growth rate of the normal to the solidification front at the liquid-solid interface.

The unit component based on the weld pool normal in the x, y and z axes are determined by

$$\vec{n}=\left(\cos\theta,\sin\theta\cos\varphi,\sin\theta\sin\varphi\right) \tag{11}$$

In order to determine the growth velocity along the [hkl] dendrite growth direction (one of the six <100> directions), the value of cosψ can be determined from the relation:

$$\cos\psi=h\cdot\cos\theta-k\cdot(\sin\theta)(\cos\varphi)+l\cdot(\sin\theta)(\sin\varphi) \tag{12}$$

The thermal gradient along the dendrite growth direction, $\vec{G}_{hkl}$, is given by the expression:

$$\left|\vec{G}_{hkl}\right|=\left|\vec{G}_n\right|\cos\psi \tag{13}$$

where $\left|\vec{G}_n\right|$ is the total thermal gradient direction in the x, y and z directions.

The expression is solved for the volume fraction of stray grains Φ as a function of G_{hkl} and V_{hkl} which yields:

$$\Phi = 1 - e^{S} \tag{14}$$

$$S = \frac{-4\pi N_0}{3}\left(\frac{1}{(n+1)\left(G^n / aV\right)^{1/n}}\right)^3 = -2.356\times10^{19}\left(\frac{V_{hkl}}{G_{hkl}^{\ 3.4}}\right)^{3/3.4} \tag{15}$$

The area-weighted average across the entire solidification interface, $\overline{\overline{\Phi}}$, is given by the expression:

$$\overline{\overline{\Phi}} = \frac{\sum_k A_k \overline{\Phi}}{\sum_k A_k} \tag{16}$$

where the variable k corresponds to the cross-section along the x axis, and A_k and $\overline{\Phi}$ are the areas and average Φ values for each cross-section, respectively. More information about this numerical model can be reviewed by the reference (Gao&Ojo,2012a).

Thermal properties of typical nickel-base alloy used in this model are directly taken from values produced in the previous literatures(Anderson et al.,2010; Banarjee&Overfelt,1999; Bonifaz&Richards,2009; Brooks et al.,1996; Luo et al.,2002; Pottlacher et al.,2002; Zeng,2006), and are presented in Table 1.

Material property	Value	Unites
Density of liquid	7578.9	Kg m^{-3}
Density of solid	7742.5	Kg m^{-3}
Viscosity of liquid	0.0074	Kg.s/m
Solidus temperature	1507	K
Liquidus temperature	1613	K
Specific heat of solid	430	J $kg^{-1}K^{-1}$
Specific heat of liquid	700	J $kg^{-1}K^{-1}$
Thermal conductivity of solid	11.4	$Wm^{-1}K^{-1}$
Thermal conductivity of liquid	28.7	$Wm^{-1}K^{-1}$
Coefficient of thermal expansion	1.33×10^{-5}	K^{-1}
Latent heat of fusion	1.45×10^{5}	J kg^{-1}
Latent heat of vaporization	6.4×10^{6}	J kg^{-1}
Surface tension	1.8	N m^{-1}
Surface tension gradient	-1.37×10^{-3}	N $m^{-1}k^{-1}$

Table 1. Nickel-base alloy properties used in this model

2.1.2. *Results and discussion*

The material adopted for this simulation is a nickel-base superalloy with a thickness of 3 mm. The focal plane of the laser beam is placed on the top surface of the base metal. Arc droplets are assumed to be steadily generated through a certain wire feed rate at an initial speed of 1.5m/s, and initial temperature 1613K. The size of the droplet mainly depends on the diameter of the feeding wire, and the generation frequency is determined by the wire feed rate and wire diameter. The distance between the arc source and laser beam is 3 mm, and the different welding conditions are listed in Table 2.

	Current (A)	Voltage (V)	Laser power (W)	Welding velocity (m s^{-1})	Droplet frequency (Hz)	Arc heat distribution parameter(mm)
Weld1	120	21	3667	0.022	217	2.23
Weld2	150	21	3667	0.022	269	2.33
Weld3	180	21	3667	0.022	304	2.45
Weld4	120	21	3667	0.028	217	2.23
Weld5	120	21	3667	0.034	217	2.23
Weld6	120	21	4000	0.022	217	2.23
Weld7	120	21	4332	0.022	217	2.23

Table 2. List of Hybrid welding parameters used for Nickel-base superalloy (Gao&Ojo,2012a)

2.1.2.1. *Role of welding parameters on microstructure development*

The properties of weld metal are affected by solidification behavior parameters, such as growth rates, temperature gradients, undercooling and alloy constitution, which are often useful to determine the development of microstructures in weld zones (Farzadi et al., 2010; He et al., 2005; Roy et al., 2006). In this study, the effect of undercooling has been simplified and thereby the solidification parameters, such as cooling rate, thermal gradient, and solidification velocity, are computed by the weld pool geometry considering only the heat transfer and fluid flow.

Since the shape of the weld pool remains constant under a steady-state condition, the solidification rate can be calculated with position along the fusion boundary. The steady state solidification rate, V_n, is related to the welding velocity in the following formula relation:

$$V_n = V_d \cos\theta \tag{17}$$

where θ is the angle between the normal to the solid/liquid boundary and the welding direction, and V_d is the welding velocity. The solidification rate increases from the edge of the weld pool and pool bottom, where θ is nearly 90°, to the weld centerline, where the velocity equates V_d along the fusion boundary (Farzadi et al., 2010).

To remove the constitutional undercooling ahead of the growth front for columnar grain growth, a high thermal gradient is required and low solidification velocities are desired

(Bussac & Gandin, 1997; Dong, 2007). The values of the temperature gradient and solidification rate vary with position along the entire solidification interface of the weld pool. The density of temperature isotherm in the maximum width is greater than that of the trailing portion of the pool boundary, so the temperature gradient is higher at the fusion line than the centerline. The ramifications of cooling rates (G×R) along the solid/liquid interface with different welding conditions are presented in Fig.3.

The cooling rate is higher at the maximum width of the fusion boundary and lower at the weld centerline for each case in Fig.3 (A). It is observed that the average range of the cooling rate is 4×10^3-2.4×10^4 K/s. The magnitude of the cooling rate decreases with an increase in the arc current and laser power, while an opposite effect will occur with a decrease in welding speed. In general, increasing the welding speed or decreasing welding heat sources power will increase the cooling rate, and the latter will result in an increase of residual stress and strain. In addition, the solidification velocities will be high and this may lead to more extensive non-equilibrium solidification and partitioning at high welding speeds (Rai et al., 2007; Vitek, 2005).

The cooling rate along the solid/liquid boundary of penetration in the conjoint zone is shown in Fig.3 (B). It is observed that the magnitude of the cooling rate is 10^5-7×10^5 K/s, which is approximate to the typical laser welding range of 10^4-10^6 K/s (Ral et al., 2009). It is higher at the bottom of the keyhole than that at the weld crown solid/liquid boundary. This distribution is consistent with higher convective heat transfer near the pool surface than the weld root. The Marangoni convection brings fluid from the weld center to the peripheral solid/liquid interface, which yields a lower thermal gradient on the weld pool surface. It can be further found that the value of the cooling rate near the bottom of the keyhole dramatically increases with increase in the arc current, welding speed and laser power, and the effect of laser power is the most obvious. Meanwhile, the magnitude of the cooling rate near the maximum depth of the weld pool is much higher than that of the rear part of the weld pool.

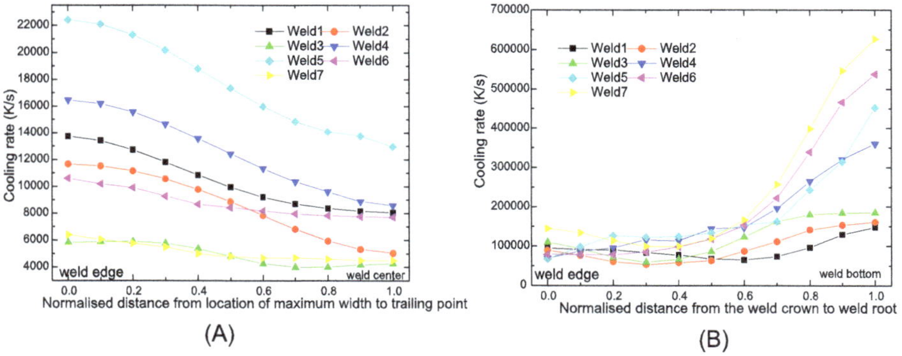

Figure 3. The calculated cooling rate along the solidification interface for different welding conditions, (A) Weld pool surface, (B) Weld penetration direction (Gao&Ojo, 2012a)

Secondary dendrite arm spacing (SDAS) is estimated from the cooling rates (G×R).The dendrite arm spacing increases from the interface of maximum weld width to the weld centerline, and decreases from the weld crown to weld root, as shown in Fig.4. These explicate that the finer arm spacing is always found near the fusion boundary on the weld crown or part of the weld root, and the coarsest structure occurs in the center of the weld crown.

As the arc current and laser power were increased, the dendrite trunk spacing along the rear of the solid/liquid interface of the weld pool increases and the magnitude of the SDAS is between 2.88 and 4.2 μm. Dendrite arm spacing decreases with increasing velocity, more obvious for higher welding speed, and the characteristic magnitude of SDAS is between 2.18 and 3.2 μm. The calculated dendrite arm spacing along the solidification interface has the same tendency as the measured values, but the former is slightly greater than the latter, especially near the fusion boundary, see Fig.4 (A).

The SDAS near the solid/liquid boundary along the weld penetration is presented in Fig.4 (B). The dendrite trunk spacing near the root part of the weld gradually decreases with increase in the arc current, welding speed and laser power. The calculated dendrite arm spacing along the direction of weld penetration agrees reasonably well with the measured value near the lower region of the keyhole, and the experiment result is higher in value than that calculated on the weld crown. Hybrid welding allows for the weld metal to remain at a high temperature for a longer time and cool at lower rates compared to approximated laser power performed at the same welding speed (Moore et al., 2004). This results in a coarser dendritic microstructure region where the laser and arc beams simultaneously interact, which is at the weld crown center rather than the solidification boundary.

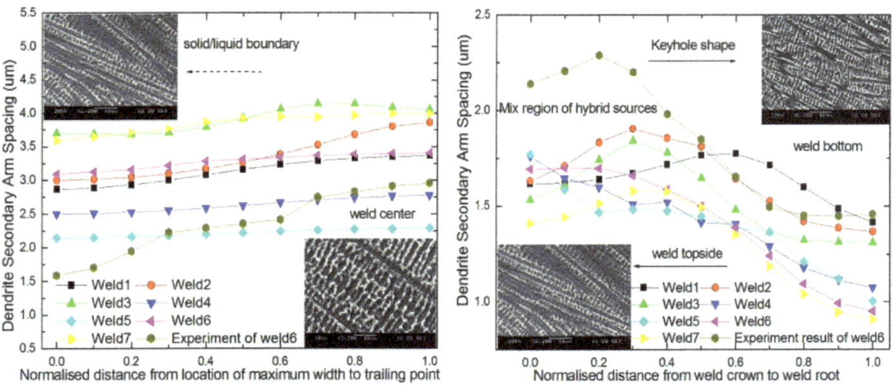

Figure 4. Predicted and experimental results of SDAS along the weld pool solidification interface, (A) Weld pool surface, (B) Weld penetration direction

Since the effect of the crystallographic orientation of the weld on the overall tendency to form stray grains is minimal, the weld direction [100] and sample surface plane (001) are specified for the sample orientation. The angles (θ, ϕ), which characterize the normal to the solidification front, are determined from three-dimension weld geometry, and dendrite growth pattern transitions can occur along the solidification interface. In consideration of the average growth velocity along the solidification interface front at each analysis cross-section, the dendrite orientation velocities of [100], [010] and [001] are presented in Fig.5. The θ, ϕ, ψ terms are used to calculate the dendrite growth velocity by Eqs. (10) and (12). The velocities of the [100] dendrites are always equal to those by welding, regardless of the weld pool shape for the (001) || [100] orientations (Rappaz et al.,1990), and average dendrite tip velocity in [010] and [001] is higher than that of the [100] orientation along the rear part of the solid/liquid interface. The average dendrite tip velocity of the [001] orientation is much larger than that of [010], and the latter is higher than that of [100] near the solidification interface. It can be underlined that $\psi > \theta$ and the dendritic orientation does not coincide with the solidification front for [010] and [001]. The discrepancy between ψ and θ is larger in [001] than [010].At the weld center, the dendrite growth velocities of [010] and [001] are lower than those at the solid/liquid boundary. For case (A), there are obvious transition positions among the solidification interface with differences in the average dendrite velocity for the [001] orientation. The average dendrite growth velocity of the [010] and [001] orientations of weld2 are lower than the other cases near the solidification interface. For case (B), the solidification time is much less at high welding speeds, and dendritic arrays do not have sufficient time to establish steady state spacing. The average dendrite tip velocity in the [010] and [001] orientations are irregularly volatile along the solidification front at the highest welding speed. It is also observed that the welding speed affects the magnitude of the dendrite growth velocity for the [001] orientation near the solidification interface. For case(C), the average dendrite tip velocity for the [001] orientation monotonically increases with increase in laser power, and its magnitude is the largest near the fusion line with the highest laser power.

2.1.2.2. Role of welding parameters on stray grain formation

The area-weighted average values of stray grain formation tendency with different welding conditions is shown in Fig.6, and partial columnar growth with the value of $\overline{\overline{\Phi}}$ is below 0.5 and above 0.0066 with increase in arc current or welding speed. These welding parameters are beneficial for avoiding CET during the entire welding process. The effect of welding speed on $\overline{\overline{\Phi}}$ is more complex. The latter initially increases with increasing welding velocity before decreasing at higher speeds. This can be explained through the changes in temperature gradient transition with different welding speeds. As the welding speed is increased, the temperature gradient is generally affected by two opposing factors (Grong, 1997). Thus, the magnitude of $\overline{\overline{\Phi}}$ will subsequently decrease with a further increase in the welding speed. The results from Anderson et al. (2010) and Vitek (2005) are similar to those illustrated here, in which the fraction of stray grain formation is generally observed to reach the maximum at an intermediate speed, and then always decreases as the speed increases.

Hitherto, literature mainly focused on wide and shallow pool shapes and higher laser power conditions were omitted due to limitations in simulation and method (Anderson et al., 2010; Gäumann et al., 2001;Vitek,2005).From the investigations of this model, the effect of higher laser power is slightly negative on SG formation. Laser power is the primary factor in weld penetration and weld pool geometry which determines the shape of the liquid/solid interface and the slope of the weld pool edges. The local angle ϕ along the solidification interface in the nail-head shaped weld contributes more to stray grain formation. Hence, the typical keyhole geometry in a hybrid pool is more prone to create CET in the weld pool

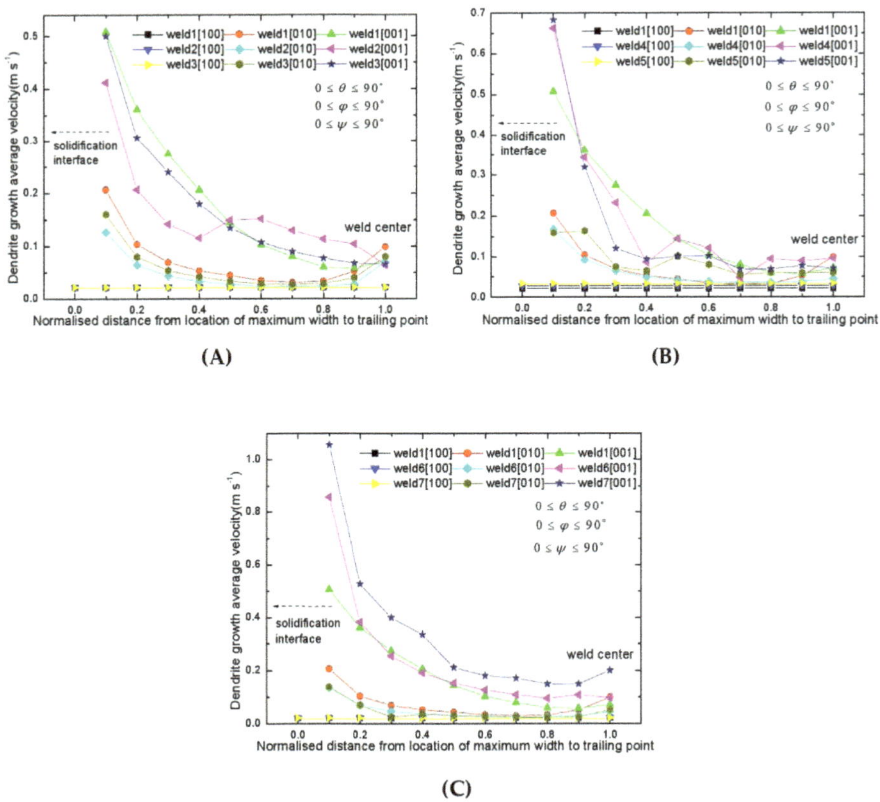

Figure 5. Dendrite tip velocity distribution around rear part of weld pool with different welding conditions, (A) variation of arc current, (B) variation of welding velocity, (C) variation of laser power (Gao&Ojo,2012a)

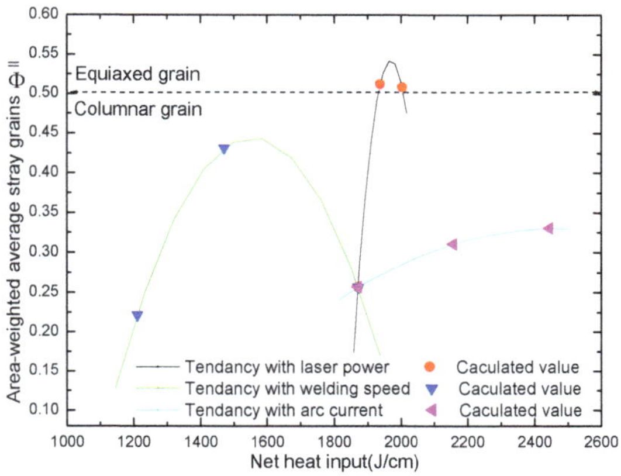

Figure 6. Area-weight average values of stray grain formation distribution as a function of heat input(Gao&Ojo,2012a)

when considering the distribution of Φ in the three directions from the above simulation, whereas the net effect of higher laser power on CET can be verified by the experimental work of Anderson et al. (2001), who used similar welding power intensity with an electron beam.

From the different tendencies of $\bar{\bar{\Phi}}$ summarized by the calculated value, higher welding speed and lower arc current or laser power ameliorate to minimize/avoid stray grain formation. In contrast, slower welding speed and higher arc current or laser power exacerbate stray grain formation, and abundant equiaxed dendrite will be yielded. Based on the above analysis, welding conditions should be optimized to minimize the overall likelihood of forming stray grains.

2.2. Liquation cracking modeling for polycrystalline superalloys welding

2.2.1. Mathematical model description

In consideration of tractable object symmetry, a 1/2 model including the weld metal, deposition filler material and base metal is adopted. A sequential thermal-mechanical finite element analysis is performed in present study based on ANSYS 13 code. Specifically, 2D thermal solid plane55 is used to simulate the temperature field and 2D structure element plane42 is then used to calculate the stress-strain field with the nodal temperature extracted from the thermal analysis as the load. All the material properties used in the model are displayed in Fig.7.

In this model, the laser-GMA hybrid process employs heating the base metal in a localized zone with sequent transient heat sources. The effect of the synergistic interaction between laser and GMA on the heat flux distribution is not considered in this simulation. The distribution of laser beam energy is assumed to be Rotary-Gauss body heat source, which is suitable for nail-head shape with large depth to width ratio of molten pool along the workpiece thickness direction (Wu et al., 2004).The distribution of the GMA energy is assumed to be a Gaussian flux over the workpiece surface.

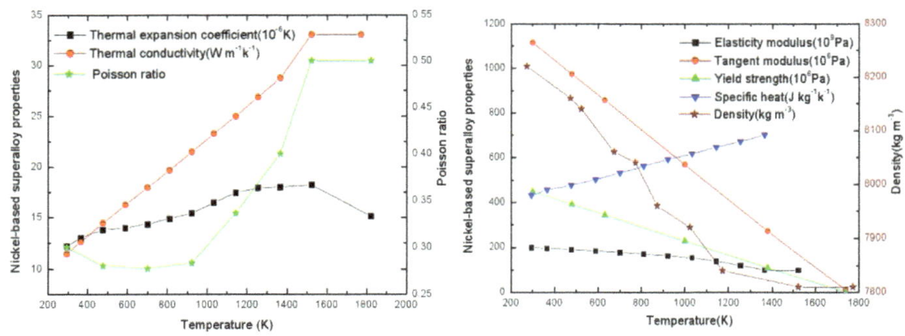

(A) **(B)**

Figure 7. The physical and mechanical properties used for the simulation (Gao&Ojo, 2012b)

The crystallographic orientation of the dendrites is not taken into account and they are assumed to grow in a direction that is always perpendicular to the growth front. Local growth velocity of dendrite along the solidification front as a function of undercooling is represented by the following function (Hunziker et al., 2000):

$$v_{kin}\{\Delta T\} = a\Delta T^n \tag{18}$$

where n=3.05, a=1.2754×10^{-7}ms^{-1}K$^{-3.05}$ and the undercooled liquid is the region between the growth front and liquidus temperature. Using growth rate R and the temperature gradient G along the fusion boundary, the cooling rate (G×R) can be calculated.

For the case of nonlinear material, the definition of total strain is given:

$$\{\varepsilon\} = \{\varepsilon^{el}\} + \{\varepsilon^{th}\} + \{\varepsilon^{pl}\} \tag{19}$$

where ε^{el} is the elastic strain vector, ε^{th} is the thermal strain vector, ε^{pl} is the plastic strain vector but the creep strain vector and swelling strain vector are ignored.

As mentioned above, the thermal elasto-plastic material model, based on von Mises yield criterion, temperature dependent mechanical properties and linear kinematic hardening rule, is considered. The incremental forms of stress-strain relation are described as (Akbari & Sattari-Far, 2009; Chang & Lee, 2009; Kong & Kovacevic, 2010):

$$[d\delta] = [D^{ep}]\{d\varepsilon\} - [D^{th}]dT \tag{20}$$

$$[D^{ep}] = [D^e] + [D^p] \tag{21}$$

where $d\delta$ is the stress increment, $d\varepsilon$ is the strain increment, $[D^e]$ is the elastic stiffness matrix, $[D^p]$ is the plastic stiffness matrix, $[D^{th}]$ is the thermal stiffness matrix and dT is the temperature increment. More information about this numerical model can be reviewed by referring to Gao (2012).

	Laser power (kW)	Arc power (kW)	Welding speed (m/min)	Time of laser heating (s)	Time of arc heating (s)	Time interval for hybrid (s)
Laser	2	-	3	0.012	-	-
Laser	4	-	3	0.012	-	-
Laser	6	-	3	0.012	-	-
Laser	4	-	2	0.018	-	-
Laser	4	-	4	0.009	-	-
Laser	4	-	6	0.006	-	-
Laser-GMA	2	4.75	3	0.012	0.05	0.02
Laser-GMA	4	4.75	3	0.012	0.05	0.02
Laser-GMA	6	4.75	3	0.012	0.05	0.02
Laser-GMA	4	4.75	2	0.018	0.075	0.03
Laser-GMA	4	4.75	4	0.009	0.0375	0.015
Laser-GMA	4	4.75	6	0.006	0.025	0.01

Table 3. Design laser and laser-GMA hybrid welding conditions (Gao&Ojo, 2012b)

Predictions for the cracking susceptibility have been estimated for a wide range of welding conditions. Two types of welding processing, three kinds of laser power and four kinds of welding speed have been investigated, which result in a total twelve laser fabrication conditions, as shown in Table 3.There is a time interval for laser-arc hybrid welding, which sequentially heats the base material due to the distance between laser beam and arc source.

2.2.2. *Results and discussion*

2.2.2.1. Weld pool solidification characteristics

The HAZ cracking in the waisted zone over the bead cross-section is attributed to the greater amount of grain boundary liquation during the cooling process. The solidification completion temperature of liquated grain boundary is around 1383K, which affects the Laves cluster dissolution and grain boundary liquid phase (Nishimoto et al., 2002). The shape of weld pool and cooling rate along localization fusion boundary derived from temperature isothermal are shown in Fig.8. The abscissa means the distance from the surface to the root of the weld bead along the fusion boundary. The left ordinate value indicates the weld geometry, which is the distance from the weld bead centerline and the right ordinate value is the cooling rate.

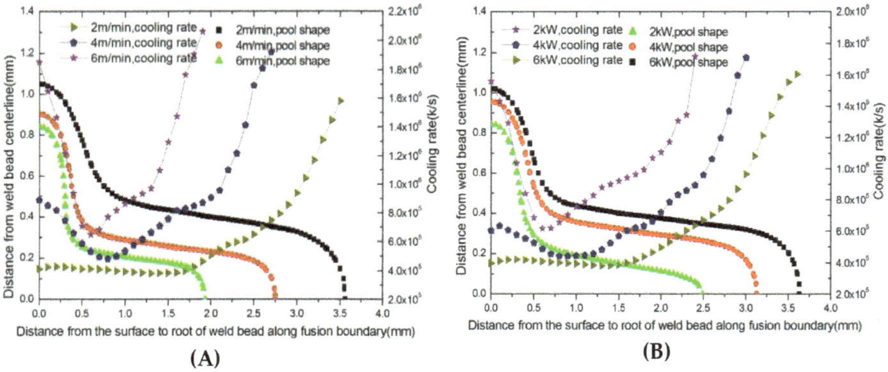

Figure 8. Correlation between the shape of weld pool and cooling rate with welding parameters for laser welding , (A) variation of welding speed, (B) variation of laser power

The results indicate that the cooling rate is a function of position along the fusion boundary among the weld fusion zone and varies simultaneously with the penetration shape. The shape decay of temperature in the surface region is attributed to the heat loss from the surface region due to heat transfer of conduction and convection in this region. Temperature decay in the initially high temperature region at the root of the keyhole during the laser beam moving away, which is due to the attainment of high temperature gradient in this region, results in faster solidification. The neck zone over the bead cross-section is more liable to heat stagnation than in other regions and undergoes lower cooling rates than other parts. That means that the maximum solidification time exists at the neck zone of the weld shape and the minimum solidification time occurs at the root of the keyhole. Thus, there is obvious solidification time transition region along the fusion boundary line. There is a steep cooling rate in the vicinity of the neck zone. With an increasing welding speed, the cooling rate in the neck zone gets steeper and weld pool geometry substantially decreases, respectively. Correspondingly, the shapes of weld pool synchronously enlarge with a laser power increase and the concomitant cooling rates considerably decrease.

Correlation between weld pool geometry and cooling rate of the hybrid pool characteristic are shown in Fig.9. The difference in these relationships between laser and hybrid laser arc welding are compared, and several conclusions can be drawn from the proposed study. First, with incorporating the effect of filler material on the weld bead cross-section, the weld pool geometry variation with welding speeds of laser hybrid welding are different from that of laser welding. Employing an additional heat and deposition metal lead to an extended region of weld width and increased neck zone radius of curvature. The volume of weld filler metal strongly depends on the welding speed and decreases with an increasing welding speed. Second, the rational link between cooling rate and weld pool geometry suggest that the neck zone over bead cross-section is more liable to heat stagnation and has larger solidification time, therefore, the minimized cooling rate occurs more readily at locations where it is underside of neck zone, which is especially observed in the neck-like waisted zone of lower heat input of laser-GMA hybrid welding as well as laser welding. Third, laser-GMA hybrid welding alters the temperature distribution of the weld fusion zone and substantially enhances the cooling rate near the weld bead neck zone. The magnitude of cooling rate is lower than that of laser welding. Cooling rate attains low value in the region of neck zone due to higher solidification time, and obviously increases near the bottom of the keyhole. The relationship between the weld pool shape and the cooling rate elucidates clearly localized solidification behavior and enables the cause of liquation cracking to be explained.

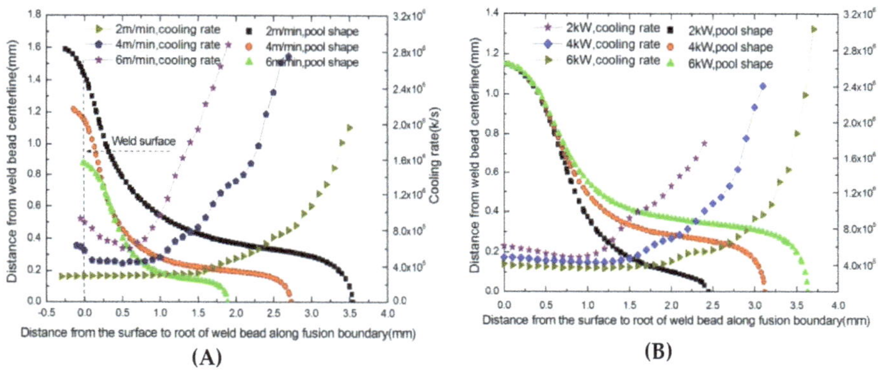

Figure 9. Correlation between the shape of weld pool and cooling rate with various welding conditions for laser-arc hybrid welding, (A) variation of welding speed, (B) variation of laser power

2.2.2.2. HAZ cracking analysis

Constitutional liquation theory and grain boundary segregation mechanism are associated with weld HAZ liquation cracking. Nishimoto et al (2002) explained the liquation cracking mechanism in laser welds of Inconel 718. The molten liquid of the Laves phase eutectically is liquated at a temperature below the solidus temperature of the matrix and infiltrates along the grain boundaries. Tensile plastic strain/stress induced by thermal shrinkage is imposed on the liquid film during the subsequent cooling process and cracking will occur with an attendant buildup of higher strain/stress across the grain boundaries.

Despite the fact that the HAZ cracking metallurgical mechanism of nickel-base alloy is revealed, the limited information is available in the literature about the explanation of the relationship between welding conditions and cracking susceptibility during the laser and hybrid laser-GMA welding.

For the purpose of comparison of crack susceptibility with various welding conditions and different welding types, von Mises stress and 1st principal strain are not sensitive to reflect these discrepancies (Gao, 2012). Quantitative evaluation of solidification crack with behavior of ductility near the solidification boundary can be calculated by crack susceptibility strain rate, which is represented:

$$\dot{\varepsilon} = \frac{\varepsilon_1}{t_s} \tag{22}$$

where ε_1 denotes total 1st principal strain along the fusion boundary, t_s denotes the lapse time of solidification boundary from liquidus temperature to crack sensitive temperature, 1383K.

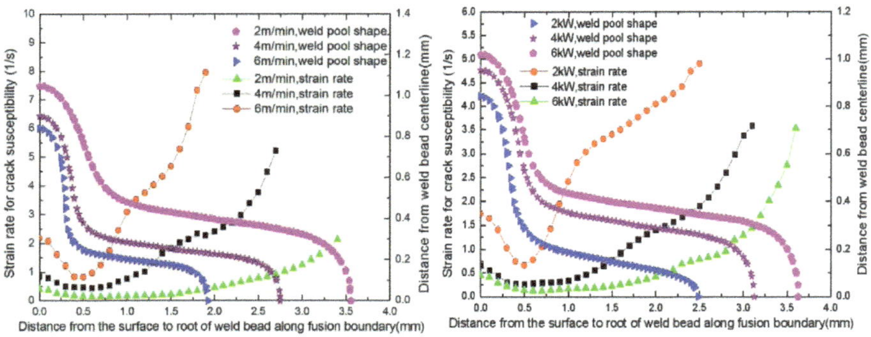

(A) **(B)**

Figure 10. Correlation between the shape of weld pool and strain rate with various welding conditions for laser welding, (A) variation of welding speed, (B) variation of laser power(Gao&Ojo, 2012b)

Relations among crack susceptibility strain rate and pool shape of laser weld are shown in Fig.10. As a whole, it is worth noting that the value of the geometry of weld pool is nearly in the same order of the crack susceptibility strain rate, and the curve shapes of crack susceptibility strain rate conspicuously change with various welding conditions. It should be noticed that there is a concave shape, which suddenly decreases and then increases in the strain rate curve adjacent to the neck region of weld pool shape that means it is more susceptible to initiate crack in the stage of solidification. After the neck zone, strain rate increases until it is at the root of the keyhole and reaches a maximum value. The reason for this gradual increase in strain rate is thought to be due to the interaction between strain distribution and solidification characteristic within the weld pool. It can be estimated that greater welding speed and smaller laser power, namely less heat input results in the steeper curve of crack susceptibility strain rate, which is prone to engender and propagate cracks in the neck zone. Thus, the crack susceptibility strain rate itself is a major parameter determining the tendency of crack susceptibility.

As mentioned before, the effects of the welding conditions on the solidification crack susceptibility are interrelated. In contrast, laser–GMA hybrid welding can minimize the risk of variation of strain rate in the neck zone of the weld bead, and the value of these ductility curves are obviously smaller than that of laser welding, as shown in Fig.11. The strain rate becomes sharp near the neck region with an increasing welding speed or decreasing laser power.

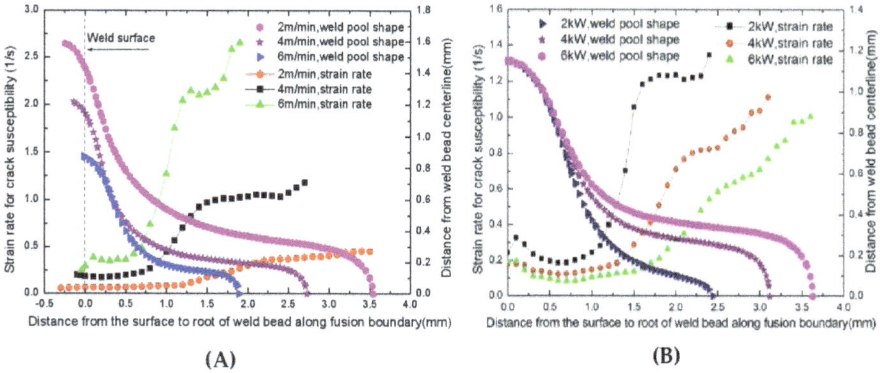

Figure 11. Correlation between the shape of weld pool and strain rate with various welding conditions for laser-arc hybrid welding, (A) variation of welding speed, (B) variation of laser power(Gao&Ojo, 2012b)

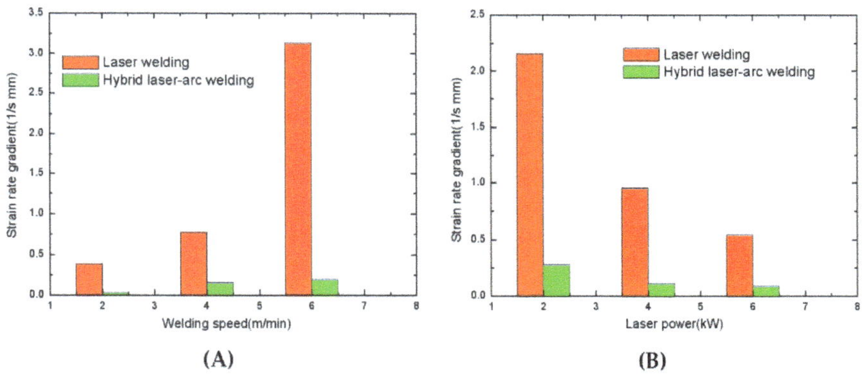

Figure 12. The strain rate gradient near the neck region with different welding conditions, (A) variation of welding speed, (B) variation of laser power

Comparison of strain rate gradient near the neck region under different welding conditions is illustrated in Fig.12. It is worth noting that the strain rate gradient decreases with heat input increase. The values of strain rate gradient of laser welding are on the whole higher than those of hybrid laser arc welding. It clearly reveals that laser welding is more susceptible to liquation cracking and the crack is located more in the neck region during low heat input. Therefore, the solidification cracking relating the liquid film stage in cracking sensitive temperature range can be explained from the view point of the contribution of strain rate gradient. From the above results, using strain rate gradient for evaluation of the susceptibility to HAZ liquation cracking provides theoretically understanding of how laser-arc hybrid welding produces less susceptibility to liquation cracking compared to laser welding as previous reported experiment results (Stelling et al., 2005).

3. Conclusions

These models provide valuable insight into the metallurgical and mechanical driving force for cracking. These results will be used to predict the effect of welding conditions on the potential for weld solidification cracking and HAZ liquation cracking and to interpret the experimental results during laser-GMA hybrid welding single-crystal and polycrystalline nickel-base superalloys. It should be noted that higher welding speed and lower laser power or arc current are beneficial for minimizing stray grain formation for laser-GMA welding single-crystal nickel-base superalloys , but generally lower heat input exacerbate HAZ liquation cracking for laser-GMA welding polycrystalline nickel-base superalloys. These promising results will also be used to identify preferred welding conditions that minimize stray grain formation, local stress-strain and associated cracking. For the future research, many efforts will focus on the different welding parameters on the cracking trend with optimized tactics and substantiate them by experiments.

For the solidification cracking modeling of welding of single-crystal nickel-base superalloys, there are some conclusions:

1. In almost all cases, the cooling rate is higher at the maximum width of the fusion boundary and at the bottom of the keyhole, while lower at the weld centerline and weld crown solid/liquid boundary. Meanwhile, it is shown that the finer arm spacing is always found near the solid/liquid boundary of the weld crown or the weld root, and the coarsest structure occurs in the center of the weld crown;
2. The average dendrite tip velocity of the [001] orientation is much greater than that of [010], and the latter is higher than that of [100] near the solidification interface for the (001) || [100] orientations. At the weld center, the average dendrite growth velocity of [010] and [001] is lower than that at the solid/liquid boundary;
3. Higher welding speed and lower laser power or arc current are beneficial for minimizing stray grain formation, when taking into consideration, on average, the tendency for stray grain formation over the entire weld pool;
   For the HAZ liquation cracking modeling of welding of polycrystalline nickel-base superalloys, there are also some conclusions:
4. The neck zone radius of curvature in the laser hybrid pool is larger than that of the laser pool, which means the neck region is more liable to heat stagnation and has a smaller cooling rate. The minimized cooling rate occurs more readily at the underside of the neck zone, which is observed in the neck-like waisted zone of laser-GMA hybrid welding as well as laser welding;
5. Laser-GMA hybrid welding alters the temperature distribution of the weld fusion zone and substantially improves the cooling rate near the weld bead neck zone; the magnitude of cooling rate is lower than that of laser welding. The cooling rate attains low value in the region of neck zone due to higher solidification time and increases near the bottom of the keyhole;
6. The values of strain rate gradient near the neck region of laser welding are on the whole higher than those of hybrid laser-GMA welding. Laser welding is more susceptible to

liquation cracking and the crack is located more in the neck region during lower heat input compared to laser-GMA hybrid welding;

7. The weld pool shape has a strong influence on the stress-strain pattern, and a neck weld pool shape can have a detrimental effect on the HAZ cracking behavior. Adding the arc heat source alters the strain distribution, and strain rate gradient provides theoretically understanding of how laser-GMA hybrid welding produces less susceptibility to liquation cracking compared to laser welding.

Author details

Zhiguo Gao
University of Manitoba, Canada

Acknowledgement

Thanks a lot to the Journal of Acta Materialia, Journal of Materials Science Research and International Journal of Advanced Manufacturing Technology, the author makes reference to some previous works in this chapter.

4. References

Akbari, D. & Sattari-Far, I. (2009). Effect of the welding heat input on residual stresses in butt-welds of dissimilar pipe joints. *International journal of pressure vessels and piping,* Vol.86, No.11, pp. 769-776

Anderson, T.; DuPont, J. & DebRoy, T. (2010). Origin of Stray Grain Formation in single-crystal superalloy weld pools from Heat Transfer and Fluid Flow Modeling. *Acta Materialia,* Vol. 58, No.4, pp.1441-1454

Banerjee, P. & Overfelt, R. (1999). Viscosity measurements of industrial alloys using the oscillating cup technique. *International Journal of Thermophysics,* Vol.20, No.6, pp.1791-1800

Bonifaz, E. & Richards, N. (2009). Modeling cast IN-738 superalloy gas tungsten arc welds. *Acta Materialia,* Vol.57, No.57, pp.1785-1794

Brooks, R.; Monaghan, B. & Barnicoat A. (1996). The Physical properties of alloys in the liquid and "Mushy" states, *International Journal of Thermophysics,* Vol.17, No.5, pp.1151-1161

Bussac, A. & Gandin, C. (1997).Prediction of a process window for the investment casting of dendritic single crystals. *Materials Science and Engineering: A,* Vol.237, No.1, pp. 35-42

Chang, K. & Lee, C. (2009). Finite element analysis of the residual stresses in T-joint fillet welds made of similar and dissimilar steels. *Int J Adv Manuf Technol,* Vol.41, No.3-4, pp.250-258

Claus,B.&Flemming,O.(2005). Review of laser hybrid welding. *Journal of Laser Applications,*Vol.17,No.1,pp.2-13

Dai, K. & Shaw, L. (2001). Thermal and stress modeling of multi-material laser processing. *Acta Materialia*, Vol.49, No.20, pp. 4171-4181

Dong, H.(2007). Analysis of Grain Selection during Directional Solidification of Gas Turbine Blades, *Proceedings of the Word Congress on Engineering 2007*, Vol.II WCE 2007, ISBN: 978-988-98671-2-6, London, U.K., July 2-4, 2007

Egbewande, A.; Buckson, R. & Ojo, O. (2010). Analysis of laser beam weldability of Inconel 738 superalloy, *Materials characterization*, Vol.61, No.5, pp.569-574

Farzadi, A.; Serajzadeh, S. & Kokabi, A. (2010) Investigation of weld pool in aluminum alloys: Geometry and solidification microstructure. *International journal of thermal sciences*, Vol. 49, No.5, pp.809-819

Gao, Z.; Wu, Y. & Huang, J. (2009). Analysis of weld pool dynamic during stationary laser-MIG hybrid welding. *Int J Adv Manuf Technol*, Vol.44, No.9-10, pp.870-879

Gao, Z. (2012).Numerical modeling to understand liquation cracking propensity during laser and laser hybrid welding (I).*Int J Adv Manuf Technol*, Doi:10.1007/s00170-012-3907-8,pp.1-13

Gao, Z. & Ojo, O.(2012a) Modeling analysis of hybrid laser-arc welding of single-crystal nickel-base superalloys, *Acta Materialia* ,Vol.60,No.6-7,pp.3153-3167

Gao, Z. & Ojo, O.(2012b) Numerical modeling of HAZ liquation cracking tendency during laser and hybrid laser-arc welding processes, *Journal of Materials Science Research* , Vol.1,No.2,pp.42-55

Gäumann, M.; Bezençon, C.; Canalis, P. & Kurz, W. (2001).Single-crystal laser deposition of superalloys: processing microstructure maps. *Acta Materialia*, Vol.49, No.6, pp.1051-1062

Grong, Ø. (1997). *Metallurgical Modeling of Welding*, The Institute of Materials, London, U.K.

He, X.; Elmer, J. & DebRoy, T. (2005). Heat transfer and fluid flow in laser microwelding. *J. Appl. Phys.*Vol. 97, No.8, pp.084909-084909-9

Hee, S.; Han, S.; You, C. & Sung, M. (2010). Analysis of residual stress on AH32 butt joint by hybrid CO2 laser-GMA welding. *Computational materials science*, Vol. 49, No.2, pp.217-221

Hong, J.; Park, J.; Park, N.; Eom, I.; Kim, M. & Kang, C. (2008). Microstructures and mechanical properties of Inconel 718 welds by CO_2 laser welding. *Journal of materials processing technology*, Vol.201, No.1-3, pp.515-520

Hu, B. & Richardson, I. (2006). Mechanism and possible solution for transverse solidification cracking in laser welding of high strength aluminium alloys. *Materials science and engineering, A* ,Vol.429,No.1-2,pp.287-294

Hunt J. (1984). Steady state columnar and equiaxed growth of dendrites and eutectic. *Materials Science and Engineering*, Vol.65, No.1, pp.75-83

Hunziker, O.; Dye, D. & Reed, R. (2000). On the formation of a centerline grain boundary during fusion welding. *Acta Materialia*, Vol.48, No.17, pp. 4191-4201

Kong, F.& Kovacevic, R. (2010). 3D finite element modeling of the thermally induced residual stress in the hybrid laser/arc welding of lap joint. *Journal of Materials Processing Technology*, Vol.210, No.6-7, pp.941-950

Liu, W. & DuPont. J. (2004). Effects of melt-pool geometry on crystal growth and microstructure development in laser surface-melted superalloy single crystals.

Mathematical modeling of single-crystal growth in a melt pool (part I). *Acta Materialia*, Vol.52, No.16, pp.4833-4847

Liu, W. & DuPont. J. (2005). Effects of substrate crystallographic orientations on crystal growth and microstructure development in laser surface-melted superalloy single crystals. Mathematical modeling of single-crystal growth in a melt pool (Part II). *Acta Materialia*, Vol.53, No.5, pp.1545-1558

Liu, F.; Lin, X.; Yang, G.; Song, M.; Chen, J.& Huang, W.(2011). Microstructure and residual stress of laser rapid formed Inconel 718 nickel-base superalloy. *Optics & laser technology*, Vol.43, No.1, pp.208-213

Long, R.; Liu, W.; Xing, F. & Wang H. (2008). Numerical simulation of thermal behavior during laser metal deposition shaping. *Transactions of Nonferrous Metals Society of China*, Vo.18, No.3, pp.691-699

Luo, X.; Shinozaki, K. & Kuroki, H. (2002). Analysis of temperature and elevated temperature plastic strain distributions in laser welding HAZ study of laser weldability of Ni-base superalloys. Welding International, Vol.16, No.5, pp.385-392

Moore, P.; Howse, D. & Wallach, E. (2004). Microstructures and properties of laser/arc hybrid welds and autogenous laser welds in pipeline steels .*Sci Technol Weld Joining*, Vol. 9, No.4, pp.314-322

Nishimoto, K.; Woo, I. & Shirai, M. (2002). Analyses of temperature and strain distributions in laser welds study of the weldability of Inconel 718 cast alloy. *Welding International*, Vol.16, No.4, pp.284-292

Park, J.; Babu, S.; Vitek, J.; Kenik, E. & David, S. (2003). Stray grain formation in single crystal Ni-base superalloy welds. *Journal of Applied Physics*, Vol.94, No.6, pp.4203-4209

Pottlacher, G.; Hosaeus, H.; Kaschnitz, E. & Seifter, A.(2002) Thermophysical properties of solid and liquid inconel 718 alloy. *Scandinavian Journal of Metallurgy*, Vol.31, No.3, pp.161-168

Rai, R.; Roy, G& DebRoy, T. (2007). A Computationally Efficient Model of Convective Heat Transfer and Solidification Characteristics during Keyhole Mode Laser Welding. *J Appl Phys.*,Vol. 101,No. 05409,pp.1-11

Ral, R.; Palmer, T.; Elmer, J. & Debroy, T. (2009). Heat Transfer and Fluid Flow during Electron Beam Welding of 304L Stainless Steel Alloy. *Welding Journal*, Vol. 88, No.3, pp.54-61

Rappaz, M.; David, S.; Vitek, J. & Boatner, L. (1990). Analysis of solidification microstructures in Fe-Ni-Cr Single-Crystal welds. *Metallurgical Transaction A*, Vol.21A, No.6, pp.1767-1782

Ribic, B.; Palmer,T. & DebRoy,T. (2009). Problems and issues in laser-arc hybrid welding. *International Materials Reviews*, Vol.54, No.4, pp.223-244

Roy, G.; Elmer, J. & DebRoy, T. (2006). Mathematical modeling of heat transfer, fluid flow, and solidification during linear welding with a pulsed laser beam. *J. Appl. Phys.*Vol.100, No.3, pp.034903-034903-7

Stelling, K.; Boellinghaus,Th.; Lammers,M. & Schobbert, H. (2005). Laser plasma power hybrid welding in vertical-up and vertical-down positions, *Proceedings of the 7th international conference on trends in welding research*, pp. 121-126,Callaway gardens resort, Pine mountains, Georgia, USA, May 16-20,2005

Vitek. J. (2005). The effect of welding conditions on stray grain formation in single crystal welds-theoretical analysis. *Acta Materialia*, Vol. 53, No.1, pp., 53- 67

Wang, N.; Mokadem, S.; Rappaz, M. & Kurz, W. (2004). Solidification cracking of superalloy single- and bi-crystals. Acta materialia, Vol.52, No.1, pp.3173-3182

Wu, S.; Zhao, H.; Wang, Y. & Zhang, X. (2004). A new heat source model in numerical simulation of high energy beam welding. *Transactions of the china welding institution*, Vol.25, No.1, pp.91-94

Yann, D.; Eric, L. & Corinne, A. (2010). Numerical modeling of Inconel 738LC deposition welding: prediction of residual stress induced cracking. *Journal of Materials Processing Technology*, Vol.210, No.14, pp.2053-2061

Yilbas, B.; Akhtar, S. & Karatas, C. (2010). Laser surface treatment of Inconel 718 alloy: Thermal stress analysis. *Optics and Lasers in Engineering*, Vol.48, No.7-8, pp.740-749

Zeng, D. (2006). Annular beam shaping and optical trepanning, In: *Doctor of Philosophy*, 22.02.2012, Available from
http://etd.fcla.edu/CF/CFE0001333/Zeng_Danyong_200612_PHD.pdf

Chapter 8

Development of a Comprehensive Process Model for Hybrid Laser-Arc Welding

Fanrong Kong and Radovan Kovacevic

Additional information is available at the end of the chapter

1. Introduction

Recently, a hybrid welding technique combining laser welding and arc welding has been finding broader applications in industry due to its unique advantages, such as higher welding efficiency and lower costs [1, 2] compared to the traditional arc welding or autogenous laser welding. Because of rapid melting and solidification occurring in the weld zone, a locally high thermal gradient inevitably exists and accompanies with the whole welding process, which really decides the final residual stress and distortion distributions of weld and affects the remained grain size in the fusion zone (FZ) and heat affected zone (HAZ). Also, the levels of residual stresses and distortions directly influence the weld quality [3]. In comparison with traditional arc welding and autogenous laser welding, the temperature field and residual stress distribution in hybrid laser-arc welding involve more variables because of the additional interaction between the laser and arc plasma [4] thus becoming much more complex and difficult to theoretically and experimentally analyze. Trial-and-error experiments are not able to fully describe those physical mechanisms involved in the hybrid laser-arc welding process. Therefore, the numerical tools have been widely used to help explain the complex welding mechanisms present in the hybrid laser-arc welding process [5, 6].

Up to now, numerical work on the welding process mostly concentrates on traditional electric arc welding, including gas tungsten arc welding (GTAW) [7 - 9], submerged arc welding (SAW) [10,11], gas metal arc welding (GMAW) [12-16], and partly on laser beam welding (LBW) [17-21]. These studies focused on the heat and mass transfer phenomena in the weld pool [8, 11-15], thermal-induced distortion and residual stresses [16, 17, 19, 20], solidification-induced dendrite growth in the FZ [21], and recrystallization in the heat affected zone (HAZ) [9]. Due to the locally rapid melting and solidification occurring in the welding process, a high temperature gradient—which inevitably exists in the weld zone—

causes a high-stress concentration in the weld zone and nearby HAZ [22], which usually exceeds the yield strength of the material. Large residual stresses presented in the welded structure can obviously reduce the fatigue strength of metal components, causing crack generation and shorten the lifetime of metal component [23], which could possibly have disastrous results. A number of mitigation procedures to reduce and/or eliminate the level of residual stress have been presented by researchers, mainly including enhancement of the material ductility of solidification zone (SZ) and HAZ, and improving the thermal and mechanical conditions in the welding processes [24].

Because of the complex physical mechanisms in the welding process—which are related to the heat source properties, material performance, and welding parameters, etc.—trial-and-error methods to optimize welding parameters takes a long time and is usually more costly. Also, understanding of the physics of the welding process is limited by only using an experimental approach. Numerical simulation as accompanied by theoretical analysis has been widely applied as a cost-efficient way to help explore the welding phenomena in different welding techniques. Eagar et al. [25, 26] spent a lot of time in developing theoretical models for GTAW processes. Dong et al. [27-29] developed numerical models to predict the residual stresses as well as fatigue life of weld obtained by the multi-pass welding process. Deng et al. [30-32] developed a series of numerical models to study the residual stress distribution in variable welding joints.

Compared to the traditional electric arc welding, laser welding has unique advantages such as high energy density, narrow HAZ, low heat input, and high energy efficiency. However, laser welding is also limited by its disadvantages like poor gap bridgeability and high equipment cost. In order to fully use the advantages of both laser and arc welding techniques, Steen et al. [33] introduced for the first time a hybrid technique by combining the laser beam and arc for welding and cutting in the late 1970s. Subsequently, researcher and engineers have presented a number of works on combining the laser and electric arc in the past decades. Considering that interaction between laser beam and arc plasma is complex, the hybrid laser-arc welding and cladding processes have not been understood fully. Most available literature on these approaches is limited at the level of the experimental study including hybrid laser-GTAW, hybrid laser-GMAW, and hybrid laser-plasma arc welding for steels, magnesium alloy, aluminum alloy, titanium alloy and dissimilar materials. In order to further study the welding mechanism of hybrid laser and arc, it is necessary to develop a comprehensive model to understand the heat and mass transfer, residual stress evolution, as well as microstructure formation in the hybrid laser-arc welding process. Zhou and Tsai [34, 35] presented heat transfer and fluid flow models to study the metal inert gas (MIG) welding and laser-MIG hybrid welding processes. Rao et al. [36] reviewed the modeling of hybrid laser-gas metal arc (GMA) welding and presented further studies on synergistic interaction between the laser beam and arc, the metal transfer features, and behavior of shielding gas. Ribic et al. [37] developed a three-dimensional (3-D) finite volume model to study heat transfer and fluid flow in the hybrid laser-GTA welding process. Considering that the microstructure formation of weld has a close relationship with the macro-scale heat transfer and fluid flow, and residual stress fields, it will be very necessary to integrate the thermal, fluid flow and mechanical modeling with the microstructure evolution like grain growth in the fusion zone and HAZ. Multi-scale and multi-physics modeling is one of most interesting simulation trends in the laser-based heat processes, especially in the hybrid laser-arc welding process.

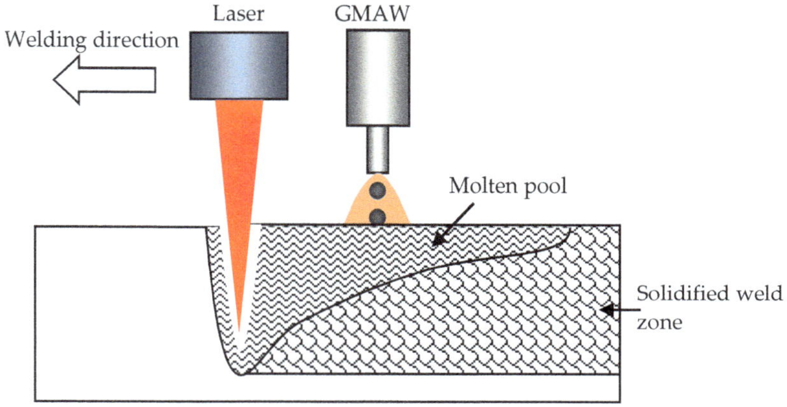

Figure 1. Schematic view of hybrid laser-GMA welding system

In this chapter, a 3-D mathematical model will be developed to numerically predict the transient temperature distributions and residual stresses in the hybrid laser-GMA welding of a thick plate of A514 steel in butt joint configuration, as shown in Figure 1. The numerical solution is achieved based on a finite element method by using a commercial numerical package, ANSYS. A Monte Carlo model is introduced to consider the grain growth and phase transformations in the HAZ. The laser and arc heat inputs and heat losses at the surface of coupons are considered by using ANSYS Parametric Designed Language (APDL). The influences of the processing parameters (including welding speed, laser power, wire feed rate, arc power, and stand-off distance from laser to arc) on the profile and geometrical size of the molten pool, residual stress distribution of the weld, and grain size in the HAZ are numerically studied. The numerically obtained results are experimentally verified.

2. Finite element modeling

2.1. Thermal analysis of hybrid laser-GMA welding

In the hybrid laser-GMA welding process, laser and GMA simultaneously heat the coupon surface in local area, which makes the thermal distribution of weld much more complex. In this study, a cylindrical volume heat-source model with a Gaussian distribution is assumed to simulate the heat input by laser, and a double-ellipsoidal volume heat source is selected to consider the heat input by GMA welding. The general thermal governing equation is shown below, in which thermal conduction-induced heat transfer is considered and temperature-dependent material properties are used [38].

$$\rho c_p \frac{\partial T}{\partial t} = \{L\}^T ([D]\{L\}T) + \ddot{q}_l + \ddot{q}_a \tag{1}$$

$$\text{where}[D] = \begin{bmatrix} k_{xx} & 0 & 0 \\ 0 & k_{yy} & 0 \\ 0 & 0 & k_{zz} \end{bmatrix}, \tag{2}$$

$$\text{and}\{L\} = \begin{Bmatrix} \frac{\partial}{\partial x} \\ \frac{\partial}{\partial y} \\ \frac{\partial}{\partial z} \end{Bmatrix}, \tag{3}$$

ρ is the density, c_p is the specific heat, T is the temperature, t is time, k_{xx}, k_{yy}, and k_{zz} are the thermal conductivity components along the x, y, and z axis, respectively; $\ddot{q}_l$ and $\ddot{q}_a$ are the volume heat generation rates due to the laser and GMA heat input, respectively.

So far, a number of heat source models have been developed to simulate the arc welding and laser welding processes. Laser welding usually consists of laser conduction welding or laser keyhole welding. The former one has lower energy density as compared with the latter one by which a keyhole is formed in the weld pool. A surface heat flux model is usually applied in the thermal analysis of laser conduction welding. However, a volume-distributed heat source model, like rotary Gaussian heat density distribution [39], is usually used for simulating a laser keyhole welding. Compared to the laser beam welding, electric arc welding has much lower energy density, and surface heat flux models with Gaussian distributions used to be applied to simulate the arc heat input in the arc welding process.

Considering that the enthalpy brought into the weld pool by melted wire in GMAW, volume-distributed heat source models are preferred, such as hemi-spherical power density distribution [40], ellipsoidal power density distribution [41], and double ellipsoidal power density distribution [42]. However, all of these heat source models are empirically derived based on the experimentally fitting data. Therefore, each heat source model mentioned above has a certain applicable range in the real production case. It is suggested that engineers in the welding process design should reasonably select a heat source model which matches well with the specific welding process. There are a limited number of publications available to numerically describe the hybrid laser and arc welding process because of lack of knowledge on complex interaction between the material, arc plasma, and laser beam [43]. Current heat source models of hybrid laser and arc including GTAW and GMAW were mostly developed with the help of experimental support [34-37].

In this study, a double-ellipsoidal heat source model is introduced to simulate the GMAW heat input, and a cylindrical heat source model with a sectional Gaussian distribution is used to consider the laser heat input. q^f_{arc} (x, y, z, t) and q^r_{arc} (x, y, z, t), depict heat input distributions inside the front and rear quadrants of the GMAW heat source, respectively, which can be expressed as follows [42, 44]:

$$\dot{q}^{f}_{arc}(x,y,z,t)=\frac{6\sqrt{3}f_f P_{arc}}{abc_f\pi\sqrt{\pi}}\exp\left(-\frac{3(x-x_0)^2}{a^2}\right)\cdot\exp\left(-\frac{3(y-L_w)^2}{b^2}\right)\cdot\exp\left(-\frac{3(z-vt)^2}{c_f^2}\right) \quad (4)$$

$$\dot{q}^{r}_{arc}(x,y,z,t)=\frac{6\sqrt{3}f_r P_{arc}}{abc_r\pi\sqrt{\pi}}\exp\left(-\frac{3(x-x_0)^2}{a^2}\right)\cdot\exp\left(-\frac{3(y-L_w)^2}{b^2}\right)\cdot\exp\left(-\frac{3(z-vt)^2}{c_r^2}\right) \quad (5)$$

where a, b, c_f, c_r are the characteristic parameters of heat sources, and a, b, c_f, and c_r are set at 4 mm, 3 mm, 3 mm and 7 mm respectively [44]. P_{arc} denotes the nominal power of the GMAW, and $P_{arc}=\mu UI$. Where μ is the energy efficiency of GMAW based on the welded metal, U denotes the arc voltage of GMAW, and I stands for arc current of GMAW. $\dot{q}_{laser}(x,y,z,t)$ denotes the laser radiation-induced volume heat input given by [45, 46]:

$$\dot{q}_{laser}(x,y,z,t)=\eta_l\frac{P_{laser}\cos\Phi}{2\pi R_l^2}\exp\left(-\frac{(x-x_0)^2+[z+(L_w-y)\tan\Phi-vt-D_{la}]^2\cos^2\Phi}{2R_l^2}\right)\cdot(y/L_w^2) \quad (6)$$

where η_l is laser absorption efficiency based on the welded material, P_{laser} stands for the nominal power of the laser beam, x_0 is the x-coordinate of the center point of laser spot at the coupon surface, L_w is the thickness of the butt joint, and R_l is the effective radius of the laser beam, Φ is the inclination angle of laser head, D_{la} is the laser-to-arc stand-off distance. Φ is set at 0° and D_{la} is set at 8 mm in this study, and v denotes the welding speed.

The boundary conditions at the sample surfaces are given by:

$$-k\frac{\partial T}{\partial n}=h_c(T-T_\infty) \quad (7)$$

where n is the normal outward vector to the surface of specimen, T_∞ is the room temperature, and h_c is the heat transfer coefficient of sample surface.

2.2. Mechanical analysis of hybrid laser-GMA welding

The mechanical analyses of hybrid laser and arc welding are similar to the previous studies on the electric arc welding and laser welding. The stress and distortion of weld are mainly caused by the thermally-induced expansion and shrinkage and the accompanying phase transformation-induced volume change. By considering the elastic-plastic material properties, stress and strain relationships in the hybrid laser-GMA weld are given by [38]:

$$\{\sigma\}=[D_e]\{\varepsilon^{el}\} \quad (8)$$

where {σ} denotes the stress vector, [D_e] denotes elastic stiffness matrix, and $\{\varepsilon^{el}\}$denotes elastic strain vector expressed by [30]:

$$\{\varepsilon^{el}\}=\{\varepsilon\}-\{\varepsilon^{th}\}-\{\varepsilon^{pl}\}-\{\varepsilon^{\Delta V}\}-\{\varepsilon^{Trp}\} \quad (9)$$

where {ε} is the total strain vector, $\{\varepsilon^{th}\}$ is thermal strain vector, $\{\varepsilon^{pl}\}$ is the plastic strain vector, $\{\varepsilon^{\Delta V}\}$ is strain vector due to phase transformed induced volume change, and $\{\varepsilon^{Trp}\}$ is strain vector due to phase transformation plasticity which is ignored in this study. The boundary conditions taken into consideration in the mechanical analysis assume that one edge of the butt joint is fixed, and the other one is only transversely shrinkage free.

2.3. Grain size prediction model by Monte Carlo method

Grain size evolution and phase transformation play a critical role in deciding the final mechanical properties of weld, and it is necessary to involve those factors in the thermo-mechanical modeling of different welding processes. Many good trials have been performed to numerically predict the grain growth in the fusion zone and heat affected zone for solidification and re-crystallization, respectively, which includes Monte Carlo (MC) model [47-50], phase field (PF) method [51], and cellular automaton (CA) model [52] combined with finite element and finite difference analyses. Here a brief introduction of MC model to predict the grain growth in HAZ will be performed. The detailed description of phase field method and cellular automaton model-based numerical prediction of grain growth in welds can be found in literature [53].

Monte Carlo model-based grain growth prediction generally includes the following several steps: (1) The representation of the considered material in a two-dimensional (2-D) or 3-D of cells, as shown in Figure 2a. The content of each cell stands for its crystallographic orientation. A region consisting of a set of consistently distributed cells with the same orientation value denotes a grain. The grain boundaries are identified by a curve in 2-D matrix or a surface in 3-D matrix between the separate planes or volumes with different orientations. (2) After selecting the matrix type and defining it by an initially random number, the free energy of a cell in the matrix with its specific crystallographic orientation based on its surroundings will be identified. (3) Randomly selecting a new crystallographic orientation for each cell. (4) Calculating the free energy of the new coming element with the new crystallographic orientation, the two energy values and their difference are then calculated. A new grain orientation that will minimize the free energy is generated with the selected transition probability [54]. These four steps will be reiterated many times at random positions in the matrix. The ultimate product is a microscopic simulation of the free energy decaying in the system, which is in fact the main driving force for grain growth. The Hamiltonian demonstrates the interaction among the closest neighbors in a particular cell, which stands for the grain boundary energy and can be calculated as follows [54]:

$$G = -J\sum_{nn}(\delta s_i s_j - 1) \qquad (10)$$

where, J is a positive constant that characterizes the scale of the grain boundary energy; s_i is one of the possible orientations, which is set between 1 and q, in the ith cell of the matrix; s_j is the crystallographic orientation of one neighboring cell; nn is the amount of neighboring cells for each element. In the Monte Carlo model, a Moore neighborhood model is selected (see Figure 2b); therefore, nn=8. δ_{ab} is the Kronecker-delta, which equals 1 when two elements in the matrix are equal, i.e., a=b, and 0 for others. As a consequence, neighboring cells with different orientations contribute a free energy J to the system and 0 with the same orientation. The total number of grain orientations, q is set at 40 in this model since it is known that the grain-growth exponent becomes almost independent of q when its value is larger than 30 [54].

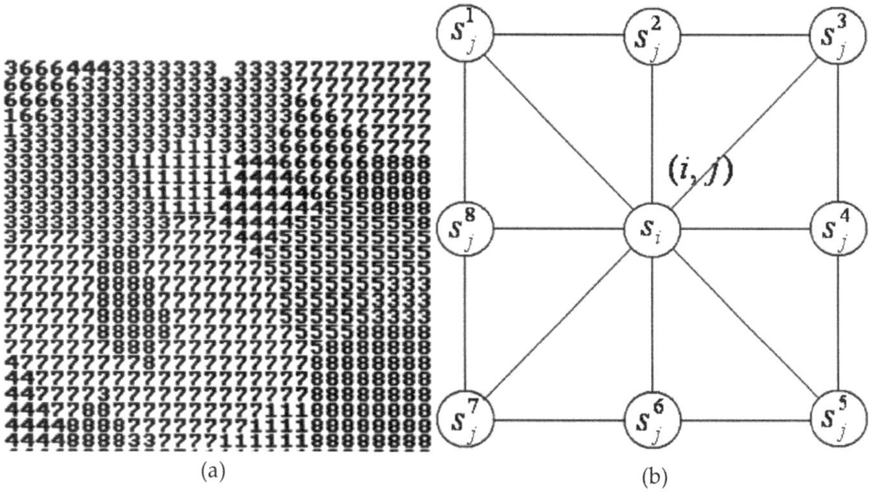

Figure 2. The grain structure in MC model with Moore neighborhood [51]
(a) The grain structure represented by a 2D square (b) Moore neighborhood

The transition probability W is given by [55]:

$$W = \begin{cases} \exp(-\frac{\Delta G}{k_b T}), & \Delta G > 0 \\ 1, & \Delta G \le 0 \end{cases} \tag{11}$$

where ΔG is the change of the free energy because of the orientation alteration, k_b is the Boltzman constant (k_b=1.38×10^{-23} m^{2}kg/s^2/K), and T is the temperature. Therefore, the speed of the moving segment can be calculated by [55]:

$$v_i = C_1[1 - \exp(-\frac{\Delta G}{k_b T})] \tag{12}$$

where C_1 is the boundary mobility. For a continuous grain growth, the final grain size can be calculated by using the following equation [55]:

$$L^n - L_0^n = f(T)t \tag{13}$$

where L and L_0 are the final and initial mean grain sizes respectively calculated by the linear-intercept method, n is the grain growth exponent and set at 1.84 in this study [56]. $f(T)$ is usually computed as an Arrhenius-type equation [55], and its expression is shown as follows:

$$f(T) = K\exp(-\frac{Q}{R_g T}) \tag{14}$$

where K is the pre-exponential coefficient, Q is the activation energy for grain growth, and R_g is the universal gas constant. In this study, K is set at 3.01×10^{-2}, and Q is set at 1.7×10^{5} J/mol [56].

The Monte Carlo method has been proven to be an effective way to simulate grain growth with slow and uniform temperature evolution such as metal casting [53]. In the hybrid welding by laser and arc, there exists a dynamic thermal process with rapid heating and cooling resulting in an abrupt temperature gradient in the HAZ and fusion zone. In the simulation of microstructure evolution, three techniques—such as the atomistic models, a grain boundary migration (GBM), and experimentally data-based (EDB) models—have been presented [53, 58-60]. The atomistic model used to be only applied to small numbers of atoms like nanocrystals [60], and it is not suitable for a large-scale FZ or HAZ simulation. The GBM model can be a good alternative for grain-growth simulation when the isothermal grain-growth kinetics is not accessible. However, the physical properties of the material in this model have to be known, and the grain size is assumed to be proportional to the square root of time. The EDB model can avoid these shortages and be applied to simulate the grain growth in HAZ when the isothermal grain-growth kinetics of metal are available. Therefore, it can be used to relate time and temperature to the Monte Carlo simulation-time step t_{MCS} [55]:

$$L = K_1\lambda(t_{MCS})^{n_1} \tag{15}$$

where λ is the discrete grid-point spacing in the Monte Carlo model, and K_1 and n_1 are constants. Through the regression computation of t_{MCS} and the Monte Carlo model predicting the grain size, the values of K_1 and n_1 are obtained as 0.715 and 0.477, respectively [56]. In the EDB model, the relationship between the t_{MCS} and the real time-temperature T(t) is further given by [60]:

$$(t_{MCS})^{nn_1} = (\frac{L_0}{K_1\lambda})^n + \frac{K}{(K_1\lambda)^n}\sum(\exp(-\frac{Q}{R_g T(t)})\Delta t_i) \tag{16}$$

where n is the grain growth exponent, T(t) is the mean temperature in a time interval Δt_i. Therefore, at any given monitoring location where the temperature is known as a function of time, t_{MCS} can be related to the real time t, which is $\sum\Delta t_i$. The t_{MCS} values at different locations calculated through Eq. (16) cannot be straightly applied to the Monte Carlo model since the selection of a grid point for updating the orientation number is stochastic in the Monte Carlo approach. Consequently, the probability of choosing each grid point is the same as in the traditional MC calculations. However, grains usually grow at higher rates in the HAZ region of higher temperature, where a sharp temperature gradient is present. This fact has to be included in any practical grain-growth calculation scheme. One solution is to develop a scheme in which grain orientations at higher-temperature locations (higher t_{MCS} locations) are updated with a higher frequency by considering a probability gradient. In other words, the site-selection probability changes with location. The larger the t_{MCS} at a location, the higher the corresponding site-selection probability [57, 60]:

$$P = t_{MCS} / t_{MCSMAX} \tag{17}$$

where t_{MCSMAX} is the maximum of t_{MCS} in the simulation domain.

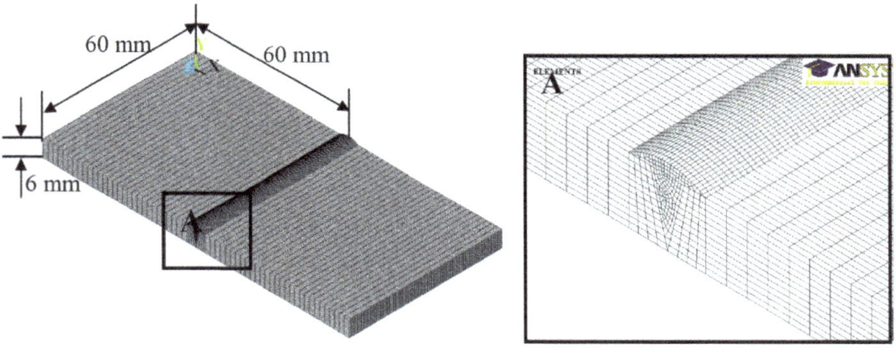

Figure 3. Finite element meshes for hybrid laser-GMA weld

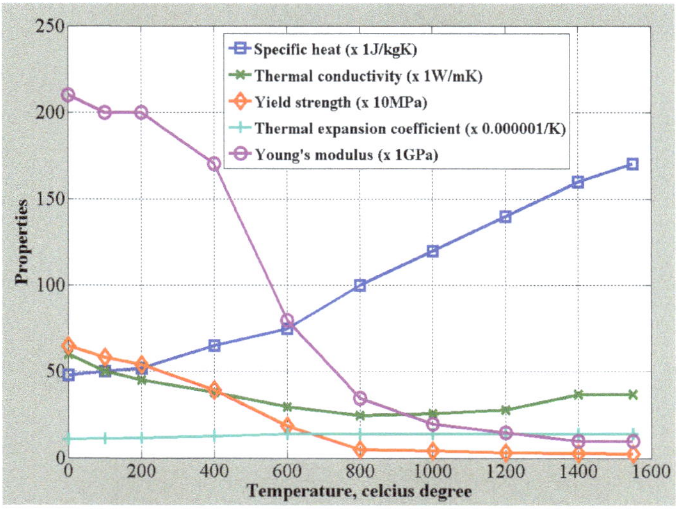

Figure 4. Temperature-dependent thermal and mechanical properties of A514 steel

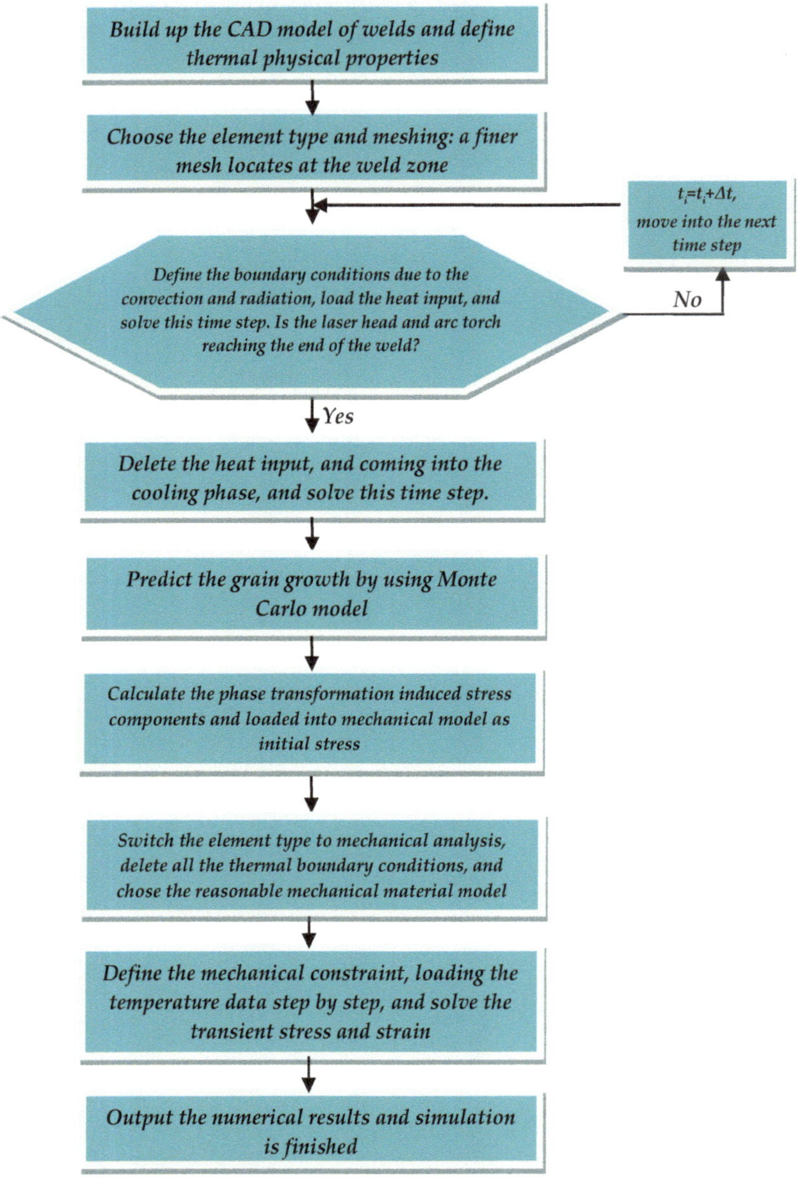

Figure 5. Numerical procedure performed in the thermo-mechanical FE analysis

2.4. Implementation of numerical procedure by using APDL

An uncoupled finite element thermo-mechanical model with considering the grain growth in the HAZ by Monte Carlo model is developed to study the temperature distribution and residual stress field in the hybrid laser-GMA welding process. A non-uniform mesh is selected in which a finer mesh is used in the weld bead and a course mesh is defined in the other region of the welded coupons (see Figure 3). A temperature-dependent material property is used in the numerical modeling, as listed in Figure 4. A thermal FE analysis is performed to achieve the temperature field of hybrid laser-GMA welding process. The wire feeding into the groove to form the weld bead has been simulated by using element kill-and-birth approach which is available in ANSYS software. The achieved geometrical size of the weld zone could be compared to the micrographs of the weld cross-section obtained by an optical microscope, by which the accuracy of thermal analysis can be verified. The numerical model is then transferred to mechanical analysis module in ANSYS by switching the element type from thermal to structural. The corresponding constraints are exerted into sample boundaries. The achieved temperature histories are subsequently loaded into the mechanical model step by step to calculate the displacement, stress/strain of the sample due to the thermal expansion or shrinkage during the welding process. A bilinear hardening principle is introduced in this study to simulate the material plastic behavior. Von Mises criterion is used for considering the yield behavior of the sample material. Figure 5 shows the numerical procedure used in this study.

3. Experimental set-up

A 4 kW fiber laser and a GMAW torch are mounted on a robotic arm to perform the hybrid welding of a thick plate for a butt joint configuration. The photo of an experimental set-up for welding is shown in Figure 6. In order to control the gap thickness, spot welding is performed at both ends of the joint before the formal welding starts. After the welding process is completed, the achieved sample will be cut into standard tensile coupon for tensile test, the left parts will be mounted for polishing, and etching to test the micro-hardness and microstructure. The base metal is high strength steel A514; its chemical composition is listed in Table 1. The wire material is ER100S-G. Its diameter is 0.9 mm, and its chemical composition is listed in Table 2. Residual stresses were measured by using the X-ray diffraction technique. Before performing the residual stress measurement, the measurement areas were cleaned by using polishing paper.

	C	Mn	P	S	Si	Cr	Mo	V	Ti	B
Min.	0.12	0.70			0.20	0.40	0.15	0.03	0.01	0.0005
Max.	0.21	1.00	0.035	0.008*	0.35	0.65	0.25	0.08	0.04	0.005

Table 1. Chemical composition of A514 [61]

Cu % max	Ni % max	Fe % max	Mn % max	Mo % max
<0.5	<5.0	Balance	<5.0	0.50

Table 2. Chemical composition of ER100S-G [62]

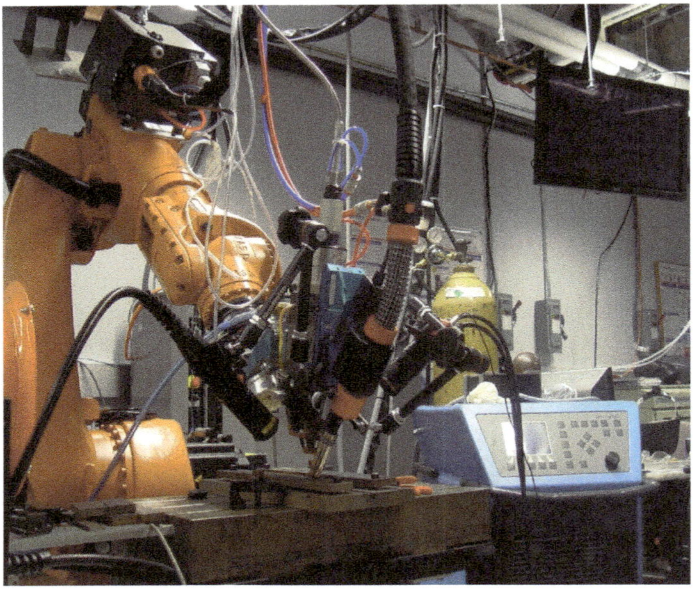

Figure 6. Photo of hybrid laser-GMA welding system used in this study

4. Results and discussion

4.1. Experimental results

Figure 7 shows the surface morphology of A514 weld obtained by hybrid laser-GMA welding, and Figure 8 presents the corresponding cross-sectional view of weld. It can be

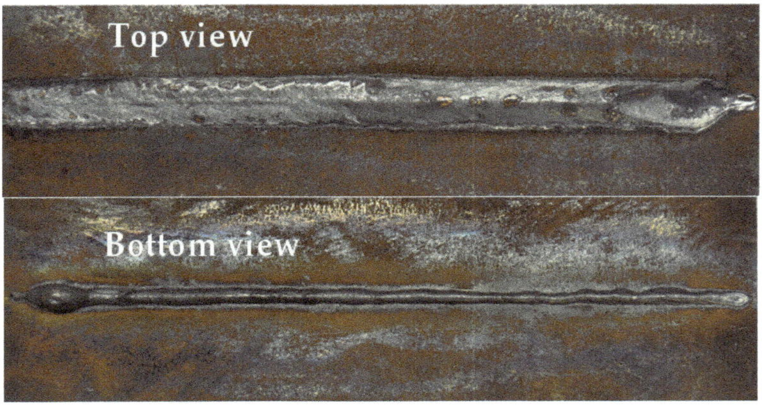

Figure 7. A514 sample achieved by 3.8 kW laser and 159A×30.5V GMAW with a welding speed of 12 mm/s and 8-mm stand-off distance between the laser and arc

seen that a sound weld quality is achieved by using hybrid laser-GMA welding and the welding-induced cracks can be effectively mitigated by reasonably selecting filler wire matched with base metal. Also, a sound mechanical property can be obtained. Figures 9a and b show the hardness distribution in the weld obtained by hybrid laser-GMA welding.

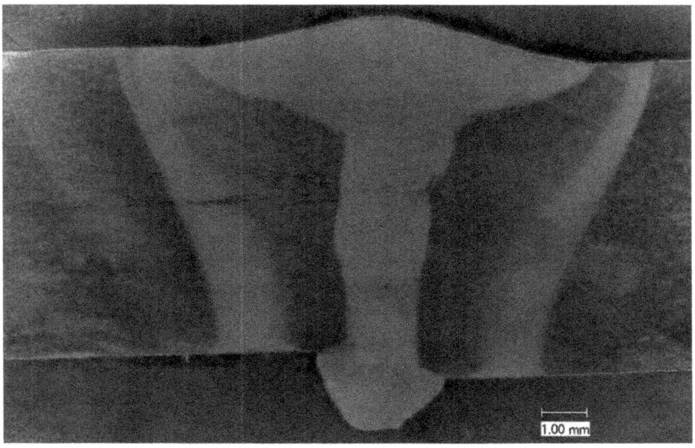

Figure 8. Cross-sectional view of A514 weld sample achieved by 3.8 kW laser and 159A×30.5V GMAW with a welding speed of 12 mm/s and 8-mm stand-off distance between the laser and arc

4.2. Thermal results and grain size prediction in the HAZ of hybrid laser-GMA welding

Finite element analyses results show the temperature at the each location of weld with respect to the welding time. The temperature evolution curves at position A, B, and C are shown in Figure 10, where position A is located at the center of weld, position B is at the heat-affected zone, and position C is in the base metal, as shown in Figure 11b. Figures 11a and b show the top and cross-sectional views of weld obtained by hybrid laser-GMA welding, respectively. It is inevitable that material heating and cooling is accompanied by phase transformation and grain size change, especially in the HAZ of weld, which is the weakest zone of the weld. In this study, a Monte Carlo-based sub-model is introduced to numerically predict the grain growth in the HAZ combined with the finite element thermal analysis. Figure 12a shows the relationship of temperature versus Monte Carlo step at Position B, and Figure 12b presents the curve of Monte Carlo step versus real time at Position B. The corresponding predicted grain size distribution in the Position B is shown in Figure 12c. The numerically predicted grain size is compared to grain size shown in the micrograph of the cross-section of weld (see Figure 12d), and a qualitative agreement is achieved.

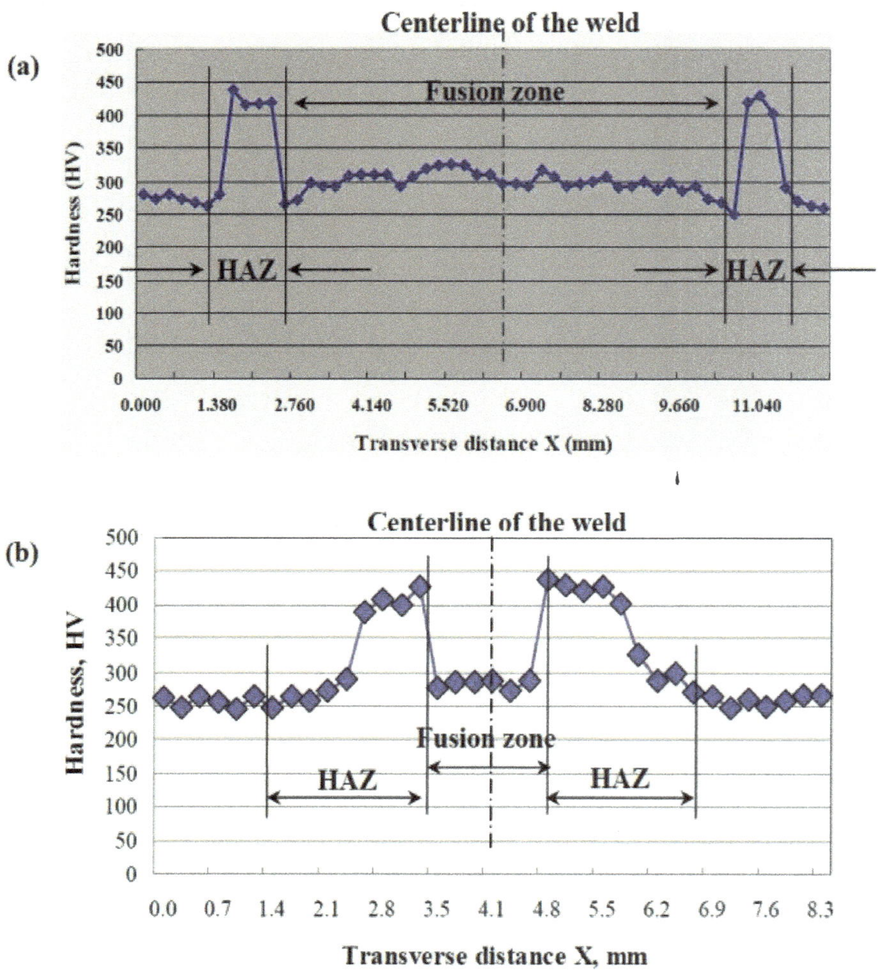

Figure 9. Hardness distribution transverse to the weld in A514 sample, (a) at the top surface and (b) at the bottom surface

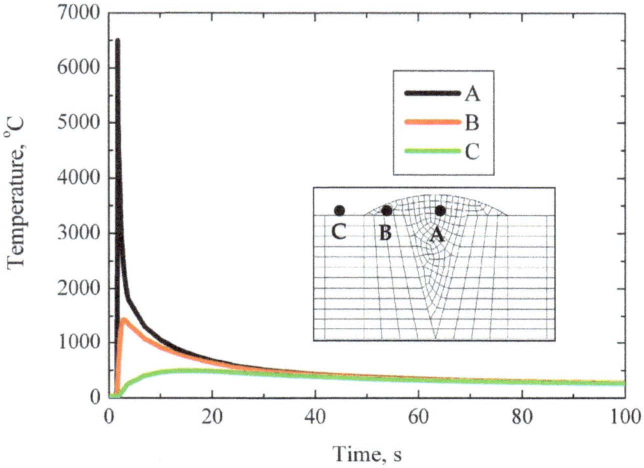

Figure 10. Temperature evolution curve at the FZ, HAZ and BM during the hybrid laser-GMA welding of A514 steel

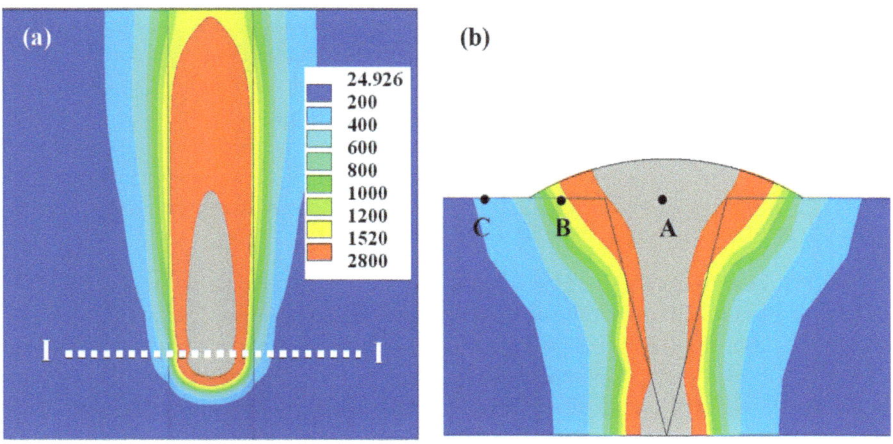

Figure 11. Numerically predicted isotherms at the top (a) and at the cross-section I-I of the weld (b) obtained by hybrid laser-GMA welding

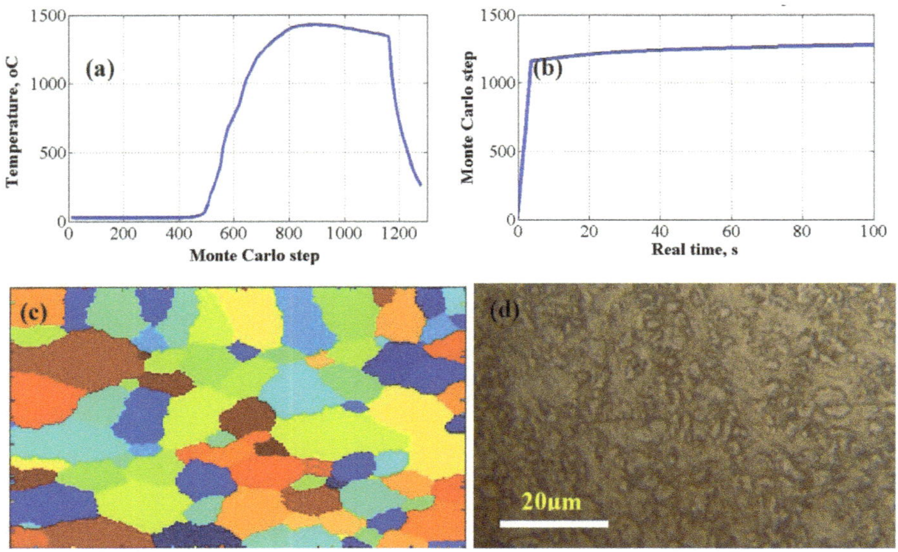

Figure 12. (a) Temperature versus Monte Carlo step, (b) Monte Carlo step versus real time, (c) numerical predicted and (d) experimentally measured grain size distributions at Position B in the HAZ of hybrid A514 weld by Monte Carlo sub-model

4.3. Thermally-induced residual stress in the hybrid laser-GMA welding

A finite element analysis is further performed to predict the thermally-induced residual stress distribution based on the previous thermal analysis results. The contours of transient stress, along thickness normal stress, longitudinal stress and equivalent residual stress of hybrid weld are shown in Figures 13a through d, respectively. It can be seen that the higher stress concentrations are located at the weld zone, which also indirectly verifies the previously experimentally obtained conclusions that the thermally induced cracks are usually generated at the weld zone, not in the base metal. The corresponding contours of stress distribution of the cross-section in the middle of weld length are shown in Figures 14a through d. A higher stress concentration is found to be located at the top region of cross-section

Figures 15a through c show residual stress distribution transverse to the weld bead at the different thicknesses in the middle of weld obtained by hybrid laser-GMA welding. It also validates the conclusion driven from Figure 14 that high tensile transverse and longitudinal stresses are located at the top and bottom regions of the weld center, high compressive transverse stresses are located at a half of the weld thickness. From the equivalent stress distribution point of view, the peak value of stress concentration is a little lower than that at the top and bottom of the weld. Figure 16 shows residual stress distribution transverse to the weld bead at the different locations along the top surface of weld obtained by hybrid laser-GMA welding. Figure 17 also shows residual stress distributions along the central line

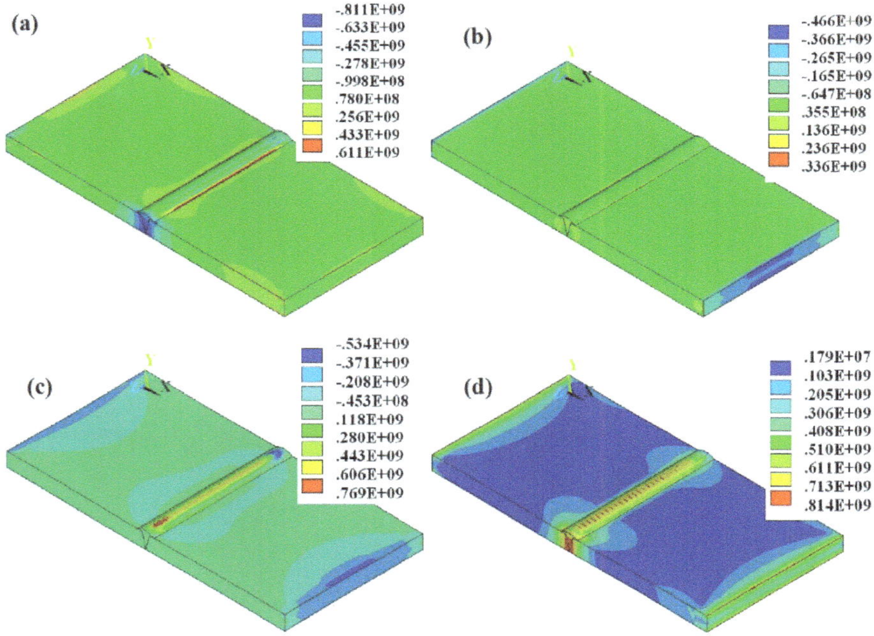

Figure 13. (a) Transverse stress SX, (b) along-thickness normal stress SY, (c) longitudinal stress SZ, and (d) von Mises equivalent residual stress SEQV mapping of weld by hybrid laser-GMA welding (unit of stress in the contour is Pa)

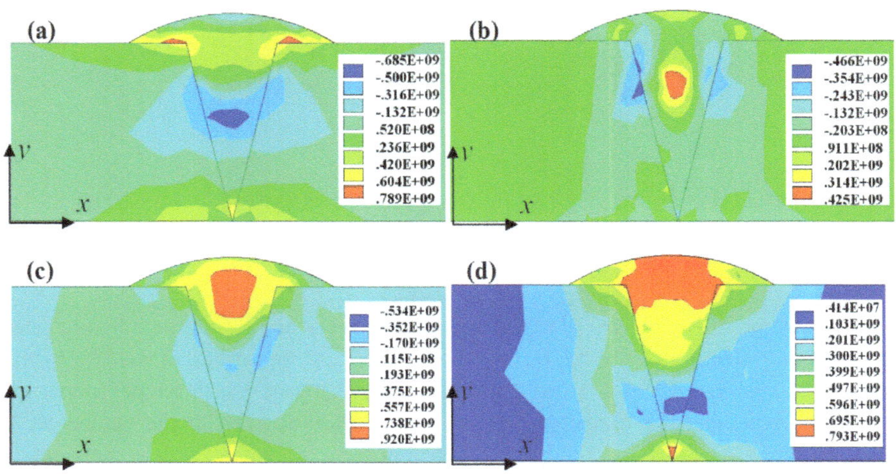

Figure 14. (a) Transverse stress SX, (b) along-thickness normal stress SY, (c) longitudinal stress SZ, and (d) von Mises equivalent residual stress SEQV mapping of cross-section of weld by hybrid laser-GMA welding (unit of stress in the contour is Pa)

at the top surface of weld centerline achieved by hybrid laser-GMA welding. It is clear that stress distribution across the weld bead is uniform along the weld; only a little drop in stress magnitude exists at the both ends of weld.

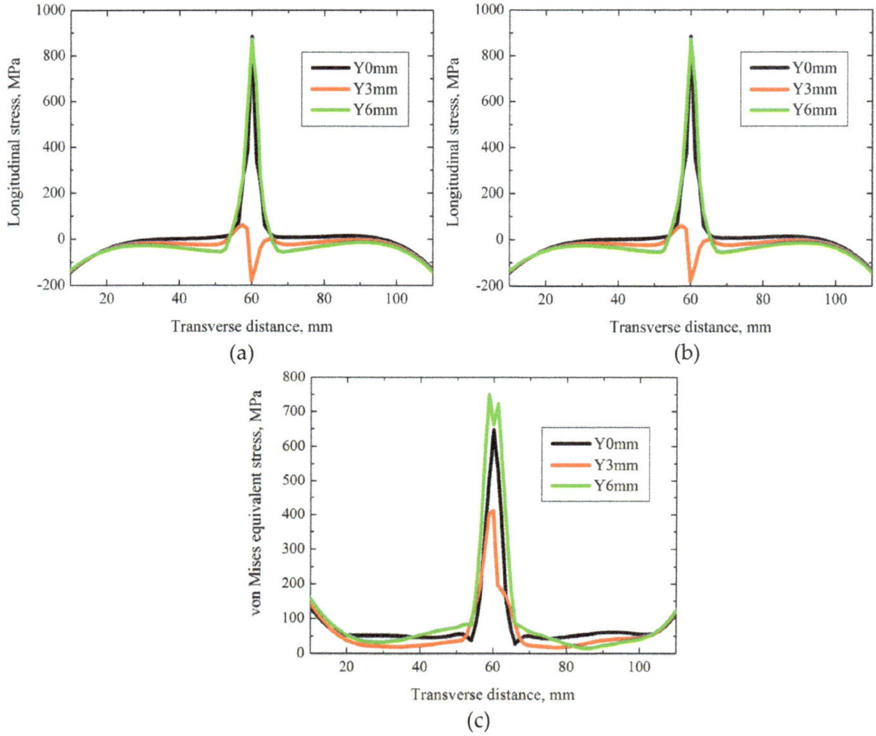

Figure 15. Residual stress distribution transverse to the weld bead at the different thicknesses in the middle of weld (z=30 mm) obtained by hybrid laser-GMA welding

Figure 18 shows a comparison of experimentally-measured and FE numerically-predicted residual stress distributions at the middle of weld length of top surface of weld by hybrid laser-GMA welding. There is a qualitative agreement between the developed numerical model and experimentally measured stress by an X-ray diffraction technique. Figures 19, 20 and 21 show the transverse, longitudinal, and equivalent stresses as well as temperature evolution with time at the positions A, B and C, respectively. It can be seen that the peak values of transient stresses at positions A, B, and C are sensitive to the temperature curve at the position A but not at the positions B and C.

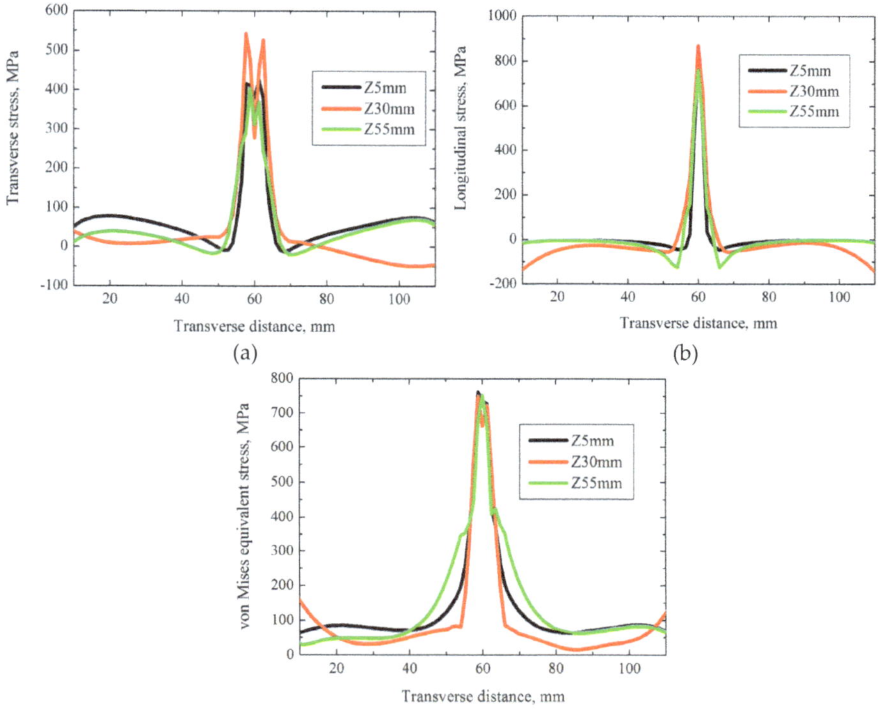

Figure 16. Residual stress distribution transverse to the weld bead at the different locations along the top surface of weld obtained by hybrid laser-GMA welding

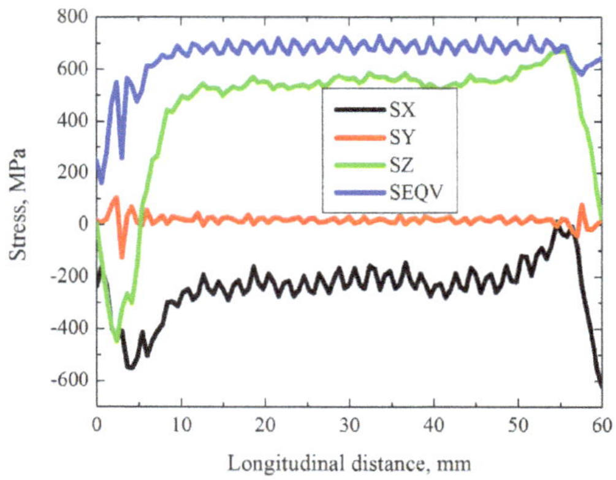

Figure 17. Residual stress distributions along the longitudinal direction of weld at the top surface of weld centerline achieved by hybrid laser-GMA welding

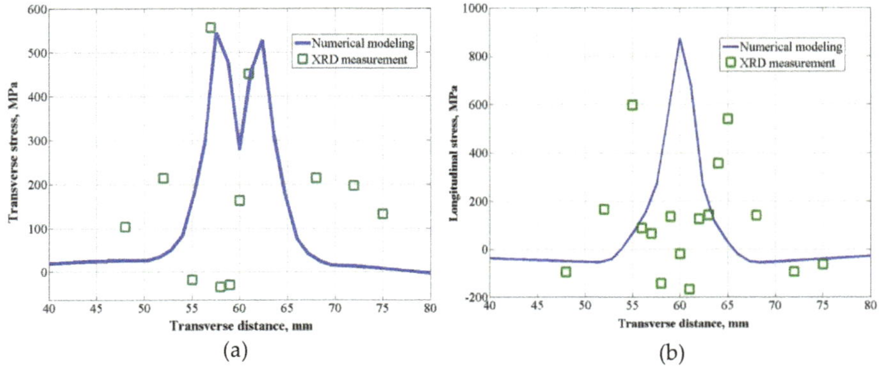

Figure 18. Comparison of experimentally-measured and FE numerically-predicted residual stress distribution at the middle of weld length of top surface of weld by hybrid laser-GMA welding

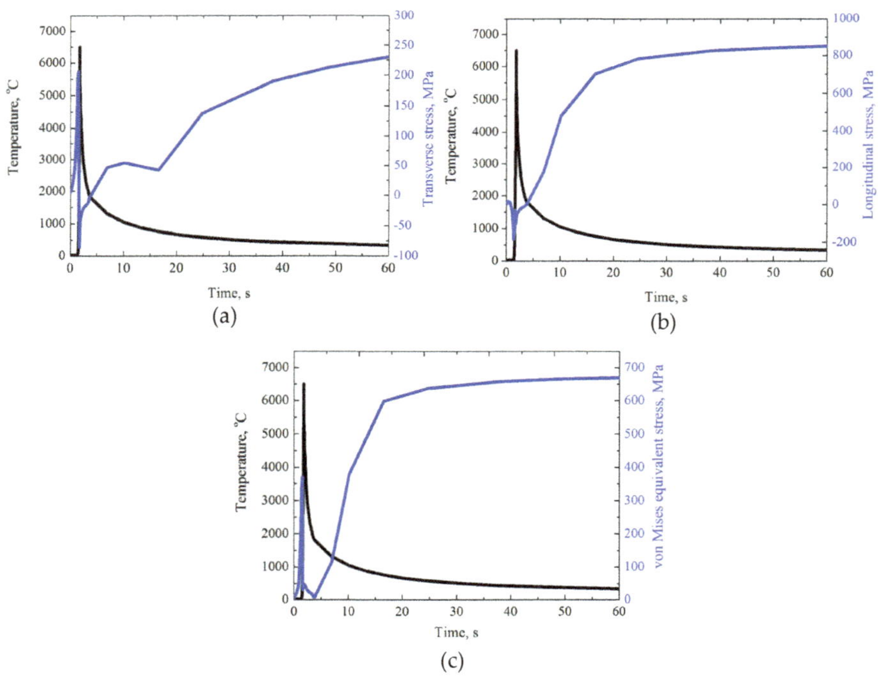

Figure 19. Stress and temperature evolution curves at the Position A in FZ of hybrid A514 weld

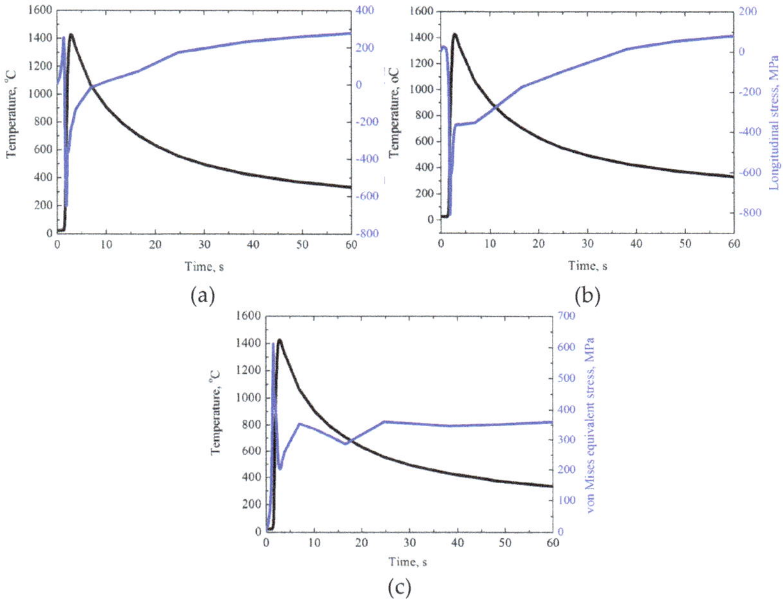

Figure 20. Stress and temperature evolution curves at the Position B in HAZ of hybrid A514 weld

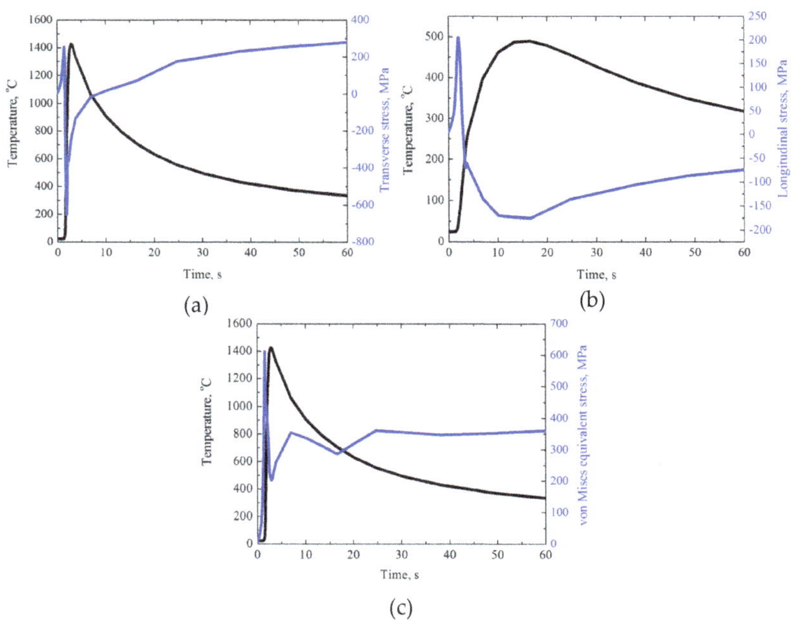

Figure 21. Stress and temperature evolution curves at the Position C in BM of hybrid A514 weld

5. Conclusions

The hybrid approach combining laser and arc has unique features which definitely help to achieve a better weld quality and to improve the production efficiency. A brief overview of modeling of hybrid laser-arc welding process has been presented in which heat transfer, fluid flow, residual stress and distortion, as well as phase transformation in the weld zone and heat affected zone, are involved. As a case study, a 3D thermo-mechanical finite element model is developed to study the thermally-induced residual stress in the hybrid laser-GMA welding process. A Monte Carlo model is introduced to numerically predict the grain growth in the heat affected zone of weld combined with finite element thermal analysis, which can be used to further understand the welding mechanisms of hybrid laser-GMA welding as well as other welding technology.

Author details

Fanrong Kong and Radovan Kovacevic*
Research Center for Advanced Manufacturing, Lyle School of Engineering,
Southern Methodist University, Dallas, TX, USA

Acknowledgement

The authors would like to thank Ph.D. candidate, Mr. Junjie Ma for his valuable help in the experiment and technical discussions and Research Engineer, Mr. Andrew Socha for his support in the experimental set-up. This work was financially support by NSF's Grant No. IIP-1034652

6. References

[1] C. Roepke, S. Liu, S. Kelly, and R. Martukanitz. Hybrid laser arc welding process evaluation on DH36 and EH36 steel. Welding Journal, 2010, 89: 140s-150s.

[2] C. Bagger and F.W. Olsen. Review of laser hybrid welding. Journal of Laser Applications. 2005, 17(1): 2-14.

[3] M. N. James, D. J. Hughes, Z Chen, H Lombard, D. G. Hattingh, D. Asquith, J. R. Yates and P. J. Webster, Residual stresses and fatigue performance, Engineering Failure Analysis, 2007,14: 384-385.

[4] P. Kah. Usability of laser-arc hybrid welding processes in industrial applications. Ph.D. Dissertation, Lappeenranta University of Technology, 2011.

[5] G. Turichin, E. Valdaytseva, I. Tzibulsky, A. Lopota, O. Velichko. Simulation and technology of hybrid welding of thick steel parts with high power fiber laser. Physics Procedia, 2011, 12: 646-655.

[6] M.H. Cho and D.F. Farson. Simulation study of a hybrid process for the prevention of weld bead hump formation. Welding Journal, 2007, 86: 174s-179s.

* Corresponding Author

[7] F.G. Lu, S. Yao, S.N. Lou, Y.B. Li. Modeling and finite element analysis on GTAW arc and weld pool. Computational Materials Science, 2004, 29:371-378.
[8] H.G. Fan, H.L. Tsai, S.J. Na. Heat transfer and fluid flow in a partially or fully penetrated weld pool in gas tungsten arc welding. International Journal of Heat and Mass Transfer, 2001, 44: 417-428.
[9] H.J. Aval, S. Serajzadeh, A.H. Kokabi. Prediction of grain growth behavior in HAZ during gas tungsten arc welding of 304 stainless steel. Journal of Materials Engineering and Performance, 2009, 18: 1193-1200.
[10] L.J. Yang, M.J. Bibby, R.S. Chandel. Linear regression equations for modeling the submerged arc welding process. Journal of Materials Processing Technology, 1993, 39(1-2): 33-42.
[11] D. V. Kiran, B. Basu, A.K. Shah, S. Mishra, and A. De. Three-dimensional heat transfer analysis of two wire tandem submerged arc welding. ISIJ International, 2011, 51 (5): 793-798.
[12] J. Haidar. A theoretical model for gas metal arc welding and gas tungsten arc welding I. Journal of Applied Physics, 1998, 84: 3518-3529.
[13] J. Haidar. Prediction of metal droplet formation in gas metal arc welding II. Journal of Applied Physics, 1998, 84: 3530-3540.
[14] J. Haidar. An analysis of heat transfer and fume production in gas metal arc welding. III, Journal of Applied Physics, 1999, 85 3448-3459.
[15] Z.H. Rao, J. Hu, S.M. Liao, H.L. Tsai. Modeling of the transport phenomena in GMAW using argon-helium mixtures. Part I- The arc. International Journal of Heat and Mass Transfer, 2010, 53: 5707-5721.
[16] C. Heinze, C. Schwenk, M. Rethmeier. Numerical calculation of residual stress development of multi-pass gas metal arc welding. Journal of Constructional Steel Research, 2012, 72: 12-19.
[17] A. Kaplan. A model of deep penetration laser welding based on calculation of the keyhole profile. Journal of Physics D: Applied Physics, 1994, 27(9): 1805-1814.
[18] R. Spina, L. Tricarico, G. Basile, T. Sibillano. Thermo-mechanical modeling of laser welding of AA5083 sheets. Journal of Materials Processing Technology, 2007, 191: 215-219.
[19] A.P. Mackwood, R.C. Crafer. Thermal modeling of laser welding and related processes: a literature review. Optics & Laser Technology, 2005, 37: 99-115.
[20] P. Martinson, S. Daneshpour, M. Koçak, S. Riekehr, P. Staron. Residual stress analysis of laser spot welding of steel sheets. Materials & Design, 2009, 30(9): 3351-3359.
[21] W. Tan, N.S. Bailey, Y.C. Shin. A novel integrated model combining cellular automata and phase field methods for microstructure evolution during solidification of multi-component and multi-phase alloys. Computational Materials Science, 2011, 50: 2573-2585.
[22] K. Asim, J. Lee, and J. Pan. Failure mode of laser welds in lap-shear specimens of high strength low alloy (HSLA) steel sheets. Fatigue & Fracture of Engineering Materials & Structures. 2011, 35: 219-236.

[23] Y. Arai, M. Kikuchi, T. Watanabe, M. Nakagaki. Residual stress due to welding and its effect on the assessment of cracks near the weld interface. International Journal of Pressure Vessels and Piping. 1995, 63(3): 237-248.

[24] Y.P. Yang, P. Dong, J. Zhang and X. Tian. A hot-cracking mitigation technique for welding high-strength aluminum alloy. Welding Journal, 2000, 79(1): 9s-17s.

[25] A.H. Dilawari, T.W. Eagar and J. Szekely. An analysis of heat and fluid flow phenomena in electroslag welding. Welding Journal, 1978, 57(1): 24s-30s.

[26] L. A. Jones, T. W. Eager, J. H. Lang. Magnetic forces acting on molten drops in gas metal arc welding, Journal of Physics D: Applied Physics, 1998, 31: 93-106.

[27] P. Dong. Residual stress analyses of a multi-pass girth weld: 3-D special shell versus axisymmetric models. ASME Journal of Pressure Vessel Technology, 2001, 123: 207-213.

[28] P. Dong, J.K. Hong, and P. Rogers. Analysis of residual stresses in Al-Li repair welds and mitigation techniques, Welding Journal (London), 1998, 77 (11): 439-445.

[29] P. Dong, J.K. Hong, J. Zhang, P. Rogers, J. Bynum, and S. Shah, Effects of repair weld residual stresses on wide-panel specimens loaded in tension. ASME Journal of Pressure Vessel Technology, 1998, 120: 122–128.

[30] Dean Deng, Yu Luo, Hisashi Serizawa, Masakazu Shibahara, and Hidekazu Murakawa. Numerical simulation of residual stress and deformation considering phase transformation effect. Transactions of Joining and Welding Research Institute (Osaka University) 2003, 32(2): 325-333.

[31] D. Deng, H. Murakawa. Numerical simulation of temperature field and residual stress in multi-pass welds in stainless steel pipe and comparison with experimental measurements. Computational Materials Science, 2006, 37: 269-277.

[32] D. Deng. FEM prediction of welding residual stress and distortion in carbon steel considering phase transformation effects. Materials and Design, 2009, 30: 359-366.

[33] W.M. Steen, and M. Eboo. Arc augmented laser welding. Metal Construction, 1979, 11(7): 332-335.

[34] J. Zhou, H. L. Tsai. Modeling of transport phenomena in hybrid laser-MIG keyhole welding. International Journal of Heat and Mass Transfer, 2008, 51: 4353-4366.

[35] J. Hu, H.L. Tsai. Heat and mass transfer in gas metal arc welding. Part I: The arc. International Journal of Heat and Mass Transfer, 2007, 50 (5-6): 833-846.

[36] Z.H. Rao, S.M. Liao, and H.L. Tsai. Modelling of hybrid laser-GMA welding: review and challenges. Science and Technology of Welding and Joining, 2011, 16 (4): 300-305.

[37] B. Ribic, R. Rai, and T. DebRoy. Numerical simulation of heat transfer and fluid flow in GTA/laser hybrid welding. Science and Technology of Welding and Joining. 2008, 13 (9): 683-693.

[38] ANSYS Inc., ANSYS 11.0 Manual, 2007.

[39] S. Wu, H. Zhao, Y. Wang, X. Zhang. A new heat source model in numerical simulation of high energy beam welding. Transaction of China Welding Institute, 2004, 25: 91–94.

[40] V. Pavelic, R. Tanbakuchi, O.A. Uyehara, and P.S. Myers. Experimental and computed temperature histories in gas tungsten arc welding of thin plate, Welding Journal Research Supplement, 1969, 48: 259s-305s.

[41] E.A. Bonifaz. Finite element analysis of heat flow in single-pass arc welds. Welding research supplement, 2000, May: 121s-125s.
[42] J. Goldak, A. Chakravarti, and M. Bibby. A new finite element model for heat sources, Metallurgical Transaction B, 1984, 15B: 299-305.
[43] P. Peyffarth, and I.V. Krivt. Welding and Allied Processes Volume I. Laser-Arc Processes and Their Application in Welding and Material Treatment. Tailor&Francis, London, 2002.
[44] F. Kong, J. Ma, R. Kovacevic. Numerical and experimental study of thermally induced residual stress in the hybrid laser–GMA welding process. Journal of Materials Processing Technology, 2011, 211(6): 1102-1111.
[45] C.S. Wu, Q.X. Hu, J.Q. Gao. An adaptive heat source model for finite element analysis of keyhole plasma arc welding. Computational Materials Science. 2009, 46: 167-172.
[46] F. Kong, R. Kovacevic. 3D finite element modeling of the thermally induced residual stress in the hybrid laser/arc welding of lap joint. Journal of Materials Processing Technology, 2010, 210 (6-7): 941-950.
[47] M. Nosonovsky, X. Zhang, and S.K. Esche. Scaling of Monte Carlo simulation of grain growth in metals. Modelling and Simulation in Materials Science and Engineering, 2009, 17: 025004 (13pp).
[48] S.Mishra, T.Debroy. Measurements and Monte Carlo simulation of grain growth in the heat-affected zone of Ti-6Al-4V welds. Acta Materialia 2004, 52: 1183-1192.
[49] S.Sista, Z.Yang, and T. Debroy. Three-dimensional Monte Carlo simulation of grain growth in the heat-affected zone of a 2.25Cr-1Mo steel weld. Metallurgical and Materials Transactions B, 2000, Volume 31B: 529-536.
[50] F. Kong, S. Santhanakrishnan, D. Lin, and R. Kovacevic. Modeling of temperature field and grain growth of a dual-phase steel DP980 in direct diode laser heat treatment. Journal of Materials Processing Technology, 2009, 209 (18-19): 5996-6003.
[51] A. Farzadi, M Do-Quang, S. Serajzadeh, A.H. Kokabi, and G. Amberg. Phase field simulation of weld solidification microstructure in an Al-Cu alloy. Modelling and Simulation in Materials Science and Engineering, 2008, 16: 165005 (18pp).
[52] Ch.A. Grandin, J.L. Desbiolles, M. Rappaz, and Ph. Thevoz. A three-dimensional cellular automaton-finite element model for the prediction of solidification grain structures. Metallurgical and Materials Transaction A, 1999, 30A: 3153-3165.
[53] Dierk Raabe, Franz Roters, Frédéric Barlat, Long-Qing Chen, Continuum Scale Simulation of Engineering Materials: Fundamentals-Mcrostructures-Process Applications, Wiley-VCH, Weinheim, 2004.
[54] B. Radhakrishnan and T. Zacharia. Simulation of curvature-driven grain growth by using a modified monte carlo algorithm. Metallurgical and Materials Transactions A, 1996, 26(1): 167-180.
[55] S. Sista, Z. Yang, and T. Debroy, Three-dimensional Monte Carlo simulation of grain growth in the heat-affected zone of a 2.25Cr-1Mo steel weld, Metallurgical and Materials Transactions B, 2000, 31: 529-536.
[56] S. Mishra, T. Debroy, Measurements and Monte Carlo simulation of grain growth in the heat-affected zone of Ti-6Al-4V welds, Acta Materialia, 2004, 52: 1183-1192.

[57] Y.H. Wei, Y.L. Xu, Z.B. Dong, and J.L. Xia, Three dimensional Monte Carlo simulation of grain growth in HAZ of stainless steel SUS316, Key Engineering Materials, 2007, 353-358: 1923-1926.
[58] M.-Y.Li, E. Kannatey-Asibu, Monte Carlo simulation of heat-affected zone microstructure in laser-beam welded nickel sheet, Welding Journal, March 2002: 37-44.
[59] Weibin Wang, Yaowu Shi, Yongping Lei, and Zhiling Tian, FEM simulation on microstructure of DC flash butt welding for an ultra-fine grain steel, Journal of Materials Processing Technology, 2005,161: 497-503.
[60] J. H. Gao, R. G. Thompson, Real time-temperature models for Monte Carlo simulations of normal grain growth, Acta Materialia, 1996, 44: 4565-4575.
[61] http://www.ssab.com/Global/SSAB/SSAB_Americas/A514%20Grade%20S%20Brochure %2006-25-08.pdf
[62] http://www.midalloy.com/html/pdf/low_alloy/TB%20ER100S-G%20Weld.pdf

Chapter 9

Numerical Simulation of Residual Stress and Strain Behavior After Temperature Modification

Farag Soul and Nada Hamdy

Additional information is available at the end of the chapter

1. Introduction

Welding, among all mechanical joining processes, has been employed at an increasing rate for its advantages in design flexibility. In addition to that, cost savings, reduced overall weight and enhanced structural performance. The highly localized transient heat and strongly non-linear temperature fields in both heating and cooling processes cause non-uniform thermal expansion and contraction. Thus, result in plastic deformation in the weld and surrounding areas. As a result, residual stress, strain and distortion are permanently produced in the welded structures. This is particular when fabrication involves the use of thin section sheet materials, which are not inherently stiff enough to resist the contraction forces induced by welding. Transient thermal stresses, residual stresses, and distortion sometimes cause cracking and mismatching of joints. High tensile residual stresses are undesirable since they can contribute in causing fatigue failure, quench cracking and stress- corrosion cracking of welded structures under certain conditions. Welding deformation is undesirable owing to the decrease in buckling strength and injures the good appearance of structures.

In addition, it causes defaults during the assembly which result in repeating the process and productivity restriction. Correction of unacceptable distortion is costly and in some cases, impossible. In welding design, the study and analysis of welding residual stresses and distortion become necessary in critical industries such as: aerospace engineering, nuclear power plants, pressure vessels, boilers, marine sector....etc. Measurement of transient thermo-mechanical history during welding process is of critical importance, but proves to be prohibitively expensive and time consuming. It often fails to provide a complete picture of temperature and stress/strain, deformation distribution in the weldment. On the other hand, detailed experimental measurements of the residual elastic strain distributions in welded parts are typically not feasible due to significant resource (man, machine and material) consumption. Mathematical modeling for residual stress evaluation provides a resource effective method in comparison to the experimental methods when all interaction fields were correctly described in the modeling process. However, development of the modeling scheme gain demands a careful experimental data. The purpose of this chapter is to develop Finite Element models that satisfy the analysis of the behavior of transient phenomena of residual stress and distortion. That can be achieved by using different methods of the mitigation technique which work as heat transfer enhancement. Approximating the mechanisms of the transient temperature and longitudinal residual stress after temperature modification can be made. The modeled welding materials are aluminum and titanium alloys concerning flat and cylindrical shapes.

2. The origin of residual stress

Residual stresses developed during most manufactured processes involving metal forming, heat treatment and machining operations deform the shape or change the properties of a material. They arise from a number of sources and can be presented in the unprocessed raw materials, and can be introduced during manufacturing or can arise from in-service loading. (Withers & Bhadeshia, 2000; Rudd, 1992; Borland, 1994; Kandil et. al. , 2001). The residual stresses may be high enough to cause local yielding and plastic deformation on both microscopic and macroscopic level, that can severely affect component performance. For this reason it is vital that some knowledge of the internal stress state can be deduced either from measurements or modeling predictions. Both magnitude and distribution of the residual stress can be critical to the performance that should be considered in the design of a component. Tensile residual stresses in the surface of a component are generally undesirable since they can contribute to the major cause of fatigue failure, quench cracking and stress-corrosion cracking.

Compressive residual stresses in the surface layers are usually beneficial since they increase fatigue strength, resistance to stress-corrosion cracking, and increase the bending strength of brittle ceramics and glass. In general, residual stresses are beneficial when they operate in the plane of the applied load and are opposite in sense (i.e, a compressive residual stress in a component subjected to an applied tensile load). The origins of residual stresses in a component may be classified as: mechanical, thermal and chemical. Mechanically generated residual stresses are often a result of manufacturing processes that produce non-uniform plastic deformation. They may develop naturally during processing or treatment, or may be introduced deliberately to develop a particular stress profile in a component (Brien, 2000). Examples of operations that produce undesirable surface tensile stresses or residual stress gradients are rod or wire drawing (deep deformation), welding, machining (turning, milling) and grinding (normal or harsh conditions). On a macroscopic level, thermally generated residual stresses are often the consequence of non-uniform heating or cooling operations. The residual thermal stresses coupled with the material constraints in the bulk of a large component can lead to severe thermal gradients and the development of large internal stresses. An example is the quenching of steel or aluminum alloys, which leads to surface compressive stresses, balanced by tensile stresses in the bulk of the component.

Microscopic thermally generated residual stresses can also be developed in a material during manufacture and processing as a consequence of the CTE mismatch between different phases or constituents. The chemically generated stresses can develop due to volume changes associated with chemical reactions, precipitation, or phase transformation. Chemical surface treatments and coatings can lead to the generation of substantial residual stress gradients in the surface layers of the component. Nitriding produces compressive stress in the diffusion region because of expansion of the lattice and precipitation of nitrides also carburizing causes a similar effect (Littmann, 1964).

3. Sheet metal fabrication

In recent years the new vision of high-tech industrial strategy looking for minimizing the cost and increasing the strength to weight ratio of critical structure such as aerospace, marine, nuclear etc. Thin walled element fabricated by welding process can promote such effect. In the last two decades the research in welding science became more vital than other manufacturing sciences in many industrial sectors. The development in welding technology is vastly increased, and the need for sheet metal fabrication by welding is necessary for many applications. Such those applications are rockets fuel tank and aircraft exhaust and engine mounts). Typical example of some components used in industries are shown in figure 1.

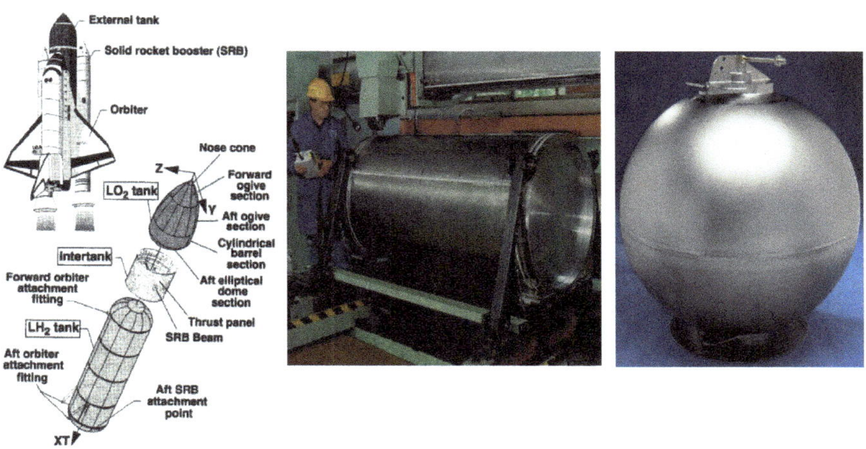

(a) (b) (c)

Figure 1. Typical thin welded component used in aerospace: (a,b,c) type of rocket fuel tank.

Other examples of sheet metal welding applications are in ships and airplanes structures. Welding can be successfully alternated to other connection processes such as riveting. Riveting has a low joint efficiency, thus a structure designed to be riveted, generally requires more materials even if joint itself is not complex. The area adjacent to the rivets is also a site of high residual stresses and stresses concentration, generating a favorable environment for stress corrosion and initiation of fatigue cracks. Welding is considered a challenge for replacing riveting in the future for airbus industries. The advantage of welding compared to the mechanical fastenings, includes cost, time and ease of fabrication. Also, simpler, lighter component design, and better joint efficiency. Laser beam welding may become a new joining technique for aircraft fuselage shells for the A318 and A380 airbuses (Airbus, 2000). Laser beam welding of the longitudinal stiffeners of the skin panels of a commercial aircraft fuselage may reduce the weight of the panels to 80% against the riveted structure production Figure 2. Welding is performed simultaneously from both sides of the stiffener. The main problem with welding is to keep the distortion as low as possible (especially the transverse deflections) and to reduce the residual stresses (especially the longitudinal tensile stresses).

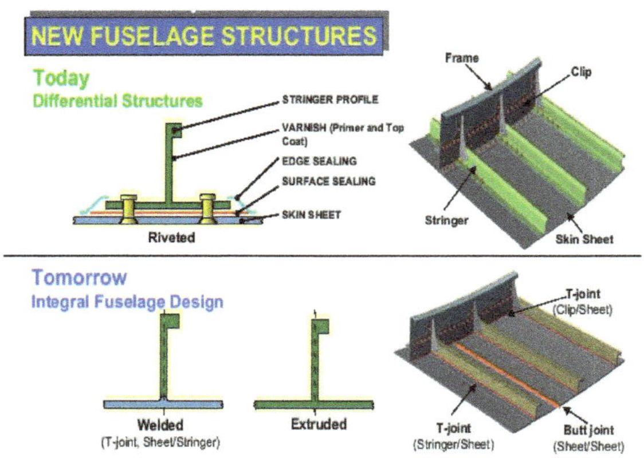

Figure 2. The fuselage structure

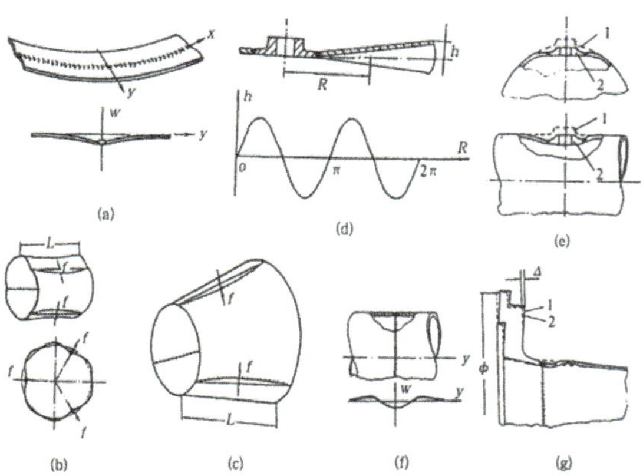

Figure 3. Typical distortion on welded sheet: longit. welds (a,b,c), circular welds (d,e), and circumference welds (f,g).

The twin problem of stress and distortion due to welding, using conventional fusion welding process, have presented fabrication problems for many years especially, in aerospace industry. Consequently, manufacturing have applied additional time consuming and costly operations. This is to remove distortions or to relieve stresses after welding that avoid variable quality problem. This is particularly so when fabrication involves the use of thin sheet section, typically in the range of 0.5-4 mm in thickness. Recently advanced aluminum and titanium alloys became very attractive materials in sheet metal fabrication in high tech- industries due to their strength to weight ratio. However, as the thickness decreases, the sheet materials are not stiff enough to resist the contraction forces induced by welding. In the aerospace fabrication, the longitudinal residual stress in addition to distortion such as longitudinal bending and buckling become more substantial. The manufacturing of welded cylinders, cones or other shaped shell elements in aerospace industries are always accompanied by distortions. Figure 3, shows typical example of those distortions produced by longitudinal or circumferential welding when fusion welding process is applied (Guan, 1999).

4. Welding and its focus research

Historically, electric arc welding appeared in the late 19th century, shortly after electric power became available. Other fusion welding processes were recently developed such as electron beam welding (EB) and laser beam welding (LB), which introduced new generations in the welding equipments and processes. Failure of welded bridge in Europe in the 1930 and the American liberty ships in world war II make the concern of welding mechanics is important. The welding researches carried out since that time, then vastly developed. Analyses of these subjects require complex computation; therefore, most early studies were primarily empirical or limited to the analysis of simple cases. A number of studies have been performed on the calculation of residual stress and deformation, but few are useful in the design process. Most of these calculation methods are limited for special purposes, e.g. (Hansen, 1968), or too complicated to be used in design and production, e.g. (Okerblom, 1955). (Puchaicela, 1997), exhibit some empirical formulas for general distortion modes in welded steel structures. Most of the formula are based on measurements of deformation and strains and cannot take in to account what really happens in the (HAZ) when the welding is cooling down.

Attempts were made to investigate changes in the bending deformation due varying pertinent parameters, see (Masubuchi, 1980). The investigation cover many practical aspects of process, but based mainly on experimental data. It is difficult to analyze the process, which is highly non-linear and involves plastic deformations and high temperatures varying in both time and space. Several reviews are, however, available such as the extensive review done by (Masubuchi, 1980; Radaj, 1992; Goldak et al., 1992) and the recent one by (Lindgren, 2001a, 2001b, 2001c). With the advancement of modern computers and computational techniques (for example, the finite-element and finite-difference method), a renewed effort has been made in recent years to study residual stresses and related phenomena. Therefore, it is now possible using computer programs to simulate the transient thermal stresses and metal movement during welding, as well as, the residual stresses and distortion that remain after welding is completed as found by (Tall, 1991 ; Hibbit & Marcel, 1972; Muraki et. al., 1975; Rybicki et. al., 1978).

4.1. The finite element method in historical perspective

The history of the finite element method is about hundred years, but it took another fifty years before the method became useful. In 1906 a paper was presented where researchers suggested a method for replacing the continuum description for stress analysis by a regular pattern of elastic bars. Later (Courant, 1943) proposed the finite element method, as we know it today, the residue of section cover the developed FEM during the years and the motivation computer tools based on FEA, and the use for those packages for simulation of welding phenomena by the researcher. Many authors have utilized the commercial finite element codes ABAQUS & ANSYS enhanced with user subroutines, to model weld simulations with great success (Dong et. al., 1998; Feng et. al.,1996; Karlsson et. al., 1989; Grong & Myhr, 1993; Voss et. al., 1999; Tenga & Linb, 1998; Li et. al.,2004). The finite element code ADINAT was used by (Karlsson & Josefson, 1990), while other authors (Junek et. al., 1999; Vincent et. al., 1999; Dubois et. al., 1984) have utilized SYSWELD to perform weld simulations. Welding is complex industrial process which often requires several trials before it can be done right. The welding is carried out by skilled workers, but in the past few years automated machines and robots are sufficiently used in the small and large industrial scales. To obtain the expected productivity through mechanization, high precision of the assembled parts must be kept. Therefore, the predictability is important in such aerospace, shipyards, nuclear and automobile industries. In order to produce a high-quality product, the accuracy control should be kept through the whole assembly line. The concept of accuracy control should be incorporated in the structural design, so that the designer can produce a better design accounting for the geometric inaccuracy. Numerical modeling and simulation of welding are a difficult and challenging problem due to the complex mechanisms involved. The wide range of problems concerned can be generalized into the fields shown in figure 4. The fields are strongly interrelated and couple in almost every possible manner. Establishment of a model accounting for all the physical effects and their couplings would be an incomprehensibly large and complex task. Hence, welding research is characterized by choice of a focal area for thorough analysis and use of suitable assumptions. Thus, the 'art' of welding research is to choose simplifications without invalidating the research focus.

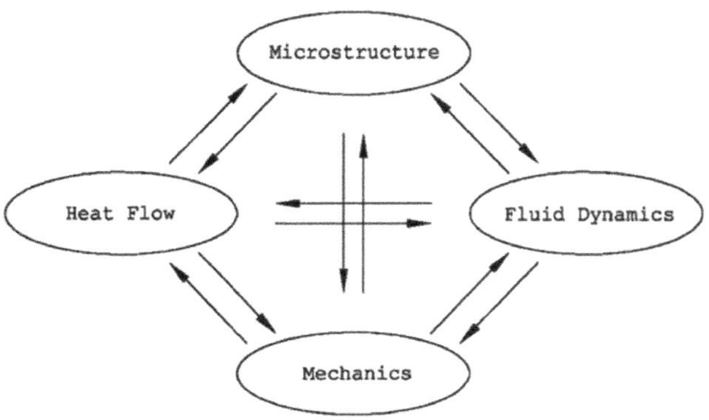

Figure 4. Coupled fields in welding analysis.

4.2. Welding induced residual stresses

Stresses arising during the welding process are referred to as internal or locked-in stresses (Radaj, 1992; Gatovskii & Karkhin, 1980). Residual stresses can be defined as those stresses that remain in a material or body after manufacture and processing in the absence of external forces or thermal gradients. Residual stress measurement techniques invariably measure strains rather than stresses, and the residual stresses are then deduced using the appropriate material parameters such as Young's modulus and Poisson's ratio. Often only a single stress value is quoted and the stresses are implicitly assumed to be constant within the measurement volume, both in the surface plane and through the depth. Residual stresses can be defined as either macro or micro stresses and both may be present in a component at anyone time. Macro residual stresses, which are often referred to as Type I residual stresses, vary within the body of the component over a range much larger than the grain size. Micro residual stresses, which result from differences within the microstructure of a material, can be classified as Type II or III. Type II residual stresses are micro residual stresses that operate at the grain-size level; Type III are generated at the atomic level. Micro residual stresses often result from the presence of different phases or constituents in a material. They can change sign or magnitude over distances comparable to the grain size of the material under analysis. To summarize, Residual stresses in the material body can be classified three type as found in (Withers & Bhadeshia, 2000; Borland, 1994; Noyan). The different types of residual stress are shown schematically in figure 5.

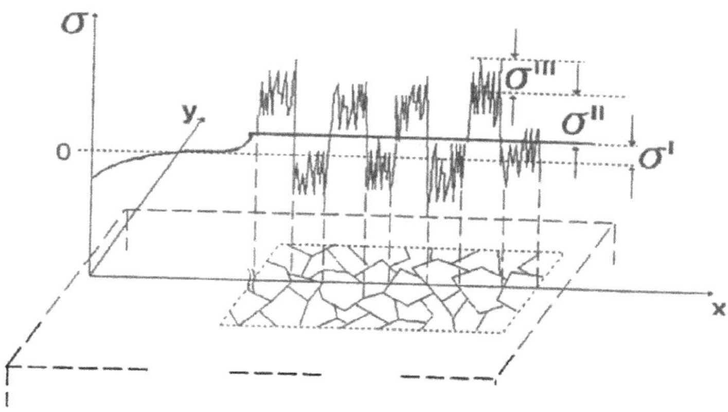

Figure 5. Categorization of residual stresses according to length scales.

Welding stresses can be classified by three characteristics: By lifetime, welding stresses can be temporary or residual, the temporary stresses do exist only in a specific moment of the non-stationary process of heating and cooling. The residual stresses can be found after the whole process of welding is completed and structure is cooled down to the room temperature. By directional the welding stresses subdivide into longitudinal (parallel to the welding direction) and transversal (perpendicular to the weld seam) and through thickness stress. By the origins, the welding stresses are subdivided into. Thermal stress, Stresses caused by the plastic deformation of the metal; Stresses caused by phase transformations.

4.2.1. *Welding induced longitudinal residual stress*

Representation of the temperature and the resulting longitudinal stress distributions that occur during welding are schematically gives in figure 6. In this example a simple bead-on-plate case is analyzed (see figure 6a). The welding arc which is moving along the *x*-axis with a speed *v*, is indicated by the arrow. Far ahead from the heat source the temperature is constant and the stress is equal to zero in all the points. Moving in the negative direction of the *x*-axis, we reach the point where the temperature starts to rise figure 6c. The points close to the weld line start to experience compression in the longitudinal direction. This deep fall changes to a fast rise of the longitudinal stress. The rate of stress change is proportional to the temperature gradient ahead of the source. It caused by the yielding point σ yield changing with temperature. As known, at elevated temperatures the material begins to soften. After some temperature (the softening temperature) the material reaches the stage when σ_Y is almost zero, and so, the points situated close to the centerline reach the softening temperature, and climb up to a zero value of the longitudinal stress. Stresses in the regions a short distance from the arc are compressive, because the surrounding metal restrains the expansion of these areas where the temperature is lower. However, stresses in the areas further away from the weld arc are tensile and balanced by compressive stresses in the areas near the weld. Going further, at some distance behind the welding arc, the temperature drops sufficiently for the material to be stiff enough to resist the deformation caused by the temperature change. Due to cooling the areas close to the weld contract and cause tensile stresses. After a certain time, the temperature change due to welding diminishes. High tensile longitudinal stresses (usually up to the yielding stress) are produced near the weld. In the regions further away from the weld, compressive stresses do exist. Figure 6d describe the final distribution of longitudinal residual stress, from literature (Masubuchi, 1980), σ_X can be approximated by:

$$\sigma_x(y) = \sigma_m[1-(y/b)^2]e^{-[1/2(y/b)^2]} \tag{1}$$

Where σ M is the maximum stress at the welding line, y is the distance from the weld line, b width of tension stress.

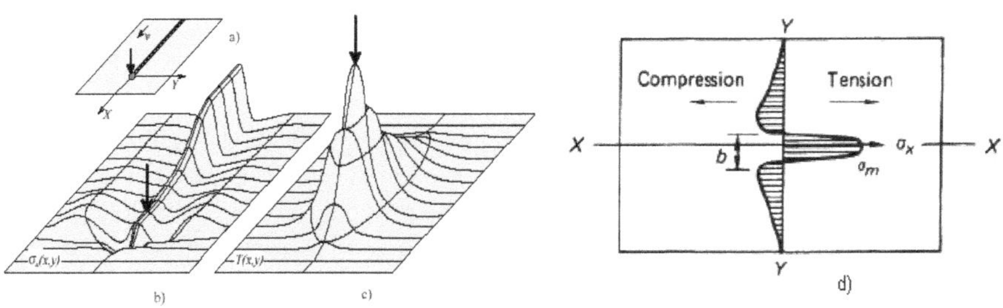

Figure 6. Schematic representation, (a, b, c) temperature vs stress during welding[Pilipenko]; d) final longitudinal residual stress.

4.3. Welding induced deformation

As in case of the stresses occurring during and after welding, welding deformation can be transient or residual. Figure 7 gives an overview of various types of welding deformations to be expected when welding plates.

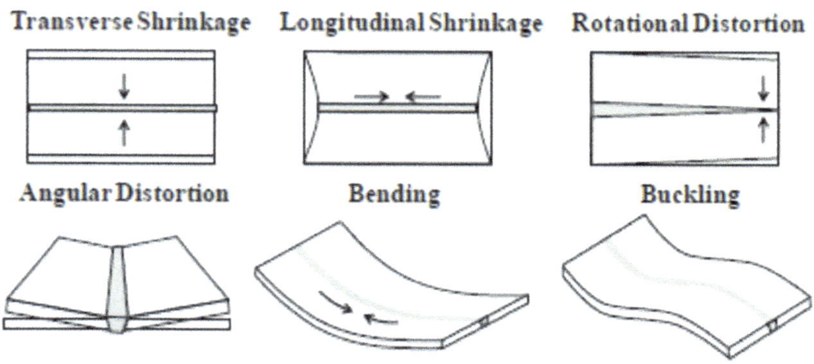

Figure 7. Different types of welding distortions. The arrows indicate the shrinkage direction of the weld metal which causes the corresponding distortion [Masubuchi 1980].

All these kinds of distortions are related to the shrinkage of the weld metal during cooling. They can be subdivided into:

1. Transverse shrinkage – shrinkage perpendicular to the weld seam
2. Longitudinal shrinkage – shrinkage in direction of the weld seam
3. Angular distortion – transverse uplift caused by a non-uniform temperature distribution in the through-thickness direction. For instance in case of butt-joints with a V-groove.
4. Rotational distortion – in-plane angular distortion due to the localized thermal expansion and contractions. Very relevant for overlap joints, for instance.
5. Bending distortion – longitudinal uplift. The same causes as angular distortion.
6. Buckling distortion – caused by compressive stresses inducing instabilities in the plates.

Driven by the need to save fuel and reduce transport and operating costs, there is a growing demand for lightweight structures, for example in the automotive and aircraft industries as well as in shipbuilding. At the very basis of this trend we find the availability of recently developed metallic alloys that actually allow the transition to more lightweight designs. Although many of the welding techniques that are currently available offer suitable material and mechanical properties, the degree of distortion remains unacceptable and residual stresses often approach component design limits. The increasing reduction in thickness will lead to a growing demand for effective solutions for residual stress and strain control during welding. An example of welding deformations in thin sheet structures can be found in the shipbuilding industry, where welding causes a typical wave-like appearance on the hull of a ship (see Figure 8). Such problem results through various stages of production have emerged as a major obstacle to the cost-effective fabrication of lightweight structures. Same situation may occurs in aerospace and aircraft assembly where the high strength to weight ratio are necessary and thin elements are used, in addition to the requirement of smooth surface to maximize hydrodynamic performance and minimize radar signature [Huang 2004]. A conservative estimation for the labor costs accumulating for post-welding distortion correction is approximately 30 % [Andersen 2000; van der Aa,2007].

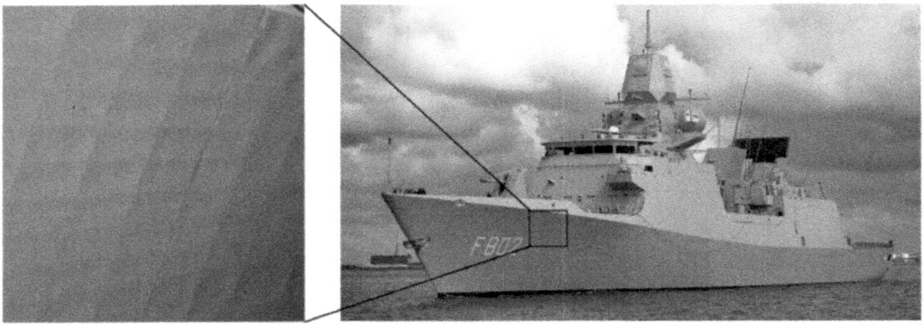

Figure 8. Ship hull defect due to distortion (van der Aa, 2007) .

4.3.1. Bending distortion in welded sheet metal

Bending distortion in sheet metal can be schematically shown in figure 9. When structure is welded, heat is supplied to melt the joint and non-uniform temperature distribution is caused owing to local melting, As a result, non-uniform thermal strains and stresses are caused and plastic strains remain after thermal cycle. Residual plastic strain around welded joints is the cause of permanent deformation. Figure 9a presents the pure cambering which may occur when the ratio of sheet length to width is high enough, but when this ratio decreases, the sheet exposes both cambering and angular distortion as shown in figure 9b. In small thickness, angular deformation is not significant because of the high homogeneity of the temperature field through the plate thickness. In some cases gradient forces countered from longitudinal shrinkage Fx, as shown in figure 9c, are more dominant and cause the cambering owing to low stiffness of thin sheet at this moment. This may occur when the longitudinal residual stress above the neutral axis of the sheet exceeds that below the neutral axis.

The possibility for minimizing or eliminating this problem is only to balance the longitudinal stress around the neutral axis otherwise, minimizing these stresses below the significant magnitude which not exceeds the component stiffness. For most welding processes, the incident surface will absorb the most energy, with the energy absorption decreasing with depth. The variation in the through-thickness heating causes variation in the longitudinal stresses through the plate thickness. This generates a bending moment, which causes the bending distortion mode.

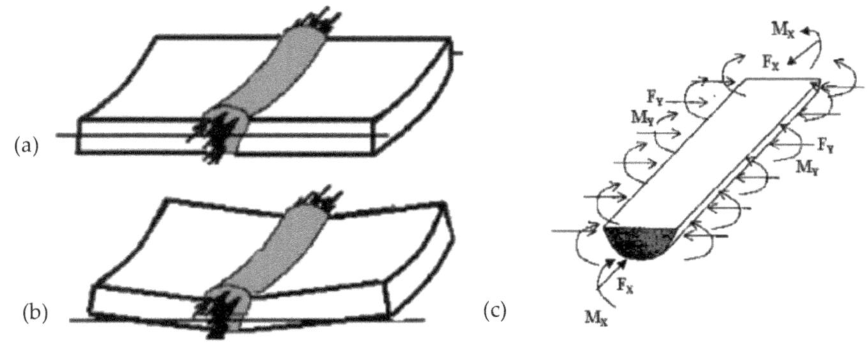

Figure 9. Schematic of: (a) cambering; (b) combined cambering and angular; (c) Shrinkage foresees in inherent strain region.

4.4. Novel control techniques for residual stress and distortion

Controlling of distortion has been investigated in series of papers by the Edison welding institute (EWI) (Conrady & Dull, 1995; Michaleris et al., 1999; Michaleris & Sun, 1997; Conrardy & Dull, 1997). Those papers concerned for the thermal tensioning technique in both static and dynamic heating during welding processes. The technique was found to be an active method in welded ship structures. Different heating sources can be used enhancing tensioning effect such as dynamic flame heating and moving laser spot heating or static heating. Another mitigation technique for controlling welding-induced stresses and distortion has been developed by Beijing Aeronautical Manufacturing Technology Research institute (Q. Guan et. al., 1994; Guan et al., 1993). The technique called dynamic controlled low stress no distortion (DC-LSND); it has been applied successfully to aerospace manufacturing for shell structures such as jet engine cases of heat resistance alloys and rocket fuel tanks of aluminum alloys (Guan et al., 1993; Guan et al., 1996). Many literature present that the residual stress can be minimized by using (DC-LSND) welding technique (Li et al , 2004a; Li et al., 2004b). many cooling media can be used in this technique such as (atomized water, compressed air, solid CO_2, liquid nitrogen, liquid argon). Beside the reduction of plastic strain, it was found that heat transfer enhancement by trailing heat sink technique work as source for balancing residual stresses above and under the neutral axis (Soul & Yanhua, 2005, 2006; Soul et al, 2010). The set-up of dynamic heating spots and trailing heat sink are represented schematically in figure 10. The studies on the temperature field characteristics and the thermal history are the foundation and prerequisite to study the stress and distortion control mechanism in welding mitigating techniques. Sometimes, it is inconvenient or even impossible to obtain the real thermal cycle at weld pool by experiment due to its limitations.

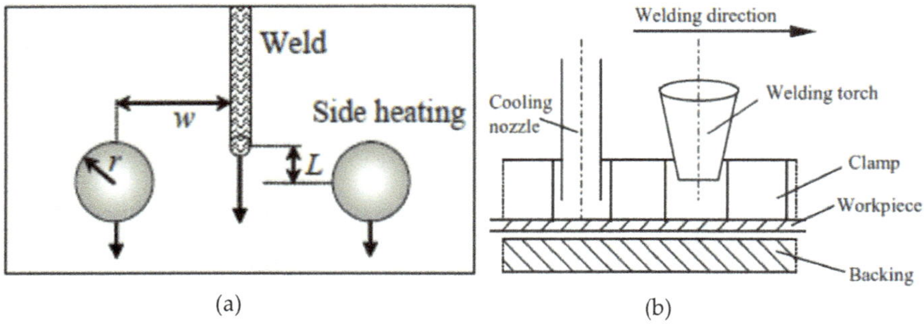

Figure 10. Schematic drawing represents: (a) moving heating spots; (b) trailing cooling spot

To overcome this disadvantage, the advanced numerical analysis technologies, such as finite element method and finite difference method, have been frequently used to obtain the whole temperature field of the welded specimen. It is necessary to develop a computer-based tool to optimize welding mitigation processes and hence minimize the expense and time incurred by extensive welding trials. In this chapter three dimensional finite element methods are employed to find qualitative analysis for the temperature field, residual stress and transient plastic strain developed during welding process. To mitigate the problem of residual stress and distortion, two different techniques were tested and compared, those are trailing heat sink and dynamic heating spots techniques. Gas tungsten arc welding process (GTAW) is used for simulation.

5. Investigation of residual stress behaviour after enhanced heat transfer

5.1. Trailing cooling spot

The proposed welding technique incorporates a trailing heat sink (an intense cooling source) with respect to the welding torch, and it is also named Low Stress No Distortion (LSND) welding. The development of this mitigation technique is based on both detailed welding process simulation using advanced finite element method and systematic laboratory trials. For understanding well LSND welding, finite element method is used to investigate the mechanism of the technique. In this chapter, 3D-FEA results from different papers done by the author have been selected. These results study the mechanism of the trailing heat sink mitigation technique and how the longitudinal residual stress was minimized based on the distance between the torch and the cooling spot. The qualitative and quantitative analysis of residual stress depend on the temperature gradient in the component during heating and cooling. So any modification of temperature topography may decrease or increase the residual stress that depends on interaction of strains. Three-dimensional models for welding, the thermal cycle and residual stress in welding are now in common use as a research tool for both academic and commercial purposes. The models use a transient 3-dimensional thermal model which is decoupled to an elastic-plastic model for calculating the stress and strain. Models were investigated with different material and dimension as shown in table 1.

Model	Size	geometry
Model 1 Al-Mn(3003)	$260mm \times 130mm \times 2mm$	Flat sheet
Model 2 Al-Mg (5083)	$240mm \times 80mm \times 3mm$	Flat sheet
Model 3 Ti-6Al-4V	$270mm \times 120mm \times 2.5mm$	Flat sheet
Model4 AL-Cu (2024	100mmOD, 96mmID, 240mm length	Cylindrical sheet

Table 1. Simulated models: material, shapes and their dimensions.

5.1.1. Establishment of heat source model

In this study, two heat source models were investigated, then correlated on fitting a practical welded sample for distinguishing the better one, so it can be used to detect other analysis. The first one is disc model proposed by (pavelic et al, 1969), the mathematical expression of the model present in Equation 2.

$$q(r) = \frac{3Q}{\pi r_0^2} e^{-\frac{3}{r_0^2} r^2} \tag{2}$$

Where q(r) is the surface flux at radius r (W/m^2), r_0 is the region in which 95 % of the heat flux is deposited, r is radial distance from center of the heat source and Q is the heat input.

The second model is double ellipsoidal power density distribution" adopted from (Goldak, 1984), the mathematical expression of the model present in Equations 3 and 4.

$$q_f(x,y,z) = \frac{6\sqrt{3}Qf_f}{abc_f \pi\sqrt{\pi}} e^{(-3x^2/a^2)} e^{(-3y^2/b^2)} e^{(-3z^2/c_f^2)} \tag{3}$$

$$q_r(x,y,z) = \frac{6\sqrt{3}Qf_r}{abc_r \pi\sqrt{\pi}} e^{(-3x^2/a^2)} e^{(-3y^2/b^2)} e^{(-3z^2/c_r^2)} \tag{4}$$

Where a, b, c are the semi-axis for the gaussian distribution in (x,y,z) direction respectively and f_f, f_r are fractions of heat deposit in front and rear of the heat source. The intensity plot for both surface heat and double ellipsoid heat source models is shown in figure 11. Variation of the semi-axis and the heat deposit fractions allows the double ellipsoid fitted to give suitable heat source especially at increased welding speed.

At the same heat input, the double ellipsoid is more reasonable for fitting the fusion boundary rather than the gaussian distribution or surface heat source model as depicted in figure 12.

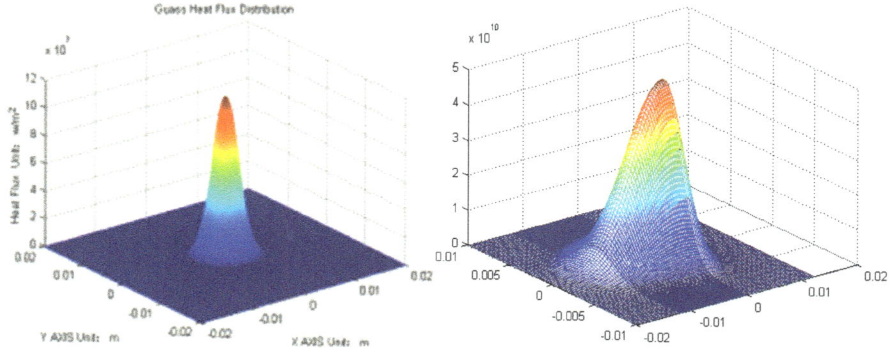

(a) (b)

Figure 11. Intensity of heat source model: (a) gaussian distribution; b) double ellipsoid

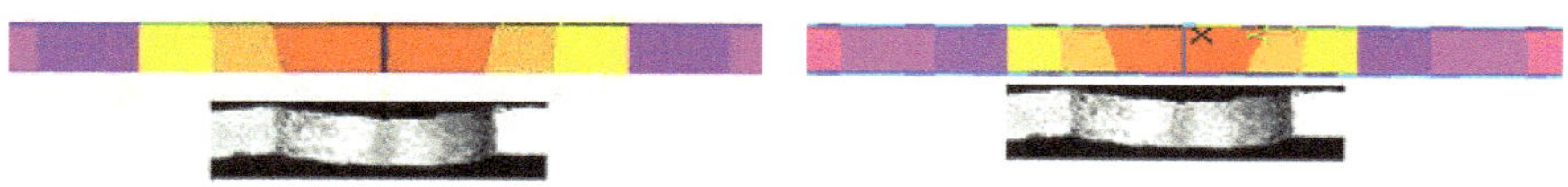

(a) (b)

Figure 12. Experimental & simulated fusion boundary fitting: (a) double ellipsoid; (b) gaussian distribution;

5.1.2. Effects of trailing cooling spot on heat transfer form

The temperature distribution resulted from the thermal analysis for the heat sink process stated above is predicted at a welding time of 25s and presented in Figure 13. The temperature decreases drastically in the zone between the arc and the heat sink, and the corresponding temperature gradient increases. Further more, in the front of the heat sink the temperature isotherms reveals the existence of high temperature gradient and therefore, some of high temperature contours are drawn back to the front of heat sink and distorted temperature distribution were formed. However, the point in and near the weld centerline may passed by more than one thermal cycle and this may expressed as [heating - normal cooling - forced cooling – surrounding heating – normal cooling]. The evidence of those cycles became more clear when different locations in the samples in both conventional and after applying heat sink processes were selected. Figure 14 reveals different location with difference cycle profiles.

The locations at or near the weld line present those cycles and exhibit significance difference in thermal profiles. The point at weld line reveals the maximum intensity of the cooling spot which indicate a steep profile and the existence of valley . The thickness of the modeled sample play an important role for the effectiveness of the technique. Figure 14c shows the thermal history of simulated welded aluminum alloy with 2mm in thickness

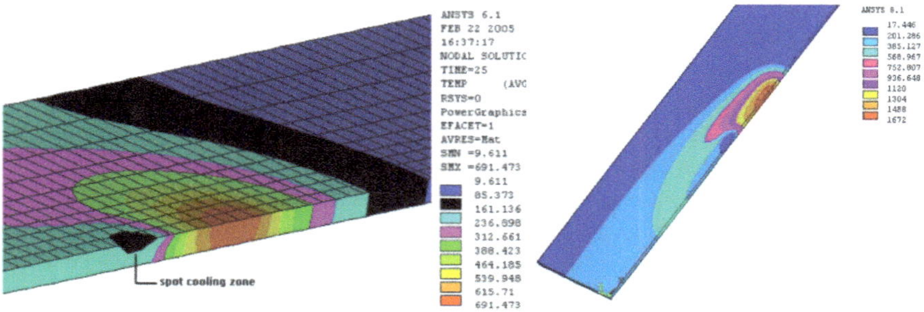

(a) (b)

Figure 13. Temperature contours in trailing cooling spot: a) Al-alloy; b) Ti- alloy (soul, 2oo5,2006)

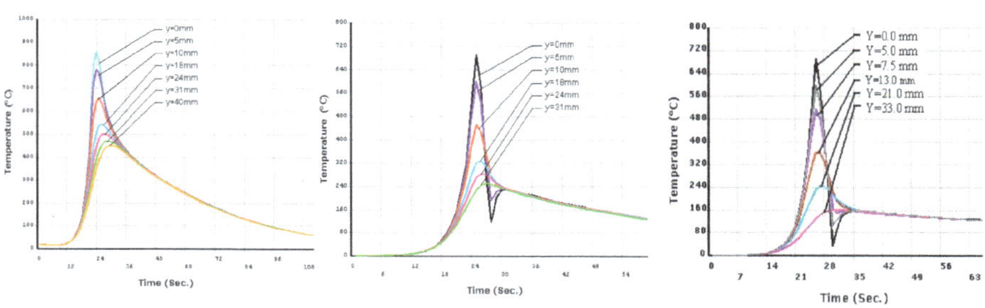

(a) (b) (c)

Figure 14. Thermal history: (a) conventional welding; (b) trailing cooling spot[t=3mm], (c) [t=2mm]

Comparing these results with that modeled with thickness 3mm as presented in figure 14b, at the point (y=0), the penetration of cooling zone is more effective in the small thickness, which reveals lower temperature magnitude at the center of the spot figure 14c. Moreover, the overall temperature tips in thermal history is decreased compared with conventional welding. This refers to some energy absorbed from the total energy by the amount action of introduced cooling.

5.1.3. Longitudinal residual stress behaviour

The result obtained from the 3-D modeling of welding a bead on plate with thickness 2.5mm for titanium alloy and 3mm for aluminum alloy using GTAW process were depicted in figure 15. Describing the process behavior, in the front of the torch, compression transient residual stress were developed which result from generated thermal strain. When the material loose the mechanical properties at high temperature such yield strength and young's modulus, no extra stresses were produced. Behind the welding pool when the shrinkage started , plastic strain was accumulated and transient residual stress changes from compression to tension in short period of time. Therefore, Its magnitude increased vastly as the temperature decreased. However, it is an important knowledge if this change with duration can be measured practically to make correlation, but it is so difficult. In fact, the final magnitude of residual stress depends on cooling rate, regarding the metallurgical science welding microstructure morphology and size depend on the cooling rate as well. So no welding residual stress developed if there is no change or homogenous microstructure was obtained.

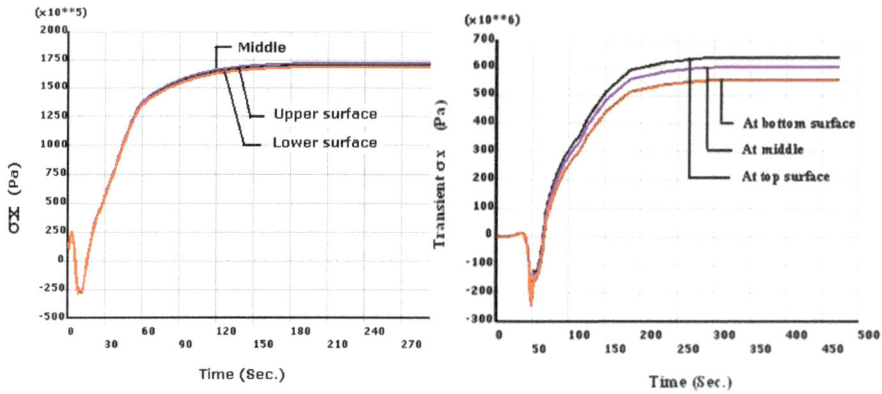

(a) (b)

Figure 15. Transient residual stress: (a) Al-Mg alloy; (b) Ti-6Al-4V alloy

In more details, after including trailing heat sink technique, which can enhance heat transfer process, the temperature was modified. The analyzed results in figure 16 obtained when the trailing heat sink is located at 30 mm behind the torch, exhibit that the stress profile has different behavior. For instance, in small area depicted in figure 16a (zone of compressed contours), the contours behind the torch were compressed in the front of the cooling spot, and its shape becomes in complicated form through this stage.

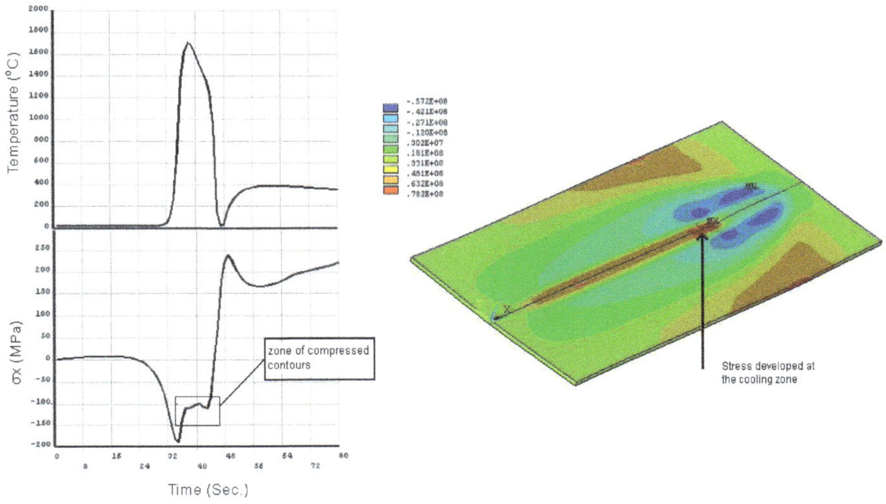

(a) (b)

Figure 16. Transient longitudinal residual: (a) stress vs thermal cycle ; (b) longitudinal residual stress contours Al-Mn alloy

The transient longitudinal residual stress stays constant for a period of time, this may also be referred to the high temperature contours intercepted between the two sources which don't affect the plastic strain. Through enough length along weld centerline behind cooling zone, the longitudinal residual stress seams to have a constant value which can be distinguished from a stable color along this length. This behavior coupled with the constant values of contours around mentioned length.

Moving a step back, far from local cooling zone, the temperature of the metal at the spot-cooling region is smaller than the surrounding, so the fast contraction of welded metal in the cooling zone and high transient longitudinal residual stress may developed as depicted in figure 16b (red color were the arrow located). Because of cooling that crossed in a short time and the hot contours appeared again due to the hot surrounding material, the residual plastic strain may released and recovery process may generate. However, both the effectiveness of the cooling at the upper surface and the abnormal temperature developed behind the heat source may affect the balance level for the front-to-rear stress pattern. Furthermore, in short distance behind the torch, high temperature contours around the cooling zone and low temperature inside, which revealed a big difference from that occurred in conventional welding process. Therefore, it may brought less stress during solidification temperature range, and reasonable longitudinal strain can be obtained. Behind the cooling zone the transient longitudinal residual stress profile seems to be decreases to somewhat value due to an increase in temperature as shown in figure 17. In this stage, the process became as heat treatment for the residual stress, and the metal may expand again due to heating which believed to be appeared in the expansion process in the rear of the cooling zone.

The drop in residual stress magnitude follows the change in temperature according to the thermal cycle may refer to the opposite change of the elastic strain when the temperature increases again by the surrounding hot metal. Therefore, the reasonable expression for this mechanism related to recovery process due to the heating process behind the cooling spot which may expand again, and then some of the plastic strain can be released. In practice, this process is quite logic during the heat treatment of materials suffering from residual stress induced by welding or other strengthening process such as cold working etc. Furthermore, the maximum longitudinal residual stress in titanium model is reduced to about (~326 MPa) from that in conventional welding, where in the aluminum model the maximum residual stress is minimized about (~55MPa) from that in conventional welding. The tested technique shows the significance influence on the final longitudinal residual stress through thickness, the clear evidence for such influence can be seen in figure 18. The residual stress at the upper surface became less in magnitude than that in the middle of plate thickness. It seems not only the over all stress minimization but the possibilities for balance the stresses around the center of the Thickness.

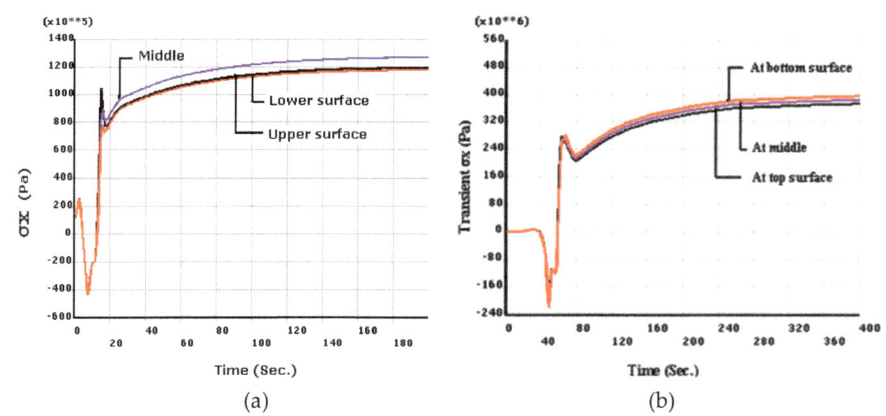

Figure 17. Transient longitudinal residual stress in moving cooling spot: (a) in Al-Mg alloy; (b) in Ti-6Al-4V alloy

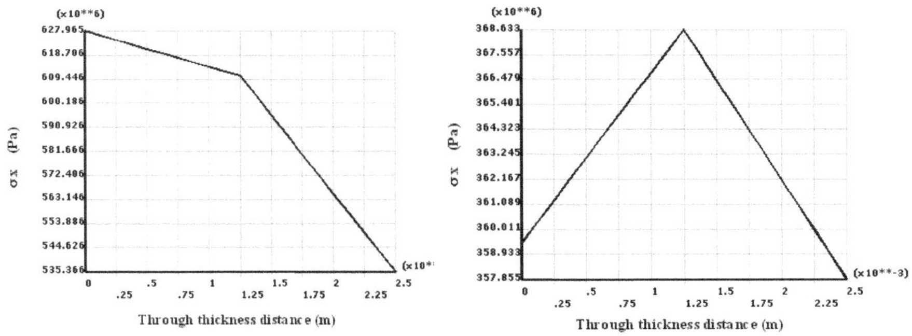

(a) (b)

Figure 18. Final longitudinal Residual Stress through Thickness: (a) In Conventional Welding; (b) In Trailing Heat Sink.

5.1.4. Longitudinal plastic strain behavior

In welding process, the residual stress magnitude based on the inherent strain or plastic strain, so the analysis of strain behavior is important to characterize the minimization of residual stress. In the work done by (soul & yan hua,2006) the transient longitudinal plastic strain profiles were analyzed at two nodes and presented in figure 19. In conventional welding the plastic strain at the centerline (Y=0mm) seams to be higher than the point a little moved a way from the center line (Y=2.5mm). However; after including dynamic cooling spot figure 19b, significant change in plastic strain behavior was obtained. The plastic strain at the centerline became lower than that at (2.5mm), it is clear that the overall plastic strain magnitude were minimized due to the used technique.

The minimization behavior refers to the complex tensioning effect based on the temperature topography. Moreover, due to the surrounding hot material, the temperature increased again behind the trailing cooling spot, and some of strains is released or partial annealing process occurs. In practical welding, it is difficult to obtain constant temperature through thickness with respect to the time whatever it is thin. This non-homogeneity refers to the three heat transfer conditions which may occurs at the upper surface rather than at the lower surface, hence the temperature is still higher above. To characterize the change in plastic strain through the thickness, simulated result of 2.5 mm modeled welded titanium sheet is analyzed. However, when no backing used in welding, the cooling rate at upper surface is still more than at lower surface. Figure 20 reveals that the maximum plastic strain occurs at the upper surface but after trailing heat sink introduced figure 20b, all the plastic strain magnitude through the thickness became close to each other due to the process effects.

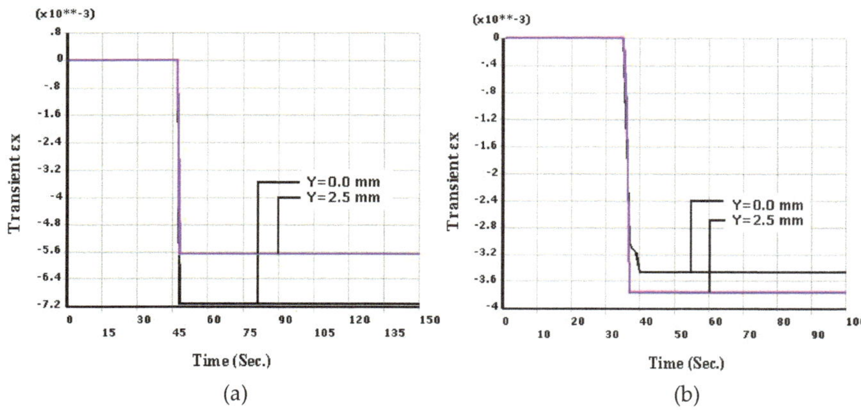

Figure 19. Transient plastic strain: (a) in conventional welding; (b) in trailing heat sink.

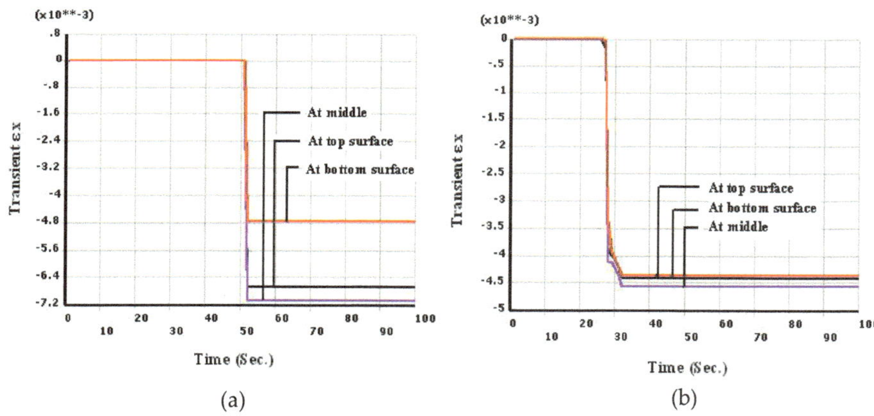

Figure 20. Plastic strain through depth: (a) in conventional welding; (b) in trailing heat sink.

Moreover, the plastic strain magnitude at the upper surface is less or equal the strain at the lower surface which cause a balance in strains around the neutral axis. Therefore, balancing of forces occurred, and bending distortion minimized or can be eliminate depends on the degree of balancing. It is conclude that the reduction in overall longitudinal plastic strain magnitude is the source of longitudinal residual stress minimization. This results can be coupled with that obtained in Figs. 15b and 17b.

5.1.5. *Effectiveness of the process on longitudinal welded cylinder model*

In the cylindrical model the trailing heat sink considered to be moving behind the heat source at the bottom side. This consideration according to the results obtained in conventional welding which indicate that the maximum axial residual tensile stress occurred at inner the surface. The temperature distribution resulted from the thermal analysis of those conventional welding and that including trailing cooling spot welding process stated in the above sections are studied at welding time of 20s. Contour plots of the temperature distribution under the two conditions are shown in figure 21. Comparing the isotherms in both process results, the contours behind the heat source were shifted in the front of trailing cooling spot due to abnormal heat transfer process. The homogeneous heat transfer in conventional welding looses its stability after cooling spot was introduced. Moreover, the temperature magnitude in compressed contours was decreased due to the absorbed heat at high temperature behind the torch by the action of cooling. The effect of change in heat transfer mechanism on the distribution of axial residual stress can be seen in figure 22. the residual stress profile is completely different after introduced dynamic cooling spot. Because of the cylindrical shape is different from the flat from point of view the stiffness resistance from point of view, the reduction in final residual stress in cylindrical model is not sufficient when it is compared with the flat model which is still thicker.

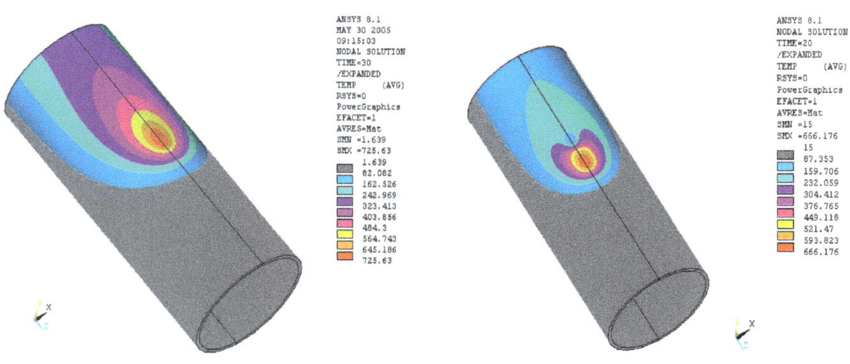

(a) (b)

Figure 21. Temperature distribution at 20 s: (a) conventional welding; (b) trailing heat sink.

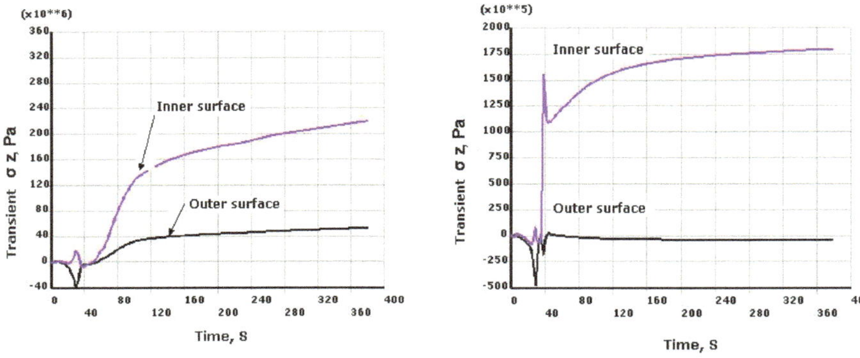

(a) (b)

Figure 22. Transient axial residual stress (a) conventional welding; (b) trailing heat sink.

5.1.6. Control of distortion by trailing cooling spot

As discussed in the previous sections, how the mechanism of heat transfer was changed after introducing heat sink. Therefore, it results in thermal tension that differs from what occurs in conventional one. The degree of tension depends on the position of the cooling spot with respect to the heat source, or the proper parameters as denoted in (Li et al, 2004, 2005; Soul et. al., 2006). All the above results show the minimization in longitudinal residual stress that refers to the decrease in responsible plastic strain. All of the previous parameters in addition of balancing stress and strain on the upper and under the neutral axis bring the elimination of the structure distortion as depicted in figure23 and figure 24.

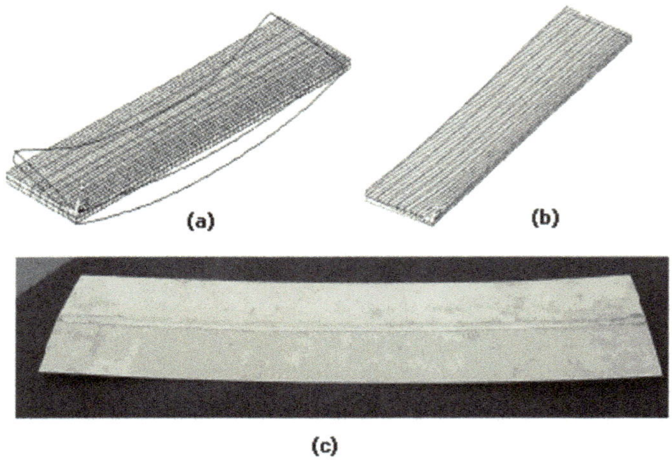

(c)

Figure 23. Final distortion: (a) conventional process; (b) with heat sink process; (c) welded experiment sample.

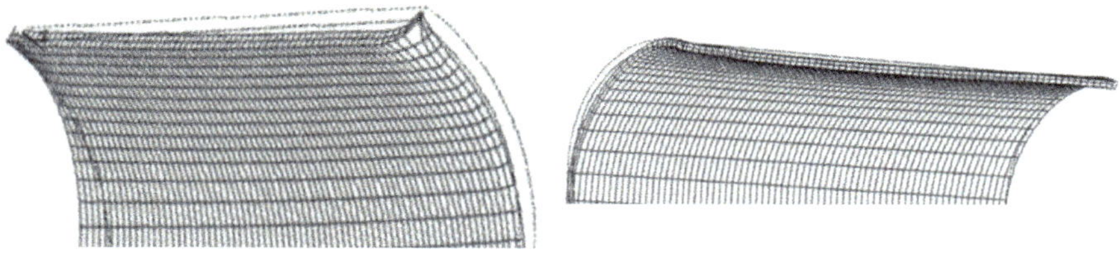

(a) (b)

Figure 24. Final deformed model: (a) GTAW process; (b) DC-LSND process.

5.2. Optimization of thermal tensioning in welded shell element

There are many different methods of welding deformation and stress reduction. At the same time, the main principles are the same. Classification of the mitigation techniques helps to understand its performance capabilities and limitations. According to the basic mechanisms on which the techniques are based.

5.2.1. Effect of the laser heating spot on the temperature distribution

The Introducing of heating spot besides the welding reveals an increase in temperature at fusion zone and other near regions comparing with the conventional welding itself. So, to obtain the similar contours of temperature at welding line, the heat input from the torch were decreased about 19%. Figure 25 shows the predicted temperature distribution in conventional welding and moving heating spots respectively. The selected separated positions for laser spots were 34 mm from the centerline. It is clear, that the heating spot not only affects the temperature distribution in the front of heat source, but also enlarges the contours in transverse direction far way from centerline. This is due to the increasing of the total energy input with respect to small specimen width. However the contours change completely at the position of spots due to the concentration of laser heating.

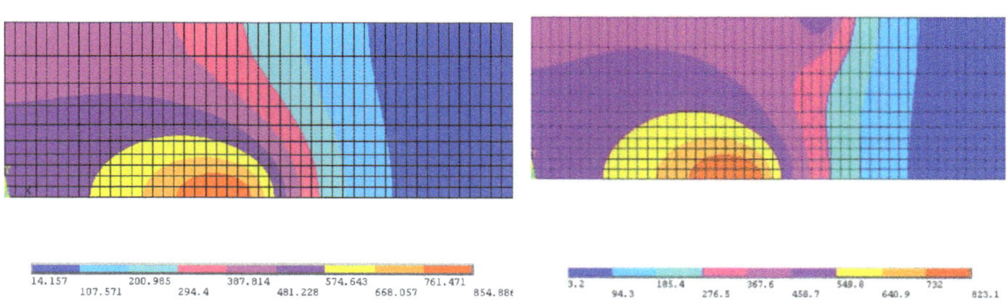

(a) (b)

Figure 25. Temperature distribution contours: (a) conventional welding; (b) moving heat spot.

5.2.2. *Effects of the laser heating spot on the stress and strain behavior*

Longitudinal residual stresses and plastic strains in transverse direction with respect to the time are presented in figure 26. The analyzed points are taken at different distance from the centerline in the mid length of the model. The distance of heat spot from weld centerline is 25 mm. The tensile stress profile developed during cooling cycle show different behavior compared to the result obtained from the conventional welding figure 26a. However, the transient longitudinal residual stress stay constant during some period of time (~40s), Thus, at that time the heat source already switched off. This behavior is related to the effect of spot heating. The heated material by laser spot were expanded to the direction of welding centerline, in the same time the metal behind the torch vastly contracted, now there are two possibilities, first, if the material is heated for enough distance in transfer direction the contraction of welded material don't meet the resistance from the heated material for mentioned period of time above, in the opposite of the conventional process were the cold material restrained the contraction. Second, when the material contacted at the welding line the heated material by spots expanded in the direction of welding line, hence the material may contracted without more tension due to the expanded material neighboring during the above mentioned period. Moreover the residual stress using this method still more than in cooling spot technique. Concerning the plastic strain behavior in this technique, the transient plastic strain result from laser heating spots reveals complicated profile too; the longitudinal plastic compressive strain is caused by the expansion of the heated material being constraint by the cooler material nearby.

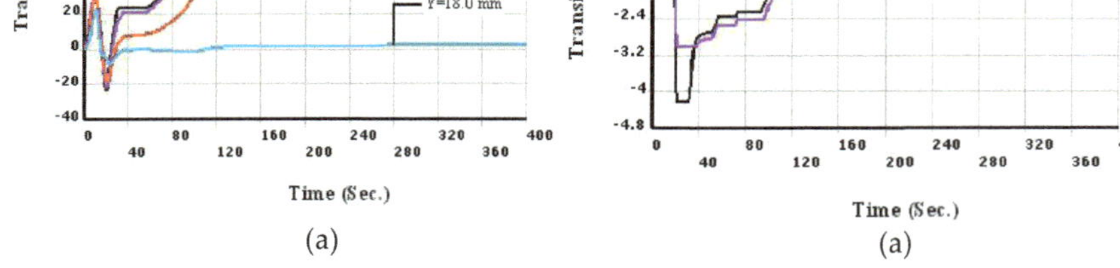

Figure 26. Simulated results in moving heating spots: (a) Transient residual stress; (b) Transient plastic strain

The rising of temperature far ways from zone induced by laser spots affects the material properties and decrease it is stiffness, hence the resistance of material to the weld metal contraction were reduced. Moreover the transient longitudinal plastic take long period of time to accomplished final magnitude due to increase of overall body temperature, and it's value less than from that obtained by trailing heat sink technique figure 17a, but the residual stress and distortion is still more. This phenomenon may refer to other strains results in transverse and through thickness direction, which need more details for their behavior under this process due to the complicated strain interaction in the three dimensional

5.2.3. *Processes parameters controlling distortion degree*

It is clear from the previous studies that any modifications in the longitudinal residual stress results from temperature modification or heat transfer enhancement. As what happen in the simulated techniques, and the main control on the twin problem is how the temperature topography looks like. However, the change of processes parameters will give changes in temperature topography. Therefore, different temperature topography generate different tensioning process or different strain interaction. According to the previous explanation, for the process optimization investigation of the effect of the distance between the heating spot and weld centerline on the cambering distortion magnitude was carried out. Figure 27 presents the degree of distortion obtained from computational results at different distance of the laser spot. The results show that the distortion increases as the distance increases. However, after introducing the heat sink process as depicted in figure 27b, the distortion may disappear and the calculated displacement is only 1.3 mm. The result is quite reasonable according to the minimized and balanced residual longitudinal stress and plastic strain or the balanced forces and moment which become insignificant to produce the curvature along welding line as mentioned in the previous sections. Moreover, the displacement in laser heating spot was reduced but still higher than that obtained in the cooling technique. At this point comparing both mitigation techniques, it can be concluded that at the same reduction of distortion, still the moving cooling spot more effective technique than thermal tensioning for residual stress minimization in thin element.

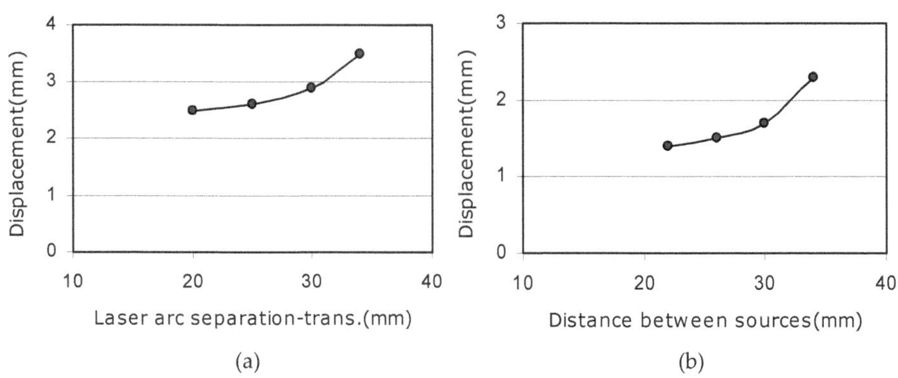

Figure 27. Parameters effects the processes performance: (a) heating spots; (b) cooling spot.

6. Conclusions

One of the major problems induced during welding of thin element structure are the twin Problem. Residual stresses reduce the performance of welded component when they have the same sense with the working stress in service. Other disadvantage of residual stress is promoting the stress corrosion cracking. Distortion from other side, impairs the structure appearance and causes misalignment during assemblies. During the years, many mitigating techniques are developed in literature to reduce the residual stress and elimination of distortion. In the end of last century, techniques introduced thermal tensioning or temperature modification were creative, that including (LSND) and heating techniques. When the dynamic action of these techniques applied, they became more suitable to approximate the solution of the problem, because it is dealing with temperature modification, and it is known that the temperature is the source of the problem. In this chapter, the two mentioned active mitigating techniques were investigated in dynamic action and correlated with each other. The main fields of interest for the investigation are temperature fields, longitudinal residual stress and strain in addition to the bending distortion. Atomized cooling water with heat transfer coefficient were proposed , also laser heating spot beside the torch with proper heat input were considered. The output results of this investigation summarized as following:

- The change of heat transfer enhancement produce temperature modification or different temperature topography.
- The modification of temperature distribution generate different tensioning inside the body, therefore new abnormal strain interaction in different direction.
- Due to complexity stiffness based-response for different shape, the transient stress behavior developed in the bead on plate model show different profile correlate with that in the cylinder model.
- One of the major advantages of FE optimization approach over welding trials was that the models enabled the transient stresses and strains during welding to be considered. This greatly simplified the understanding and optimization process.
- During the DC-LSND process, the thermal history for metal in the weld centerline and at closed region passed more than one thermal cycles, as (heating-cooling-heating-cooling) and the duration time for staying the metal at high temperature is shorter than in conventional process. This is found to reveal different stress and strain history profiles at this region compared with the profiles obtained in the conventional process.
- The significant reduction in the residual stress obtained at the top of surface and the maximum stress at the middle of the thickness. This helps to balance the shrinkage forces above and under the weld metal center, and distortion can be prevented
- The suggested techniques gave a reduction of the peak residual stress. Optimization of both DC-LSND and thermal tensioning can give similar reduction or elimination of bending distortion of such used models thickness. At this optimization still the DC-LSND method has a better effectiveness on longitudinal residual stress minimization

Future work analysis need for strains in other (Y and Z) directions, to see it is behavior after applying the above techniques and correlate with the longitudinal direction.

Author details

Farag Soul
University of Elmergeb, Faculty of Engineering, Libya

Hamdy Nada
University of Menoufia, Faculty of Engineering, Egypt

7. References

Withers P. J. & Bhadeshia H. (2000), Residual Stress – II: Nature and Origins, *Mat. Sci. Tech.* Rudd C. O. (199), Residual stresses and their measurement, *Proc. 1st Int. Conf. on Quenching & Control of Distortion, Chicago,* Illinois, USA, 193-198.

Borland D. W.(1994), Residual stress measurement – Methods, limitations and significance, *Proc. 2nd Australian Int. Conf. Surface Engineering,* Adelaide, 114-121.

Kandil F. A. et. al. (2001); A Review Of Residual Stress Measurement Methods -*A Guide To Technique Selection,* NPL Report MATC(A)O4.

Brien E. O. (2000), Residual Stress – Friend and Foe: *A Civil Aircraft Perspective,* ICRS-6, Oxford, UK, 10-12 July, vol. 1, 13-21.

Evans E. B.(1986), "Residual stresses in processing." *Encyclopaedia of Materials Science & Engineering,* Vol. 6, 4183-4188

Littmann W. E.(1964), Measurement and significance of residual macrostress in steel", SAE 793A, *Proc. of the Automatic Eng. Cong.,* Detroit, MI, 13-17.

Airbus 2000, http://www.airbus.com /gmf99_1.html

Guan Q. (1999), A survey of development in welding stress and distortion controlling in aerospace manufacturing engineering in china *Welding in the world,* Pergamon press Ltd, 43(1).

Hansen B.(1968) , Welding stresses and deformation. *Svejse Copenhagen* report no. 68.07 edition, 1968. in Danish.

Okerblom N. O. (1955), *The Calculations Of Deformations Of Welded Metal Structures.* Mashgiz, Moscow-Leningrad, 1955.

Puchaicela J. (1997), Control of distortion of welded steel structures, *Welding J.,* pp.49-52.

Masubuchi K. (1980), *Analysis of welded structures,* Pergamon Press, London.

Radaj R. (1992), *Heat Effects of Welding,* Spring-Verlag.

Goldak J. et. al. (1992), Coupling heat transfer, microstructure evolution and thermal stress analysis in weld mechanics, *Proc. of IUTAM Mechanical effects of welding,* Springer Verlag, Berlin.

Lindgren L. (2001), Finite element modeling and simulation of welding part 1: Increased Complexity, *Journal of thermal stress,* Vol. 24, pp 141-192.

Lindgren L. (2001), Finite element modeling and simulation of welding part 2: Improved materials modeling, *Journal of Thermal Stresses,* Vol. 24, pp 195-231.

Lindgren L. (2001), Finite element modeling and simulation of welding part 3: Efficiency and Integration, *Journal of thermal stress,* Vol. 24, 2001, pp 305-334.

Tall L. (1991), Residual Stresses In Welded Plates--A Theoretical Study, *Weld. J.*, Vol 43 (No. 1), P 10s-23s

Hibbit H.D. & Marcel P.V. (1972), A Numerical Thermomechanical Model For The Welding And Subsequent Loading of A Fabricated Structure, Nsrdc Contract No. N00014-67-A-019-0006, Tech. Rep. No. 2, Department Of The Navy.

Muraki T. et. al. (1975), Analysis Of Thermal Stresses And Metal Movement During Welding-- Part I: Analytical Study; Part Ii: Comparison Of Experimental Data And Analytical Results, *J. Eng. Mater. Technol. (Trans. Asme)*, 1975, P 81-84, 85-91

Rybicki E.F. et al. *(1978)*, A Finite-Element Model For Residual Stresses And Deflection In Girth-Butt Welded Pipes, *J. Pressure Vessel Technol. (Trans. Asme)*, Vol 100, P 256-262

Courant R. (1943), Variation Methods For The Solution Of Equilibrium And Vibration, *Bulletin Of The American Mathematical Society*, 49. Pp 1-23.

Dong P. et al. (1998), Residual Stress Analysis Of A Multi-Pass Girth Weld, *Asme Pvp-Fatigue, Fracture, And Residual Stresses*, 373, Pp. 421-431.

Feng Z. et al. (1996), A Finite Element Model For Residual Stress In Repair Welds, *Asme Pvp-Residual Stresses In Design, Fabrication, Assessment And Repair*, 327, Pp. 119-125.

Karlsson L. et al. (1989), Residual Stresses And Deformatins In A Welded Thin-Walled Pipe, *Asme Pvp- Weld Residual Stress Andplastic Deformation*, 173, Pp. 7-10.

Grong Q. & Myhr O. R. (1993), Modelling Of The Strength Distribution In The Heat Affected Zone Of 6082-T6 Aluminum Weldments" Mathematical Modelling Of Weld Phenomena, The Institute Of Materials, London, Pp. 300-311.

Voss O. et. al. (1999), Consideration Of Microstructure Transformations In The Calculation Of Residual Stresses And Distortion Of Larger Weldments, *Mathematical Modeling Of Weld Phenomena 4*, The Institute Of Materials, Pp584-596.

Tenga T. L. & Linb C. (1998), Effect Of Welding Conditions On Residual Stresses Due To Butt Welds, *International Journal Of Pressure Vessels And Piping* 75, 857–864.

Li J. et. al. (2004), Studies On Characteristics Of Temperature Field During Gtaw With A Trailing Heat Sink For Titanium Sheet, *J. Of Mater. Proc.Tech.* 147,328–335.

Karlsson R. I. & Josefson B. L. (1990), Three-Dimensional Finite Element Analysis Of Temperatures And Stresses In A Single-Pass Butt-Welded Pipe, Asme-*Journal Of Pressure Vessel Technology*, 112, Pp. 76-84.

Junek L. et. al. (1999), Residual Stress Simulation Incorporating Weld HAZ Microstructure, *Asme Pvp- Fracture, Fatigue And Weld Residual Stress*, 393, Pp. 179-192.

Vincent Y. et. al. (1999), On The Validation Of The Models Related To The Prevision Of The Haz Behavior, *Asme Pvp-Fracture, Fatigue And Weld Residual Stress*, 393, Pp. 193-200.

Dubois D. et. al. (1984), Numerical Simulation Of A Welding Operation: Calculation Of Residual Stresses And Hydrogen Diffusion, *Asme Fifth International Conference On Pressure Vessel Technology, Materials And Manufacturing*, Ii, San Francisco, Ca, Pp. 1210 - 1238

Gatovskii K. M. (1980) & Karkhin V. A. (1980), *Theory Of Welding Stresses And Deformations*, Leningrad Shipbuilding Institute, *Rus.*

Withers P. J. & Bhadeshia H. K. (20000), Residual Stress, Measurement Techniques, *Mat. Sci. Tech.*, 2000.

Borland D. W. (1994), Residual Stress Measurement Methods, Limitations And Significance, *Proc. 2nd Australian Int. Conf. Surface Engineering*, Adelaide, March 1994, 114-121.

Noyan I. C. , Defining Residual Stresses In Thin Film Structures", *Advances In X-Ray Analysis*, Vol.35, 461-473.

Guan Q. et al. (1986), Transient Welding Strain Measurement By Moire Technique. *Transactions Of Chinese Welding Society*, 7(1)

Artem P.(2001). Ph.D thesis, computer simulation of residual stress and distortion of thick plates in multi-electrode submerged arc welding.

Soul F. & Yanhua Z.(2005), Numerical Study Of Residual Stress Field During Arc Welding With Trailing Heat Sink. *Fluid Structure Interaction And Moving Boundary Problem, Transaction Engineering Science*,Vol.84,683-692, Witpress.

Soul F. & Yanhua Z. (2005), Thermo-Mechanical Analysis of Longitudinal Welded Aluminum Alloy Cylinder Using 3D FEM, *The Fifth International Forum on Aluminum Ships*, Tokyo.

Huang T. D. et. al. (2004), Fabrication And Engineering Technology For Lightweight Ship Structures, Part 1: Distortions And Residual Stresses In Panel Fabrication , *Journal of Ship Production*, 20(1), P.43-59,

Andersen L. F. (2000), Residual Stress And Deformations In Steel Structures - *PhD, Technical University Of Denmark, Department Of Naval Architecture And Offshore Engineering.*

Conrady C. & Dull R. (1995), Prediction Techniques For Distortion Control, *Report-Edison Welding Institute.*

Michaleris P. et al. (1999), Minimization Of Welding Residual Stress And Distortion In Large Structures. *Welding Journal, Welding Research Supplement*, 361-366.

Michaleris P. & Sun X. (1997), Finite Element Analysis Of Thermal Tensioning Techniques Mitigating Weld Buckling Distortion. *Welding Journal*,Vol.76, 451-457.

Conrardy C. & Dull R. (1997), Control Of Distortion In Thin Ship Panels. *Journal Of Ship Production*, Vol.13, 83-92.

Guan Q. et a (1994) . Low Stress No-Distortion (Lsnd)Welding A New Technique For Thin Materials. Iiw Doc. X-1190-89, *Welding In The World*, Vol.33.

Guan Q. et al. (1993), Dynamically Controlled Low Stress No-Distortion Technology And Its Apparatus. *Chinese Patent* 9310690.8.

Guan Q. et al. (1996), Low Stress No Distortion Welding For Aerospace Structures. *China Welding*, Vol.5,1-9.

Li J. et al (2004). Stress And Distortion Mitigation Technique For Welding Titanium Alloy Thin Sheet. *Sci.Technol. Weld. Join.*, Vol.9(5),451-45

Soul F. & Yanhua Z. (2006), Numerical Study On Stress Induced Cambering Distortion And Its Mitigation In Welded Titanium Alloy Sheet. *J. Sci. Technol. Weld. Join.*, Uk, Vol.11, 688- 693.

Soul1 F. et. al. (2010), Residual Stress Correlation In Two Different Mitigation Techniques Using FEA, *Advanced Materials Research Vols. 83-86 (2010) Pp 1254-1261*

Pavelic V. et al. (1969), Experimental And Computed Temperature Histories In Gas Tungsten Arc Welding, *Welding Journal Research Supplement*, 48, 295s-305s.

Goldak J. et. al. (1984), A New Finite Element Model For Welding Heat Sources, *Metallurgical Transactions B*, Vol. 15b, 1984, P 229-305.

E.M. van der Aa(2007), PhD thesis, Delft University of Technology

Chapter 10

Principles and Thermo-Mechanical Model of Friction Stir Welding

Jauhari Tahir Khairuddin, Jamaluddin Abdullah,
Zuhailawati Hussain and Indra Putra Almanar

Additional information is available at the end of the chapter

1. Introduction

Friction stir welding (FSW) is a solid-state welding process that gained much attention in research areas as well as manufacturing industry since its introduction in 1991 [1, 2]. For almost 20 years, FSW has been used in high technology applications such as aerospace to automotive till high precision application such as micro welding. The main feature of a solid-state welding process is the non-melting of the work material which allows a lower temperature and a lower heat input welding process relative to the melting point of materials being joined. This is advantageous over the conventional fusion welding where excessive high heat input is required to melt the work material. Much less heat input required for FSW translates into economic benefits, safer and less complicated welding procedures. The friction stir welding make it possible to join light weight materials such as aluminium alloy, magnesium alloy, copper and titanium alloys which are very difficult to weld by conventional welding. These clear advantages have greatly increased the usage of these materials in structural applications [3, 4]. In addition, FSW also makes possible to produce sound weldment in 5000 and 7000 series aluminium alloys that are not possible to be welded using conventional method. FSW does not produce sparks or flames. Thus, safety, environmental and legislation issues are not of major concern. FSW process provides proven good quality and strong weldment with inexpensive and lesser number of equipment, eliminates the use of filler metal and improved weldability. Due to these factors FSW has successfully been employed in aerospace, automobile and ship building industries. The need to further understand and improve FSW process continues to propagate in many applications. Many researchers have looked into several methods including mathematical modelling of the process, aiming at understanding the physical-material interaction [8 – 14]. However, there is lack of recorded work in the literature on a system or method to quantitatively measure the welding parameter such as force and torque in FSW process.

This chapter aims to introduce FSW process, its parameters, the applications and a thermo-mechanical model of the process. From the mathematical model derived, a measurement apparatus was developed for force and torque determination during FSW process and experiments were performed to validate the model.

2. Friction stir welding process

FSW set up consists of (1) cylindrical rotational tool, (2) two or more work materials of similar or dissimilar material combinations (3) backing plate and finally (4) clamping or holding fixture as shown in figure 1.

The rotating tool design consists of a combination of two cylinders of a specific radius ratio known as shoulder and smaller radius pin or probe, where the height of the pin or probe is usually more than half of the work material thickness but not equal to its overall thickness. The work materials to be joined may be arranged as conventional welding method but the most common configurations used in FSW are abutted and lapped configuration. For any configuration, FSW has the capability to join thick plate without the need for special preparation prior to welding process. Meanwhile, the backing plate is to ensure the establishment of confine volume and it becomes a must when welding with a pin penetration approaching the bottom of the work materials. The most crucial part of the work materials set up is the clamping or holding fixture. Improper clamping may jeopardize the mated surfaces to be welded and will generate gaps leading to the formation of worm hole or cavities in the weldment. FSW is the non-filler process; hence no substitute material to fill in gaps created by the separation of the work materials is required.

Figure 1. Friction stir welding experimental set up [5].

2.1. Stages in friction stir welding process

FSW process involves four phases which are (1) plunging phase, (2) dwelling phase, (3) welding phase, and finally (4) exit or retract phase. Briefly, the process starts with rotating tool pin or probe thrusting onto the configured work materials under a constant axial load to generate friction heat. This process will continuously increase the temperature at the immediate contacting surface of the rotating tool and work material. The process continue until the temperature at the immediate contact of the rotating tool and the work material increased to a temperature which causes the work material to soften, plasticized and significantly lose its strength. Consequently, these conditions allow the rotating tool to penetrate to a certain depth usually almost to the thickness of work material. The plasticized material is subjected to displacement by the rotating tool pin plunge, effectively being flashed out with a portion of the generated heat, thus introducing new immediate lower temperature and harder surface of work material. This event explains the transient heat generated through pure mechanical friction work at the tool and work material interface. The end of the plunging phase is signified by the sound contact of the rotating tool shoulder with the immediate work material surface.

At this moment, the process enter the dwelling phase where the rotating tool is allowed to dwell for a period of time, causing the temperature to increase further, up to its hot working temperature. The heat generated from frictional work is greatly dependent on the relative increase of contact surface area as well as the relative speed. The heat generated causes the affected area under the shoulder to expand considerably. Phenomenally, the heat causes the work material close to the immediate contact to lose its strength and becoming plastic. Once this condition is reached, thin soft material layers is produced and would stick to the dynamic rotating tool surfaces (pin and shoulder) and being forced to be displaced along. Instantly the mechanical friction heat generation is partially turned into plastic dissipation heat generation. It is explained by the energy dissipated from the internal shearing of different velocity between the displaced soften work material layers to static more solid surface. Ideally, intermittent heat generation mechanisms due to friction work and plastic dissipation take place because of the transient heat transfer effect and the material ability to regain its strength as heat is lost to the ambient. In addition, the other role of these frictional work and plastic deformation mechanisms are to induce soft material displacement and causes the stirring action or severe material deformation which later produce the amalgamated joint.

The dwelling phase is followed by welding phase. After the local temperature of work material under the rotating tool approaches its hot working temperature and is soft enough to be stirred and displaced, the rotating tool is moved transversely along the welding line. This traverse motion caused the plasticized soft material at the leading edge of the rotating tool being squeezed and sheared through a small slit formed by the displaced soft material at the side or lateral of the tool, preferably in the direction of tool rotation. The displaced soft material is then deposited to the gap at the trailing edge left by rotating tool pin or probe. The soft plasticized material is forcedly displaced by the rotating tool along its rotating direction under a closed encapsulation of harder solid work material wall and rotating tool shoulder. The soft material is forged to the trailing edge in layers, forming weld nugget. At each traverse increment of the rotating tool motion, the displacement of soft plasticized work material to the trailing edge will introduce new solid, lower temperature work material at the leading edge. Thus it reintroduces friction work heat generation mechanism prior to plastic deformation mechanism and continuously repeating the heat generation process all over again at each traverse displacement of the tool. This produces cyclic transient heat generation. This cyclic process takes part throughout the welding phase and strongly affected by the combination of the rotating tool's rotational and traverse speed. Recap, during the welding phase the plasticized material is subjected to displacement, extrusion and shearing mechanisms facilitated by the tool rotation, thrust and transverse movement under cyclic heat generation along welding line and finally consolidating welding nugget in the trailing side.

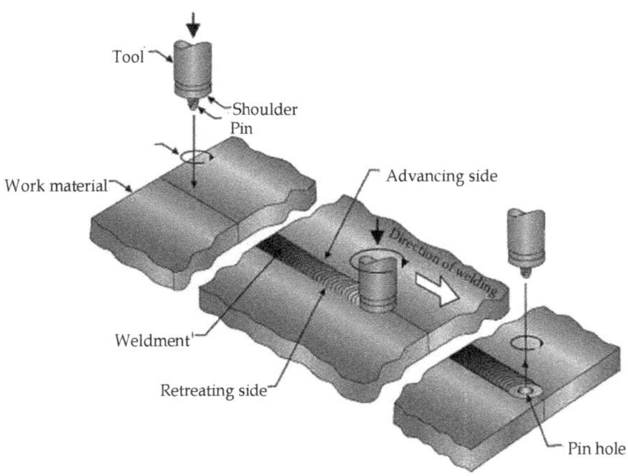

Figure 2. Friction Stir welding process phases of a butted work material configuration [6].

At the end or exit phase of FSW process, the rotating tool is retracted away from the work material leaving a cylindrical hole mark that once occupied by the tool pin. For cosmetic reason, the cylindrical hole may be filled with filler material at the end of the welding process but the most common method used is by introducing dummy material prior the exit phase. Dummy material is of the same material used for the work material to be weld and placed at the end of welding line where the rotating tool is allowed to traverse to and exit within it. The dummy material is later to be cut away leaving good surface finish. Though, this cosmetic issue would remain in the application of innovated friction stir spot welding.

These process phases in FSW are dependent to one another to produce a good amalgamated weldment and are strongly affected by the welding parameters. The assurance of good weldment is determined by proper control of varying measureable welding parameters such as rotational speed, axial plunge force and torque, traverse speed, tool geometry and orientation in the form of heat energy. Similar to other conventional welding methods, heat energy notably determines the quality of the joint.

2.2. Thermo-mechanical dynamic of friction stir welding

To understand the working principle of friction stir welding, it is necessary to explore the physics related to heat generation. Friction stir welding process started with initial mechanical friction between tool and the surface to be welded which results in heat generation. Rotating tool rotates and advances in trust motion against the work material followed by heat generated through plastic dissipation or deformation of sheared and strained layers of soften material close to the rotating tool surfaces. The heat generation mechanisms occur intermittently or in cycle throughout the welding process.

2.2.1. Friction heat generation

Mechanical friction work is initiated when rotating tool surfaces are in contact with the immediate stationary surface of work material under a normal load. While the rotating tool is sliding, it introduces velocity difference between the dynamic rotating tool and the static work material surface and thus creates friction work and subsequently, heat. This mechanical friction work is described based on the Amonton's laws which firstly explain that the friction between two separate bodies is directly proportional to the normal load applied onto the bodies. In this law, coefficient of friction of static friction is a constant variable and temperature dependent but only to be considered as kinetic friction when the contact condition is non-sticking or sliding. Secondly, the friction force is independent on the apparent contact area [7].

In general application, friction work is assumed based on Coulomb's dry friction model between solid bodies, which is also at the same time conforms the aforementioned Amonton's law. In details, the mechanical work and heat generation relationship in the presence of sliding friction is explained by contact conditions between hard and soft metallic material interaction. It involves very small scale asperities at the contact surfaces of hard and soft material. As normal force is acting on the rotating tool, it is being distributed onto smaller area asperities at the contact surfaces which resulted in a very high pressure per unit force. Due to relative velocity difference, the normal force causes ploughing of the soft material by the hard materials' asperities. The soft material gets agitated, deformed and finally broken, releasing the stored energy in the form of heat [8].

The released energy which is a very high local thermal energy causes the temperature to rise. The heat energy is eventually transferred and stored into the rotating tool and the lump work material. The heat causes the work material to soften, reduces its strength, broken and deforms into soft material layer in between the rotating tool and the work material. This soft layer is gradually displaced by the rotating tool pin revealing new surface contact condition for another cycle of ploughing action by the rotating tool. The process promotes another cycle of mechanical friction heat generation.

2.2.2 Plastic dissipation heat generation

This phenomenon typically occurs at higher temperature resulting from friction heat mechanism but only significant during dwelling and actual welding phase where work material is confined under the rotating tool. Through friction heat mechanism, work material temperature under the rotating tool is increased to a degree where work material layer at rotating tool interface started to lose it strength, yielded, stick and move with the rotating tool. This phenomenon increases the thermal softening effect from friction heat and causes shears at the work material to work material layers interface. It reduces friction heat mechanism but at the same time introduce high strain rate plastic deformation. As a result, highly localized heat is generated internally within the work material itself away from the rotating tool to work material interface due to dynamic velocity differences and boundary sliding condition [9, 10, 11].

Due to the heat transfer within the process, the plastically deformed material tend to recover its strength thus establishing new lower temperature level and reinitiate mechanical friction heat generation mechanism [12]. Furthermore at the welding phase, heat generation from plastic dissipation mechanism is increased at cost of the travelling rotating tool and the shearing of the work material to higher extend toward trailing edge. It further increases the process temperature as the rotating tool is reintroduced to new contact condition at the leading edge and reinitiates the mechanical friction heat generation mechanism [13].

These heat generation mechanisms throughout FSW process occur in cycles due to the instance of slip and stick contact conditions and the alternating boundary conditions at the material-rotating tool interface [14]. In this regards, heat generation mechanism has most influential effects on the process introducing high strain rate and thermal effect which soften work material adjacent to the tool to work material interface, to be encompassed the at the trailing edge and produces welding nugget which posture the main characteristic of the joint [15].

2.2.3. Heat transfer

Simultaneously, heat generated is constantly being transferred within the system; portions of heat are distributed within the work material, the rotating tool, the holding fixture or backing plate and finally to the ambient. In depth, heat generated is subjected to three dimensional (3-D) heat flows away from the heat source under boundary conditions. Heat input and heat transfer at the rotating tool are indirectly coupled at the rotating tool to work material interface into the work material through heat conduction and incorporated with convective heat transfer effect around the pin in the deformation zone [12, 13, 16, 17]. For the work material, convective and radiative heat transfers are considered for heat exchange at the top work material surface, past the shoulder peripheral while at the same time only convective condition is considered for the bounding surfaces of work material [18, 19]. For the case of backing plate, appropriate variable gap conductance is considered for work material to backing plate interface depending on specific thermal contact resistance condition; of temperature dependent or contact pressure dependent or surface contour of work material or any of the combinations [13, 18, 20].

Ultimately 100% energy generated from mechanical work throughout the process is converted to heat and physical deformation with approximately 88% of heat is conducted and distributed globally through to the lump work material, backing plate and else to the rotating tool [12]. Heat plays a very significant role not only toward the physical success of the joint but also towards the process temperature profile, heat transfer and the work material internal strain and stress distribution. It causes direct influences on the final weldment microstructure and residual stress resulting properties which are strongly correlated to the welding variables and temperature-dependent material properties [21].

2.2.4. *Friction stir welding mechanism*

Weldment is produced in the course of welding phase where layers of materials are forced to move along with and around the rotating tool surfaces at a specific contact condition; fully sliding or sliding and sticking or fully sticking. The soft material layers motion are heavily deformed and driven by the tool rotational direction. It is forced through the retreating side toward trailing edge and downward closed to the pin before finally being forged and deposited at the once occupied volume of the rotating tool pin at the trailing edge under severe plastic deformation [17, 21, 22, 23].

The mechanical torque of the rotating tool causes mechanical shearing to the immediate work material close to the rotating tool and work material interface, forcing soft work material layers to motion and strained, creating flows of fine layers of materials prior progression into weldment. The flow motion or velocity is visually estimated through the grain size and shape, correlated to internal strain rate [24]. The material flow and joining mechanisms are described by the region formed by the FSW process known as the flow zones. It describes the zones where shear layers are visibly distinguished by material characteristics and exhibit the evident of non-melting but severely deformed soft material deposited into amalgamated weldment in layers and flowing manner.

2.2.5. *Welding characteristics*

Temperature profile and history of FSW process are resembled by the distinct regions at the weldment. These regions are characterized by discrete microstructure sizes, shapes and varying properties, produced by the significant thermal effect and mechanical deformation. Under the heat generation, lump effect and heat transfer of the process, thermal profiles are being distributed from the crown shaped heat source around the rotating tool to work material interface toward the peripheral work materials surfaces and edges [22, 25]. These regions are known as; 1) weld nugget, the product of plastic deformation due to the stirring effect deposited behind the rotating tool pin at the trailing edge, 2) thermo-Mechanically Affected Zone (TMAZ) of internally sheared plastic deformation within the work material away from rotating tool to work material interface, 3) heat Affected Zone (HAZ) of structurally altered and thermally affected region due to intense temperature different between TMAZ and base metal temperature region, and 4) base metal of work material which is not physically affected by the thermal effect [26].

Temperature profile within the FSW process portrayed direct relationship of the heat generation, torque generated through the rotating tool, loads exerted throughout the work material and power consumed by the welding process. Both thermal and mechanical effect from heat generation and stirring effect engender welding characteristic in term of stresses, tensile and hardness properties. For which, the rotating tool rotational and traverse speed interchangeably influenced the temperature profile and affectively manipulate the material flow behaviour, weld material chemical composition, microstructure orientation, strain, residual stress, thermal stress, material hardness and strength of the weldment [27, 28, 29]

2.3. Applications of friction stir welding

Recently, FSW won attractive attention by manufacturing industries because of the ability that outperformed other welding technique such as tungsten inert gas (TIG) to weld aluminium. Besides, it has the ability to be adapted in advanced automation system such as robotic which is very predominant in aluminium components and panel fabrications of highly rated technology readiness level [30]. For example, it is an applicable technique adapted for rail cars for fabrication of floor panel part of a Type 700 Shinkansen or bullet train, as well as the aluminium roof, side wall and floor panel for suburban train and other more recent commuter or express rail cars [31]. This is due to its low distortion and its suitability to produce large welded products including prefabricated panels as well as tailored blanks and joints. These advantages are also shared in marine application when it is first commercially used for ship building in 1996 with the ability to joint thick panels, sandwiched, honeycomb panels and corrosion resistance material panels [32]. FSW is well favoured for performing butt join in comparison to conventional arc welding application which also turns out to be significantly viable in terms of low labour cost and shorter welding cycle time [33].

The other important application of friction stir welding is in aeronautical and aerospace industries where aluminium alloy is used as primary material for their construction. The welding process enables manufacturers to completely replace the riveted joints and assemblies of lapped and abutted configuration that are used mainly for fuselage sections, propellant and fuel tanks of commercial air craft as well as space launch vehicle [34]. Thus, the process allows total elimination of the use of thousands of rivets. This has resulted in better quality, stronger and lighter joints at reduced assembly cost for aviation industry. Meanwhile in automotive application, FSW and its recently innovated process of friction stir spot welding (FSSW) are introduced to replace conventional resistive spot welding (RSW) in the transition when aluminium alloy application is ready and becoming ideal to be used for panels in passenger and commercial vehicle bodies. Aluminium application in automotive industry are sought for its prefabricated and tailored panel, good strength-to-weight ratio, potential for reducing fuel consumption, its ease of recycling as well as marked reduction in production cost. This has compelled car manufacturers to use the same concept not only for the body panels, but for other parts as well [35].

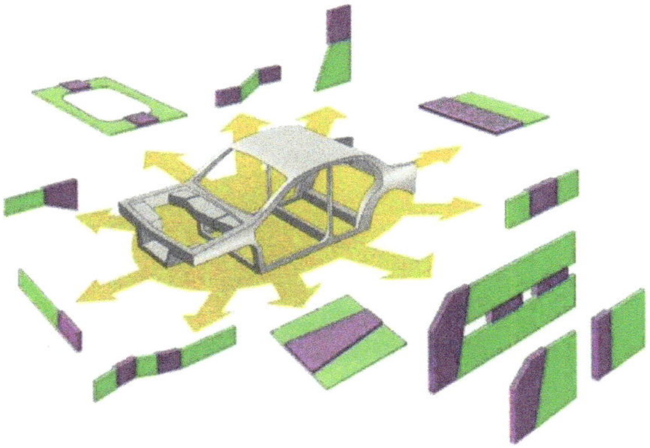

Figure 3. Tailored welded blanks in a passenger vehicle [36].

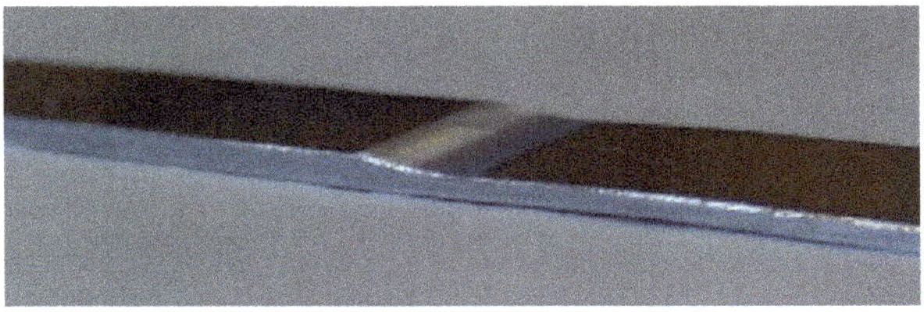

Figure 4. A component of friction stir tailored welded blank [36].

Since its introduction, friction stir welding has matured significantly with widespread use of wide range aluminium alloys for structural applications with its compatibility to be used in ferrous, stainless steel, nickel, copper and titanium alloys. However, only a small percentage of world welding and joining market has implemented the process. It is still relatively underutilised and this is varied among industries, university researches, collaborations and other niche applications. With all the challenges to joints wide variety hard to weld materials, the motivation of the application and adaptation of this noble technology in manufacturing not only to meet quality joints, but for the potential economic value and also for its environmental friendliness.

2.4. Recent development & future outlook

Since the last two decades, there are considerably vast lab and industrial work done on FSW process which leads to the emerging advancement of new materials application and combinations, process improvement, tool designs, welding configurations, tailored blank application and adaptation to automation. Though, the most important innovation in the process itself is its variations; (1) high speed FSW (HS-FSW), (2) ultrasonic stir welding (USW), (3) thermal stir welding (TSW), (4) friction stir spot welding (FSSW), (5) friction stir joining (FSJ), (6) friction stir processing (FSP), (7) friction bonding (FB) [37].

The recent advancement shows the introduction of HS-FSW aiming at reducing the process forces by means of increasing heat generation rate. This reduction in forces is to realize the idea of permitting manual handheld welding work. HS-FSW process is also designed to achieve lighter and portable device or equipment. This will make possible manual handheld welding and eliminates the need for rigid fixturing of the work material and rotating tool.

USW is another variation of the process where ultrasonic energy is used to assist in initial heat generation. The objectives are to reduce process forces and welding time. The major boost of USW is it reduces dependency on the tool shoulder to generate heat and instead, make use of the coupled ultrasonic vibration to further agitate the rotating tool at the contact interface, amplifying the mechanical friction effect.

Different from USW that utilize ultrasonic energy to generate heat, TSW decouples heat generation from mechanical friction of conventional FSW and instead, utilizes external induction heating to increase work material temperature at the welding spot or welding line. Though induction heating may ultimately increase the rate of heat generation but it reduces strain rate of plastic deformation, a prevalent characteristic in FSW. Reduced strain rate will significantly affect the final welding properties [38].

The most common variation of all is the FSSW, which is well known to be used in transportation industries to replace rivet and as well as adhesive joining method. Though this method produces cosmetic defect in the form of pin hole created by the rotating tool, the weldment has sufficient mechanical strength for joining method [39].

While the other FSW variants are dedicated for metallic materials, FSJ method is dedicated for joining thermoplastic materials. It has been used for polypropylene, polycarbonate and high density polyethylene materials [40. The advancement of FSJ method may possibly makes way for mechanical joining method involving plastic matrix composites.

FSP is a unique non-joining method variant of FSW where it utilizes the stirring process to alter microstructure derived from FSW method to be super fine, modified with improved physical properties and at the same time suppressed defects such as porosity and short crack [34]. The introduction of foreign particle into base material during FSP creates new near quality to metal matrix composite or MMC structure and thus improving the material properties. This provides a new platform to produce improved future materials.

The consequent of FSP method emerges as a modified method of FB that allows bonding of overlapping thin plates through locally modified microstructure. The process utilizes less stirring effect due to the short pin geometry and thus differs from common FSW method.

The majority of these FSW variants stress on their ability to increase welding time and at the same time to reduce forces exerted from the welding process. The other purpose of these methods is to provide means of producing joining method involving different type of materials and configurations. More or less, these method share the same mechanism of mechanically manipulating soften work material and formation of fused material at the trailing edge.

3. Friction stir welding parameters

Independent process variables play significant effect on the welding process and the process control. The process variables entail the axial force for plunging, rotating tool rotational speed, rake angle, welding speed and tool geometry. The aforementioned variables strongly affect the heat generation rate, temperature profile within the work material, mechanical power required by the process, material evolution of the weldment and also loads distributed within the work material. These variables are extensively discussed in this chapter to understand the mechanics of joining, process and final weld properties optimizations, where direct measurement can be done experimentally and predicted numerically.

3.1. Effect of welding parameters on joint quality

It has been reported that rotational and traverse speeds have both direct and indirect influence to the final weldment. Its direct influences to the mechanics of joining suggesting the degree of stirring based on the contact condition and multi component loads. Whilst, it's indirect influences to mechanical properties of the weldment is heavily derived from the combination of temperature exposure and tool design [41]. Low rotational speed induces low stirring effect and finally low heat generation rate. At low traverse speed, it increases exposure to heat source, facilitate more material flow and reduced multi component loads, vice versa to high traverse speed but only to an extent. Extreme high rotational speed results too much heat while extreme high traverse speed results less heat, low stirring effect and increase travel resistance due to heavy loads on the rotating tool and hard work material at the leading edge. At this moment, the material is sheared to the lateral side instead of moving around the rotating tool direction [42]. Excessive heat which is generated from either high rotational speed or low traverse speed, or combination of both conditions, significantly reduces the mechanical properties of joint due to microstructure evolution of the regions exposed to excessive heat [43]. Albeit, the appropriate traverse and rotational speed below critical speeds might result optimum heat generation rate and reduced thermal exposure that produce good strength and hardness of weld joints [44, 45].

Torque produced during the welding process depends heavily on the contact conditions which are determined by the rotating tool rotational speed, degree of softness of work material or plasticity at the rotating tool to work material interface, axial load exerted and the tool design [22, 46, 47]. Any changes in the traverse speed at a constant tool rotational speed do not significantly affect the temperature profile compared to changes done in the rotational speed. These make the torque to be insensitive to the traverse speed. In addition, optimum rotating tool design influences torque produced through the effect of sum of contact areas and contact conditions at the interface where it plays major role in plastic deformation or strain work distribution, material transportation or flow and toward process loads and work material temperature [41, 48, 49, 50]. As temperature increases, work material temperature dependent shear stress plump and no longer display full solid properties thus reduces the torque at the rotating tool to work material interface and further reducing power and energy required to produce heat within the process.

3.2. Optimizing process parameters

Considerably vast works has been made in pursue to understand and optimize the physical process of FSW that influenced by the associated variables using both empirical as well as numerical models for the heat generation, material interaction and flow.

3.2.1. Thermal modelling

Heat generation is modelled based on the torque required to rotate a circular shaft relative to the material surface. The model is made by assuming a constant coefficient of friction, pressure distribution and 100% conversion of the shearing work to heat. It is also assumed that the net power required is directly proportional to the tool rotational speed and the tool shoulder radius i.e. $q \approx \omega R$ [51].

Friction work principle used for heat generation model is coupled with plastic work principle to model three-dimensional heat and material flow based on temperature dependent coefficient of friction and temperature dependent pressure distribution for aluminium alloys [26]. In detail, three-dimensional visco-plastic flow and heat transfer have been investigated through solving the equations of conservation of mass, momentum and energy by considering the heat source at two separate conditions. Firstly, visco-plastic flow and heat transfer at the tool to work material interface due to the mechanical friction or due to the plastic dissipation heat generation mechanisms and secondly, the visco-plastic flow and the heat transfer within the layers of soft work material away from the tool to work material interface under the combination of both friction and plastic dissipation heat generation mechanisms [11, 52, 53, 54].

The model is defined by the contact area, the radial distance of the rotating tool pin and shoulder, the material shear stress or the spatially variable coefficient of friction or their combinations, the angular velocity and the exerted normal pressure acted on the work material surface. Then, the model is validated through comparison of the computed heat generation, peak temperature and the total torque exerted on the tool with the experimental results. Contact conditions at the interface are described as sliding, sticking or partial sliding or sticking, known as slip factor is also being adapted. The slip factor is derived experimentally, determined by the plunge force and torque from the welding process. The slip factor yield a proportional relationship between plunge force and heat generation where Coulomb's law of friction is applied to describe the shear forces at the interface [10, 55]. Slip factor is also used to evaluate the welding energy and temperature, utilizing torque based heat input [53]. These works allow the prediction of the welding temperature from the transverse speed, tool rotational speed and the applied force. Ultimately, fully coupled thermo-mechanical model with adaptive boundary condition which applied both thermal and mechanical model is used to predict transient temperature profile, active developed stresses as well as the three-dimensional force components [13].

3.2.2. *Model validation*

The force and temperature measurement experiments are conducted under different welding parameters for model verification by differing the transverse and the tool rotational speed whilst maintaining the constant vertical force. The result is later used for the calculation of the heat input into the tool and workpiece. In relation, numerical model is developed to take account the process effects on the work material such as the material plasticity properties, thermal expansion and stress, cooling effect, stress stiffening, stress distribution, material strain, residual stress as well as the thermal history o the welding process [56, 57, 58].

The mechanical effect is visualised and simulate as material flow model to describe the process parametric effects on the soft work material flow and the welding mechanisms [59, 60]. The physical material flow highlights the particular material strain, its distribution and flow motion around the rotating tool, at the leading edge of advancing and retreating sides and at the trailing edge as well as the actual bond that might occur in the FSW process. Thus, thermo-mechanical model is required to exhibits the importance of the three-dimensional loads and torque exerted by the rotating tool material to determine the best and optimized parameter for the welding process of any materials which is greatly related to the joint and process properties such as residual stresses, process temperature, joint strength and productivity. The parameter optimization is related to variables such as welding speed, rotating tool rotational speed as well as rotating tool design. Proceeded, welding power has been modelled to determine the overall heat input for FSW process based on the traverse speed and tool rotational speed as well as its effect on the material properties. The model has been derived based on the relationship between the rotating tool rotational speed and the represented rotating tool torque with association to the key parameters aforementioned [61, 62].

Significantly all of these works and models represent the correlation of the mentioned independent variables of the associated welding parameters as well as material properties serve the heat generation, temperature distribution and joining mechanism that contribute to the success of the welding process. In summary, the relationship between the independent process variables and the dependent process output to the heat generation mechanisms is best described as in a closed and correlated system, explained by the interaction of process welding variables and key process conditions; physical, metallurgical, heat generation and heat transfer effect [63].

4. Mathematical modelling of welding forces & torque

FSW is not widely available in general application rather than conventional fusion joining methods especially in automotive industry due to shorter process and heat generation cycle. Thus, the true understanding of FSW process is still in far-reaching; understanding the physics and nature of the process, lacking of standards guideline and practice, optimization of the process for typical material usage or application as well as still typically not suitable for small scale robotic or manual handling. The mathematical model on the welding forces and torque is proposed for optimizing FSW process and to study the effects of varying its parameters. The model is to be compared with experimental work [64, 65].

4.1. Model development & assumption

Initial heat generation takes place at the first contact of the rotational tool pin surface and continue throughout the plunging phase where the temperature distribution of the work material is asymmetrical at the leading, trailing edge as well as advancing and retreating side. It is based on the assumption that the interface heat generation is constant with the consideration of the constant rotating tool angular speed, ω, temperature-dependent pressure distribution function, $P(T)$, heat capacity, ρc_p, thermal conductivity, k and constant coefficient of friction, μ_k. Based on Fourier's 2nd law [66];

$$\rho c_p \frac{\partial T}{\partial t} = k\left(\frac{\partial^2 T}{\partial x^2} + \frac{\partial^2 T}{\partial y^2} + \frac{\partial^2 T}{\partial z^2}\right) + \dot{q} \tag{1}$$

Where ρc_p is the heat capacity, x, y, and z are the space coordinate and is the heat source term correspond to heat generated from the welding process. The function of the heat generation is directly related to friction work of the contacting surface thus accounts the sum of contact surface area of the tool. The sum or contact area is represented as function of rotating too plunge depth, h. The data used is as in table 1. Given;

$$A_{(h=0)} = \pi r_p^2 \tag{2}$$

$$A_{(0<h<5)} = A_{(h=0)} + \pi\left[\left(\frac{R_p}{h_p} h_{(0\to5)} + r_p\right)\left(\frac{R_p - r_p}{\sin 2}\right) - r_p s_2\right] \tag{3}$$

$$A_{(h=5)} = A_{(0<h<5)} + \pi(R_s - R_p)^2 \tag{4}$$

For A is the contacting surface between the rotating tool and the work material surface, r_p is the bottom pin radius, R_p is the top pin radius, h_p is the total height of the pin and $\pi r_p S_2$ is the minor cone area.

Furthermore, the torque required to rotate the rotational tool relative to the static workpiece surface under $P(T)$ represents the conversion of mechanical work of the rotating tool. Given;

$$M = \int_0^{M_R} dM = \int_0^R \mu P(T) 2A(h) r dr = \mu P(T) A(h) R^2 \tag{5}$$

Or by simplification of the sum of contact area through cylindrical approximation, given [67, 68];

$$M = \int_0^{M_R} dM = \int_0^R \mu P(T) 2\pi r^2 dr = \frac{2}{3}\mu\pi P R^3 \tag{6}$$

Where M is the interfacial torque of the in contact workpiece surface and rotating tool surface, μ_k is the coefficient of friction, R is the contact surface radius, and $P(T)$ is temperature dependent pressure distribution across the interface.

For fully sliding contact condition and with assumption of all the friction work is converted into frictional heat, the average heat input per unit area and time becomes;

$$q_0 = \int_0^{M_R} \omega dM = \int_0^R \omega\mu P(T) 2\pi r^2 dr = \frac{2}{3}\omega\mu\pi PR^3 = \frac{4}{3}\mu\pi^2 nPR^3 \tag{7}$$

Where q_0 is net power in Watt (Nms^{-1}) and ω is rotational speed ($rads^{-1}$). Apparently, in equation 7, heat input depends on the normal pressure distribution function, contact surface radius, temperature dependent function of coefficient of friction and the rotational speed of the rotating tool which produce transient heat generation, distributed into the work material and thus characterized the process variables for the friction stir welding process.

In order to understand the physics of the welding process in the expression of the mechanical loading associated to the welding process, a mathematical representation is derived based on figure 5;

$$r_1 = -(x_1 + x)i + y_1 j - z_1 k \tag{8}$$

$$r_2 = -(x_2 + x)i - y_2 j - z_2 k \tag{9}$$

$$r_3 = (x_3 - x)i - y_3 j - z_3 k \tag{10}$$

$$r_4 = (x_4 - x)i + y_4 j - z_4 k \tag{11}$$

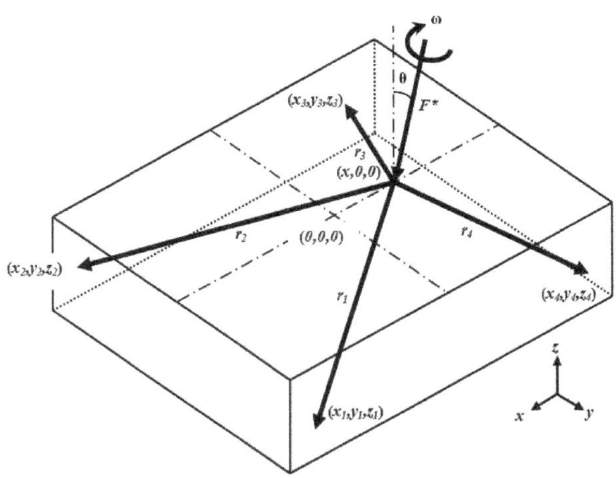

Figure 5. Free body diagram of multi-component load measuring device [65].

$$\sum F = 0 \tag{12}$$

$$F^{*}_{x,y,z} = \sum_{n=1}^{4} Fn_{x,y,z} \tag{13}$$

$$F^{*}\sin\theta i + F^{*}\cos\theta k = (F^{*}_{1x} + F^{*}_{2x} + F^{*}_{3x} + F^{*}_{4x} - F_{1x} + F_{2x} + F_{3x} - F_{4x})i + (-F_{1y} - F_{2y} + F_{3y} - F_{4y})j + (F^{*}_{1z} + F_{1z} + F^{*}_{2z} - F_{2z} + F^{*}_{3z} - F_{3z} + F^{*}_{4z} + F_{4z})k \tag{14}$$

Where F^*, is the plunge force as the function of contact surface area and under the temperature dependent pressure distribution $P(T)$ for a desired depth of penetration. The torque exerted on the work material by the rotating tool mechanical friction work is calculated based on equation 6 and coupled by the moments reacted at each of the measuring references on the workpiece as in figure 5;

$$\sum M = 0 \tag{15}$$

$$M^{*}_{x,y,z} = \sum_{n=1}^{4} rn_{x,y,z} \times Fn_{x,y,z} = 4(Rn_{x,y,z} \times Fn_{x,y,z}) \tag{16}$$

$$M^{*}_{x,y,z} = \frac{2}{3}\mu F^{*}_{x,y,z} R \tag{17}$$

Where;

$$M_1 = M_2 = M_3 = M_4 \tag{18}$$

$$r_1 \times F_1 = r_2 \times F_2 = r_3 \times F_3 = r_4 \times F_4 \tag{19}$$

In figure 5, the moment exerted on the measuring references is in equilibrium and the acting forces exerted by the rotational tool mechanism are determined from equation 18, generally;

$$\frac{1}{4}M^{*}_{x,y,z} = Rn_{x,y,z} \times Fn_{x,y,z} \qquad ;n = 1,2..,4 \tag{20}$$

$$r_n \times F_n = (y_n Fn_z + z_n Fn_y)i - (-x_n Fn_x)j + (-x_n Fn_y - y_n Fn_x)k \tag{21}$$

The general equation can be presented in a matrix form as;

$$\begin{bmatrix} \frac{1}{4}M^{*}_{x} \\ 0 \\ \frac{1}{4}M^{*}_{z} \end{bmatrix} = \begin{bmatrix} 0 & z_n & y_n \\ -z_n & 0 & x_n \\ -y_n & -x_n & 0 \end{bmatrix} \begin{bmatrix} Fn_x \\ Fn_y \\ Fn_z \end{bmatrix} \tag{22}$$

At any static equilibrium where the slip contact condition between the rotating tool and the workpiece surface remain constant, the moment at each of the measuring references remain the same but dependent on the radius of the rotating tool and the function of pressure distribution *P*(*T*) for the respective rotating tool position. The values for the moment are also remained constant at any Cartesian coordinate (*x*, *y*) location of the rotating tool on the surface of the workpiece. Although the basic assumption for the constant contact condition for the function of pressure distribution *P*(*T*) in modelling the process is not physically correct but appropriate to be acceptable in the context of numerical model as average value used throughout the investigation.

Properties/parameter	Value
Work material dimension, mm	200 X 200 X 7
Shoulder radius, mm	9
Tool radius, mm	3
Pin radius, mm	2.852
Pin height, mm	5
Pin conical angle, °	2
Tool angle, °	2
Workpiece material	6061 T-6
Tool material	M42
Coefficient of friction	Figure 7
Plunge forces	Figure 8

Table 1. Summary of data used for loads and torque calculation [5].

4.1.1. Co-planar analysis

A closer approximation is made on the geometry of the work material is to measure the reaction forces and moment caused by the rotating tool from the friction welding process. Equation 22 is reduced by performing co-planar analysis at measuring points of the workpiece as in figure 6 for the initial heat generation until the full penetration of the stirrer tool.

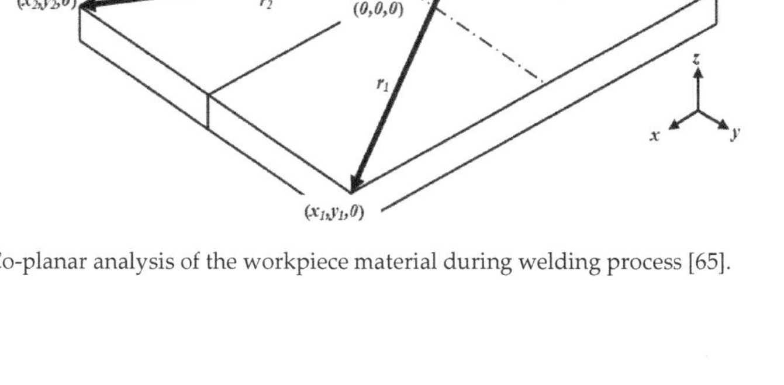

Figure 6. Co-planar analysis of the workpiece material during welding process [65].

Parametric analysis are carried out based on the work material temperature dependent tensile, yield and shear strengths properties, welding configuration, work material dimension and also tool design to investigate loads and torque distributed within the work material. The temperature dependent material properties data used in the calculation are as in figure 7 and 8 used to determine the theoretical plunge forces for the FSW process. In addition, a constant plunge force is proposed as to simulate the experimental work as comparison to numerical analysis.

Theoretically in this work, the mechanical properties suggest the controlling parameters of the welding process especially regarding the plunge force which is based on the contact area and controlled pressure parameter on the tool, heat energy consumed and the mechanical loads applied.

The details of the parametric analysis are explained based on the case studies as carried as;

Case 1: Theoretical plunge force based on temperature dependent tensile strength. This approach is based on the assumption that contact condition is fully sliding follows the temperature dependent coefficient of friction curve as in figure 7.

Case 2: Theoretical plunge force based on temperature dependent yield strength. This approach is also based on the assumption that contact condition is fully sliding.

Case 3: Theoretical plunge force based on temperature dependent shear strength. The assumption for fully sticking contact condition is that the rotating tool which is trusting and work material is separated by a thin layer of plasticised material at the contact interface.

Case 4: Experimental plunge force based on manual force and plunge depth control method [5]. Force and plunge depth control methods are common control method for FSW process [69, 70].

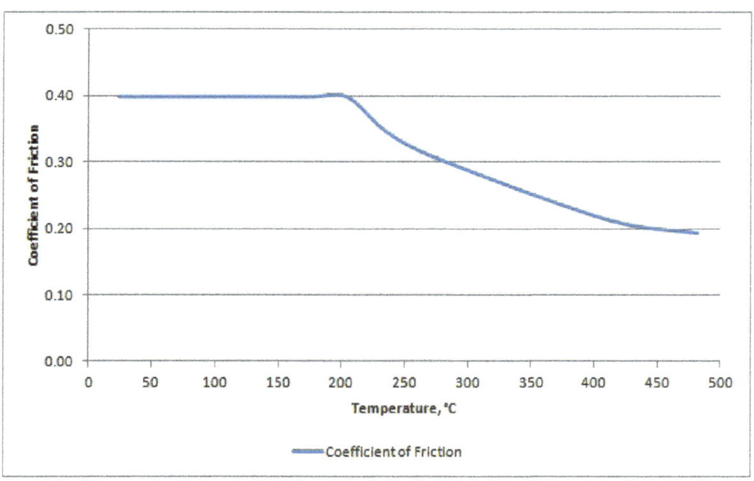

Figure 7. Temperature dependent coefficient of friction [5].

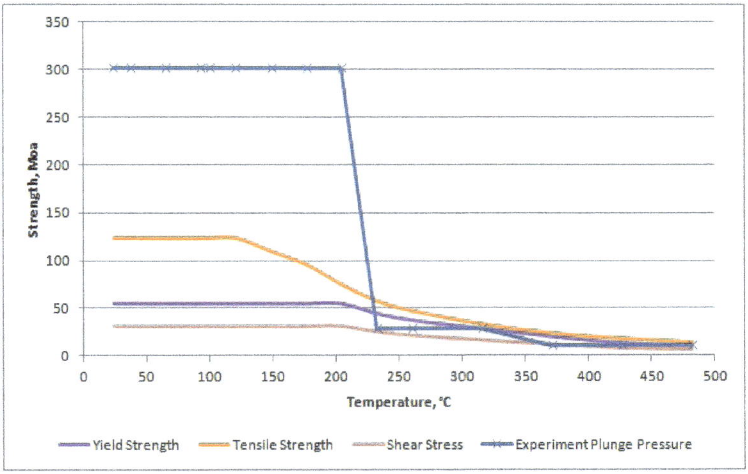

Figure 8. Variable temperature dependent material properties of aluminium alloy 6061 and plunge pressure variation for FSW process cases 1 – 4 [5, 6].

Though in this work, only case 1, 3 and 4 are being considered to visualize the maximum and minimum limits of the parametric effect in comparison to the experimental work. Case 1 encompasses the maximum strength value of the work material prior failure due to the mechanical effect of the FSW process. As for case 3, it represents the minimum requirement of the FSW process to initiate mechanical or physical effect on the work material. Finally, case 4 is the corresponding load suggested based on the work holding fixture as in figure 1 [5, 65].

4.2. Model validation with experimental data

The plunge variation profiles in FSW processes based on the temperature dependent work material properties and experimental plunge pressure value are as exhibited in figure 8. It is noticeable that the cases are significant at temperature range approximately 120°C to 220°C, where the theoretical plunge pressure started to decrease. The work material becomes soft and lost its strength, allowing the rotating tool to displace axially into the work material. The corresponding rotating tool plunge force theoretical calculation based on the plunge pressures schemes aforementioned are compared to experimental result and exhibited in figure 9.

During the plunging phase, the rotating tool thrusting under the plunge pressure profiles and constant rotational speed produces mechanical torque. It acts on the work material and initiates the mechanical friction work thus results the friction heat. The variations of mechanical torque exerted at the rotating tool and the work material interface during the plunging phase are as in figure 10.

Sticking and sliding contact conditions referred as case 1 and 3, exhibit significant torques built up beyond 300°C. For case 4, high torque exerted at initial plunge phase abruptly decrease as the temperature approaches 220°C before it increase back at temperature beyond 300°C. The decrease in torque indicate the softening effect of the work material while the immediate increase of the torque values are due to the rotating tool shoulder surface comes into contact to the work material.

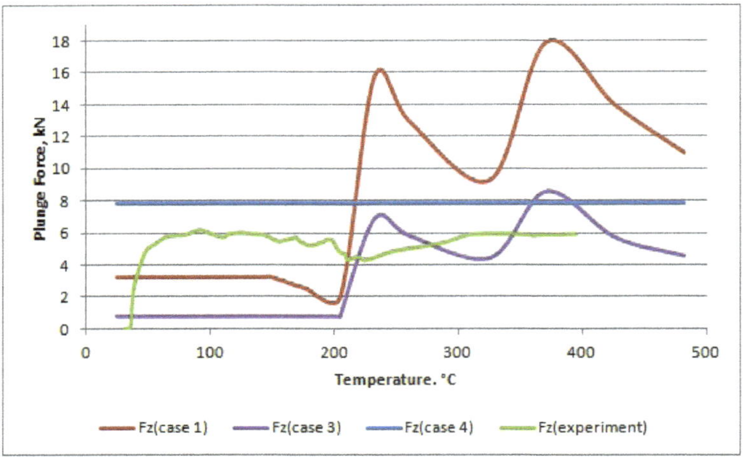

Figure 9. FSW plunge forces as controlling parameters [5].

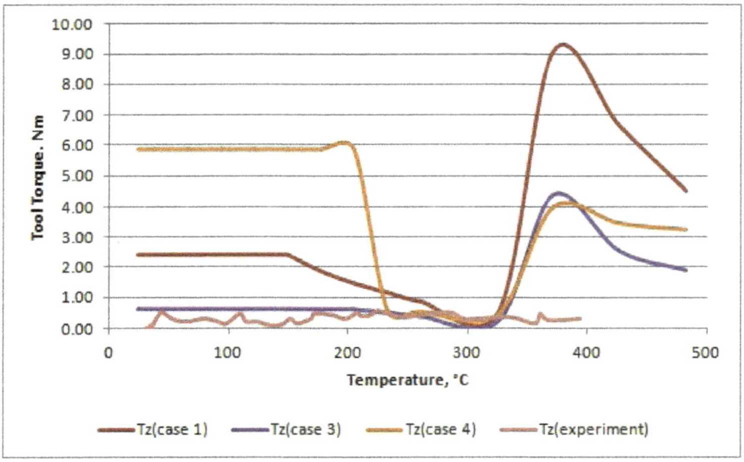

Figure 10. Tool torque exerted at the tool - work material interface for case 1, 3 & 4 [5].

An approximation of the process is based on case 4 where plunge pressure is based on manual control of constant plunge force. The comparison of the theoretical plunge forces and experimental work of the FSW process reveal distinct variation based on the assumption used. However, the values act as references for the actual FSW process that agree to the experimental result as shown in figure 9. The experimental result shows that its profile is approximate to case 4 plunge force scheme only at lower value. As for the rotating tool torque, the experimental value remains closed to the value exerted by case 3 scheme at almost constant trend. It described the nature of material in respond to the heat and mechanical friction during the FSW process.

5. Conclusion

FSW process benefits solid state joining method that has great advantage on light weight material such as aluminium alloy due to its thermal properties which make it difficult to be joined using conventional methods. Similarly to the other welding method, heat generation and heat transfer play major role in determining the success of the joining process as well as predominantly establish the joint characteristics and properties. Though the detail of the process mechanism and the effect on the welding has been widely studied in lab scale, good understanding of the process mechanism provides a better view on choosing the best parameter for the process and finally to achieve the best result in practice.

The future outlook of the process is very promising with new interest on its recent development that allows broader application in term of material used as well as process improvement. In addition, the development of mathematical analysis provides the ability to predetermine the effect of parametric study of the process effect on the work material at a shorter time as well as to be adapted to process automation.

Author details

Jauhari Tahir Khairuddin and Jamaluddin Abdullah
School of Mechanical Engineering, Universiti Sains Malaysia, Nibong Tebal, Penang, Malaysia

Zuhailawati Hussain
School of Materials and Mineral Resources Engineering, Universiti Sains Malaysia, Nibong Tebal, Penang, Malaysia

Indra Putra Almanar
Mechanical Engineering Department, Universiti Teknologi Mara, Permatang Pauh, Penang, Malaysia

Acknowledgement

This present work is supported by Universiti Sains Malaysia through RU-Grant (814084), Universiti Sains Malaysia Institute of Postgraduate Studies Graduate Research Fund (IPS-GRF) and USM-Fellowship scheme, which are greatly acknowledged.

6. References

[1] Thomas WM, Nicholas ED, Needham JC, Murch MG, Templesmith P, Dawes CJ. Friction Stir Butt Welding, GB Patent No. 9125978.8; 1991.

[2] Dawes CJ, Thomas WM. Friction Stir Welding of Aluminium Alloys. TWI Bulletin; 1995; 6 : 124-127.
[3] Hwang YM, Kang ZH., Chiou YC, Hsu HH. Experimental Study on Temperature Distributions Within the Workpiece During Friction Stir Welding of Aluminium Alloy. Int J of Mach Tools Manu; 2008; 48 : 778-787.
[4] Kim D, Badarinarayan H, Kim JH, Kim C, Okamoto K, Wagoner RH, Chung K. Numerical Simulation of Friction Stir Butt Welding Process for AA5083-H18 Sheets. Eur J Mechanics A-Solid; 2010; 29 (2) : 204 – 215.
[5] Jauhari TK. Development of Multi-Component Device for Load Measurement and Temperature Profile for Friction Stir Welding Process [M.Sc Thesis]. Penang: Universiti Sains Malaysia; Unpublished. 2012.
[6] American Welding Society. Welding Handbook, 9th Edition, Volume 3: Welding Process; AWS; 2007.
[7] Popov VL. Contact Mechanics and Frictions: Physical Principal and Application. London: Springer; 2010.
[8] Sherwood BA, Bernard WH. Work and Heat Transfer in the Presence of Sliding Friction. Am J Phys. November 1984; 52 (11) : 1001-1007.
[9] Ravichandran G, Rosakis AJ, Hodowany J, Rosakis P. On the Conversion of Plastic Work Into Heat During High-Strain-Rate Deformation. Shock Compression of Condensed Matter Meeting. American Physical Society; June 24-29, 2001 . p 557 – 562.
[10] Schmidt H, Hattel J, Wert J. An Analytical Model for the Heat Generation in Friction Stir Welding. Model Simul Mater Sc. 2004; 12 (1) : 143-157.
[11] Nandan R, Roy GG, DebRoy T. Numerical Simulation of Three-Dimensional Heat Transfer and Plastic Flow During Friction Stir Welding. Metal Mater Trans A. April 2006; 34A : 1247 – 1259.
[12] Schmidt HB, Hattel JH. Thermal Modelling of Friction Stir Welding. Scripta Mater. 2008; 58 : 332 – 337.
[13] Soundararajan V, Zekovic S, Kovacevic R. Thermo-Mechanical Model With Adaptive Boundary Conditions for Friction Stir Welding AL 6061. Int J Mach Tool Manu. 2005; 45 (14) : 1577-1587.
[14] Schneider J, Beshears R, Nunes Jr. AC. Interfacial Sticking and Slipping in the Friction Stir Welding Process. Mat Sci Eng A. 2006; 435 – 436 : 297 – 304.
[15] Colligan KJ, editors. Friction Stir Welding: From Basic to Applications – 2. The Friction Stir Welding Process: An Overview. CRC Press; 2010. p 15 – 38.
[16] Khandkar MZH, Khan JA. Thermal Modelling of Overlap Friction Stir Welding for Al-Alloys. J Mater Process Manuf Sci. 2001; 10 (2) : 91 – 105.
[17] Schmidt H, Hattel J. Modelling Heat Flow Around the Tool Probe in Friction Stir Welding. Sci Technol Weld Joining. 2005; 10 (2) : 176 – 186.
[18] Khandkar MZH, Khan JA, Reynolds AP. In: Prediction of Temperature Distribution and Thermal History During Friction Stir Welding: Input Torque Based Model. Sci Technol Weld Joining. 2003; 8 (3) : 165 – 174.

[19] Song M, Kovacevic R. Thermal Modelling of Friction Stir Welding in A Moving Coordinate System and its Validation. Int J Mach Tools Manuf. 2003; 43 : 605 – 615.
[20] Shi QY, Dickerson T, Sherchliff HR., Thermo-mechanical Finite Element Modelling in Friction Stir Welding of AL – 2024 Including Tool Loads. In: Proceeding of the 4th International Symposium on Friction Stir Welding. 2003.
[21] Nandan R, DebRoy T, Bhadeshia HKDH. Recent Advances in Friction Stir Welding – Process, Weldment Structure and Properties. Prog Mater Sci. 2008; 53 : 980 – 1023.
[22] Arora A, Nandan R, Reynolds AP, DebRoy T. Torque, Power Requirement and Stir Zone Geometry in Friction Stir Welding Through Modelling and Experiments. Scripta Mater. 2009; 60 : 13 – 16.
[23] Arora A, Zhang Z, De A, DebRoy T. Strains and Strain Rates during Friction Stir Welding. Scripta Mater. 2009; 61 : 863 – 866.
[24] Jata KV, Semiatin SL. Continuous Dynamic Recrystallization During Friction Stir Welding of High Strength Aluminium Alloys. Scripta Mater. 2008; 43 (8) : 743 – 749.
[25] Song M, Kovacevic R. Numerical and Experimental Study of the Heat Transfer Process in Friction Stir Welding. In: Proceedings of the Institution of Mechanical Engineers, Part B: J Eng Manu. 2003; 217 (1) : 73 – 85.
[26] Dong P, Lu F, Hong JK, Cao Z. Coupled Thermomechanical Analysis of Friction Stir Welding Process Using Simplified Models. Sci Technol Weld Joining. 2001; 6 (5) : 281-287.
[27] Elangovan K, Balasubramanian V, Babu S. Developing an Empirical Relationship to Predict Tensile Strength of Friction Stir Welded AA2219 Aluminium Alloy. J Mater Eng Perform. 2008; 17 (6) : 820 – 830.
[28] Moreira PMGP, Santos T, Tavares SMO, Richter-Trummer V, Vilaca P, de Castro PMST. Mechanical and Metallurgical Characterization of Friction Stir Welding Joints of AA6061-T6 with AA6082-T6. Mater Design. 2009; 30 : 180 – 187.
[29] Tutum CC, Hattel JH. Optimisation of Process Parameters in Friction Stir Welding Based on Residual Stress Analysis: A Feasibility Study. Sci Technol Weld Joining. 2010; 15 (5) : 369 – 377.
[30] Smith C, Hinrichs J, Crusan W. Robotic Friction Stir Welding: The State-of-the-Art. In: Proceeding of the 4th International Friction Stir Welding Symposium. Salt Lake City, UT; May 2003.
[31] Kumagai M, Tanaka N. Application of friction stir welding to welded construction of Aluminum alloys. J Light Met Weld Constr. 2001; 39 : 22-28.
[32] Kallee SW, John D, Nicholas ED. Railway Manufacturers Implement Friction Stir Welding. Weld J. 2002; 81 (10) : 47 – 50.
[33] Delany F, Kallee SW, Russell MJ. Friction Stir Welding of Aluminium Ships. In: International Forum on Welding Technologies in the Shipping Industry. Shanghai, China; June 2007.
[34] Arbegast WJ. Friction Stir Welding – After a Decade of Development. Weld J. 2006; (3) : 28 – 35.

[35] Kallee SW , Kell JM, Thomas WM, Wiesner CS. Development and Implementation of Innovative Joining Process in The Automotive Industry. In: DVS Annual Welding Conference. Essen, Germany; September 2005.
[36] ESAB. Friction Stir Welding – the ESAB Way, XA00123720 [Internet]. 2004 [cited 24 May 2012]. Available from:
http://www.esab.co.kr/esab/mc_data/pdf/Friction%20Stir%20Welding.pdf
[37] Nunes AC. Friction Stir Welding. In: Schwart MM, Innovation in Materials Manufacturing, Fabrication, and Environmental Safety, Boca Raton, FL . CRC Press. 2011: 88 – 119
[38] Ding J, Carter R, Lawless K, Nunes AC, Russell C, Suits M. Friction Stir Welding Flies High at NASA. Weld J. 2006; 85 (3) : 54 – 60.
[39] Gerlich A, North T. Friction Stir Spot Welding of Aluminum Alloys. Canadian Weld Assoc J, IIW Special Ed. 2006; 57 – 58.
[40] Mishra RS, Ma ZY. Friction Stir Welding and Processing. Mat Sci Eng. 2005; 50 (1-2) : 1-78.
[41] Elangovan K, Balasubramanian V. Influences of Tool Pin Profile and Tool Shoulder Diameter on the Formation of Friction Stir Processing Zone in AA6061 Aluminium Alloy. Mater Design. 2008; 29 : 362 – 373.
[42] Zhang Z, Liu YL, Chen JT. Effect of Shoulder Size on the Temperature Rise and the Material Deformation in Friction Stir Welding. Int J Adv Manuf Tech. 2009; 45 : 889 – 895.
[43] Liu HJ, Shen JJ, Huang YX, Kuang LY, Liu C, Li C. Effect of Rotation Rate on Microstructure and Mechanical Properties of Friction Stir Welded Copper. Sci Technol Weld Joining. 2009; 14 (6) : 577 – 583.
[44] Ren SR, Ma ZY, Chen LQ. Effect of Welding Parameters on Tensile Properties and Fracture Behavior of Friction Stir Welded Al-Mg-Si Alloy. Scripta Mater. 2007; 56 : 69 – 72.
[45] Patil HS, Soman SN. Experimental Study on the Effect of Welding Speed and Tool Pin Profiles on AA6082-O Aluminium Friction Stir Welded Butt Joints. Int J Eng Sci Technol. 2010; 2 (5) : 268 – 275.
[46] Zhang Z, Zhang HW. Numerical Studies on the Effect of Transverse Speed in Friction Stir Welding. Mater Design. 2009; 30 : 900 - 907
[47] Cui S, Chen ZW, Robson JD. A Model Relating Tool Torque and Its Associated Power and Specific Energy to Rotation and Forward Speed During Friction Stir Welding/Processing: Inter J Mach Tools Manuf. 2010; 50 : 1023 – 1030.
[48] Buffa G, Hua J, Shivpuri R, Fratini L. Design of Friction Stir Welding Tool Using the Continuum Based FEM Model. Mater Sci Eng A. 2006; 419 : 381 – 388.
[49] Hattingh DG. Blignault C, van Niekerk TI, James MN. Characterization of the Influences of FSW tool Geometry on Welding Forces and Weld Tensile Strength Using an Instrumented Tool. J Mater Process Technol. 2008; 230 : 46 – 57.
[50] Arora A, De A, DebRoy T. Toward Optimum Stir Welding Tool Shoulder Diameter: Scripta Mater. 2011; 64 : 9 – 12.

[51] Frigaard Ø, Grong Ø, Midling OT. A Process Model for Friction Stir Welding of Age Hardening Aluminum Alloys. Metall Mater Trans A. May 2001; 32 (5) : 1189-1200.
[52] Buffa G, Hua J, Shivpuri R, Fratini L. A Continuum Based FEM Model for Friction Stir Welding – Model Development. Mater Sci Eng A. 2006; 419 : 389 – 396.
[53] Nandan R, Roy GG, Lienert TJ, DebRoy T. Three-Dimensional Heat and Material Flow During Friction Stir Welding of Mild Steel. Acta Mater. 2007; 55 : 883 – 895.
[54] Ulysse P. Three-Dimensional Modeling of the Friction Stir-Welding Process. Int J Mach Tools Manuf. 2002; 42 : 1549-1557.
[55] Colegrove PA, Shercliff HR, Zettler R. Model for Predicting Heat Generation and Temperature in Friction Stir Welding From the Material Properties. Sci Techno Weld Joining. 2007; 12 (4) : 284-297.
[56] Chen, CM, Kovacevic, R. Thermomechanical Modeling and Force Analysis of Friction Stir Welding by Finite Element method. In: Proceeding of the Institution of Mechanical Engineers, Part C. J Mech Eng Sci. 2004; 218 : 509-519.
[57] Hamilton C, Dymek S, Sommer. A Thermal Model for Friction Stir Welding in Aluminum Alloys. Int J Mach Tool Manuf. 2008; 28 : 1120-1130.
[58] Chen CM, Kovacevic R. Parametric Finite Element Analysis of Stress Evolution During Friction Stir Welding. In: Proceeding of Institute of Mechanical Engineers Part B. J Manuf. 2006; 22 (28) : 1359 – 1371.
[59] Fratini L, Buffa G, Palmeri D, Hua J, Shivpuri R., Material Flow in FSW of AA7075 – T6 Butt Joints: Numerical Simulations and Experimental Verifications. Sci Technol Weld Joining. 2006; 11 (4) : 412 – 421.
[60] Zhang W, Zhang Z, Chen JT. 3D Modeling of Material Flow in Friction Stir Welding Under Different Process Parameter. J Mater Process Techn. 2007; 183 : 62-70.
[61] Pew JW, Record JH, Nelson TW, Sorensen CD. Development of A Heat Input for Friction Stir Welding. In: ASM proceeding of International Conference: Trends in Welding Research. 2005: 247 – 251.
[62] Pew JW. A Torque-Based Weld Power Model for Friction Stir Welding [Msc thesis]. Brigham Young University; December 2006.
[63] Colligan KJ, Mishra RS. A Conceptual Model for The Process Variables Related to Heat Generation in Friction Stir Welding of Aluminum. Scripta Mater. 2008; 58 : 327 – 331.
[64] Benedyk JC: Aerospace Structural Metals Handbook – Nonferrous Alloys; CINDAS LLC. 2008: 68.
[65] Jauhari TK, Indra PA, Zuhailawati H. Mathematical Model for Multi Component Forces & Torque Determination in Friction Stir Welding. Adv Mater Res. 2011; 230-232 : 1255 - 1259.
[66] Incropera, DeWitt: Fundamentals of Heat and Mass Transfer, 2th Edition; Singapore; Wiley. 1990.
[67] Midling OT, Grong Ø. A process model for friction welding of Al-Mg-Si alloys and Al-SiC metal matrix composites; 1: HAZ temperature and strain rate distribution. Acta Metall Mater. 1994; 42 : 1595-1609.
[68] Crossland B. Friction Welding. Contemporary Physics. 1971; 12 (6) : 559-574.

[69] Longhurst WR, Strauss AM, Cook GE. Enabling Automation of Friction Stir Welding: The Modulation of Weld Seam Input. J Dyn Sys Meas Control. 2010; 132: 1-11.
[70] Davis TA, Shin YC, Yao B., Observer-Based Adaptive Robust Control of Friction Stir Welding Axial Force. IEEE-ASME Trans Mechatron. 2011; 16 (6) : 1032-1039

Analytical Model for Estimating the Amount of Heat Generated During Friction Stir Welding: Application on Plates Made of Aluminium Alloy 2024 T351

Miroslav Mijajlović and Dragan Milčić

Additional information is available at the end of the chapter

1. Introduction

Two decades after its invention, friction stir welding (FSW) has the status of a novel and promising welding technique that has not yet been fully described, investigated or utilized in industry [1]. It is a process of joining two metals/alloys, with relatively simple equipment, that initializes complex physical processes in/around the parts that are being joined, resulting in the monolith structure of these parts.

FSW is a patented [2-4] solid state welding technique that needs no consumables or shielding gasses (except in some special cases) for the creation of a welded joint. Welding is performed with a specialized, usually cylindrical, welding tool which mounts into the spindle of a machine that can rotate the tool around its axis (Fig. 1, a).

During the last few years, FSW has been standardized for the welding of aluminium [6, 7] and its principle of operation has been fully described for single welding tool application [7]. Welding parts (recognized as a base metal, workpieces) mount on the backing plate (anvil) [7] and rigidly clamp in such a manner that abutting or dilatation of the base metal is prevented. The rotating welding tool is inserted into the base metal by (axial) force at a start point on the joint line and travels along it [7]. While it travels, the welding tool "machines" material in the base metal in the zone near the traveling path and confines it in the working zone in a mixture that is deposed beside the welding tool as a weld. When the welding length is reached, the welding tool retracts from the base metal and welding is completed.

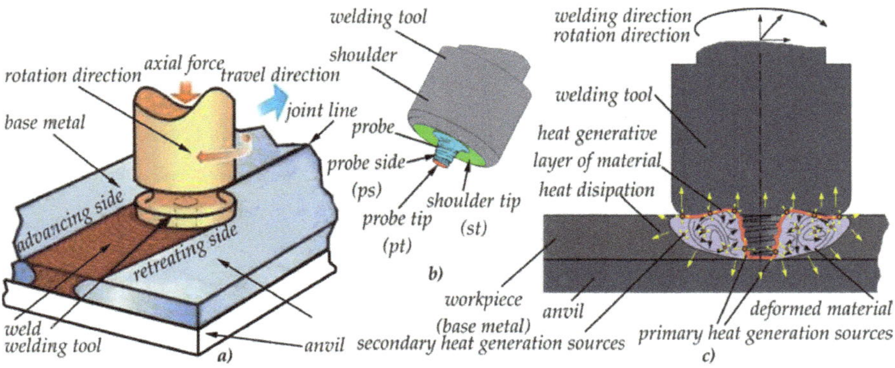

Figure 1. Schematic of FSW: a) principle of operation, b) welding tool, c) schematic of heat generation during FSW [5]

2. Heat generation during FSW

During FSW, the welding tool (Fig. 1, b) slides over the base metal, stirring, deforming, and mixing it. The base metal, anvil, and welding tool increase in temperature due the influence of the welding tool on the base metal. This change in temperature is a sure sign of heat generation caused by frictional contact that takes place during the welding process.

Thermodynamics recognizes several different types of heat transfer from a hotter to a colder body [8, 9]. Both the heat transfer and heat as an energy type have been investigated for a number of cases. However, a challenge appears when heat generation occurs as a result of the contact of two bodies. Heat generation is a process of energy transformation that takes place when one form of energy transforms into heat [8, 9]. This transformation is complex and it depends on the nature of the contact between the bodies, delivered loads, what materials are in contact, the surroundings, movement of the bodies etc. [9, 10]. Heat generated during FSW is the product of the transformation of mechanical energy delivered to the base metal by a welding tool. The transition of mechanical energy from the welding tool to the base metal happens between these bodies [10, 11]. Understanding that heat generates when a metallic body receives an "energy boost" and recognizing the dominant physical processes involved in the contact between the welding tool and base metal (friction, wear, adhesion, deformation, recrystallization of material, etc. [5, 11]), some might say that heat during FSW is primarily generated due to friction and deformation processes that appear during FSW [11]. Friction processes always appear in boundary layers and, therefore, the frictional heat generates in the boundary heat generative layer. Deformational heat appears wherever the deformation of base metal appears: in the boundary layer as well as in zone of deformed material around the welding tool [5, 11-13]. Heat generates due to other processes (e.g. infrared radiation, vibrations) but at a much lower intensity than results from friction and deformation.

Mechanical energy primarily transforms into heat when the welding tool contacts the base metal, while secondarily it transforms in deformed material around the welding tool (Fig. 1, c). That is the reason why the heat generative layer is the primary heat generation source while deformed material around the welding tool is the volume of material where secondary heat generation sources appear [5]. Understanding the process of heat generation and estimating the amount of heat generated during FSW are complex and challenging tasks that requires a multidisciplinary approach. Estimating the amount of heat generated during FSW aids in understanding the appearance of thermal stress and structural changes in base metals inflicted by heat. Understanding the heat generation in FSW might help in selecting the optimal technological parameters of FSW (e.g. rotation frequency, travel rate) from aspects of minimal thermal stresses and deformations, energy consumption etc.

2.1. State of the art

Heat generation and heat transfer became a topic of research related to FSW during mid 1990s. However, understanding heat generation and heat transfer processes within FSW requires understanding several other physical processes: material flow around the welding tool, contact pressure inflicted by the welding tool, the friction coefficient, wear, change of thermo-mechanical properties and heat transfer coefficients etc. Nandan et al. [14] gives a review of thermal processes in FSW, from the invention of FSW until 2008.

Chao and Qi [15] have introduced a 3-D heat transfer model in FSW with constant heat input. Constant heat flux at the shoulder of the welding tool, constant contact pressure and pure Coulomb's friction law for estimating shear stress, and heat were the main assumptions of the model. The experimental welding of plates made of aluminum alloy 6061-T6 was performed and the temperature history of welding plates was estimated. Heat input was adjusted ("trial and error" principle) until numerical and experimental temperatures were matched. As such, this model is the first model developed for estimating the amount of heat generated during FSW. Frigaard and Grong [16] presented a process model for heat flow in FSW, where they assumed that heat is generated only by friction on the tops of shoulders and probes. Heat input and friction coefficients were adjusted during the welding process to keep the calculated temperature below the melting point of base metal material. Heat input was a moving heat source with a linear distribution of heat flux at the contact surface. Gould and Feng [17], and later Russell and Shercliff [18], have applied the Rosenthal equation [19] for describing the moving heat source, heat flux distribution, and heat transport within base metals, welding tools and the surrounding area. Models consider friction heat only at the shoulder and use a finite difference method for a numerical solution of the heat equation. Russell and Shercliff [18] based the heat generation on a constant friction stress equal to the shear yield stress at elevated temperature, which is set to 5% of the yield stress at room temperature. The heat input is a pure point or line source. Colegrove et al. [20] have used an advanced analytical estimation of the heat generation on the welding tool with a threaded probe to estimate the heat generation distribution. The results show that the fraction of heat generated by the probe is about 20% of the total amount. Shercliff and Colegrove [21] developed a material flow model that investigates the influence of threads on the probe on material flow. An advanced viscous material model is introduced and the influence of different contact conditions prescribed as the boundary condition is analyzed. A thorough presentation of analytical estimates of the heat generation in FSW and influence of material flow on heat generation is given, as well. Khandkar et al. [22] introduced a torque based heat input model where experimentally estimated torque is a heat source. Khandkar modeled advanced heat transfer within the FSW process with frictional and deformational heat input into the process. Song and Kovačević [23] investigated the influence of the preheating period on the temperature fields in FSW. A sliding condition of the welding tool over the base metal was assumed and an effective friction coefficient and experimental plunge force are input into the heat source expression. Schmidt and Hattel [12] have defined an analytical model for estimating the amount of heat generated during FSW that recognizes the shoulder and the probe of the welding tool as heat sources and concludes that about 89% of heat is generated at the shoulder. Heat has friction and deformation components and the total heat is a sum of both with influence of the contact state variable [12, 24]. The effective value of the friction coefficient was used in calculations. Reliability of the previously proposed ideas and principles of heat generation were summarized by Nandan et al. [25]. Nandan has performed FSW of dissimilar aluminum alloys and his results have shown that a constant state variable (also referred as an extent of slip) gives values close to sticking.

Further advances in heat generation and modeling included finite element analysis (FEA) in FSW. Ulysse [26] presented a 3-D visco-plastic FEA model using the commercial software FIDAP. The heat generation was determined to be a product of the effective stress and the effective strain-rate. Results show that the model consistently over predicted the measured temperatures probably from an inadequate representation of the constitutive behavior of the material used in FSW. Steuwer et al. [27] used the experimentally observed mechanical power as input in the model. The influence of tool loads on residual stresses was investigated. Chao et al. [28] recognizes two boundary value problems in heat transfer: heat transfer in the welding tool as steady state while considering heat transfer within workpieces as transient. The amount of the heat that flows to the tool dictates the life of the tool and the capability of the tool in the joining process. In discussions, it is shown that the majority of the heat generated from the friction, about 95%, is transferred into the workpiece and only 5% flows into the tool and the fraction of the rate of plastic work – deformation, dissipated as heat is about 80%. Heurtier et al. [29] presented a 3-D model based on the fluid-velocity fields where the tool shoulder and the plastic strain of base material near the welding tool were heat sources. The model has shown good agreement regarding the numerical and experimental results. Santiago et al. [30] introduced a model with rigid and visco-plastic materials in which the plates move towards the rotating tool and the material flow at the interface is specified as a boundary condition. The results estimated from the model correspond to the steady state of the FSW process that has been proposed by Chao [28]. Schmidt [31] has adopted a fully coupled thermo-mechanical dynamic analysis model also aiming to achieve the steady welding state in ABAQUS/Explicit.

Colligan [32] gave a conceptual model that describes dominant parameters affecting heat generation including a detailed description of the existing literature and the principles of specific physical processes in FSW, e.g. friction coefficient. Kalya et al. [33] investigated torque, specific energy and temperature during FSW and gave a correlation between torque, angular frequency and travel rate. The conclusion was a simple correlation between the friction coefficient and mathematically modeled torque. Kumar et al. [34] proposed an experimental setup for estimating the friction coefficient for different contact pressures, temperatures and materials of workpieces. A number of results showed that the friction coefficient varies during FSW from 0.1 up to 4. Setup and the mathematical model are limitedly applicable due to the design of the setup, kinematical complexity of the FSW and approximations involved in the mathematical model. Colligan [35] has investigated the material flow around the welding tool using a tracer embed and "stop action" technique. The results gave deep insight in the material flow patterns, influence of the thread on material flow, influence of rotation direction, as well as some metallurgic and details about heat transport in the zone of the deformed material. Nandan et al. [36] gave some numerically estimated results about viscoplastic material flow, heat transport, viscosity changing and material deposition behind the welding tool. Ouyang and Kovačević [37] have shown that material flow is always dissimilar along the joint line when welding either similar or dissimilar materials. Material flow-patterns can be easily traced in the vortex-like microstructure known as the welding nugget. Material in the zone of the nugget suffers significant plastic deformation, thermal recrystallization and acts like a quasi heat source. Lorrain [13] has explained in detail the shear layer of material and material flow patterns around the unthreaded probe. The existence of the rotating layer in the material near the welding tool is pointed out and it is concluded that material flow is significantly lower when a welding tool without threads is used. Lack of material flow had no influence of the strength of the weld.

2.2. Analytical model for estimating the amount of heat generated during FSW

Heat generation is an unavoidable process following the friction stir weld-creation process. Since FSW is a welding procedure that uses a welding tool as an initiator of the joining process of workpieces, the welding tool delivers activation energy [38, 39] to workpieces and the joining of the workpieces is achieved while heat generates.

This study presents an analytical model for estimating the amount of heat generated during FSW [5]. The model recognizes geometrical, kinematic, physical and energetic possibilities of heat generation during FSW, recognizes dominant parameters affecting the heat generation process and uses them to estimate how much mechanical power is transformed into heat. Existing models for estimating the amount of heat generated during FSW [12, 15, 17, 20, 22, 23, 25, 26] recognize many parameters affecting the heat generation process. Some of them are topology and geometry of the welding tool, technological parameters (tool rotation speed n [rpm] or angular velocity ω [rad/s], travel rate v_x [mm/s], tilt angle, etc.), loading (axial force F_z, torque M_t etc.), physical phases of FSW, duration of the welding procedure, duration of certain phases of the welding procedure, etc. Furthermore, these parameters initiate other parameters that affect heat generation process: friction coefficient μ, contact pressure p, shear stress τ, temperature T, mechanism of heat generation (defined over e.g. the constant state variable δ), etc. However, presented models simplify FSW assuming e.g. constant friction coefficient [12], constant contact pressure [15], pure frictional heat generation [12, 15, 17, 23], heat generation only due work of the largest part of the welding tool [20, 23, 25, 26], no heat generation when temperature in the workpiece reaches melting point [22, 23] etc. Such assumptions are affecting the usability and the precision of results derived by developed models.

The model presented here considers many of the previously analyzed parameters. Special care in the model is given, beside estimating values of parameters, to the mutual dependences between parameters and their influence on the heat generation process. Such dependences are numerous and it is not possible to recognize all of them. Furthermore, many of them are too difficult to be explained analytically and require the numeric calculations and the experimental estimation/validation. These are the reasons why analytical model considers only the most important dependences (Fig. 2, a).

Because of the nature of this approach, the proposed analytical model relies on three major elements: analytic algebra, numerical calculations and experimental data [5]. The analytic algebra is based on existing research and results but includes some improvements. The algebra is developed for a complete welding tool, involves more dominant parameters in the calculations than in previous models, recognizes more dependencies between parameters, neglects fewer parameters and has a shorter calculation time. One of the improvements of the algebra is the implementation of a numerical material flow model with respect to energy balance in workpieces. The numerical calculations use adequate numerical procedures to give good precision and convergence during a short-computing time.

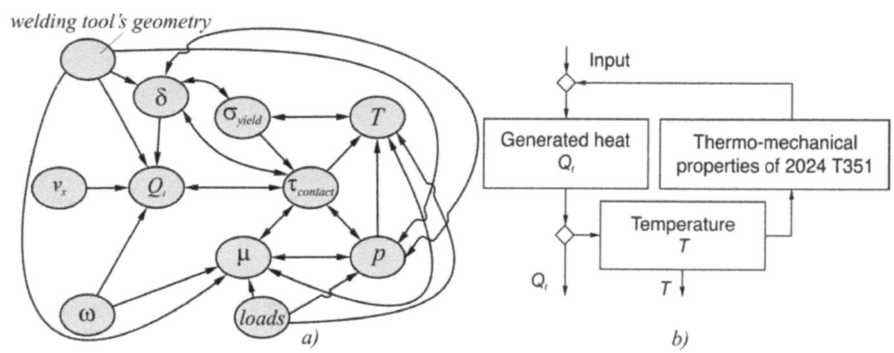

Figure 2. a) Schematic of mutual dependencies between generated heat and dominant influencing parameters [5, 44], b) Partial algorithm for generated heat estimation [5]

The analytical model gives precise results only if experimentally estimated parameters are involved in the model. Furthermore, verification of the analytical model can be done by comparing the results from analytical model with experimentally estimated results. Experimental data is obtained during the welding of the workpieces made of Al alloy 2024 T351 and is used as input and for verification of the analytical model. Mutual dependences of the parameters affecting the heat generation are derived in iterative work regime of the analytical model: time and space are discretized and conditions numerically estimated for the present discretized moment of time are the input for the future discretized moment of time (Fig. 2, b).

Since welding tool is the main initiator of the welding and the heat generation process, it s important to analyze the welding tool and its influence on workpieces as well as the physical engagement of the welding tool while welding.

2.2.1. *Active surfaces, active surfaces engagement and physical phases of the FSW process*

A number of different types of welding tools have been introduced from 1992 until the present [1, 40, 41]. They differ in shape, dimension, mechanical properties etc., and every tool is applicable to a specific material and limitedly applicable to some others. However, all welding tools have the same basics: they consist of at least one shoulder carrying at least one probe directly involved in welding. Recently advanced bobbin tools [7] have appeared when the welding tool has two shoulders. No matter how complex or simple the welding tool is, a limited portion of the welding tool is in constant contact with the base metal and performs the welding. The welding contact region (WCR) on the welding tool consists of three areas called the active surfaces of the welding tool (ASWT). There are always three of them: probe tip (pt), probe side (ps) and shoulder tip (st) (Fig. 1, b). Complete welding and all physical processes following it appear on these surfaces or close to them [5]. Probe tip is usually the smallest ASWT located at the top of the probe. It can be flat, curved or flanged, and rounded at the corner where it connects with the probe side. The probe side is cylindrical or conned ASWT sharing the same rotational axis with the probe tip. The area of the probe side is enlarged by various threads or flanges that help in more intensive mixing of material into the weld [41]. The root of the probe side connects with the shoulder. The shoulder tip is the largest ASWT, usually flat or curved in a manner that creates a "reservoir" for flashed material that come from the workpieces. It is confined to the top surface of the weld and has a role in imperfection-free weld creation [11].

At the beginning of the FSW process, the welding tool is positioned above the workpieces and the rotation axis of the welding tool is (nearly) perpendicular to the joint line.

After positioning, the welding tool starts to rotate at a constant rate (n revolutions per minute, angular velocity of ω [rad/sec]) and slowly plunges into the workpieces in the direction of the $-z$-axis. That is the start of the welding process starts $t=t_0$. Plunging stops when the plunging depth is reached ($t=t_1$, duration of the plunging is $t_{pl}=t_1-t_0$). The plunging depth is equal to the height of the workpieces or slightly smaller and it is achieved at a constant travel rate of the welding tool of v_z. The welding tool continues to rotate and dwells until $t=t_2$. During this time period ($t_{dw1}=t_2-t_1$) the workpiece material is being prepared for the welding: it heats and softens in the area near the welding tool. Afterwards the welding tool begins a translational movement along the joint line (x-axis) at a constant travel rate of v_x. The rotation and translation of the welding tool make the workpiece material (near the welding tool) deform, stick and mix into a monolith composition (weld) that is deposed in the area behind the welding tool. Movement of the welding tool along the joint line lasts until the welding length l is reached, at $t=t_3$. This period ($t_w=t_3-t_2$) is the productive phase of the welding process. Translation of the welding tool stops and the tool dwells at the end point until $t=t_4$ ($t_{dw2}=t_4-t_3$). The welding tool then moves in z direction and leaves the weld and workpieces. When the welding tool is completely removed from the workpieces ($t=t_5$, $t_{po}=t_5-t_4$) the welding process is over. The physical phases of FSW are shown in Fig. 3. In certain circumstances, dwelling can be excluded from the welding process, however, a full FSW process consist of the plunging phase (t_0 to t_1), first dwelling phase (t_1 to t_2), welding phase (t_2 to t_3), second dwelling phase (t_3 to t_4), and pulling out phase (t_4 to t_5) [5].

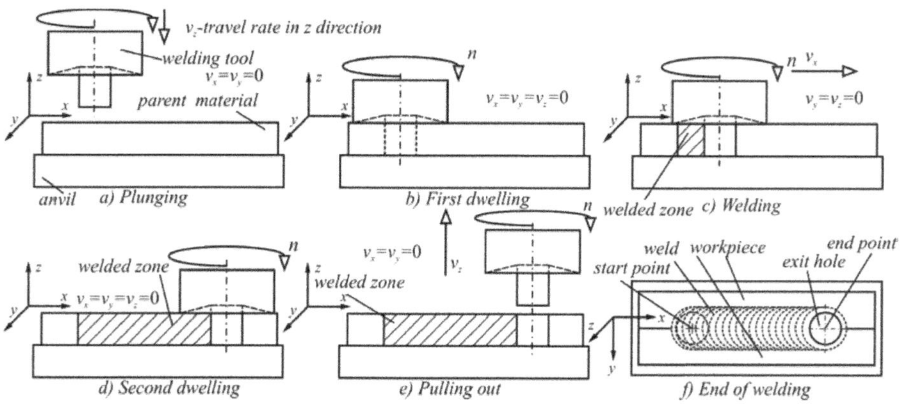

Figure 3. Physical phases of the friction stir welding process

Active surfaces of the welding tool are involved differently in the welding process during a complete cycle of welding, and engagement of every active surface varies during the cycle (Fig. 4, b). The probe tip is involved in welding from the beginning of the welding process until the end of the second dwelling phase (t_0 to t_4). Since the complete probe tip is fully involved in welding, engagement of the probe tip is considered as maximal. The probe side is involved in the welding process when intensive plunging appears (at $t_{pl'}$)[5, 42-46]. Engagement of the probe side rises with the rise of the probe plunge into workpieces. With the end of the plunging phase (t_1), engagement of the probe side reaches a certain value and keeps it during the complete first dwelling phase. When the welding phase starts (at t_2), engagement of the probe side rises toward maximal. This value is reached when the welding process stabilizes – travel rate (s) v_x reaches a steady value (at $t_{pl''}$) and it keeps constant until the end of the second dwelling phase (t_4) and afterwards it decreases. The shoulder tip is involved in the welding process before the end of the plunging phase (t_{st}) and it reaches full engagement when plunging stops (t_1) and keeps it until the end of the second dwelling phase.

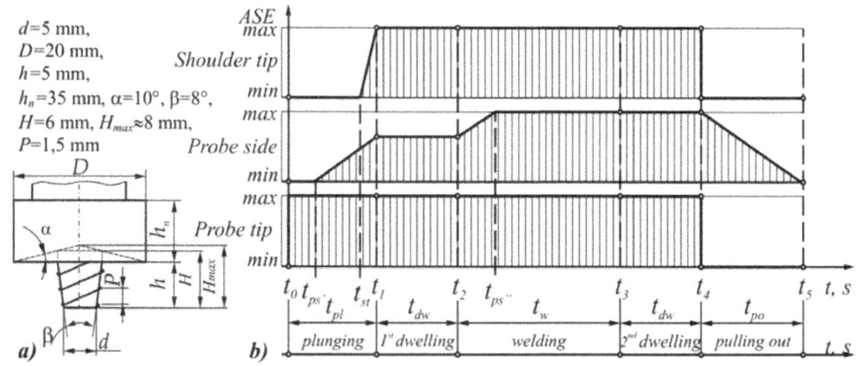

Figure 4. a) Welding tool used for experiments [5], b) Active surface engagement (ASE) [44]

2.2.2. Estimating the amount of heat generated during FSW

As previously mentioned, the heat generation process within FSW is a process that transforms mechanical energy (power) into heat. If η_Q represents a heat transformation [5], the total amount of heat generated during FSW - Q_t is a function of mechanical power P_a delivered to the welding tool:

$$Q_t = \eta_Q P_a \text{ [W]}, \quad \eta_Q = (0,\ 1) \tag{1}$$

The welding tool performs dual movement: translation (tr) and rotation (rot), and the total amount of generated heat is the sum of translation Q_{ttr} and rotational-generated heat Q_{trot}:

$$Q_t = Q_{ttr} + Q_{trot} = \cancelto{0}{Q_{ttr}} + Q_{trot} \tag{2}$$

The amount of translation heat is significantly smaller than amount of rotational heat [5, 12] and it can be neglected in analysis.

Heat is generated at or near the ASWT [5, 10, 12] and the total amount of generated heat is the sum of heat generated on every ASWT:

$$Q_t = Q_{pt} + Q_{ps} + Q_{st} \tag{3}$$

where Q_{pt} – the amount of heat generated at probe tip, Q_{ps} – the amount of heat generated at probe side and Q_{st} – the amount of heat generated at shoulder tip.

Simplifying the analysis and assuming that the total amount of mechanical power transforms into heat (η_Q =1), the total amount of heat becomes:

$$Q_t = Q_{tr} = P_a \tag{4}$$

Mechanical power depends on angular frequency ω and torque M_t and the total amount of generated heat is:

$$Q_t = \omega M_t \tag{5}$$

and

$$\mathrm{d}Q_t = \omega \mathrm{d}M_t = \omega r \mathrm{d}F_t = \omega r \tau_{cont} \mathrm{d}A \tag{6}$$

where $\mathrm{d}F_t$ - infinitesimal force, r - distance of a infinitesimal segment, $\mathrm{d}A$ - infinitesimal area, τ_{cont} - contact shear stress in material.

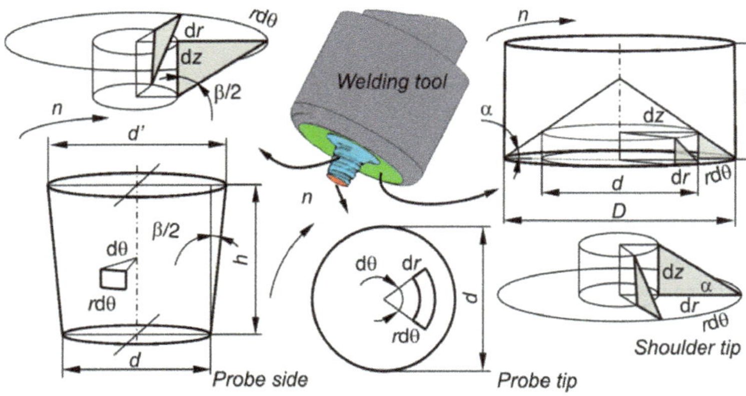

Figure 5. Active surfaces of the FSW welding tool

Different topologies of active surfaces result in different amounts of heat generated on them that give different expressions for estimating the amount of generated heat (Fig. 5). After the integration of Eq. 6, the expressions for the analytical amount of heat generated on every ASWT are, respectively:

$$Q_{pt} = \int_0^{2\pi}\int_0^{d/2} \omega r^2 \tau_{cont} \mathrm{d}\theta \mathrm{d}r = \frac{2}{3}\pi\omega\tau_{cont}\left(\frac{d}{2}\right)^3 \tag{7}$$

$$Q_{ps} = \int_0^{2\pi}\int_0^{h} \omega\left(\frac{d}{2}\right)^2 \tau_{cont}\left(1+\mathrm{tg}\frac{\beta}{2}\right)\mathrm{d}\theta \mathrm{d}z = 2\pi\omega\tau_{cont}\left(\frac{d}{2}\right)^2 h\left(1+\tan\frac{\beta}{2}\right) \tag{8}$$

$$Q_{st} = \int_0^{2\pi}\int_{d/2}^{D/2} \omega r^2 \tau_{cont}\left(1+\tan\alpha\right)\mathrm{d}r\mathrm{d}\theta = \frac{2}{3}\pi\omega\tau_{cont}\left[\left(\frac{D}{2}\right)^3-\left(\frac{d}{2}\right)^3\right]\left(1+\tan\alpha\right) \tag{9}$$

where: d - nominal diameter of probe, D - diameter of shoulder, h - height of probe, α - cone angle of shoulder, β - cone angle of probe.

There is heat generated by friction (frictional heat) and heat generated by deformation (deformational heat) [5, 10-12]. Both types of heat appear simultaneously on every ASWT and both influence one another. Considering both types of heat and their mutual influence on one another, the total amount of heat generated on the probe tip, probe side, and shoulder tip are, respectively:

$$Q_{pt} = \left(1-\delta_{pt}\right)Q_{pt}^{fr} + \delta_{pt}Q_{pt}^{def} \tag{10}$$

$$Q_{st} = \left(1-\delta_{st}\right)Q_{st}^{fr} + \delta_{st}Q_{st}^{def} \tag{11}$$

$$Q_{ps} = \left(1-\delta_{ps}\right)Q_{ps}^{fr} + \delta_{ps}Q_{ps}^{def} \tag{12}$$

where heat indexed with *fr* represents frictional heat, heat indexed with *def* represents deformational heat, δ_{pt}, δ_{ps}, δ_{st} – a dimensionless contact state variable (extension of slip) at the probe tip, probe side and shoulder tip, respectively.

The frictional and deformational amount of heat in equations 10, 11 and 12 for every ASWT, using Equations 7, 8 and 9 with respect to the contact shear stress [12, 5] is:

$$\tau_{cont} = \begin{cases} \mu p & \text{, for frictional heat generation} \\ \tau_{yield} & \text{, for deformational heat generation} \end{cases} \tag{13}$$

where: μ-friction coefficient, p-contact pressure, τ_{yield}-shear yield strength of workpieces.

Beside geometrical dimensions of the welding tool (d, D, h, H, α, β, etc.) and technological parameters of the process (ω, v_x), all other parameters (μ, p, τ_{cont}, τ_{yield}, T, $F_z(t)$, $M_t(t)$, δ_{pt}, δ_{ps}, δ_{st}, t_1, t_2, $t_{ps'}$, t_{st}, etc.) necessary for the analytical model have to be estimated analytically, numerically, experimentally or combining the estimation procedures.

Estimating the friction coefficient: Due to the complex kinematics of the FSW, it is difficult to establish a straightforward procedure for estimating the friction coefficient in FSW. Previous research recognizes the friction coefficient as a variable in FSW, but neglects the variation and assumes a constant value throughout the complete cycle of FSW. Usually, the friction coefficient within FSW, for a welding tool made of steel and workpieces of aluminium is equal to 0.3-0.4 [12, 34].

Kumar et al. [34] proposed an experimental model for estimating the friction coefficient during FSW. The model is based on the experimental estimation of the momentum of friction and axial force, which are necessary for estimating the friction coefficient. Figure 6 gives the functional schematic of the measuring place for the estimation of the friction coefficient. To estimate the coefficient of friction during FSW, it is necessary to estimate the momentum of friction and axial force [5]. The momentum of friction is the multiplication of the tangential force $F_t(t)$ (measured at force sensor 10, Fig. 6) and length of the pole (friction pole) L_t. If the diameter of the welding tools probe in contact is $d(t)$ and axial force is $F_z(t)$, the friction coefficient μ can be estimated as [14, 34]:

$$\mu = \frac{3F_t(t)L_t}{F_z(t)d(t)},\ t_2 \geq t \geq t_0 \tag{14}$$

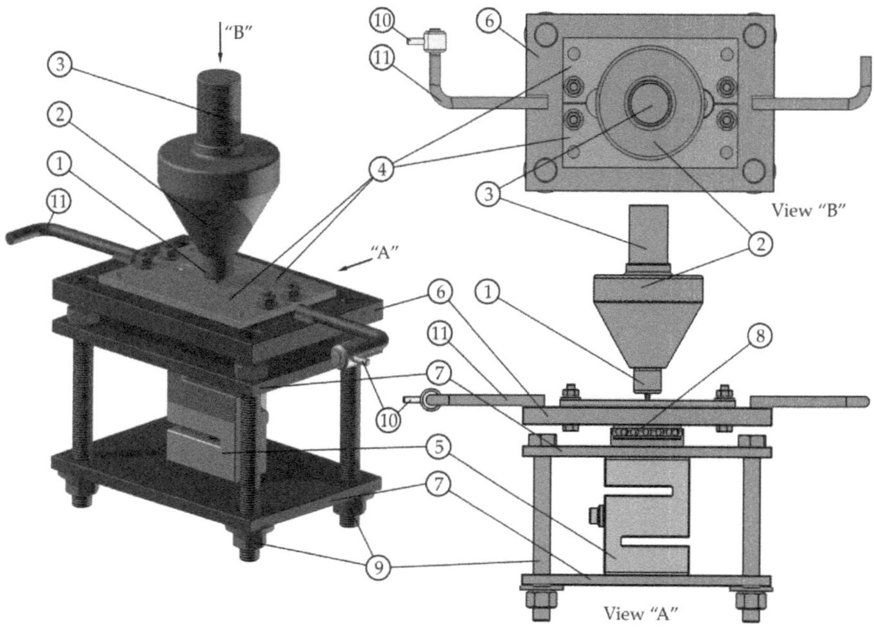

Figure 6. The measuring configuration for the momentum of friction and axial force: 1-welding tool, 2-welding tool's spindle, 3-shaft, 4-workpieces, 5-force sensor (axial force), 6-anvil, 7-backing plate, 8-ball bearing, 9-fundamental bolts, 10-force sensor (tangential force), 11-pole

The proposed model gives approximate results and only for the first two phases of FSW – plunging and first dwelling. The model is not applicable to the welding phase because the measuring system loses its stability when the welding tool travels along the join line and the momentum of friction cannot be measured [5]. Without proper model for estimating the friction coefficient during welding phase, friction coefficient has to be modeled. Friction coefficient used for the analytical model is estimated regarding the experimental results.

Estimating the contact pressure: Contact pressure p appears at the beginning of the plunging phase as a result of axial load $F_z(t)$ on the welding tool. Hertz [45] has proposed the first model for distributing contact pressure if a cylinder with a flat base punches into the plane, while Munisamy et al. [46] and Levytsky [47] have proposed models describing contact pressure distribution and heat generation when the axis of the cylinder is tilted.

Distribution of the contact pressure $p(r,t)$ delivered by the flat probe tip (Fig. 7, a) is [48]:

$$p(r,t) = \frac{2F_z(t)}{d\pi\sqrt{d^2 - 4r^2}},\ t_0 \le t < t_{st},\ 0 \le r \le \frac{d}{2} \tag{15}$$

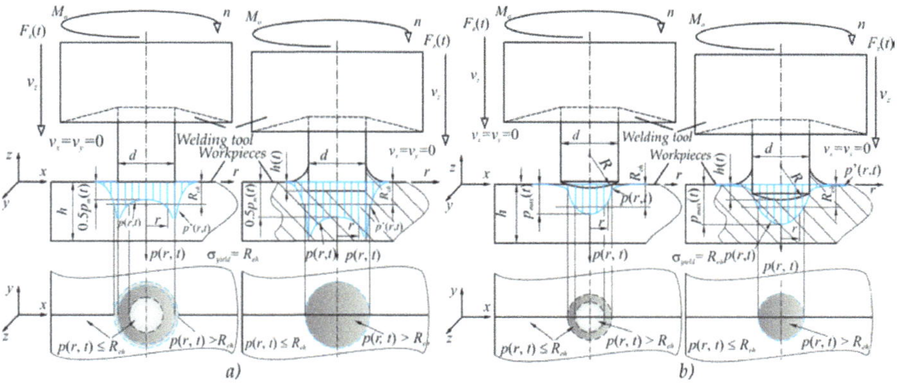

Figure 7. Contact pressure distribution for a) flat and b) spherical probe tips [4, 42, 43]

If the probe tip has a spherical shape, contact pressure is distributed (Fig. 7, b) as [48]:

$$p(r,t) = \frac{2}{\pi}\sqrt{d^2 - 4r^2}\sqrt[3]{\frac{3F_z(t)\overline{E}^2}{d^5}},\ t_0 \le t < t_{st},\ 0 \le r \le \frac{d}{2} \tag{16}$$

where $\overline{E}$ – represents the median modulus of elasticity estimated as:

$$\frac{1}{\overline{E}} = \frac{1 - \nu_{wt}{}^2}{E_{wt}} + \frac{1 - \nu_{wp}{}^2}{E_{wp}} \tag{17}$$

and E_{wt} – the modulus of elasticity of welding tool's material, ν_{wt} – the Poison's ratio of welding tool material, E_{wp} – the modulus of elasticity of the workpieces's material, ν_{wp} – Poison's ratio of the workpieces's material.

For engineering practice median contact pressure $p_m(t)$ gives good results:

$$p \approx p_m(t) = \frac{4F_z(t)}{d(t)^2\pi},\ d(t) = d,\ t_0 \le t < t_{st},\ 0 \le r \le \frac{d}{2} \tag{18}$$

Research [5, 10, 42-44] has shown that contact pressure distributed to workpieces reaches different values in different zones – in some zones it overcomes the yield strength of workpieces, while in other zones it has values lower than the yield strength (Fig. 8, c).

Existence of such zones multiplies the resistance of workpieces and plunging and intensive plunging appears when [5, 10, 42-44]:

$$p_m(t) > k_{eh}\sigma_{yield}(T);\ k_{eh} = 1.5\text{-}3 \tag{19}$$

where: $\sigma_{yield}(T)$ – yield strength of workpieces in function dependence with temperature T.

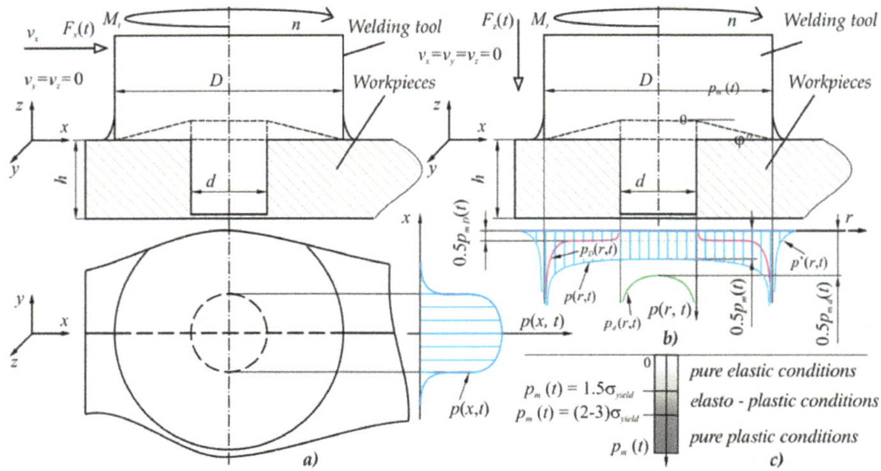

Figure 8. Contact pressure distribution [4, 42, 43]: a) probe side, b) shoulder tip, c) contact pressure defining contact conditions [5, 10, 42]

Contact pressure delivered by the shoulder tip is similarly distributed (Fig. 8, b) with a flat probe tip [5]. It appears smoothly since the shoulder tip is continuously involved in welding. Superposed contact pressure delivered by the probe tip and shoulder tip [5] is:

$$p \approx p_m(t) = \frac{4F_z(t)}{d(t)^2 \pi}, \quad d(t) \begin{cases} \approx \dfrac{d-D}{t_{st}-t_1} \cdot (t - t_{st}) + d, & t_{st} \le t < t_1 \\ = D, & t_1 \le t < t_4 \end{cases} \tag{20}$$

Contact pressure delivered by the probe side (Fig. 8, a) is a case of a modified "cylinder in cylinder" contact problem [45, 49]. Threads on the probe side increase the complexity of the analysis of the contact pressure distribution, however, with or without threads, median contact pressure on the probe side is:

$$p \approx p_m(t) \begin{cases} \approx \dfrac{F_x(t)}{dh}, & t_2 < t < t_3 \\ \approx 0, & t \le t_2,\ t \ge t_3 \end{cases} \tag{21}$$

where: $F_x(t)$ – force in welding direction, h – height of the probe/workpieces.

Estimating the tangential shear stress: When the deformation of workpieces appears, the rotational layer of the softened material travels around the welding tool [5, 12, 35, 37]. This is possible only if loads delivered by the welding tool inflict tangential stresses larger than the shear yield strength. The boundary value of such tangential shear (contact) stress, from von Misses yield criterion in uniaxial tension and pure shear [5, 12, 41, 42], is:

$$\tau_{cont} = \tau_{cont}(T) = \tau_{yield}(T,\varepsilon) = \sigma_{yield}(T,\varepsilon) / \sqrt{3} \tag{22}$$

where: $\tau_{cont}(T)$ – tangential contact stress in function of temperature T, $\tau_{yield}(T,\varepsilon)$ – shear yield strength of workpieces' material in function of temperature T and strain rate ε, $\sigma_{yield}(T,\varepsilon)$ – yield strength of workpieces' material in function of temperature T and strain rate ε. Yield strength of material is highly dependent on the temperature and strain rate, and the analysis of tangential stresses within FSW requires the full temperature and strain history in of the workpieces in a wide zone around the welding tool [5, 11, 14, 17, 23, 24, 27, 30]. However, analysis of heat generation in FSW can neglect the influence of strain on the decrease of yield strength and still maintain sufficient precision [12]. Neglecting is possible since the maximal temperatures of the material reach about 80% [38] of the melting temperature when strain has significant values due to near melting conditions in the material [18, 22]. Tangential contact shear stress is:

$$\tau_{cont} = \sigma_{yield}(T) / \sqrt{3} \tag{23}$$

where: $\sigma_{yield}(T)$ –yield strength of workpieces' material in function of temperature T. Thermo-mechanical properties of Al 2024 T351 are given in [5, 12, 50-53].

Estimating the contact state variable: The contact state variable or extent of slip is a parameter defining the influence of slipping in the heat generation process following the difference in the velocity of the welding tool and material, and relates frictional vs. deformational heat. It is obtained after curve fitting the experimental data regarding measured velocities [12, 14, 54, 55]:

$$\delta = \delta_{\min} + \left(1 - \delta_{\min}\right)\left(1 - e^{A_R}\right),\ A_R = -\delta_0 \frac{\omega r}{\omega_0 R} \tag{24}$$

where: δ_{min} – minimal measured slip, δ_0 – adjustable parameter depending on the material of the workpieces, R – maximal radius of the welding tool, ω_0 – normalized angular frequency of the welding tool (often the mid-point of the diapason of the measured angular frequencies).

Early works [12] considered the extent of slip as a single value for a complete welding tool. Experiments [5] have shown that the decomposition of the welding tool provides more precise results for the extent of slip if estimated for every ASWT separately. For example, when welding Al alloys with a steel welding tool with a threaded probe [12], with concern for the ASE of ASWT, the partial extent of slip is:

$$\delta_{pt} = \begin{cases} 0, & t_0 \le t < t_2,\ t_3 \le t < t_4 \\ \delta_{pt\ \min} + \left(1 - \delta_{pt\ \min}\right)\left(1 - e^{A_d}\right),\ A_d = -\delta_0 \dfrac{2\omega r}{\omega_0 d}, & t_2 \le t < t_3 \end{cases} \tag{25}$$

$$\delta_{ps} = \begin{cases} 0, & t_1 \le t < t_2,\ t_3 \le t < t_4 \\ \delta_{ps\ \min} + \left(1 - \delta_{ps\ \min}\right)\left(1 - e^{A_d}\right),\ A_d = -\delta_0 \dfrac{2\omega r}{\omega_0 d}, & t_{ps'} \le t < t_1,\ t_2 \le t < t_3 \end{cases} \tag{26}$$

$$\delta_{st} = \begin{cases} 0, & t_{st} > t \geq t_4 \\ \delta_{st\ \min} + \left(1 - \delta_{st\ \min}\right)\left(1 - e^{A_D}\right),\ A_D = -\delta_0 \dfrac{2\omega r}{\omega_0 D}, & t_{st} \leq t < t_4 \end{cases} \tag{27}$$

where: $\delta_{pt} = 0.1$, $\delta_{ps} = 0.2$, $\delta_{st} = 0.1$, $\delta_0 = 0\text{-}1$ from [5].

Estimating the temperature history of workpieces: Estimation of workpiece temperature requires knowing how much heat is generated during welding since heat influences temperature increase and it has to be done in an iterative regime . An iterative regime requires the discretization of time and space (Fig. 9, b), numeric calculations and significant computing time [5]. Temperature history of workpieces and welding tool can be estimated solving heat equations:

$$\rho_w c_w \frac{\partial T}{\partial t} = \lambda_w \left(\frac{\partial^2 T}{\partial x^2} + \frac{\partial^2 T}{\partial y^2} + \frac{\partial^2 T}{\partial z^2} \right) + q_v \quad \text{(for workpieces)} \tag{28}$$

$$\rho_{wt} c_{wt} \frac{\partial T}{\partial t} = \frac{\lambda_{wt}}{r} \frac{\partial}{\partial r}\left(r \frac{\partial T}{\partial r} \right) + \frac{\lambda_{wt}}{r^2} \cdot \frac{\partial}{\partial \varphi}\left(\frac{\partial T}{\partial \varphi} \right) + \lambda_{wt} \frac{\partial}{\partial z}\left(\frac{\partial T}{\partial z} \right) + q_v \quad \text{(for welding tool)} \tag{29}$$

where: ρ_w – density of the workpiece, c_w – specific heat capacity of the workpiece, λ_w – thermal conductivity of the workpiece, ρ_{wt} – density of the welding tool, c_{wt} – specific heat capacity of the welding tool, λ_{wt} – thermal conductivity of the welding tool.

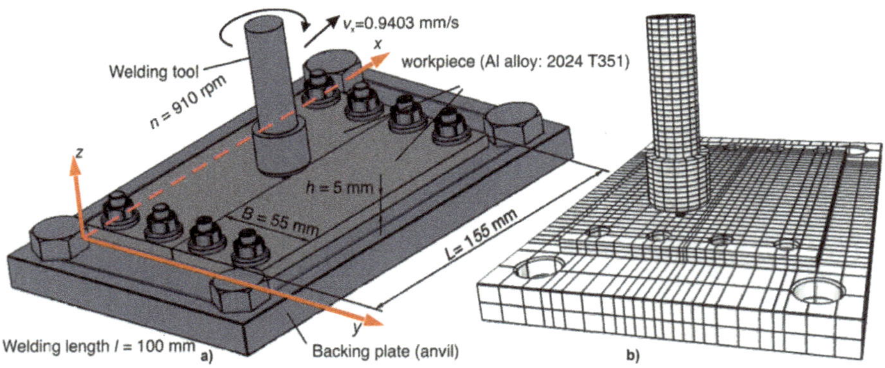

Figure 9. Workpieces, welding tool, bolts and anvil positioned for welding: a) realistic view with dimensions and some technological parameters used in experiment [5], b) discretized view (primarily and secondarily meshed with adaptive grid) [5]

Thermal energy generation source q_v is directly affected by generated heat Q_t and volume receiving generated heat V_t:

$$q_v = Q_t / V_t \tag{30}$$

Initial conditions for such a system include the recognition of the initial temperature:

$$T(x, y, z, t_0) = T(r, \varphi, z, t_0) = T_0 \tag{31}$$

Boundary conditions are complex due to the complex geometry of the welding tool and complex kinematics. As an example, boundary conditions for the top surface of workpieces involve convective and radiation heat transfer:

$$\lambda_w \left(\frac{\partial T}{\partial z} \right)_{z=h} = \alpha \left(T_0 - T_{i,j,k} \right) + \sigma \bar{\varepsilon} \left(T_0^{\ 4} - T_{i,j,k}^{\ \ 4} \right) \tag{32}$$

where: α - heat transfer coefficient, σ - Stefan-Boltzmann constant, $\bar{\varepsilon}$ - thermal emissivity of workpieces, T_0 – ambient temperature.

Boundary conditions on the contact between workpieces and anvil:

$$\lambda_w \left(\frac{\partial T}{\partial z} \right)_{z=0} = \lambda_a \left(\frac{\partial T}{\partial z} \right)_{z=0} \rightarrow \lambda_w \left(\frac{\partial T}{\partial z} \right)_{z=0} = \alpha_{\text{aprox}} \left(T_{i,j,k} - T_0 \right) \tag{33}$$

where: λ_a – the thermal conductivity of anvil, α_{aprox} – approximated heat transfer coefficient.

Boundary conditions between the welding tool and workpieces involve conduction between the parts. Such condition is decomposed to a classical conductive boundary condition:

$$\lambda_w \left(\frac{\partial T}{\partial r} \right)_{r=d/2} = \lambda_{wt} \left(\frac{\partial T}{\partial r} \right)_{r=d/2} \tag{34}$$

and influence of heat transfer due to material flow [5]. The material of the workpieces in the welding zone travels around the welding tool and partially carries its energy balance with it. If analyzed in discretized space and during discretized time, nodes "travel" from one discretized position to another and they "carry" its temperature, and while traveling they get and lose a burst of heat. This model of material travel is based on research concerning material flow around the welding tool [5, 13, 14, 20, 25, 26, 30, 34, 37], and applied in numerical calculations of temperature and heat flow. The model is named "node substitution and replacement method - NSRM" [5].

Results can be obtained analytically and numerically – for temperature estimation, a finite difference method, an explicit scheme with an adaptive grid were used, with the application of algorithm NSRM. The numerical solution of Eqs. 29-30 with the application of Taylor series for approximation of 2nd order derives and node positioning in discretized space is:

$$T_{i,j,k}^{m+1} = \frac{\Delta t}{\rho_w c_w} \left[\lambda_w \left(K_{x''} + K_{y''} + K_{z''} \right) + q_v \right] + T_{i,j,k}^{m} \quad \text{(for workpiece)} \tag{35}$$

$$T_{i,j,k}^{m+1} = \frac{\Delta t}{\rho_{wt} c_{wt}} \left[\lambda_{wt} \left(\frac{K_{r'}}{r_i} + \frac{K_{\varphi''}}{r_i^2} + K_{z''} \right) + q_v \right] + T_{i,j,k}^{m} \quad \text{(for welding tool)} \tag{36}$$

where:

$$K_{x''} = \frac{T^m_{i+1,j,k} - T^m_{i,j,k}}{(x_{i+1} - x_i)(x_i - x_{i-1})} - \frac{T^m_{i,j,k} - T^m_{i-1,j,k}}{(x_i - x_{i-1})^2}, K_{y''} = \frac{T^m_{i,j+1,k} - T^m_{i,j,k}}{(y_{i+1} - y_i)(y_i - y_{i-1})} - \frac{T^m_{i,j,k} - T^m_{i,j-1,k}}{(y_i - y_{i-1})^2}, \quad \text{(a)}$$

$$K_{z''} = \frac{T^m_{i,j,k+1} - T^m_{i,j,k}}{(z_{i+1} - z_i)(z_i - z_{i-1})} - \frac{T^m_{i,j,k} - T^m_{i,j,k-1}}{(z_i - z_{i-1})^2}, K_{\varphi''} = \frac{T^m_{i,j+1,k} - T^m_{i,j,k}}{(\varphi_{i+1} - \varphi_i)(\varphi_i - \varphi_{i-1})} - \frac{T^m_{i,j,k} - T^m_{i,j-1,k}}{(\varphi_i - \varphi_{i-1})^2}, \quad \text{(b)}$$

$$K_{r'} \approx r_{i+\frac{1}{2}} \frac{T^m_{i+1,j,k} - T^m_{i,j,k}}{(r_{i+1} - r_i)(r_i - r_{i-1})} - r_{i-\frac{1}{2}} \frac{T^m_{i,j,k} - T^m_{i-1,j,k}}{(r_i - r_{i-1})(r_i - r_{i-1})}, r_{i+\frac{1}{2}} \approx r_i + \frac{r_{i+1} - r_i}{2}, \quad \text{(c)} \qquad (37)$$

$$K_{r''} \approx \frac{T^m_{i+1,j,k} - T^m_{i,j,k}}{(r_{i+1} - r_i)(r_i - r_{i-1})} - \frac{T^m_{i,j,k} - T^m_{i-1,j,k}}{(r_i - r_{i-1})^2}, r_{i-\frac{1}{2}} = r_i - \frac{r_i - r_{i-1}}{2} \quad \text{(d)}$$

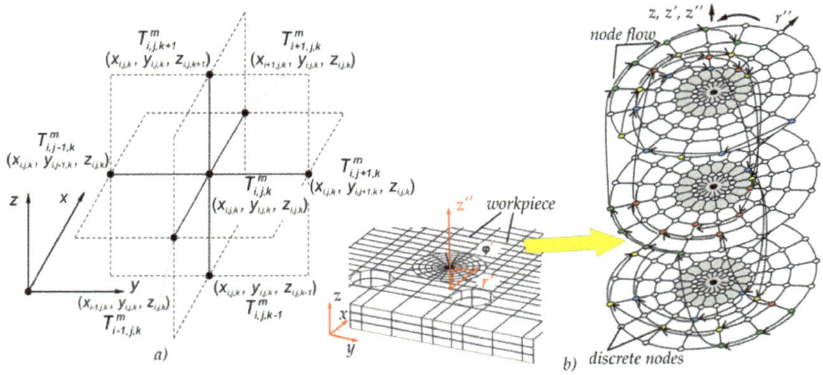

Figure 10. a) Discrete nodes with coordinates and temperatures, b) discretised space with a schematic of "node replacement and substitution" method [5]

3. Experimental procedure applied on aluminium alloy 2024 T351

The analytical model for estimating the amount of heat generated during FSW [5] is closely bound to the experimental research:

The analytical model gives realistic results only if some inputs into the model (axial force, torque, momentum of friction etc.) are obtained during realistic welding;

Validation of the analytical model and verification of gained results is possible only if some outputs from the model (e.g. numerically estimated temperature history of workpieces) are compared with the experimentally obtained results.

Such demands of the analytical model require a workplace with measuring equipment. Figure 11 gives a model of a realized workplace [5] where the welding of plates of Al 2024 T351 (some details given in Fig. 10, a) with the welding tool (given in Fig. 2, a) was performed.

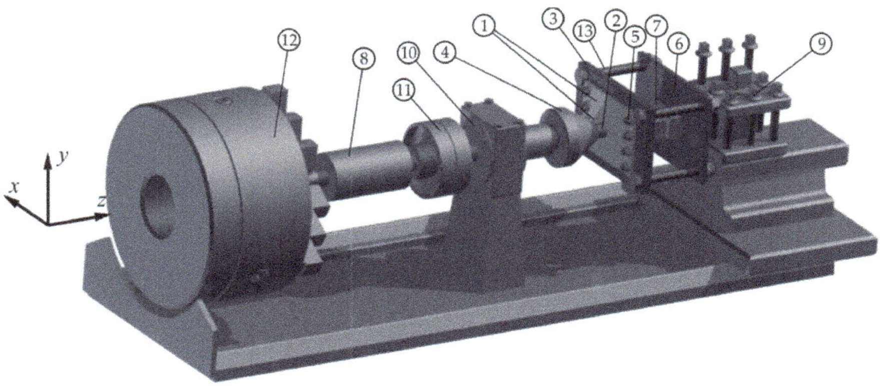

Figure 11. Workplace for FSW with measuring equipment: 1-workpiece, 2-welding tool, 3-anvil, 4-welding tool's spindle, 5-bolts, 6-backing plate, 7-force sensor, 8-torque sensor, 9-machine's tool rest, 10-bearing house, 11-clutch, 12-machine's spindle, 13-fundamental bolts

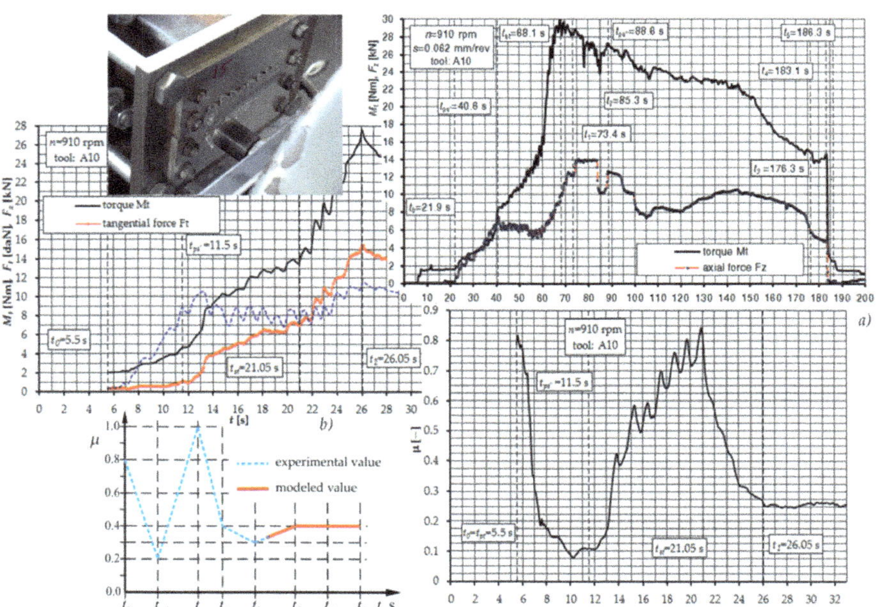

Figure 12. a) Typical diagram of measured torque and axial force, b) Typical diagram of measured torque, axial and tangential forces during FSW (plunging, first dwelling), c) Calculated friction coefficient, d) Modeled friction coefficient [5]

Welding was performed on a universal lathe with a horizontal operational axis (z-axis). Anvil, force sensor and backing plate are assembled and mounted on the machine's tool rest. Holes in the anvil (where fundamental bolts assemble force sensors) and the anvil and backing plate are larger than the diameter of the bolts and allow for the axial translation of the force sensor and measurement of axial force. Workpieces (dimensions given in Fig. 9, a) are bolted to the anvil and tilted over the machine's tool rest at an angle of 1°. The welding tool and its spindle are clutched to the torque sensor that is mounted into machine's spindle. A rigid axial-radial bearing house between the clutch and the welding tool disables the transmission of axial and radial forces from the welding tool to the torque sensor and machine's spindle. Such a design provides the correct measuring of axial force on the welding tool. Second working/measuring configuration (given in Fig. 6) is used for measuring the momentum of friction, axial, and tangential force at the welding tool in order to estimate the experimental friction coefficient. The machines used for this measuring configuration are vertical milling machines. The anvil and workpieces are mounted on an axial bearing above the axial force sensor. The anvil carries a tangential friction pole engaged in measuring the tangential force delivered by the torque of the spindle. As previously mentioned, such a design of the measuring equipment provides usable results only for plunging and the first dwelling phases. An infrared camera was used for measuring both the configurations that were used to estimate the thermal history of workpieces and the welding tool.

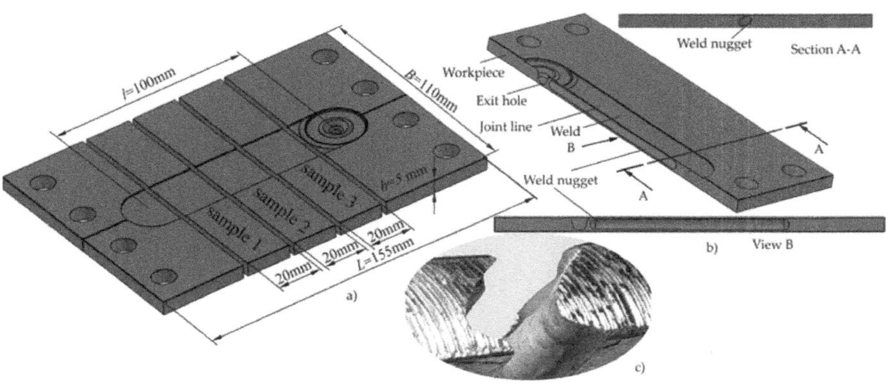

Figure 13. a) Schematic of sample extraction; b) Schematic of weld nugget's position; c) Tension sample destroyed at boundary of weld nugget

The experimental procedure of welding had three stages: to get familiar with the welding process, to get optimal technological parameters and to measure parameters necessary for the analytical model while creating qualitative welds [5]. Figure 12, a represents a typical diagram of torque and axial force measured during welding resulting in a qualitative weld.

Using a second measuring configuration, torque, axial and tangential force are obtained for numerous conditions. Figure 12, b shows a typical diagram of these parameters, obtained with optimal technological parameters. These values are used for estimating the experimental value of the friction coefficient (Fig. 12, c). However, since the proposed method gives limitedly usable values of the friction coefficient, based on experimental results, the friction coefficient is modeled for usage in the analytical model (Fig. 12, d).

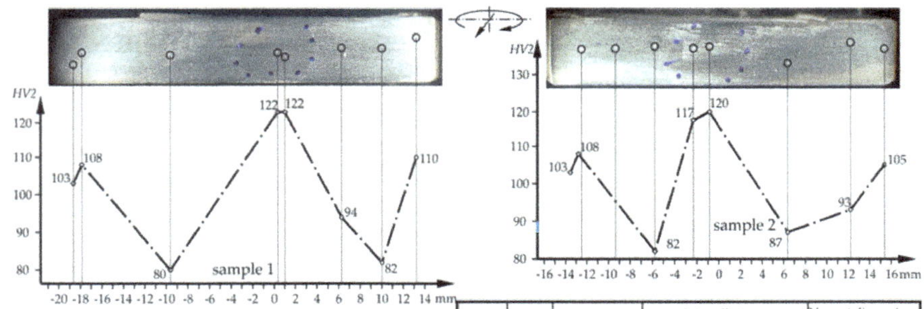

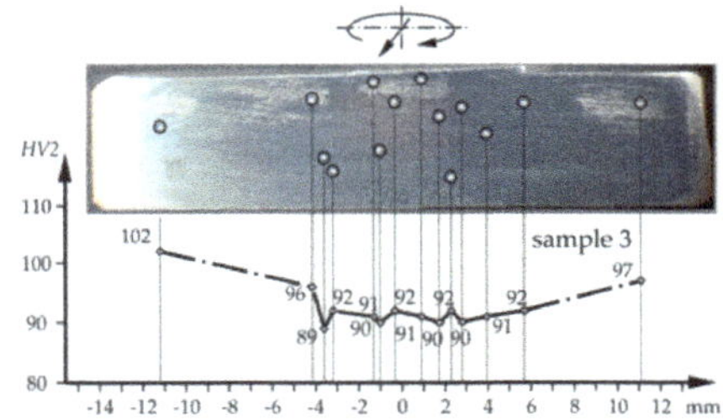

Welding No.	Sample	Ultimate tensile strength of joint σ_{joint} [N/mm²]	Joint efficiency $E_f = \sigma_{joint}/\sigma_{workpiece}$ 100% $\sigma_{workpiece}$ = 410 N/mm² – experimentally estimated ultimate tensile strength of workpiece	Nugget dimensions $b_f \times h_f$ [mm×mm]
#1	1	255	62.19	6 × 4.5
	2	266.7	65.05	4.8 × 3.8
	3	335,8	81.90	3.8 × 3.5
#2	1	245.8	59.92	6 × 4.2
	2	295	71.95	5.2 × 3.9
	3	350	85.36	4.2 × 3.5
#3	1	220.8	53.85	5.8 × 4.5
	2	316.7	77.24	4.6 × 4.3
	3	344.2	83.95	4 × 3.8

Figure 14. Hardness, joint (tensile) efficiency and observed dimensions of the welding nugget in samples 1, 2, 3 extracted from welded workpieces

In order to investigate the properties of welds (tensile and bending efficiency, hardness, metallic structure etc.), test samples were extracted from welded workpieces. Figure 13, a gives a schematic of the sample dimensions and positions of samples in workpieces. All welded joints used for sampling had a vortex-like structure of material called weld nugget, located along the joint line (Fig. 13, b). All tested samples (1-3) from all tested welds have a crack in the weld zone at the border of the nugget (Fig. 13, c).

Tested samples have shown a bending efficiency of about 12% (reaching a bending angle of about 11° while samples from parent material have reached an angle of about 89°). Results of joint (tensile) efficiency, the samples' hardness and the dimensions of the weld nugget in test samples are shown in Fig. 14.

4. Results and discussion

Whereas the experimentally obtained results included in analytical model makes use of some the necessary parameters affecting the heat generation process within FSW (contact pressure, shear stress etc.) and the amount of generated heat is relatively easily estimated, the numerical estimation of the temperature field of the wokpieces requires some computational time [5]. Experimentally estimated temperature can be easily inputted into analytical model in order to estimate the amount of heat generated during FSW and computational time can be significantly shorten. However, temperature change is the main product of heat generation and the verification of analytical model can be done via temperature comparison [56].

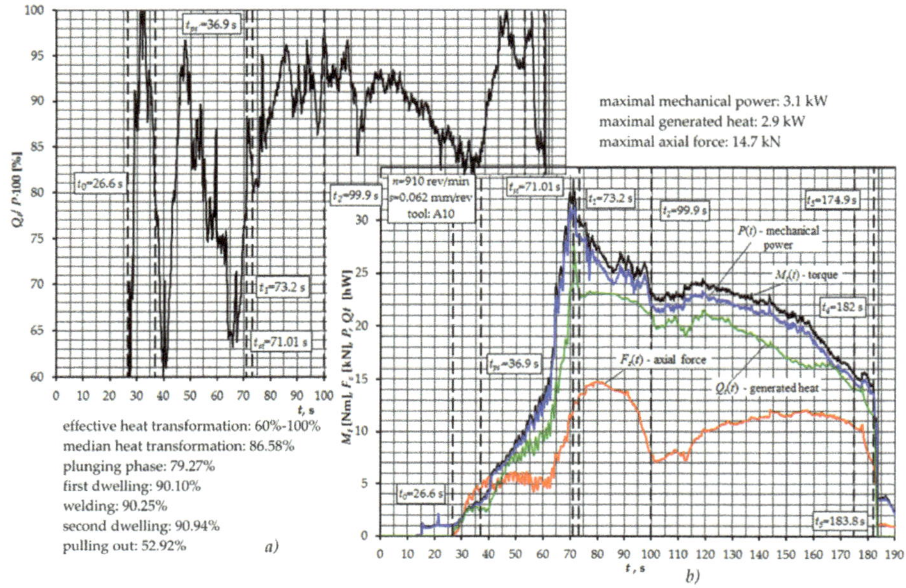

Figure 15. a) Effective heat transformation, b) Analytically estimated amount of generated heat [5, 56, 57]

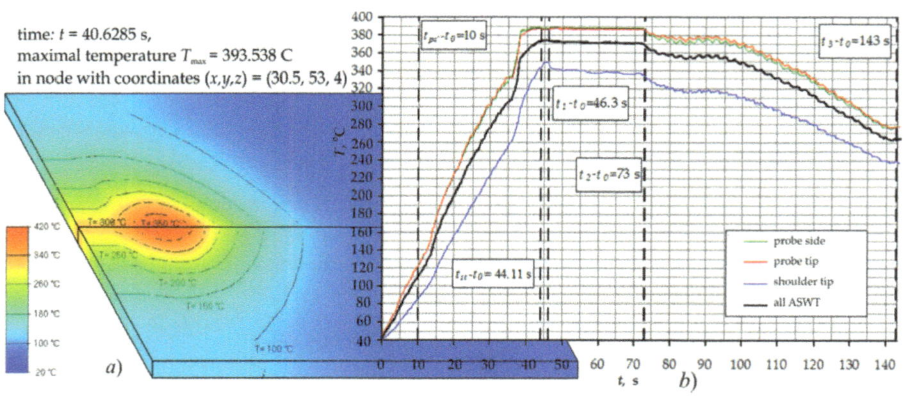

Figure 16. a) Numerically estimated temperature T of welding plates (rendered image); b) Median temperature of the material on contact with ASWT [5]

Figure 15, b gives the analytically estimated amount of generated heat during the welding of plates made of Al 2024 T351 (the dimensions of plates and technological parameters of the process are given in Fig. 9, a) with the welding tool with conned, threaded probe (given in Fig. 3, a). There were six repetitions of the welding procedure and the same number of applications of the analytical model. Characteristic moments of the process, operational technological parameters, values of measured axial force and torque delivered to the welding tool are given in Fig. 15, b, as well. The ratio of generated heat and engaged mechanical power is effective heat transformation η_{Qef} (Fig. 15, a) and it varies between 60%-100% in this application. The median value of effective heat transformation is 86.58% while it reaches 90.25% during welding phase. The maximum generated heat is 2.9 kW, reached during the plunging phase (when the shoulder tip is involved in the welding process at t_{st}).

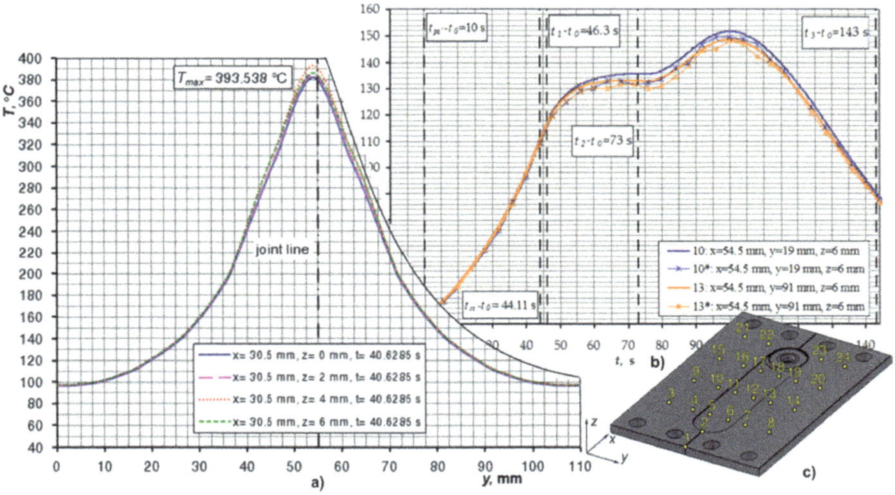

Figure 17. a) Temperature distribution in workpieces in plane normal to the joint line, b) temperature history (experimental and numerical) of specific discrete nodes, c) location of selected discrete nodes [56, 57]

Figure 16, a gives a rendered image of numerically estimated temperature in workpieces. Rendering is done for the moment the temperature reached maximal value – a few seconds before the tool shoulder was involved in the welding process. Figure 17, a gives the temperature distribution in a plane normal to the joint line for the same moment of time. The temperature peak is slightly dislocated from joint line to the advancing side of the weld. Analysis of heat generation required information about temperatures on contact between ASWT and the workpieces. Figure 16, b gives the median temperatures of all ASWT during the welding cycle. A comparison of the numerically and experimentally derived temperatures is done for 24 discrete points on the workpieces with adequate points from the numerical simulation (Fig. 17, c). Figure 17, b gives a diagram of both temperatures for two selected points. The largest difference between temperatures was about 11% in point 1 (absolute of 12°C) while other points had differences of 0.5-3.5% (absolute of 2-13°C) [5].

If the engaged mechanical power and generated heat are compared, it is clear that they have the same trend and notable changes in both are mostly connected with the characteristic moments of time. Mechanical power consumption and heat generation are more intense in the plunging and the first dwelling phases than in other phases. With a drop in the axial force (at the beginning of the first dwelling phase), torque drops as well, which results in the stabilization of the heat generation process and a drop in the temperature of the workpieces. The welding process is always in a quasi-equilibrium state: increase in temperature lowers the generated heat, engages mechanical power and axial force; decreases of the temperature, and increases power consumption and heat generation (almost sinusoidal temperature character, Fig. 16, b). At the end of the first dwelling and the beginning of the welding phase, the system (tool-workpieces-anvil) has reached nearly the maximal observed temperatures. During the welding phase, the system constantly decreases the temperature – probably workpieces were "overheated" due to the long lasting first dwelling phase ($t_{dw1}\approx27$ s). Power consumption and heat generation drop as well while axial force steadily rises until the moment when effective heat transformation reaches app. 85% ($t\geq150$ s, Fig. 15, b) and axial force reaches a value of 11-12 kN. Analyzed results from tensile and hardness testing (Fig. 14), best distribution of hardness, smallest weld nugget and the most efficient welds of app. 80% are reached in test samples no. 3 – at the end of welding, where heat transformation has a value near 85%. Samples 1 have the worst joint efficiency (app. 55%), exhibiting W-shaped hardness distribution with great variation of hardness and largest weld nugget. Welding in the zone of samples 1 is performed at highest temperatures and with an effective heat transformation of app. 90% and the lowest axial force (7-9 kN).

5. Conclusions

The analytical model developed here utilizes analytical algebra, experimentally gathered data and numerical calculations to estimate how much of the mechanical power delivered to the welding tool is transformed into heat. The novel approach in the model focuses on the recognition of the active surfaces involved in welding, engagement of active surfaces during welding, recognition of dominant parameters involving heat generation and their estimation, recognition of mechanisms of heat generation and their utilization, and implementation of a new numerical model for defining the material flow around the welding tool. Experiments and temperature based validation of the model are done on Al 2024 T351 5 mm thick plates.

Results from the model have shown that 60-100% of the mechanical power of the welding tool transform into heat during FSW with a median of 86.58% in a complete welding cycle. The median value of heat transformation during welding is about 90% which is in agreement with previous results. Comparison of workpiece temperatures from numerical calculations and experiments has shown a maximal relative error of 11% (about 13°C) while the maximal temperature of workpieces reached a maximal temperature of app. 394°C (79% of Al 2024 T351 melting temperature).

Heat generation appeared to be extreme in the welding-procedure-dependent process: preheating of welding plates, derived by a long dwell of the welding tool at the beginning of the welding, has significant influence on heat generation and the quality of the welded joint. Too long dwelling "overheats" workpieces and the welding phase happens during continuous transient cooling. In such conditions, heat transformation is near (and above) 90%, joints have 50% tensile efficiency and a large weld nugget. For conditions of heat transformation of app. 85%, joints reach app. 80% of tensile efficiency and have the smallest weld nugget. In both situations, it was necessary to have minimally 11kN of axial force to acquire a qualitative weld. If the optimal technological parameters of FSW and the welding tool are selected, heat management in FSW is of great importance for the quality of the weld. Due to the low bending efficiency of investigated welding, alloy 2024 T351 is legally considered to be a tough and a limitedly weldable alloy.

Author details

Miroslav Mijajlović* and Dragan Milčić
University of Niš, Faculty of Mechanical Engineering Niš, Serbia

6. References

[1] TWI, official web page: http://www.twi.co.uk/technologies/welding-coating-and-fabrication/friction-stir-welding/ (01.12.2011.)
[2] Thomas W M et al., Friction stir butt-welding, Int. Patent App., PCT/GB92/02203, 1991.
[3] Thomas W M et al., Friction stir butt welding, GB Patent App., No 9125978.8, 1995.
[4] Thomas W M et al., Friction stir butt welding, UP Patent, No 5 460 317, 1995.
[5] Mijajlović M. Investigation and Development of Analytical Model for Estimation of Amount of Heat Generated During FSW (in Serbian), Ph. D. thesis, Faculty of Mechanical Engineering Nis, University of Nis, Nis, Serbia, 2012.
[6] AWS D17.3/D17.3M:2010 Specification for FSW of Al Alloys for Aerospace Applications.
[7] ISO 25239: Friction stir welding – Aluminium, parts: 1-5, 2011.
[8] Maxwell JC. Theory of Heat. Dover Publications, Inc. ISBN 0-486-41735-2, 1871
[9] Ilić G. Radojković N. Stojanović I. Thermodynamics II: Basics of heat distribution (in Serbian), Vranje, Yugoslavia, 1996.
[10] Stamenković D. Djurdjanović M. Tribology of the Press Fit Joints (in Serbian), Faculty of Mechanical Engineering, Nis, Serbia, ISBN–86–80587–48–6, 2005.
[11] Djurdjanović M. Mijajlović M. Milčić D. Stamenković D. Heat Generation During Friction Stir Welding Process. Tribology in Industry. 2009; 31(1-2):8-14.
[12] Schmidt H.J. Hattel J.W. An analytical model for the heat generation in friction stir welding. Model. Simul. Mater. Sci. Eng. 2004; 12 (2004): 143–157.
[13] Lorrain O. Favierb V. Zahrounic H. Lawrjaniec D. Understanding the material flow path of friction stir welding process using unthreaded tools, Journal of Materials Processing Technology, 2010; 210 (2010): 603–609
[14] Nandan R. DebRoy T. Bhadeshia H.K.D.H. Recent advances in friction-stir welding – Process, weldment structure and properties. Progress in Mat. Sci. 2008; 53 (6): 980–1023.
[15] Chao YJ. Qi X. Thermal and thermo-mechanical modeling of friction stir welding of aluminum alloy 6061-T6, J. Mater. Proc. Mfg. Sci. 1998; 7 (1998): 215–233.

* Corresponding Author

[16] Frigaard O. Grong O. Modeling of the heat flow phenomena in friction stir welding of aluminum alloys, Paper presented at: INALCO '98. Proceedings of the Seventh International Conference on Joints in Aluminum, Cambridge, 15–17. April, 1998.
[17] Gould J.E. Feng Z. Heat flow model for friction stir welding of aluminum alloys, Journal of Material Processing and Manufacturing Science. 1998; 7(1998): 185-194.
[18] Russell M.J. Shercliff H.R. Analytical modeling of microstructure development in friction stir welding, Proceedings of the first International Symposium on Friction Stir Welding, Thousand Oaks, CA June 1999.
[19] Rosenthal D. The theory of moving sources of heat and its application to metal treatments [J]. Transaction ASME, 1946, 43(11): 849–866.
[20] Colegrove P. Painter M. Graham D. Miller, T. 3 Dimensional Flow and Thermal Modeling of the Friction Stir Welding Process, Proceedings of the Second International Symposium on Friction Stir Welding, June 26–28, 2000, Gothenburg, Sweden.
[21] Shercliff H.R. Colegrove P.A. Modelling of friction stir welding, Mathematical Modelling of Weld Phenomen. Maney Publishing, London, UK, 2002; 6(2002): 927-974.
[22] Khandkar M.Z.H. Khan J.A. Reynolds A.P. Prediction of temperature distribution and thermal history during friction stir welding: input torque based model. Science and Technology of Welding & Joining. 2003; 8(3): 165-174.
[23] Song M. Kovačević R. Thermal modeling of friction stir welding in a moving coordinate and its validation, Int. J. Machine Tool Manufacturing. 2003; 43 (6): 605–615.
[24] Hamilton C. Dymek S. Sommers A. A Thermal Model of Friction Stir Welding Applied to Sc-Modified Al-Zn-Mg-Cu Alloy Extrusions. International Journal of Machine Tools and Manufacture. 2008; 49 (2008): 230 – 238.
[25] Nandan, G.G. Roy, T.J. Lienert, T. Debroy, Three-dimensional heat and material flow during friction stir welding of mild steel, Acta Materialia. 2007; 55 (2007): 883–895.
[26] Ulysse, P. Three-dimensional Modelling of the Friction Stir-welding Process. International Journal of Machine Tools & Manufacture. 2002; 42(2002): 1549-1557.
[27] Steuwer A. Peel M. Withers, P.J. Dickerson, T.L. Shi, Q. Shercliff, H.R. Measurement and prediction of residual stresses in aluminium friction stir welds. Journal of Neutron Research. 2003; 11(2003): 267-272.
[28] Chao Y. J. Qi X. Tang W. Heat transfer in friction stir welding-experimental, and numerical studies. Transactions of the ASME. 2003; 125(2003): 138–145.
[29] Heurtier P. et al. Mechanical and thermal modelling of Friction Stir Welding Journal of Materials Processing Technology. 2006; 171 (2006): 348–357
[30] Santiago D.H et al. Numerical Modeling of Welded Joints by the Friction Stir Welding Process. Materials Research. 2004; 7(4): 569-574.
[31] Schmidt H. Hattel J. A Local Model for the Thermo-mechanical Conditions in Friction Stir Welding. Modelling and Simulation in Mat. Science and Eng. 2005; 13(2005): 77-93.
[32] Colligan KJ. Mishra RS. A conceptual model for the process variables related to heat generation in friction stir welding of aluminium. Scripta Mat. 2008; 58(2008): 327–331.
[33] Kalya P. Krishnamurthy K. Mishra RS. Baumann J. Specific Energy and Temperature Mechanistic Models for Friction Stir Processing of Al–F35. Presented at: 2007 TMS Annual Meeting Friction Stir Welding and Processing IV. Orlando, FL; 2007, pp. 113-126

[34] Kumar K. Kalyan C. Kaias SV. Srivatsan TS. An Investigation of Friction During Friction Stir Welding of Metallic Materials. Mat. and Man. Proc. 2009; 24(4): 438-445.

[35] Colligan K. Material Flow Behavior during Friction Stir Welding of Aluminum. Weld. J. Suppl. 1999; 78(7): 229s–237.

[36] Nandan R. et al. Improving reliability of heat transfer and materials flow calculations during friction stir welding of dissimilar aluminum alloys. Welding Journal, 2007; 86(10): 313S-322S.

[37] Ouyang JH. Kovačević R. Material Flow and Microstructure in the Friction Stir Butt Welds of the Same and Dissimilar Aluminum Alloys. The Journal of Materials Engineering and Performance. ASM International. 2002; 11(1): 51-63.

[38] Djurdjanović M. Tribology, Welding, Friction Welding. [unpublished materials used in studies at University of Nis, Faculty of Mechanical Engineering Nis, Serbia]. 2000-.

[39] Vukicevic M. Petrovic Z. Djuric S. Bjelic M. Oxy-fuel welding (in Serbian). Faculty of Mechanical Engineering. Kraljevo. Serbia. 2007

[40] Živković A. Influence of Friction Stir Welding Tool Geometry on Properties of Welded Joint of Alloys Al 2024 T351 (in Serbian) Ph. D. thesis. Faculty of Mechanical Engineering University of Belgrade, Belgrade, Serbia 2011.

[41] Rai R. De R. Bhadeshia HKDH. DebRoy T. Review: friction stir welding tools. Science and Technology of Welding and Joining. 2011; 16(4): 325-342.

[42] Mijajlović M. Milčić D. Anđelković B. Vukićević M. Bjelić M. Mathematical Model for Analytical Estimation of Generated Heat During Friction Stir Welding Part 1. Journal of Balkan Tribological Association. 2011; 17(2) 179-191.

[43] Mijajlović M. Milčić D. Anđelković B. Vukićević M. Bjelić M. Mathematical Model for Analytical Estimation of Generated Heat During Friction Stir Welding Part 2. Journal of Balkan Tribological Association. 2011; 17(3): 361-370

[44] Mijajlović M. et al. Study About Friction Coefficient Estimation in Friction Stir Welding. 2011; Presented at: Balkantrib 11, The 7th International Conference on Tribology, Proceedings, Thessaloniki, Greece, pp. 323-330.

[45] Hertz HR. On Contact Between Elastic Bodies(in German), Collected Works, 1(1885). Leipzig. Germany. 1895.

[46] Munisamy RL. Hills DA. Nowell D. The solution of the contact between a tilted circular rigid punch and an elastic half-space. Wear. 1995; 184(1995): 93-95.

[47] Levytsky VP. Interaction of Rigid Cylinder with Elastic Half-Space by Heat Generation on The Contact Area. Int. J. Engng. Sci. 1994; 32(11): 1693-1702.

[48] Galin LA. Contact Problems; The legacy of L.A. Galin. Series: Solid Mechanics and Its Applications (in Russian). Nauka. Moscow, Russia. 15(2008). 2008.

[49] Goryacheva I. Sadeghi F. Contact characteristics of a rolling/sliding cylinder and a viscoelastic layer bonded to an elastic substrate. Wear. 1995; 184(1995): 125-132.

[50] EN 573-3:2007, Al and Al alloys Part 3: Chemical composition and form of products.

[51] EN 485-2:2008-10, Al and Al alloys —Part 2: Mechanical properties.

[52] Metals Handbook: Properties and Selection: Non Ferrous Alloys and Special – Purpose Materials. 1992; 10(2). The American Society for Metals, ASM International, 1992

[53] ASM Aerospace Specification Metals Inc. Aluminium 2024 T351 Data Sheet. c2012 [Cited 27 August 2012]. Available from http://asm.matweb.com.

[54] Deng Z. Lovell MR. Tagavi KA. Influence of material properties and forming velocity on the interfacial slip characteristics of cross wedge rolling. Man. Science and Eng. 2001; 123(2001) :647–653, 2001.

[55] Khandkar H. Zahedul M. Khan J. Thermal modeling of overlap friction stir welding for Al-alloys, J. Mater. Process. Mfg Sci. 2001; 10 (2): 91–106.

[56] Milčić D. et al. Temperature Based Validation of the Analytical Model for Estimation of the Amount of Heat Generated During FSW. Thermal Science. In press. 2012.

[57] Mijajlović M. *et al.* Experimental Studies of the Parameters Affecting the Heat Generation in the Friction Stir Welding Process. Thermal Science. In press. 2012.

Section 3

Sensing of Welding Processes

Chapter 12

Visual Analysis of Welding Processes

Yoji Ogawa

Additional information is available at the end of the chapter

1. Introduction

Fusion welding process is very convenient method to join metal structures. It plays the most important role on industrial production. The process itself is quite simple. The generated heat sources such as arc and/or laser melt the limited parts of work pieces to be joined, and the joint zone is unified after solidification of melting metal in the groove. However fusion welding process contains much interesting research targets(Ogawa,2011). For example, metal contains many elements those thermal properties are quite different. The surface is oxidized. When the concentrated heat source strikes on the metal surface, some elements are evaporating away from the original work piece. If the shielding is insufficient, some amount of hydrogen and oxygen invades into the hot metal. Hydrogen reduces mechanical property of the structure. And quick re-solidification produces quite different structure compared to the original base metal which was qualified by thermal refining process. Usual fusion welding process contains four phases such as solid, liquid, gas and plasma. Plasma is quite hot, maximum temperature exceeds 10,000K. The temperature of metal in solid phase to be joined is normally in room temperature. The temperature of molten pool in the base metal is raised up more than melting point. But, this temperature is about 3,000K or less. Then the temperature gap between liquid parts and plasma is tremendous. The physical reactions in those regions include many unknown factors(Zacharia & David,1993). These reactions occur in quite small area about one centimetre cubic space. Therefor the moving speed of target in observing area is quite high. This is the main reason why high-speed imaging technique is necessary on visual analyssis of fusion welding processes. Almost all of welding processes are carried out in factories on the earth. However, some welding processes must be carried out in deep water or in space. These are another interesting area to study.

2. Basics on imaging for fusion welding phenomena

2.1. General instruction for Gas Tungsten Arc Welding process

Basic concepts to study arc welding process had been constructed before 1970s(Pattee et al.,1973). The most useful signals to understand arc welding process is arc voltage and current. However, these two electrical signals are insufficient to understand actual process. Then, high-speed filming technique was introduced. The most important matter to get reasonable image of arc welding process is how to eliminate the meaningless light from the process. Very strong light is radiated from the arc. This strong light protects to observe actual welding procedure.

Figure 1 shows a fundamental concept of Gas Tungsten Arc welding (GTAW) process. GTAW is one of the simplest of fusion welding process. Electrons emitted from tungsten cathode impinge a base metal. Some parts of argon atoms, those are normal inactive shielding gas to avoid oxidation of molten metal, ionize in an arc column. The arc column is a channel of current between cathode and anode, and it includes the same number of ions and electrons. The strong light is radiated by recombination reaction with ions and electrons. Understanding of current density distribution is important to know thermal transportation, and the distribution of this radiation suggests the temperature of arc column. The GTAW process is materialized by the tight energy balance among the cathode, arc column and weld pool. And there exist very complicated physical and chemical reactions in each region.

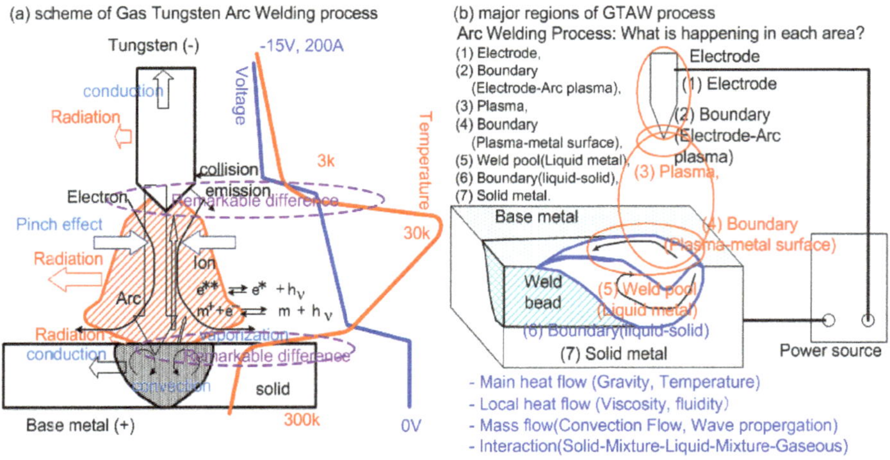

Figure 1. Scheme of Gas Tungsten Arc Welding(GTAW) process.

Figure 1(b) shows the main typical regions of interests during GTAW process. There are three major zones such as base metal, arc column and the tungsten electrode. The base metal has three different regions. They are melting zone, solid zone and boundary between solid and liquid. The boundary zone is the most important on metallurgical point of view. Crystal structure in boundary becomes large, and the mechanical property of the metal changes as a result. Some defects like blowhole or undercut occur in this region. A melting area as called weld pool is complicated. Electrons and atoms hit on the molten surface, and some atoms invade into the pool. A surface of the base metal in front of the pool contains much oxide. The melted oxide also invades on the pool. Many physical and chemical reactions occur on and in the pool. Rapid liquid metal flow also occurs on and in the pool.

The tungsten electrode is expected as a non-consumable cathode. The cathode is mainly heated by Joule heating by current passing through the electrode. The surface of the cathode is heated by collision of atoms and ions from the outer space, and the top area, which called cathode spot, is cooled by electron emission. Normal tungsten electrode contains a few per cents of thorium oxide to improve emission ability. Thorium oxide on the cathode surface evaporates during arc process. A cool surface of the electrode is normally oxidized. And tungsten oxide is much easier to melt and evaporate compared to pure tungsten, then some amount of tungsten oxide, 3 or more millimetres from the top, melt and move toward the top where the arc generates. And the temperature near the top is hot enough to evaporate the invading tungsten oxide. Evaporated tungsten oxide is easy to dissociate to tungsten and oxygen. A temperature of middle region between the top and the area where tungsten oxide evaporates is cool enough to crystallize the impinged tungsten atom from the outer space. Very complicated physical and chemical reactions are occurred on the cathode surface. A microscopic observation is required because normal diameter of the tungsten electrode is 1.6-3.2mm.

The arc column existing between cathode and anode is hot enough where ionization, dissociation, and recombination frequently occur. Particles in this region are heated and forced by electro-magnetic field. Much power is conducted to the work piece by moving electrons. Some power is lost to open air by radiation and conduction. Arc light is the radiation loss of arc, and the frequency distribution of radiation has important information of arc characteristics. It contains temperature distribution information in arc column. The approximate power of arc radiation is very strong, and it protects precise observation of cathode and anode surfaces. The energy transfer from the arc is crucial in the weld pool formation. And the energy transfer due to convection becomes important in the weld pool compared to heat conduction. Total thermal radiations from the weld pool and solid surface are very weak compared to arc light. How to observe actual motion in and on the weld pool under luminous arc column is the issue to study(Shaw,1975; Inoue,1981; Ogawa,2011) .

2.2. How to get good image

2.2.1. Spatial effect

Figure 2 shows basic principles to show spatial effect of arc light. Suitable selections of exposure time, diaphragm and filtering set-up of camera are essential to get high quality image. An external light is usually used to improve image quality of target which has a strong light source in it. When the arc region is very small as shown in figure 2(a), normal light source is enough. When the arc region becomes large as shown in figure 2(b), stronger light source is required to get clear image of the whole apparatus. Some halation occurs on and near the arc. However, the arc region is also small enough to get clear image of experimental set-up. When the arc region becomes larger as shown in figure 2(c), normal light source is insufficient because arc radiation is quite high. The size of this figure is normally used for observation of high speed imaging. We need some special technique to improve image quality. Details of how to get high quality image of fusion welding process are described in section 3.

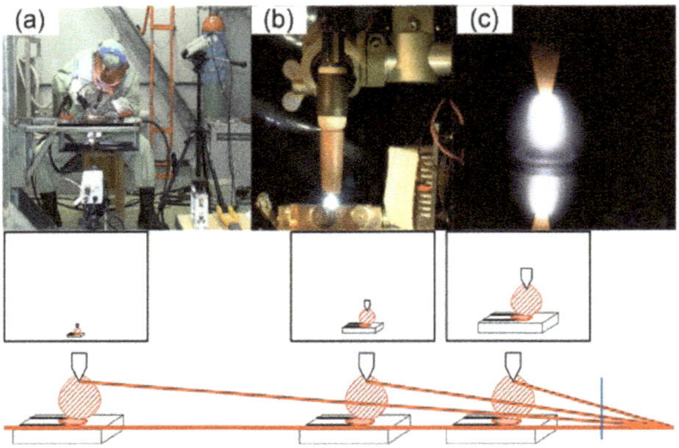

Figure 2. Size effect on image quality.

2.2.2. Fundamentals of high speed imaging

High speed video camera is a digital processor with huge memory. Captured high speed image data are processed as digital data inside of the camera. Depth of the captured data in the camera is normally 10 to 14bits. However, data depth of output image/movie is normally less than 8bits to fit normal picture/movie standards. Objects in three dimensional space are captured onto two dimensional digital data. Many two dimensional data sets are stored time by time as shown in figure 3(a). In case of monochrome camera, only one colour information on each pixel is stored. In case of colour camera, it captures three kinds of colour information on each pixel. They are red, green and blue. The arc light itself contains from ultra violet to infrared regions. Normal colour camera is set to fit human eyes, so light which is out of visible range of human eyes are avoided. But, ultra violet and infrared lights from arc are so strong compared to normal scene, so it affects the colour tone of the captured image(Ogawa,2008).

When the camera condition is set to get clear arc image, almost all back ground become black as shown in figure 2(c). One simple solution to capture clear background image is using a strong external light. Another good solution is using narrow banded interference filter which protects arc light and pass the light from back ground. Spectroscopic data is also useful to understand arc characteristics. Factors of interest to understand arc welding process are three dimensional space information, wavelength information and temporal variations as shown in figure 3(b). Dynamic range of camera device or data depth of image is also important factors for precise analysis.

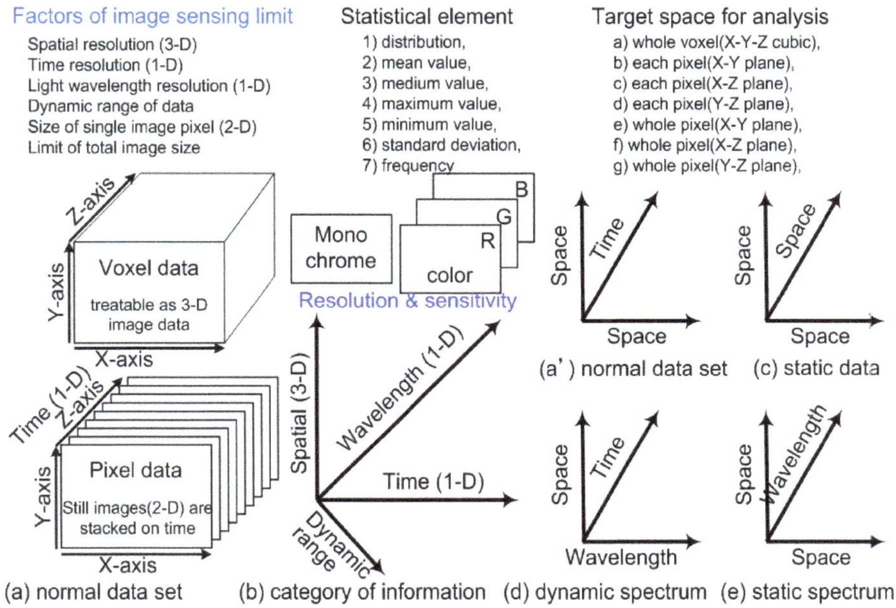

Figure 3. Data structures of high speed imaging.

The image of object just on the focal plane is clear, and the brightness of captured data is higher than those from out of focus position. When the objects is static, range information can be detected by changing the camera position or focal distance as shown in figure 3(c). Spectroscopic imaging is also important to understand the process behaviour as shown in figure 3(d). If the object is stable and constant, static spectrum information can be detected as shown in figure 3(e). Two dimensional digital data of each still images are stacked time by time in case of high speed imaging. The data itself can be treated as voxel data as shown in figure 3. Then statistical analysis becomes possible, and it becomes easy to choose numerous points of view to analyse the whole process. This is the remarkable benefit of high speed imaging to study on welding process.

2.2.3. Non-linearity on brightness

However, we need careful selection and understanding of observation conditions. Usual camera is fabricated to fit human sense. Figure 4 shows effects of exposure time on data brightness. The same static object is captured by different exposure time. One important fact to realize is an effect of dark noise. Some special camera has special cooling system for sensor device to avoid dark noise. However normal high speed camera has no special cooling system, then some dark noise is added on the data. And the values are affected by temperature and exposure time. Another important fact is special fitting process on too bright objects. There are no linearity in the highest range as shown in figure 5. Halation is protected by this process, and the picture looks very clear. However, quantitative analysis cannot guaranteed. The relation of actual brightness on object and captured data value depends on camera. One typical relation is shown in figure 5. When we try to use high speed camera for quantitative analysis like detecting temperatures, we need precise correction of several different temperature object before actual observation.

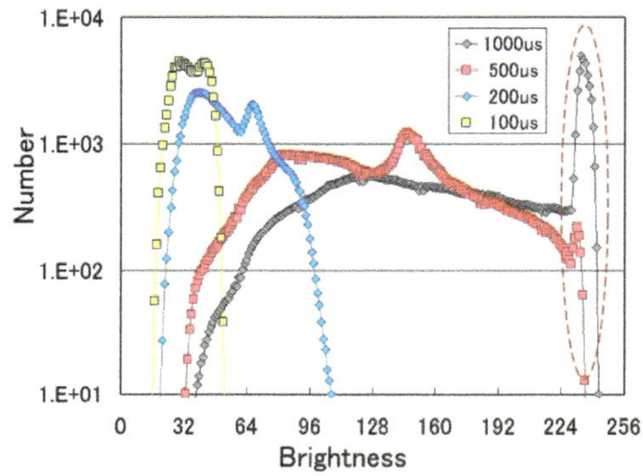

Figure 4. Effect of exposure time on brightness distribution of data.

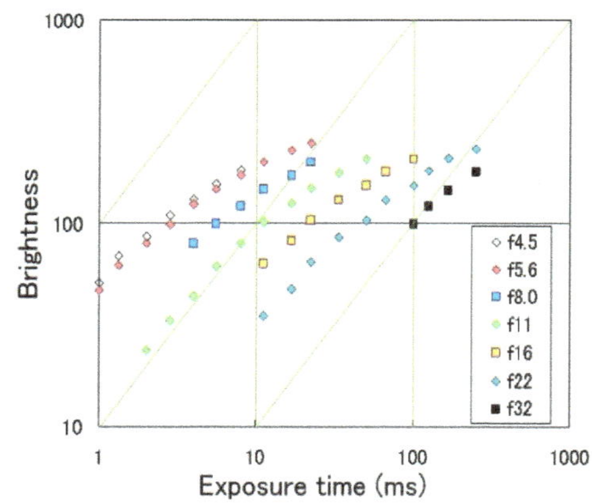

Figure 5. Typical relation between actual brightness of objects and data value.

2.3. Radiation effect

Surface conditions of a molten pool and a solid metal are quite different. The surface of the molten pool is smooth like mirror, because of surface tension, and a light illuminated on the molten pool reflects totally to the geometrical direction. On the contrary, solid surface has rough surface, and the light illuminated on the solid surface is scattered to wide directions.

This is an important difference for imaging of arc phenomena. Another important difference is radiation efficiency by surface condition. Radiation from oxides is greater than that from pure surface.

One of the major purposes of an imaging of arc is to provide a monochromatic two dimensional image from a polychromatic object. Optical frequency distribution at appropriate point has important information on arc temperatures at that point, such as electron temperature, atom temperature and ion temperature. Normal colour picture is a mapping of colour information from three dimensional scene on to a two dimensional frame. In case of normal picture of arc column, one point of the picture contains integrated intensity information in depth and integrated intensity on wavelength information. A single colour picture which means a picture at appropriate wavelength, as called monochromatic photography, and/or emission spectroscopy which measures information on wavelength distribution at appropriate point are required to estimate the temperature at that point. They are two dimensional data, and each point has information of intensity. The situations are also a matter of time. Static or quasi-static phenomena is not usual for arc welding processes, they usually changes time by time. So the time domain analysis is also required. Time domain information is easily recorded as time series of two dimensional data as shown in figure 3.

The intensity of the data show integrated values across the depth of the space. The image density measurements, as integrated intensity in depth, are input data for the Abel Integral Equation. Radiance (at each wavelength) is obtained as a function of position. The local ratio of radiances for two spectral lines then yields the temperature as a function of position. Monochromatic imaging also has more direct value. Arc light is an exponential function of radiation temperature. Total power of radiation is more than 1000 times of thermal radiation from molten metal. This is the major difficulty to get clear image and weld pool simultaneously. The highest value of radiation is estimated by Wien's law. The highest radiation just near the boiling temperature of iron is in near infrared region between 950 and 1000nm. And radiation from arc at this region is low enough to get clear image of the molten metal. Following formula are used to calculate temperature from image.

$$E_\lambda = \frac{2\pi hc^2}{\lambda^5} \frac{1}{e^{hc/kT^\lambda} - 1} \qquad (1)$$
$$h = 6.6256 \times 10^{-34}\ Js \quad (\text{Planck's Constant})$$

$$E = \sigma\, T^4 \quad (\text{Stefan-Boltzmann's Law}) \qquad (2)$$
$$\sigma = 5.67 \times 10^{-8}\ Jm^{-2}s^{-1}\deg^{-4} \quad (\text{Stefan-Boltzmann's Constant})$$

$$\lambda_m T = 2.898 \times 10^{-3}\ m\ \deg \quad (\text{Wien's Transition Law}) \qquad (3)$$

$$I = A\, T^2 \exp(-e\phi / kT) \quad (\text{Richardson's Theory})$$
$$e\phi = \text{work function} \qquad (4)$$
$$k = 1.387 \times 10^{-23}\ JK^{-1} \quad (\text{Boltzmann's Constant})$$

2.4. Spectroscopic and monochromatic imaging

Figure 6 shows examples of argon arc images on mild steel and SUS304 stainless steel. Left hand-side pictures are over exposure images. Information of arc column is saturated, but image of cathode is clear. Weld pool and base metal can be detected. In right hand-side pictures, exposure time is too short to identify whole cathode surface, but plasma shape and metal vapour from the weld pool can be recognized. The pictures on middle part are suitable camera condition to recognize whole parts.

One good way to improve image quality is using circumferential filter of narrow band. Figure 7 shows the effect of band-pass filter on arc images. Figure 7(a) shows normal colour image, which reduce the intensity of all wavelength by ND(Neutral Density) filter. Images of different wavelength are quite different, but all of each image are captured in the same welding condition.

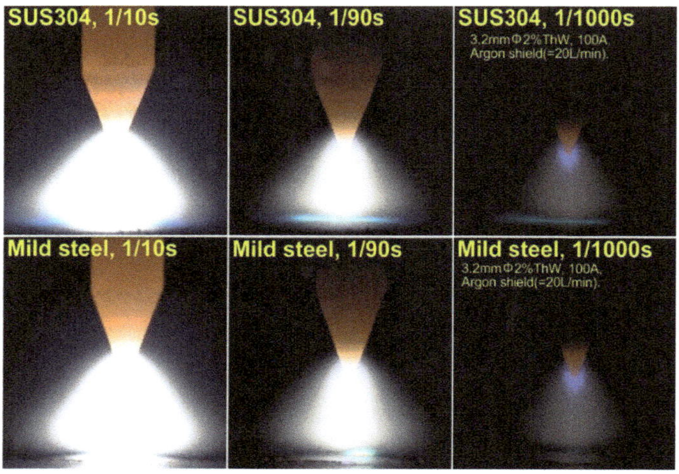

Figure 6. Example of argon arc image on metal plates.

The spectroscopic measurement of plasma condition is essential to understand plasma temperature. Plasma temperature is a key element to understand energy and mass balances in arc welding process. The problem for spectroscopic study on arc welding process is existence of metal vapour from work piece and electrode. Ionization potential of metal is much less than shielding gas. The metal vapour influences not only on ionization of plasma but also an energy transfer in boundary regions between plasma and electrodes. A grating monochromator is usually used to record spectra from representative arcs. Radiation strength is a function of wavelength depends on particle temperature. Typical line spectra due to transition of energy level indicate probabilistic number of temperature. In case of argon shielded arc, suitable lines in the Ar I (neutral) and Ar II (singly ionized) spectra are identified, measured and used to determine mean temperature of the species from the Saha's equation, Boltzmann distribution and the radiation law, by the two line method. The ratio of typical two spectra indicates the temperature, and this method is a good way to reduce measuring error caused by measuring device such as transmission loss of lens and sensitivity of sensor device.

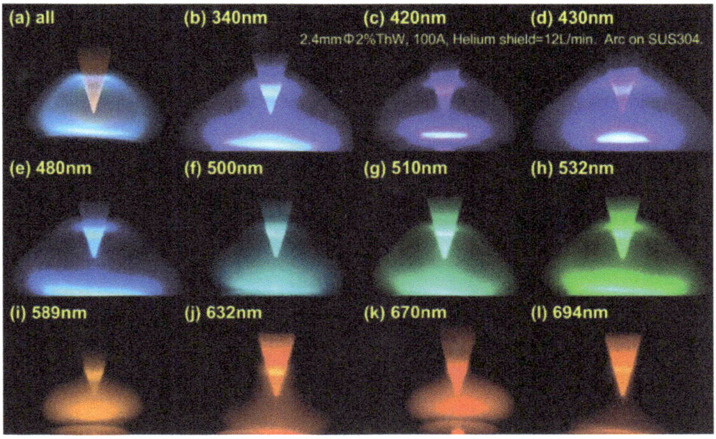

Figure 7. Example of single colour images.

The tendency is the same for thermometry. A monochromatic imaging is used to identify the measuring point of spectroscopy, and it is also used to identify spatial distribution of radiation. Spectroscopy by a prism is the simplest way to measure an interaction between radiation and matter as a function of wavelength, but resolution is much less compared to the grating monochromator. Figure 7 shows typical colour images, single colour images and spectroscopic images for several shielding gases. A single colour image at 694nm and spectroscopic image along the central line of the torch centre are captured by the same lens simultaneously. Image through the lens is divided by a prism. One half of light is passing through the band-pass filter of 694nm. The other half image is passing through the special prism which is called Imspector, and it reaches to the sensing device. Monochrome sensing devices are used for both cameras. Spectroscopic images are shown by pseudo colour system to identify the quantitative differences. Helium has a few spectra, so it can be used to identify the actual wavelength on horizontal position. Sensitivity of the image sensor is affected by wavelength. Normally, it is the highest between 500nm to 700nm range. The sensitivity over 900nm is one fourth of visible range or less. The sensitivity in the short wavelength less than 400nm is also low. Reduction loss of lens is affected by the wavelength also. Therefor precise quantitative comparison is difficult without correction of sensitivity and resolution. However, qualitative consideration becomes easy by spectroscopic imaging.

Dynamic ranges of data depth and optical frequency range are the most important factors to use measurement. A dynamic range in computer vision means data depth, and common data depth is 8 bits. The data depth of 8 bits is completely small for optical diagnostic. It usually requires more than 12 bits. Another important dynamic range is sensitivity of the solid imaging devices such as CCD (Charge Coupled Device) and CMOS (Complementary

Metal Oxide Semiconductor). Normal dynamic range of CCD camera is about 10^3. This value is much less for accurate measurements. A CMOS camera with specially tuned electronic device for sensitivity as a wide dynamic range of $10^{10\text{-}12}$ is available to use. However data depth is still a major matter to protect scientific use. Practical methods to use a solid devices for imaging are (1) using an well designed optical filter, (2) controlling the shutter timing and duration, (3) using a wide dynamic sensor, and (4) correction of image data those taken under different capture conditions as shutter speed, iris, and different neutral density filters.

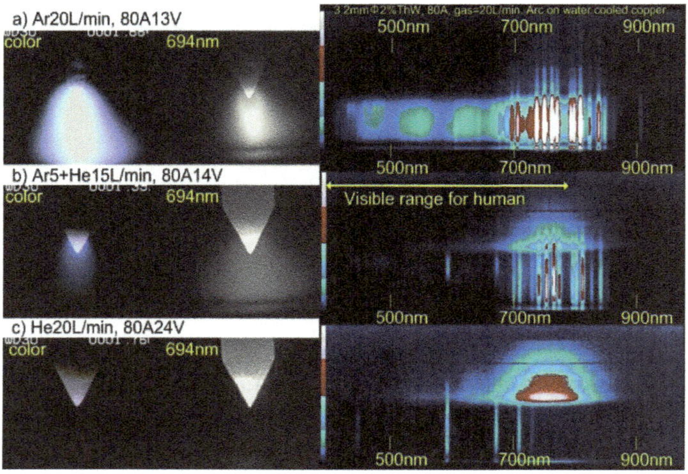

Figure 8. Effect of gas contents on arc plasma condition.

3. Observation of transient response

3.1. Observation of meta-stable stage of arc ignition

Argon and helium are usually used in GTAW process. Much smut covers on and around the welding bead in case of helium arc welding. This is a quite simple evident of metal vaporization during arc welding. Metal has much less ionization potential compared to argon or helium. Therefore, effects of metal vapours on arc properties and subsequently on weld bead configuration have been studied since 1950s(Pattee, 1973). Emission spectroscopy and monochromatic imaging were used to determine the prominent metal species present and distribution in arcs.

Temperatures of base metal and tungsten electrode are the same as room temperature until welding process is starting. Temperatures of base metal and tungsten electrode are quickly raised high. However they have some amount of heat capacity, so there are time rags to be stable temperatures. High speed video is a good tool to capture the dynamic behaviour. But, it is not so good to show movies to understand the dynamic behaviour. Capturing the video finishes very short time. But watching the captured movie takes a quite long time. Showing the behaviour of dynamic response is another problem. Showing picture cue in time order is one answer, but numbers of pictures to show and time step are difficult to decide. Using streaking image on important location is good answer to understand the dynamic behaviour at glance as shown in figure 9. Vertical line on the centre line of the electrode is drawn by time. The top time chart shows the dynamic behaviour of centre line between arc start and 15seconds later. The behaviour at first 1 seconds changes much. It takes about 4-5 seconds to be a stable condition. The second top picture shows the detail behaviour of transient state. The third and the fourth pictures show the behaviour of the horizontal lines at cathode tip and 0.5mm below the cathode tip. These pictures show the behaviour of plasma behaviour. These streak images are good way to understand time response qualitatively. The pictures shown on the right are single colour pictures at 694nm. They are taken simultaneously. Taking pictures of different wavelength is a good way to consider actual behaviour of the process. When the cathode becomes stable condition, upper range of the cathode becomes brighter. Tungsten oxide on the surface becomes wet in this region, because temperature of the tungsten electrode becomes high enough to melt tungsten oxide on and near the surface. This high speed image shows from arc ignition to stable stage. However, video rate and image resolution are insufficient to understand actual dynamic behaviour.

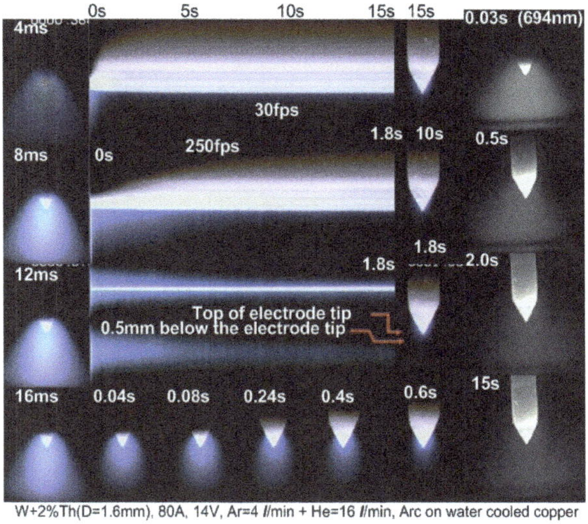

Figure 9. Typical example of GTAW at arc ignition stage.

3.2. Qualitative methods to show transient phenomena

Figure 10 shows examples of ultra-high speed pictures taken at 54k frames per seconds (fps). A picture before No.1 is absolutely in black. There is a big change between No.1 and No.2 pictures. There is brown coloured channel between the electrode and the base metal in No.1 image. This brown colour, which envelops the electrode top, is brighter than the other area. And three bright spots are shown on the base metal. These three spots have blue colour, which is a colour of recombination from metal vapour. One bright spot is remained in No.2 picture. Blue colour zone looks like aurora occupies almost all space between the electrode and the base metal. Brown coloured zone is disappeared. One bright spot appears on the upper place of the electrode. There is a drastic change on arc behaviour at the ignition moment. Using higher video rate is better to analyse in this arc ignition period, but this video rate is the highest in this space resolution (320x256pixels) of this high speed camera. Pictures among No.3 and No.12 look almost similar. This is another problem, when we consider the process from captured high speed video. Capturing times of ultra-high speed video is a few seconds, but it takes too much time to watch whole video. Much time to save the data into a hard disk is another troublesome issue. One solution to recognize typical transient behaviour is reconstructed the stacked images into streak image which contains important information of time response.

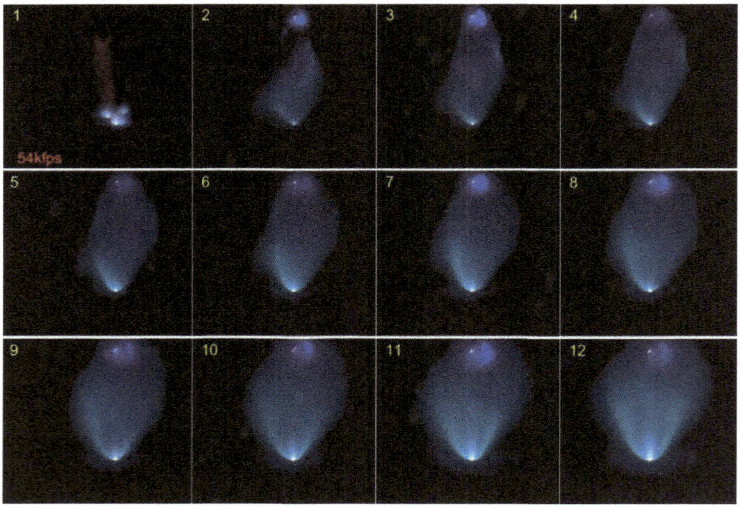

Figure 10. Time series of high speed video at arc ignition stage.

Figure 11 shows an example how to reconstruct the movie to a still picture. Left picture is a typical one shot at ignition stage. This is a typical case of arc ignition. Arc ignites in air, there is no shielding gas. Cathode and anode are easily oxidized, and their reactions produce much heat. Arc ignition in good shielding is shown in figure 12. Arc is very stable in this case, but some drastic behaviour occurs in first 5 ms.

The video rate of 250fps as shown in figure 9 is insufficient to understand actual arc ignition, because only two pictures in arc ignition stage can be captured in this video rate. Ultra high video rate as shown in figure 11 and 12 indicates that quite complicated response is appeared during first 5ms. High video rate is required for understanding of precise transient behaviour, but transient states of welding process happens only short time range. Welding process usually continues order of 10 to 100 minutes. And some unexpected transient behaviour during steady condition also occurs in short time. Therefore, high speed video needs very huge memory size.

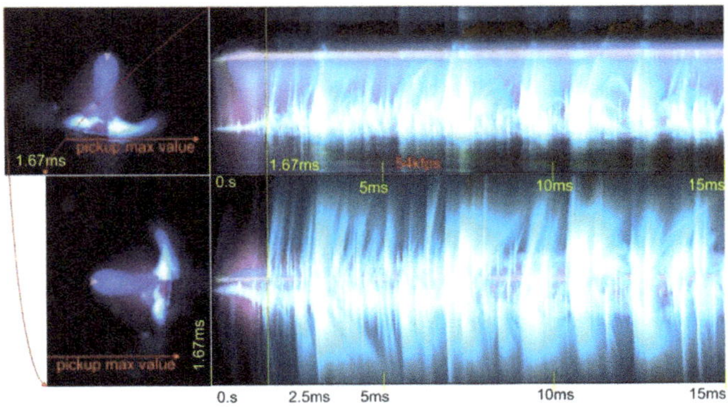

Figure 11. Example of time charts to recognize arc ignition stage in air.

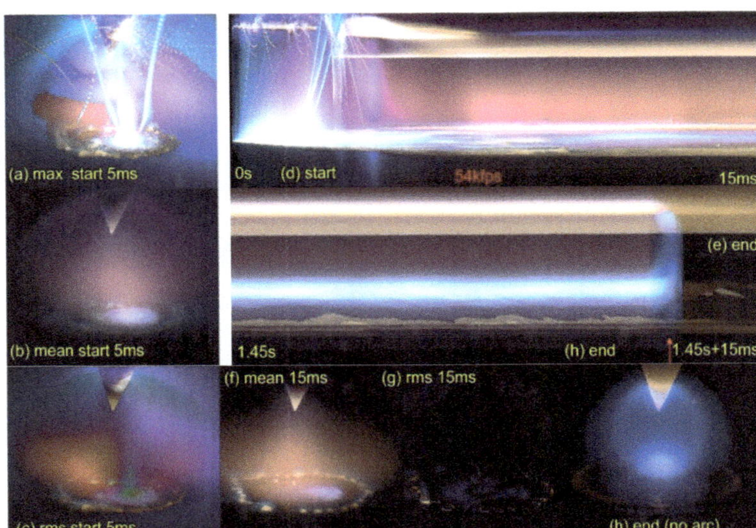

Figure 12. Methods to show process behaviour GTAW in helium shield.

Using statistical data on time axis is another good way to understand spatial behaviour. High speed camera captures many images in time series. Picking up the brightest value, mean value and/or deviation for each pixel gives us good information. Figure 12(a) shows the maximum value during arc ignition stage of 5ms. A whole time chart in this arc start duration along 15ms is shown in figure 12(d). The picture of maximum values gives us typical quantitative information on spatial behaviour. A locus of spatters indicates particle size, velocity and flying direction with its origin. Sizes and positions of anode and cathode are apparently appeared as bright zone. A picture of mean value is almost the same as normal still picture. An image of standard deviation indicates the active regions. The image of maximum values emphasizes singular situation like sputtering. So, this image is useful to identify the place somewhat abnormal situation exists. On the other hand, image of standard deviation usually hides one time irregular event.

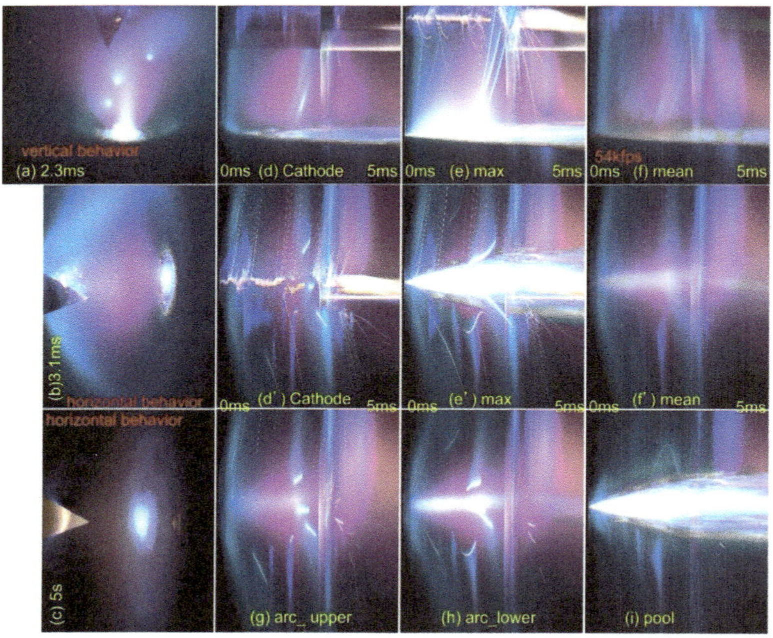

Figure 13. Examples of drawing methods to show time response.

Streak image is convenient for quick understanding of time response. And there are many points of view to get valuable information on time response as shown in figure 13. One is to check the typical line like centre line of the cathode as shown in figure 13(d). The arc process usually assumes as cylindrical symmetry, however this assumption is always wrong during arc ignition stage and improper welding condition. These conditions are main cases when we need the analysis by high speed camera. However, difference between the streak image of max value and it on the centre line suggests much good information for analysis. A streak image of mean value is also important to recognize global time response of the process. Vertical streak image is simple compared from the horizontal streak image. There are typical three different area such as the cathode region, plasma region and the anode region on the horizontal streak image. They are shown in figure 13(d', g, h and i), respectively. Pictures on typical moments are essentially important to understand the process. We can reconstruct the three dimensional behaviour of the process to use typical features of streak images in our brains. However, we need some other pictures for reconstruction of spatial features.

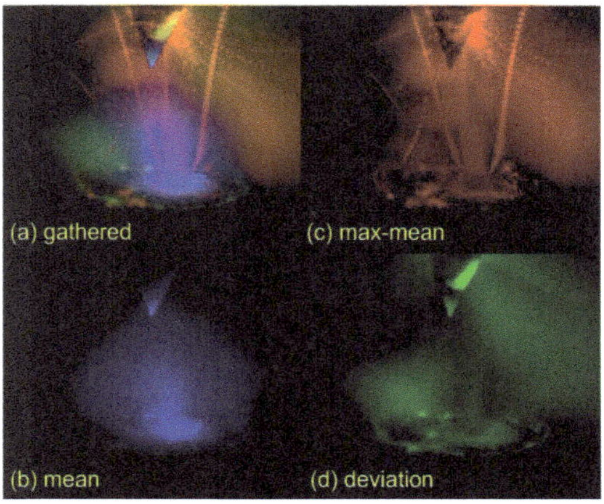

Figure 14. Examples of methods to show typical characters.

Figure 14 shows a synthotic image assembled different feature images. Figure 14(b) shows mean values of stacked images in blue colour. Figure 14(c) stresses one time feature in red, used values are maximum values minus mean value. Figure 14(d) shows deviation in green colour. Figure 14(a) is a gathered image among these three colours of different features. This kind of synthotic image which contains different feature helps to understand the spatial property at any periods. Discussions of this paragraph are somewhat sensitive or qualitative. Merit of using high speed camera is that captured huge data is digital, and quantitative analysis is expected in this field.

Figure 15 shows time response of unsuitable welding condition. Behaviour of arc ignition stage is almost the same as suitable condition. Because high frequency power source assists arc ignition. But current is too low to heat up the cathode temperature to keep steady arc. When the high frequency power is shut off, cathode spot (arc ignition area) becomes to move irregularly. It becomes stable from after 243ms of arc ignition. However, arc spot exists on the shoulder of the cathode. A single colour image of 957nm is captured in this case to get in good contrast to show arc and cathode spot. Reactions on cathode and anode are high-lighted in green, and reaction in plasma is high-lighted in red. Colour tone suggests the location of happening.

Colour image contains much information. But, colours captured by cameras have different characteristics. Captured image by usual colour camera is set to fit human sensitivity on natural scene. Radiation from arc has discrete line spectra. Therefore colour tone of captured arc image is very different by the makers and sensor kinds. We usually use (R,G,B) colour sets on pictures. Some cameras use (Y,U,V) colour system to proceed data in the camera. YUV colour system is reasonable on natural scene which has continuous property on colour frequency. Colour tones on arc welding process are different by camera kinds. Some good cameras have ability to correct colour tones, but it is very difficult to make the same colour tone from different cameras.

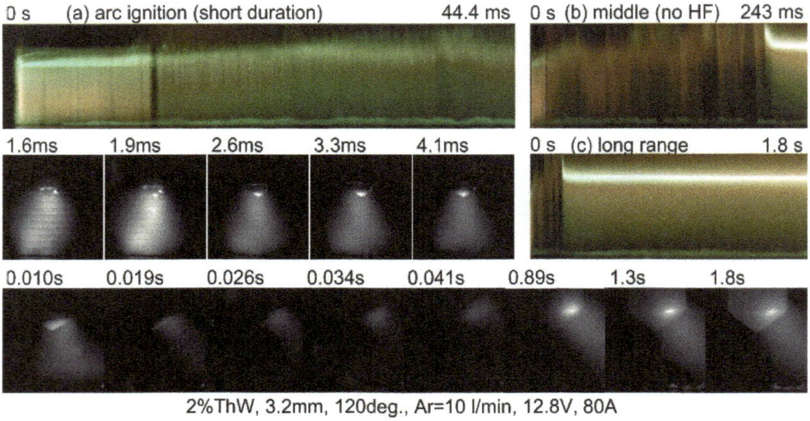

Figure 15. Examples of time response on not suitable cathode geometry.

3.3. Quantitative methods to show transient phenomena

Brightness is only apparent quantitative data at the first stage of analysis. Many quantitative elements such as cathode size and brightness distribution of cathode spot, maximum brightness of arc, arc area size, brightness distribution in arc, pool size, mean brightness of pool, metal flow speed on pool, etc.. The brightness is a good indicator to pick up unusual feature. One problem for analysis of captured data is a data depth. Normal data depth is 8bit, this means that digital range of brightness value is from 0 to 255. This data range is so small for actual arc welding process. When we focus the analysis in arc ignition period, brightness value of major target on this stage is less than those in steady state.

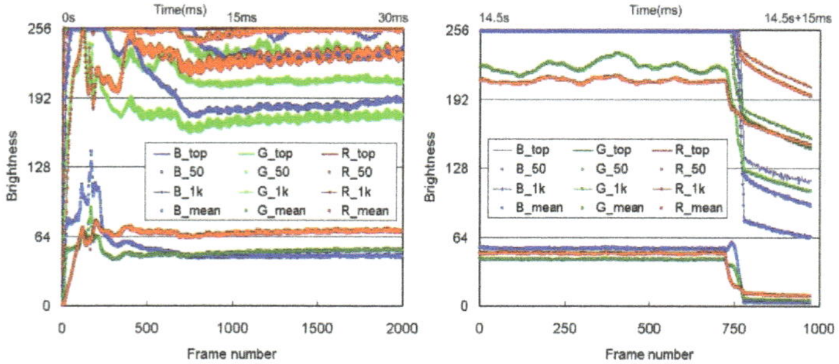

Figure 16. Time response of brightness in each frames.

Figure 16 shows the example of brightness values. This figure show maximum value(top), 50th value from the top, 1000th value from the top and mean values for each colour. Camera condition is set to get clear image of sputters during arc igniton period, so rather over exposure condition is used, and some data in bright area are saturated. Top 1000th data is appeared in the graph, and many of top 50th data also appeared in the figure. Data size of this video is 320x256(=81,920elements), then top 1000th data means top 3% value. In this meaning, capturing condition of this figure is almost the best to understand on arc ignition stage. Figure 16 also shows brightness on the end stage. Brightness increases with time until being on the steady state. So, all top values are saturated. Radiation of plasma suddenly stops with arc expiration, however many hot particles remains in space. Radiation from hot metal vapour is only appeared on this ending stage as shown in figure 12(h), so mean value of blue is increased at this ending moment.

Mean value is low compared to top values, because it contains low values in dark back ground. Figure 17 shows the difference of mean level for total image and it on arc and relating area. Tendency of both data is almost the same except sensitivity on alteration. A power source uses inverter control at high frequency. Brightness of arc changes with this frequency, and the frequency is close to image capturing rate of 54kfps. So some interference occurs as beat plotted in figure 17(b). However area size of target is almost the same, so mean value for whole image has some sense. Spectroscopic high speed imaging is also carried out, but video rate is 2kfps. Time charts of horizontal distribution shows the difference. Radiation by recombination of atom is essential at arc ignition duration until ion produces.

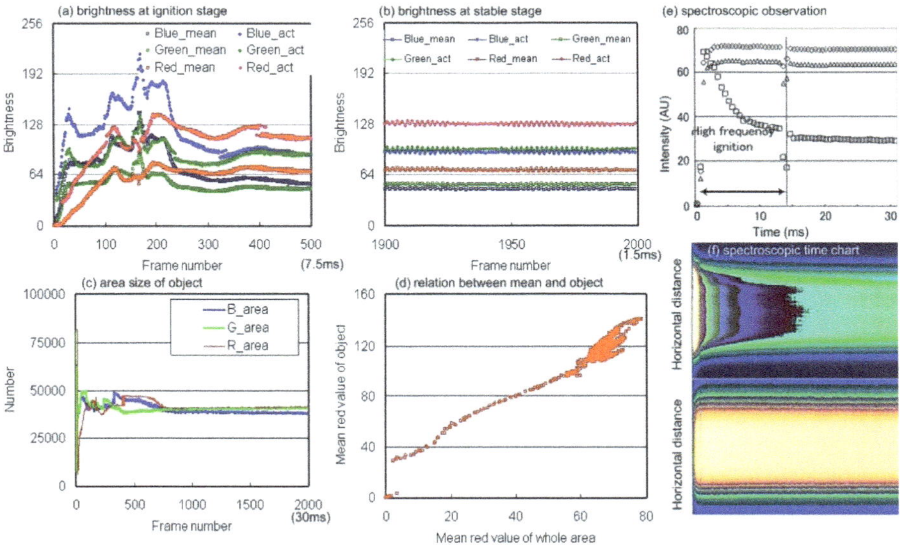

Figure 17. Examples of brightness characteristics.

Estimation of temperature on cathode and anode is important to understand welding process. Using thermal radiation value is good way to estimate temperature distribution even it usually over estimates the temperature by arc influence. Another way to estimate cathode temperature is using cathode spot area size. Electron density is a function of temperature, so estimation of mean temperature of cathode spot becomes possible when we count this area size from image. Figure 18 shows effect of gas contents on cathode temperature. Cathode spot in argon is concentrated on tip. Spot area size increases with helium addition. Cathode spot size suddenly increases when helium content exceeds 25%. Cathode temperature becomes high with cathode size enlargement. This sudden change causes by cooling effect of electrode. When the cathode spot is small and it locates on the top, current passing through the cone shaped electrode is heated efficiently by concentrated current in the cone shaped electrode. Cooling effect acts also efficiently by conduction in the cone. When cathode spot is enlarged by helium addition in argon, heating efficiency decreases. Heating by hot helium collision onto the upper position of the electrode also increases.

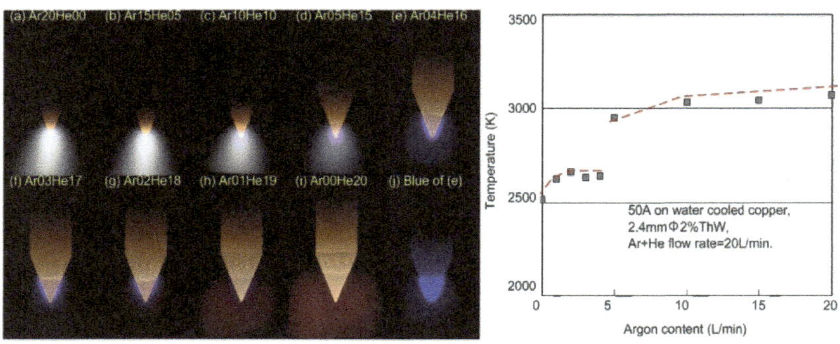

Figure 18. Effect of gas contents on arc behaviour.

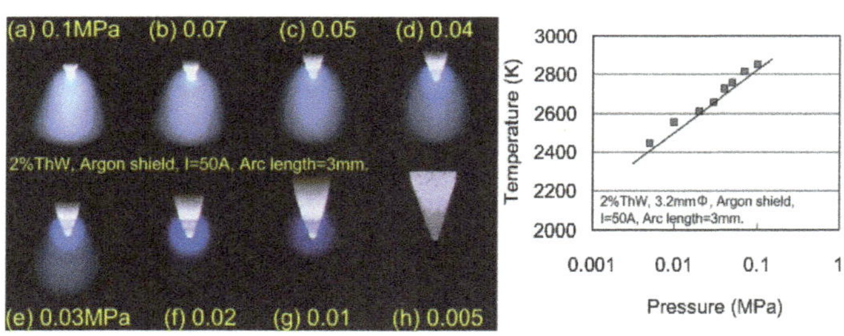

Figure 19. Effect of pressure on arc behaviour and electrode temperature.

Temperature change by ambient pressure is also estimated as shown in figure 19. Shielding gas is pure argon in this case, but argon is fulfilled in an experimental pressure chamber in this case. So, there is no flow of argon along the electrode, cooling action by shielding gas is not existed. And cooling system of cathode is different, those are reactions that cathode temperature at atmospheric condition is higher than the temperature shown in figure 18. Argon arc in low pressure looks like helium arc at atmospheric condition.

Figure 20 shows dynamic response of thermal radiation. Both electrodes are tungsten. Upper one is cathode, and lower one is anode. This movie is captured through interference filter of 532nm. Pseudo colour display is chosen to show brightness difference, because data depth of this high speed camera is 10bit. Much heat is lost by emission of electron at cathode. Anode is heated by impingement of electrons. This is the reason why anode is much brighter than cathode. Figure 21 shows effects of wavelength on brightness. Upper pictures show original monochrome image, and lower pictures show pseudo colour image. Thermal radiation at short wavelength is much less compared to long wavelength, because the highest radiation occurs at about 950-1100nm. Capturing conditions are set in proper value for each wavelength. Brightness is also normalized, so contour line shows approximate brightness distribution.

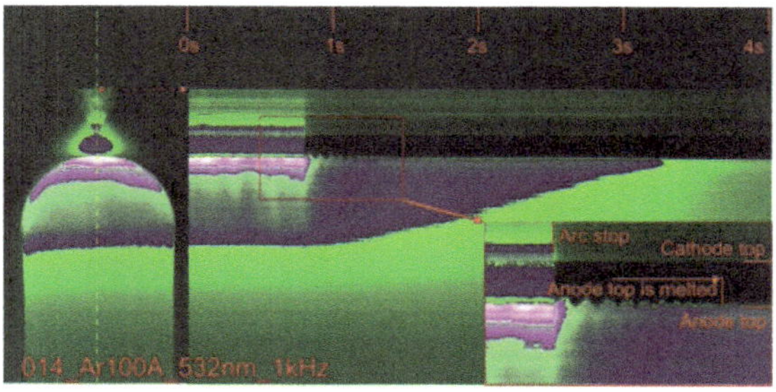

Figure 20. Dynamic response of thermal radiation.

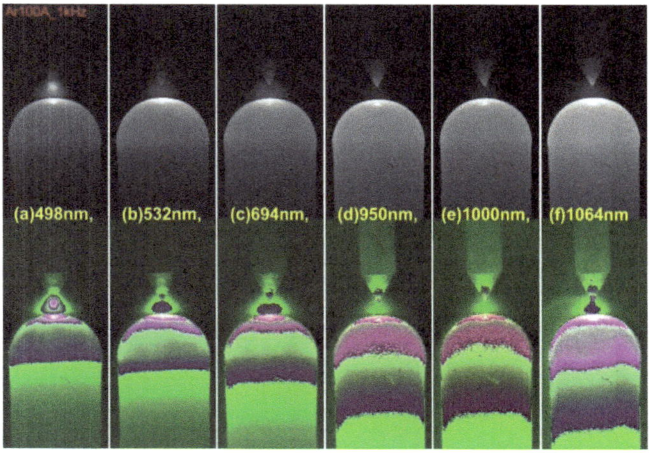

Figure 21. Effect of wavelength on radiation.

Estimation of temperature from brightness is simple method. But correction of obtained data is difficult. There are many unknown factors to correct. One simple way to correct the data is using the data at solidification area. Latent heat of solidification causes some typical feature around this area. When the brightness data is arranged as time chart, the same value continues at solidification stage as shown among point b and c in figure 22. Period between 'a' and 'b' is melting stage without arc influence. Dropped values from arc stage to no arc stage are about 1500 at point D, 1000 at point C and 750 at point A. These values are affected by radiation from arc. Point A is not melted. Solidification starts at b, and it ends at c when almost all metal near the point c is solidified. Data during melting period, which is between 'a' and 'b', are almost the same. Brightness data increases when solidification starts, surface of solid state is rough and it is covered by oxide, so radiation efficiency is high than the liquid state. One problem is why brightness data at C and D are different. Time period during solidification is longer at Point D, this is reasonable because heat capacity on fat area is higher.

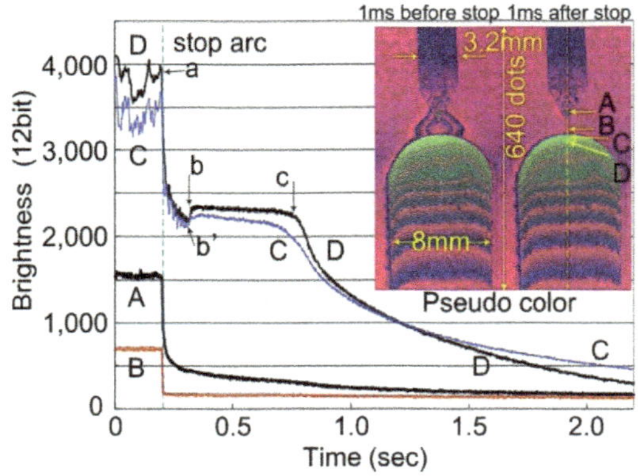

Figure 22. Measurement of surface temperature.

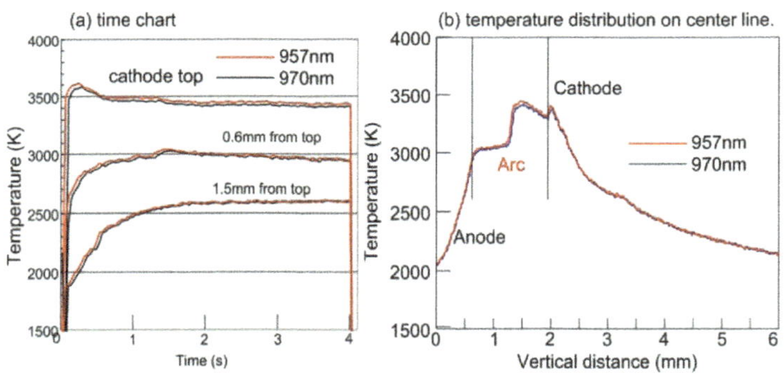

Figure 23. Example of single colour video for arc ignition stage.

Figure 23 shows temperature distribution calculated by brightness data. Argon is used as shielding gas in this case, and bead on plate welding is carried out. Used band-pass filters are 957nm and 970nm. Capturing condition is set to fit the brightness on tungsten electrode becomes just below the saturation. The reason why near wavelength is used is to estimate the influence from arc radiation. Calculated results for both cases are almost the same. Temperature at top position becomes high at early stage of arc ignition. The arc is concentrated at the top, therefor the temperature becomes high. Temperature becomes stable about 2 to 3 seconds later. Values of temperatures are higher compared to those as shown in figure18. Reasonable correction of brightness data is necessary.

Figure 24 shows temperature distribution of bottom surface of the welding pool. There is no affection by arc radiation, and boundary between liquid and solid is clearly recognized in this case. Melting temperature can be used to correct brightness data to temperature. Brightness on solid area is higher than that on solid area, therefore different fitting formula are used to determine the temperature value. Inside of red colour region is molten pool, and precise temperature distribution on welding pool is drawn in right. Upper picture shows early stage of welding, and lower one is the distribution in steady state. Length of welding pool becomes longer.

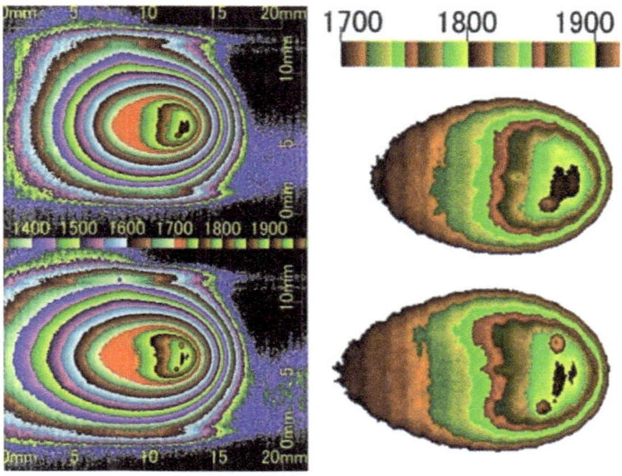

Figure 24. Pseudo colour image of temperature distribution.

4. Analysis of steady state

4.1. Effect of active flux on arc behaviour

Figure25 shows effect of active flux on behaviour of welding process. Upper pictures are captured on slant position. Lower pictures are capture in horizontal position to watch metal vapor on the pool. Major difference is size of anode area. Anode area for normal welding is wide. Anode area for active flux is narrow, and some vapor jet is shown on anode area. Next difference is position of metallic vapour color on the cathode. Metal vapour in arc decreases plasma temperature, because ionization potential of metal is much lower than argon and helium. And metal vapour is fully ionized in arc. Metal ion moves to cathode by electric field. Anode area for active flux is very narrow. This means that electron is constricted to this size, and almost all electrons impinge on this area. On the contrary, electron for normal case is wide spread to broad anode area.

Figure 26 shows the dynamic response of arc behaviour from normal area to an area where active flux is painted. Arc starts from left(normal area) to right(active flux area). Anode area on normal case is wide. When an welding pool reaches to active flux area, melted active flux invades on to the pool. And anode area is pushed to rear side by invaded flux layer. Upper pictures and lower pictures were captured by different maker's camera. A camera captured images in figure 26 is also made by different maker. Cone angle of electrodes, and surface treatment, and captured date are, but welding condition and material are the same. Tones of colour are quite different for these pictures. Another difference between normal arc and active flux is pool behaviour. Front position of the pool becomes closer on active flux case. And there is no change on pool length, so pool end moves to rear on active flux. Vibration level increases on active flux, many small ripples are produced.

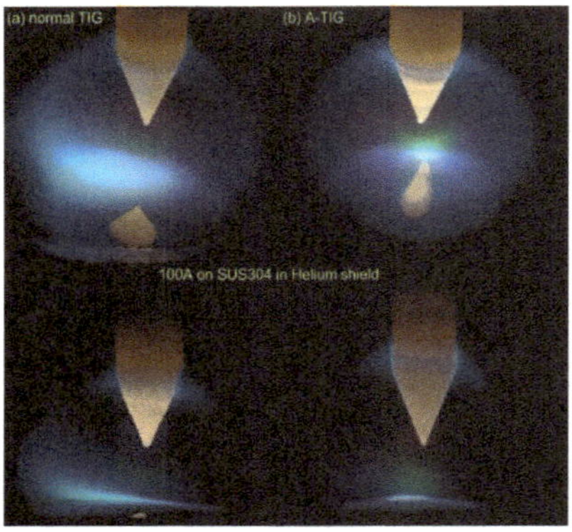

Figure 25. Effect of active flux on arc process.

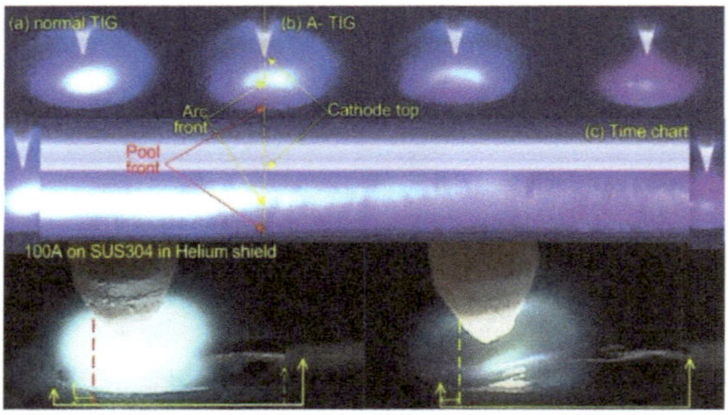

Figure 26. Effect of active flux on arc and pool behaviour.

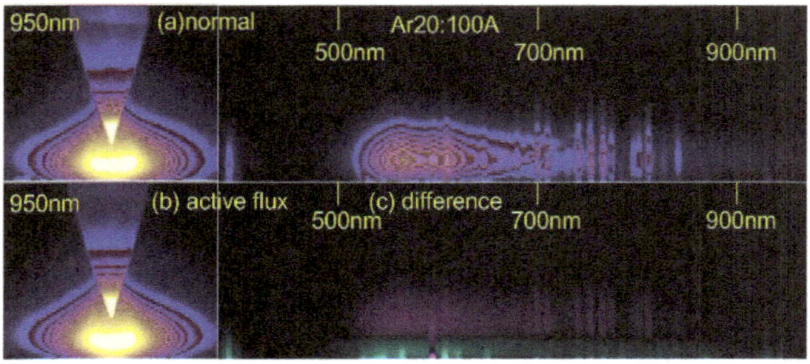

Figure 27. Effect of active flux on arc plasma.

Figure 27 shows typical pictures of plasma configuration and spectrum distribution in pseudo colour. Arc is generated in argon. Left pictures are single colour images at 950nm, which range has no strong spectra for argon, for normal stainless steel and stainless steel with active flux. These pictures are displayed in pseudo colour to intense the difference of both conditions. These are single colour images, and brighter point indicates higher temperature. Upper right picture shows spectrum distribution along central line of cathode for normal welding. Lower right picture shows the difference between normal case and active flux. The region where the brightness for normal plate is higher than it for active flux, displayed in green colour. Red colour shows the opposite case. The intensity shows amount of the difference of brightness. Radiation from normal plate near the molten pool is larger. On the contrary, the brightness for active flux is higher in the electrode surface and outer space of main plasma as shown in red colour. Radiation from active flux area near surface is very low.

Actual physical and chemical process acts on brightness. Brightness on cathode is low. It is difficult to recognize the difference to watch normal video by human eye. However there is some difference of frequency on space and/or time. Pseudo colour display is good method to show spatial difference as shown in figure 28(a,b). Small difference due to chemical reaction also can be extracted as shown in figure 28(c,d). Melted thorium oxide on electrode moves from upper side to top position. Behaviour of this chemical and physical reaction becomes visible by some numerical treatment. These reactions remain the evidence on the electrode. These evidence can be watched by SEM and EDM.

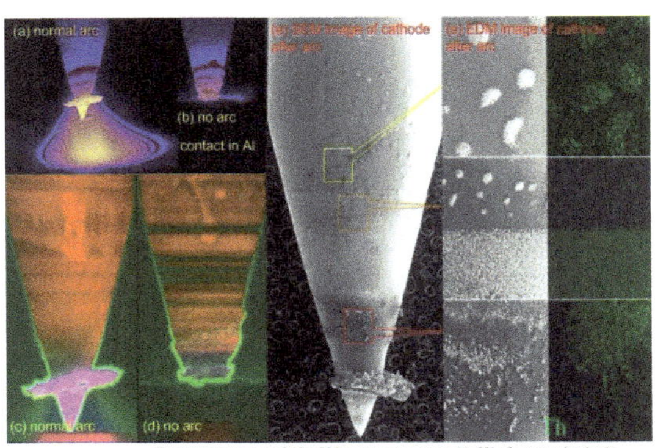

Thoriated tungsten of 15V-200A arcing in argon with 2% of oxygen for 1 min

Figure 28. Effect of oxygen for reactions on cathode.

4.2. Effect of gravity on arc behaviour

Figure 29 shows effects of gravity on welding process. High speed video is captured by drop tower experiment. Height of free drop zone is 10m, and time duration of micro gravity is 1.3sec. This time period is short, but it is enough to detect transient motion from normal gravity condition to micro gravity condition. There is no time limit to continue welding process before drop trial. Shape of molten metal is clearly affected by gravity. There is no apparent change on molten metal flow on the weld pool. However, inertial force acts on metal flow, and the time response of inertial force is unknown. Reconstruction period to balance static forces is finished in 10ms, and some vibration by overshoot motion remains around 10ms.

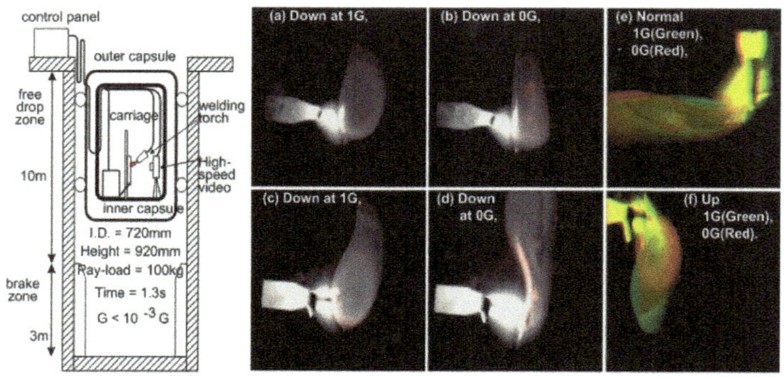

Figure 29. Results of drop tower experiment.

4.3. High speed imaging of metal transfer on GMAW process

High speed imaging of wire melting and droplet transfer phenomena have been carried out for long time. Silhouette imaging by using of strong external light was essential for observing metal transfer during Gas Metal Arc Welding (GMAW) process. GMAW process has some periodical vibration on metal transfer. Using band pass filters in near infrared wavelength becomes convenient because it contains information on temperature(Ogawa, 2004).

Figure 30 shows the typical metal transfer of GMA welding by interference filter of 950 nm without external lighting. Streak imaging is also useful to show dynamic behaviour. The reactions at the plasma/metal interface include oxygen removal at the anode and the discharge and pick-up of oxygen ions at the cathode. Figure 30 shows the reaction of oxygen invaded into the molten metal in the melted wire. A combination of invaded oxygen and carbon in wire makes carbon oxide gas. The gas is abruptly expanded by high temperature and exploded on the way to the work piece. Streak image shows points for spattering and unusual situation. Spatters are recognised as spike lines to the outsides from the wire centre, and the flying speed is recognised as its locus angle. Abrupt expansion of the droplet is recognised as an irregular knot. This scene is automatically detected by image processing of this figure. The merit of the high speed imaging system is that it uses digital data; therefore, effective analysis can proceed automatically(Ogawa et al., 2003).

A single colour video is a kind of thermal image, and it is presented in pseudo colour to emphasise the physical changes. Cathode spot exists in the first frame. The melting part at the top of the wire is growing, and the region between this melting part and the solid wire becomes slender. Current density of this slender portion becomes high, and temperature of this region increases quickly. Pinch force is also acted in this region. The combination of these forces drives to release the metal. Arc is soon generated between new tip and droplet with some small spatters. This set of images gives a visual representation of transfer modes in GMAW.

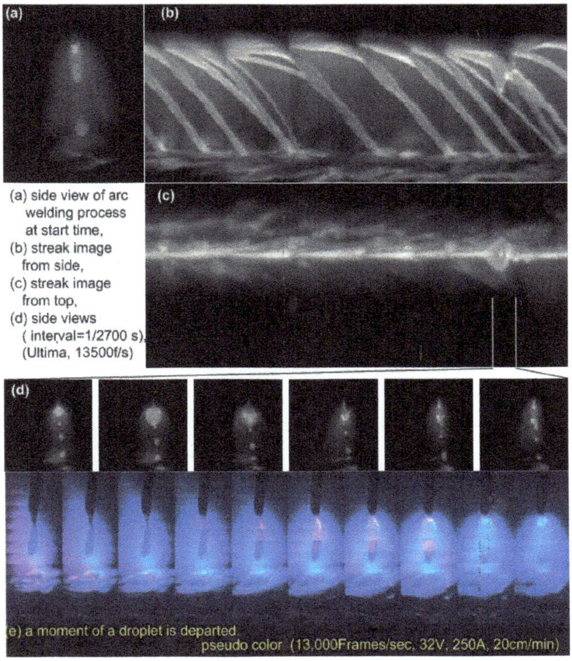

Figure 30. Time chart to pick up abnormal condition.

4.4. High speed imaging of Laser Welding process

Laser welding is very high speed welding method. Hybrid system such as Laser Arc hybrid and Laser Hot wire hybrid are often used to improve joining efficiency of large structures. Normal observation of arc welding process uses fixed torch system. Work piece moves during welding. High speed camera is heavy and big to carry on laser torch. When moving object is captured by fixed camera, object to watch moves in the scene as shown in figure 31(a). Reconstructing the scene as static torch system is easy as shown in figure 31(b). However, this system needs wide range of image size. Static coordinate observation system saves image size and/or improve spatial resolution as shown in figure 31(c).

Torch moving system has good points to present information on quality of whole welding result as shown in figure 32. These pictures are produced by using of histogram information. Lights have several infomations of their origin, and they appear in statistical features. Behaviours of arc, fume and sputters are apparently drawn by statistical features. Laser itself is invisible. But laser acts on fume and plume, and some statistical change appeares on its value. So laser channel can be shown from reconstructed image as shown in figure 33. Figure 33 shows pseudo color display of mean value image, original mean image and deviation images, respectively. Laser beam channel is apparent in this figure.

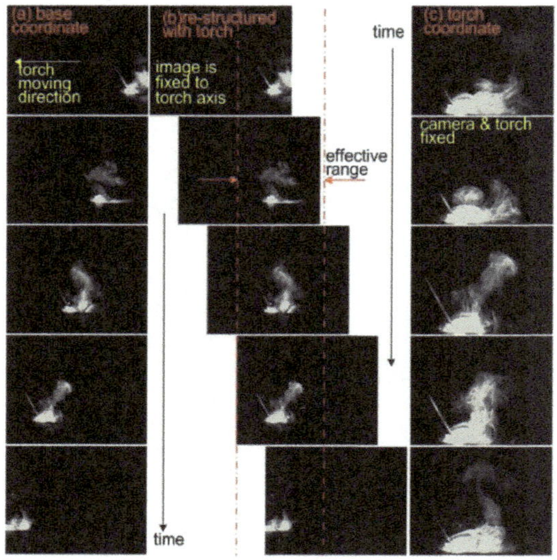

Figure 31. Comparison of coordinate system.

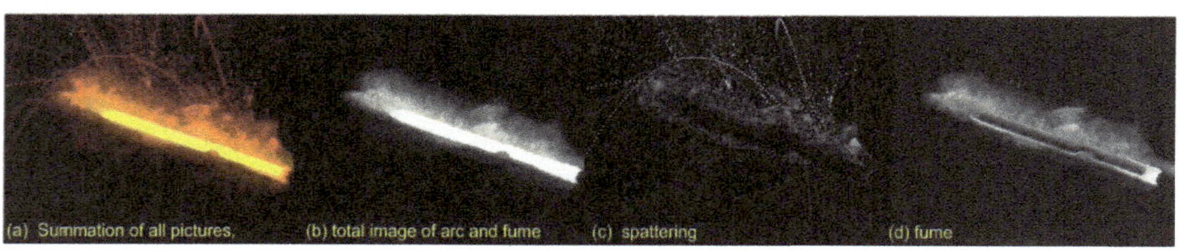

Figure 32. Exsample of imaging methods..

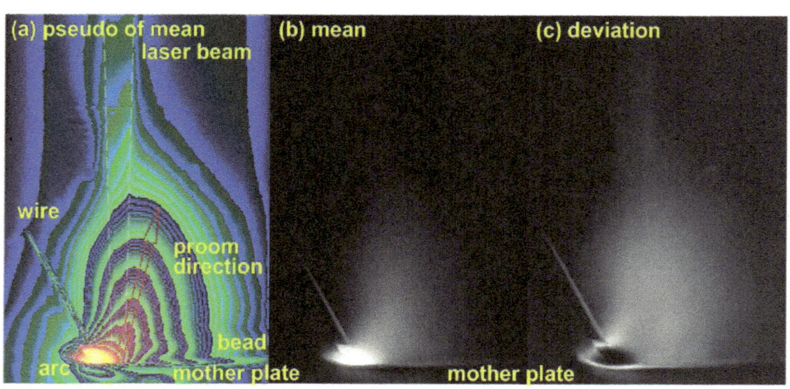

Figure 33. Statistical image of laser arc hybrid welding.

4.5. Monitoring and evaluation of GTAW process

Figure 34 shows a training system for weldrs of one side butt welding of thin stainless steel. This system uses four cameras to identify welder's skill. Welding trials are carried out by manual operation. A torch camera captures weld pool and arc. Camera 2 and 3 are fixed on carriage to capture surface and bottom situations of welded plate. Camera 4 is a fixed camera to observe welders motion. The camera 3 is the most important camera to identify the weld quality. This camera captures the conditon of bottom pool, and the system indicates analysed status of penetrated condition by sound on real time. Five tones are used to notice the actual condition to the welder. The welder can watch bottom situation by small LCD monitor inside the cover face. This monitor indicates only the image from camera 3. Voltage and current signals are also recorded and shown on the screen with colour. When the signal is out of suitable range, normal green colour changes to red. The collected data is stored in the data folder. In the same moment, a mentor of the welder tells to the welder about important points, and this voice is also recorded in the system. The welder can watch his own operation to select his data. Almost the same features are reproduced by the system with mentors indicated voice. Several reference information on mentors' operation and editorial videos are stored in the reference folder. The trainee can watch the reference video at any time. So he can learn his skill without any stress of actual welding operation. Another purpose of this system is to study relationship among voltage and amperage signals, torch camera image, and penetrated situation. Data of more than 200 welders of several welding companies were acquired to improve evaluation algorithm from torch camera image and electrical signals of manual welding. The same data on automatic welding operation are also acquired in various welding conditions. Evaluation of weld quality by one camera system becomes possible, when the feature of the welder was stored in the data base. One camera system is used on actual production process, and whole manual process is recorded to evaluate the quality of the products.

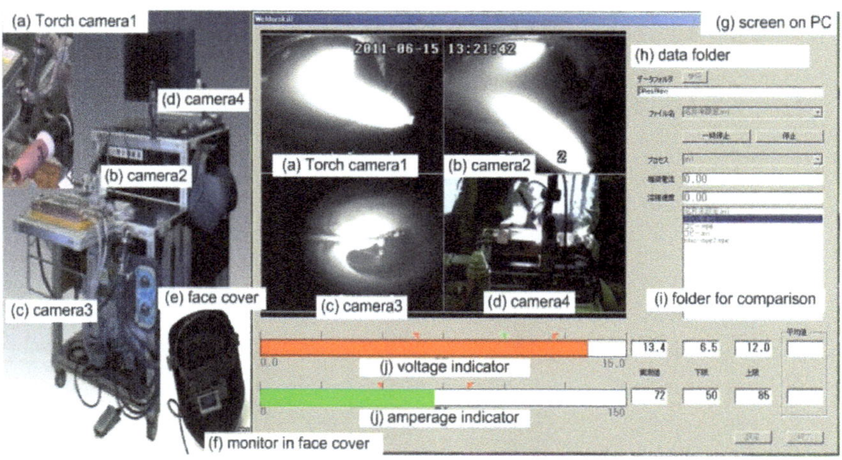

Figure 34. Training machine of GTA welders of one sided full penetration butt welding.

5. Conclusion

High speed imaging on welding process is a powerful tool to understand their nature. Deep knowledge on principle of target process is necessary to take and analyse succesfully. Some fundamental technique on high speed imaging are described in this paper.

Author details

Yoji Ogawa
Ogawa Giken Co.ltd. & AIST, Japan

6. References

Inoue, K. (1981) Image processing for on-line detection of weld process, Trans. JWRI, No.10, pp.13-18

Ogawa, Y. (2008) Accurate three dimensional measurement of arc welding phenomena, J. Soc. Instrum. Control Eng., Vol.47, pp.65-68 (in Japanese)

Ogawa, Y. (2011). High speed imaging technique Part 1, Science and Technology of Welding and Joining, Vol.16, No.1, pp.16-26

Pattee, H. E. et al. (1973). Effect of arc radiation and heat on welders, Welding Journal, Vol.52, pp.297-308

Shaw, C. B. Jr (1975). Diagnostic studies of the GTAW arc. Part 1. observational studies, Welding Journal, Vol.54, pp.33s-44s

Zacharia, T. & David, S.A. (1993). Mathematical Modeling of Weld Phenomena, Welding Journal, Vol.52, pp.3-23

Chapter 13

Monitoring of Arc Welding Process Based on Arc Light Emission

Marek Stanisław Węglowski

Additional information is available at the end of the chapter

1. Introduction

A welding process may be expressed as a system shown in Figure 1 whose outputs depends on the welding conditions or their nominal constants (which determine the dynamic model of the system), controlled by the variables or inputs (adjustable welding parameters), and affected by the disturbances (fluctuations or variations of welding conditions from their nominal constants), (Zhang Y.M., 2008).

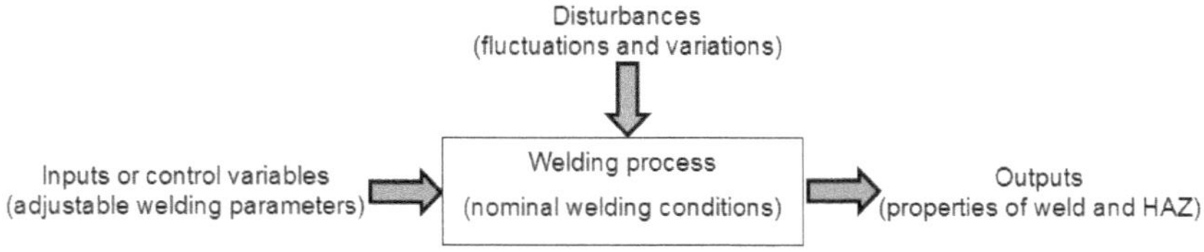

Figure 1. The welding process as a system (Zhang Y.M., 2008)

One of the major goals of the process monitoring in welding is to assure that the required welding parameters are being applied into the process to make a quality of welds. In case abnormal welding parameters are detected, the resultant segment of welds may be post-examined using the more precise methods, (Blakeley P. J., 1990; Siewert T., et al, 1992; Chen X.Q., 2002; Pan J., 2003). This would help to reduce the need of strict/expensive process controls and reduce an extensive use of the costly post NDT (Non-destructive Testing) of all welds. To this end, the monitoring devices are required to be fully automatic and the data analysis of sensed signals including welding parameters and signal generated from the welding arc need to be optimized. In addition, the monitoring devices must also incorporate the criteria so that they can judge if the welds are acceptable or need the additional examinations/repairs.

Monitoring of the welding processes can be divided into the traditional and non-traditional methods (Fig. 2). The traditional methods are based on the monitoring of the electrical and other direct welding parameters, (Kim J.W., et al, 1991; Johnson J.A., et al, 1991; Modenesi P. J., Nixon J.H., 1994; Luksa K., 2006). The non-traditional ones use many different signals, for example: x-ray radiation (Guu A.C., et al, 1992), IR and UV emission (Fan H., et al, 2003), ultrasonic wave (Carlson N.M., et al, 1992), acoustic emission (Taylor-Burge K.L., Harris T.J., 1993) and sound (Saini D., Floyd S., 1998; Luksa K., 2003) to analyse and detect the process.

The traditional methods have been effectively used in the welding process monitoring and control. For example, the measurements of the welding current and arc voltage can be used to estimate the stability of the welding processes, especially with the advanced methods of the signal analysis and AI methods, (Smith J., Lucas B., 1999). The so-called "through the arc sensing", which is based on the measurement and analysis of the welding current and arc voltage, is a widely used traditional method which has been accepted as one of the effective methods for the weld seam tracking. The synergic control of GMAW machines (Amin M., Naseer A., 1987) is also based on the measurements of the current and arc voltage. One interesting case where on-line control of the weld quality is based on the characteristics of the welding arc signal is the narrow groove GMAW with an electromagnetic arc oscillation (Kang Y.H., Na S.J., 2003). The relatively complex plasma arc welding process can be also monitored by using of an electrical signal from the pilot arc (Lu W., Zhang Y.M., 2004). In addition, traditional methods are also useful for detecting of disturbances of the welding process in the form of surface impurities and insufficient shield in GMAW. Monitoring is carried out using specialized monitoring equipment or universal measurement cards.

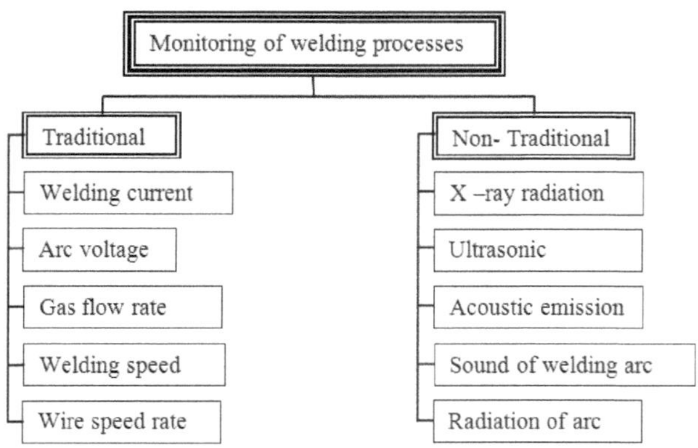

Figure 2. Monitoring methods of welding processes and typical signals

While the traditional methods have advantages of being low-cost and have achieved many successes as aforementioned, many existing issues may require the use of the signals more than the welding current, arc voltage, and the other direct welding parameters. For example, monitoring and control of a weld penetration is an important issue in welding, (Zheng B., et al, 2009) which may require the use of the non-traditional methods. The real-time vision systems take the lead in the non-traditional monitoring methods especially on the robotic welding applications. The CCD (Charge Coupled Devices) video cameras which can be used with the fast algorithms can give us real-time estimates of stability of the process, quality of welds, for example depth of the penetration, (Zhang Y.M., et al, 1993). However, the investment for non-traditional methods is typically high. Cost effective non-traditional methods such as the arc light radiation monitoring which can measure and analyse the intensity of the whole range of the arc light spectrum or intensity of a single emission line (Li P.J., et al, 2001; Wang Q.L., et al, 1997; Yoo C.D., et al, 1997; Ancona A., et al, 2004; Sadek C.A., et al, 2006; Li Z.Y., et al, 2009; Mirapeix J., et al, 2008) are thus desired. This method was first used to determine the length of the arc in the method of MAG in 1966 (Johnson C.A., et al, 1966). The spectroscopy methods are also finding application in the other welding processes such as laser welding (Stabillano T., et al, 2009; Bruncko J., et al, 2003; Kong F., et al, 2012). However, new methods of monitoring require the use of sophisticated measuring equipment, which in most cases need to be adapted for the measurement in welding.

2. Arc light radiation

They are many sources of an electromagnetic radiation of the welding arc area. It can be: the arc column, the regions close to the electrodes, the liquid metal transported across the welding arc, the molten pool, the heated region of the base material around the molten pool, the heated end of the electrode wire ((Pattee H.E., et al, 1973). The welding parameters strongly influence on the range of the wavelength of the electromagnetic radiation and their spectral composition (Hinrichs J.F., 1978). The intensity of the radiation produced by the welding arc is a function of the welding process itself and of the welding variables. The welding arc spectrum can be divided according to wavelength as shown in Table 1. The welding arc in the TIG method are shown in Figure 3.

Type	Wavelength [nm]
Extreme ultraviolet	4 – 200
Ultraviolet	200 – 400
Visible	400 – 750
Infrared	750 – 1300
Far infrared	1300 – Hertzian wavelength

Table 1. Radiation from welding arcs (Hinrichs J.F., 1978)

Figure 3. The shape of the TIG arc (welding current 100 A, arc length 2 mm)

The energy dissipated in the arc column is mainly dissipated by the conduction and convection. Emission of an electromagnetic radiation is 10 to 15% of the energy supplied to the arc (Marzec S., Janosik E., 1995). Thermal radiation, whose source is a body of high temperature, is characterized by a continuous spectrum of radiation. The source of the continuous spectrum in the arc is mostly a liquid weld pool (Quigley M., 1977). Radiation characteristics of ions and atoms in the arc is a discrete (Glickstein S., 1976). This type of radiation is analysed in the literature as a plasma radiation.

The plasma at a temperature within the range between several tens of eV and keV (the energy scale 1 eV = 11 600 K) emits an infrared radiation, visible, ultraviolet or X-rays, which, due to the emission mechanism can be divided into the three basic types (Huddlestone R., 1965):

- the line radiation of atoms or ions sent during the move from the one discrete energy level to another (transition between states related);
- the recombination radiation associated with the free-electron uptake by one of the discrete levels of atoms or ions (the transition between the state of the free and associated states);
- braking radiation in the free-electron zone's ion (transitions between free states).

The total radiation of the plasma arc is the sum of a continuous radiation and line radiation of the spectral lines (Szymański A., 1991). This sum can be written as:

$$\varepsilon_{\lambda} = \varepsilon_{\lambda,c} + \sum \varepsilon_{\lambda,L} \tag{1}$$

where: $\varepsilon_{\lambda,c}$ - intensity of radiation with a continuous spectrum, $\varepsilon_{\lambda,L}$ - intensity of spectral lines; sum appearing on the right side of the equation is carried out after all lines lying in the area.

For the optically thin plasma radiation intensity with the continuous spectrum can be written as (Szymański A., 1991):

$$\varepsilon_{\lambda,c} = k_{\lambda,c}(T) \cdot B_{\lambda}(T) \cdot \left(1 - \exp\left(-\frac{hc}{\lambda kT}\right)\right) \tag{2}$$

where: $B_{\lambda}(T)$ - the Planck function for blackbody, $k_{\lambda,c}(T)$ - the total absorption coefficient, T – arc temperature [K], h - Planck's constant ($6{,}6262 \cdot 10^{-34}$ [Js]), c - speed of light in vacuum ($2{,}9979 \cdot 10^{8}$ [ms^{-1}]), λ - wavelength [nm], k - Boltzmann constant $1{,}38 \cdot 10^{-23}$ [JK^{-1}].

The formula for the intensity of the continuous radiation is valid regardless of whether the plasma is in a state of the local thermal equilibrium (LRT) or not. The intensity of a spectral line $\varepsilon_{\lambda,L}$ data is an expression (Szymański A., 1991):

$$\varepsilon_{\lambda,L} = \frac{hcg_q A_{qp} N_e N_i}{8\pi\lambda U_i(T)} \left(\frac{h^3}{2\pi mkT}\right) \cdot \exp\left(\frac{E_{i,q} - \Delta E_i}{kT}\right) \cdot P_{qp}(\lambda) \tag{3}$$

where: g_q – the statistical weight of upper level; A_{qp} – probability of transition; $E_{i,q}$ – ionization energy of the upper level; ΔE_i – decrease of the ionization potential; P_{qp} –line profile, N_e – electron density, N_i – joins density, U_i - the statistical sum of the ion, m – mass of particle.

A spectral distribution and an intensity of the thermal radiation depends on the body temperature. Black bodies with the temperatures up to 500 K emit mostly the infrared radiation at a wavelength > 2 μm. Body with a temperature above about 1000 K, in addition to a long-term infrared radiation also emit the infrared radiation close to the wavelength range 0,78 ÷ 1,4 μm, and very little, because less than 1% of visible radiation. Only body at a temperature higher than 3000 K emits infrared and a visible radiation is also a slightly (0,1%) long-term ultraviolet radiation. Only the body with a temperature above 4000 K emits ultraviolet radiation shorter than 315 nm (Marzec S., Janosik E., 1995). The welding arc radiation intensity is the greatest at the wavelengths between 200 and 1300 nm [16]. The share of the infrared radiation, visible light and ultraviolet radiation in the spectrum depends on the welding method, and the welding parameters (H.E. Pattee, et al, 1973).

The highest intensity of the visible radiation of the arc welding processes is observed in the MIG/MAG and next MMA, TIG and plasma welding. It was also found that the intensity of the ultraviolet radiation increases with the square of the welding current and the intensity of visible radiation is not growing so vigorously (H.E. Pattee, et al, 1973). The intensity of the ultraviolet radiation and visible light emission when welding with the coated electrodes and cored wires (MIG/MAG and self-shielding wires) in the presence of welding fumes is less than that in case of TIG process (for similar welding current). Under the same conditions, the intensity of the infrared radiation does not change dramatically. When a submerged arc welding process is used the visible and ultraviolet radiation is absorbed by a layer of a flux.

Radiation characteristics of the ions and atoms in the arc has a discrete character, and the source of radiation is mainly argon atoms and ions, iron, oxygen and nitrogen. The intensity of radiation of the other elements is much lower. In the wavelength range of the visible radiation spectrum iron, oxygen and nitrogen lines, and only partially spectrum of argon, which the ionization potential is much higher are mainly composed (Petrie T.W., Pfender E. 1970). This also means, that the emission of the light by atoms and ions of argon occurs at the higher temperatures than the temperatures reached at the arc welding for example, in a mixture of $Ar+CO_2$.

The discrete spectral studies provide information about the temperature of the radiation-emitting particles, because the excitation of particles required to provide it with a certain amount of the energy, and for this reason the temperature can be measured. The source of this type of radiation in the arc welding is mainly plasma arc column, but also the metal transported by the arc, slag, and the surface of the welded components (Etemadi K., Pfender E., 1982). The energy regions close to the anode and cathode arc are consumed for heating and melting of the electrode and the base material. It is known that the potential and kinetic energy of the electrons are converted into the surface of the anode heat causing it to intense heat (Petrie T.W., Pfender E., 1970).

The arc radiation is a complex phenomenon and dependent on a number of the welding parameters. To apply for monitoring the radiation of the arc welding process with a high accuracy and reliability is necessary to create the model of the binding intensity of the visible radiation from the arc welding parameters (Yoo C.D., et al, 1997). The welding arc can be regarded as a point source of the radiation. However, this approach in many applications seems to be insufficient. A better approach is to treat the arc as a cylindrical source of the radiation. This model accurately reproduces the actual shape of the welding arc and makes examination of the arc radiation easier. For this reason, it will be elaborated. Cylindrical model can also be simplified and presented as a half-sphere of the welding arc, which is used in the design of the monitoring systems of automated welding process, based on the machine vision systems (Lee C.W., Na S.J., 1996; Yu J.Y., et al, 2003).

Arc column consists of the three types of particles: electrons, ions and neutral atoms. It is assumed that the arc column is in the state of a local thermodynamic equilibrium, in which the electron collisions play an important role in the excitation and ionization.

Equation 2 describes the arc emission of the radiation with a continuous spectrum. Given the relationship between the wavelength and frequency $c / \lambda = \nu$, and the Planck function for the black body, as well as when $h\nu / kT \ll 1$, the Rayleigh-Jeans'a law is performed. Then the equation 2 can be simplified to:

$$\varepsilon_{\nu} = k'(\nu)\frac{2\nu^2}{c^2}kT_e \tag{4}$$

where: υ - frequence [Hz], T_e - the kinetic temperature of electrons [K].

Compatibility equations 2 with 4 is better than 5% for $\lambda_L T > 4{,}3$ cmK, where λ_L is wavelength [cm]. The right side of equation 4 has a value equal to 1 (approximately) for the infrared and visible radiation. Also, at atmospheric pressure and a normal range of the welding current, the electron temperature is close to the temperature of the arc. Considering the above and apart from the differences in the temperature, it can be eq. 4 write as:

$$\varepsilon_{\upsilon} = k'(\upsilon)\ \frac{2\upsilon^2}{c^2}kT \tag{5}$$

where: T the arc temperature [K].

To simplify the discussion, the gradient of the temperature along the axis of the arc can be omitted. By combining the emission coefficients for the different areas of the arc, the energy radiated from the entire arc can be expressed as:

$$B_{i\upsilon} = \iiint \varepsilon_{\upsilon} d\upsilon \tag{6}$$

After the calculation of the emission factors in the whole arc welding, and after assuming that electrical conductivity and voltage gradient are constant and taking into consideration the impact of visible light weld pool (Zhang Y.M., Li P.J., 2001):

$$B_{iv} = G_1 L I^{\gamma}\left(e^{\frac{G_2}{I}} - \frac{1}{2}\right) + G_3 I^2 + G_4 \tag{7}$$

where: γ, G_i – constants, L – arc length, I – welding current.

Equation 7 gives a relationship between the radiation and the visible arc welding parameters, including the current intensity and the arc length. The authors of the model (Zhang Y.M., Li P.J., 2001) show that this equation is satisfied for arc welding with a current of 150 A, because at the higher currents the current density is not constant over the entire volume of the welding arc.

3. Investigation of the arc electromagnetic radiation

The investigation has been carried out to date focused mainly on examining the luminance of the arc, impact of radiation on the health of the welders (Hinrichs J.F.,1978) and systems to protect them, and to development the tracking systems (torch position). Analysis of the visible light spectrum emitted by the arc welding is used to study the distribution of a temperature in the arc (Farmer A.J.D., Haddad G.N., 1984), calculate the average temperature of the welding arc, an amount of a hydrogen in the shielding gas (Grove L., et al, 1970), and the temperature of molten metal weld pool. The analysis of the arc light emission may help to develop the technique of taking photographs of the welding arc. Spectroscopic methods are a useful tool for studying turbulent shielding gas after leaving the gas nozzle in the TIG and MIG/MAG methods, relationship between the spectral distribution of radiation and the type of a base material and the electron density distribution.

It should be emphasized that the study of the visible radiation in the method of the arc welding MIG / MAG was also used to monitor the metal transfer process in the arc (Wang Q.L., Li P.J. 1997). Methods that use the electrical signals (measure the welding current and arc voltage) are effective only to track the short arc and globular metal transfer welding process. When the metal is being transferred by the spray mode, the signal / noise ratio is too small, and the greater accuracy is achieved by measuring the intensity of the visible radiation arc (Wang Q.L., Li P.J., 1997). Optical methods are also applied to scan the length of the welding arc in the TIG and the MIG/MAG methods.

In parallel, a wide range of plasma research is conducted. First of all, emission spectroscopy and scattering of a laser radiation (laser spectroscopy) were used. These methods allow the calculation of plasma parameters such as a temperature and concentration of atoms (ions, electrons).

The emission spectroscopy is a passive method in which the electromagnetic radiation from the plasma (one or many spectral lines) is recorded and analysed. The advantage of this method is particularly simple measurement. This requires an optical focusing system, a monochromator or a spectrometer and detector, which can be photomultiplier or CCD. The disadvantage of this method is that the recorded radiation is a total emitted from the plasma. In order to obtain measurement data from one particular point of the measurement, it is necessary to use the Abel transformation (Cho Y.T., Na S.J., 2005). Another disadvantage is the need to run the calculation assumptions that the plasma is in a state of a local thermodynamic equilibrium and is optically thin.

Laser spectroscopy is a more universal method. However, it requires a laser light source and a detection system. The method of the laser spectroscopy allows for determination of the plasma parameters at a given point. In some cases, calculation of the plasma parameters without the assumption that the plasma is in thermodynamic equilibrium allowed. This technique uses: the Rayleigh scattering, Tomson scattering, laser induced fluorescence (LIF) and two photon laser induced fluorescence.

Plasma radiation recorded in the measurements perpendicular to the axis of discharge (called side-on) is the sum of the smaller contributions from the various layers of plasma (Figure 4). The known Abel transformation (Cho Y.T., Na S.J., 2005) allows to determine ε(x) knowing I(x).

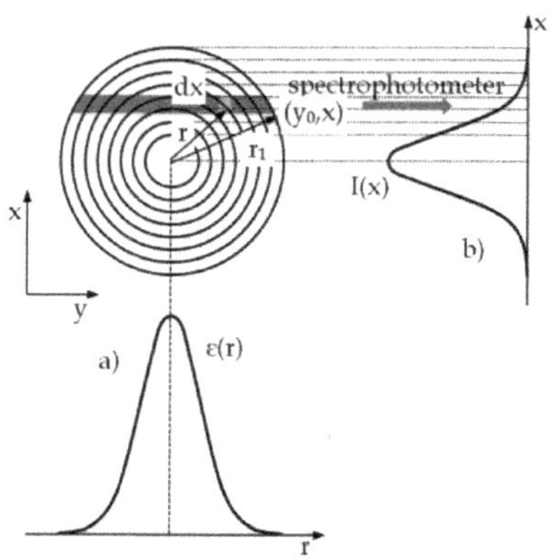

Figure 4. Cross section of the plasma column, the discharge axis is perpendicular to the plane of the paper, A - radial distribution of the emission factor, B - distribution of intensity observed on the side, I(x) - radiation distribution of intensity in the plane perpendicular to the direction in which the plasma is observed, x – distance from the direction of observation of plasma (Cho Y.T., Na S.J., 2005)

If the plasma in the observed cross-section is cylindrically symmetrical and the phenomenon of self-absorption does not occur, the radiation distribution of intensity in the plane perpendicular to the direction of observation of the plasma can be determined by the formula (Cho Y.T., Na S.J., 2005):

$$I(x) = 2 \cdot \int_{x}^{r_0} \frac{\varepsilon(r) \cdot r}{\sqrt{r^2 - x^2}} dr \tag{8}$$

where: ε(r) – intensity of radiation emitted by the plasma per unit thickness or distant from the axis of the discharge, x – distance from the direction of observation of plasma (Fig. 4), $2r_0$ – diameter of the area in which the plasma occurs.

Up to now, the main goals of the plasma investigation in the welding arc were creation the mathematical and physical models of the arc (Fan H.G., et al, 1997). These models will be very useful to design new welding machines. Very important aspect of experiments is to find the correlation between the electric welding parameters and the properties of the welding arc. Many experiments concern studies about the influence the composition of the shielding gas on the plasma properties. Also some investigation concern a magnetic arc deflection (Kang Y.H., Na S.J., 2002). The most important aim of investigation is to calculate the arc efficiencies. Many experiments were focus on the physical properties of the plasma welding arc, for example the temperature distribution, velocity fields of the electrons and ions, electrode work functions, and the local thermodynamic equilibrium in free-burning arcs in argon. Modern methods of welding, including A-TIG method, prompted the author (Ogawa, Y. 2004) to study the effect of the additional elements and compounds intentionally introduced to the area of the welding arc on properties.

In a study of the arc radiation it is important to determine the influence of individual factors on the width of the spectral peaks. Typical spectral line profile is shown in Figure 5 together with the characteristic values: x_c – wavelength of the center line, FWHM - Full Width at Half Maximum, I_{max} – maximum value for the radiation intensity of spectral line. The natural width of the spectral lines (Huddlestone R.H., Leonard S.L., 1965) is due to the finite lifetime of the energy levels and is higher, the lifetimes are shorter. Emission line profile resulting from natural broadening is the Lorentz distribution.

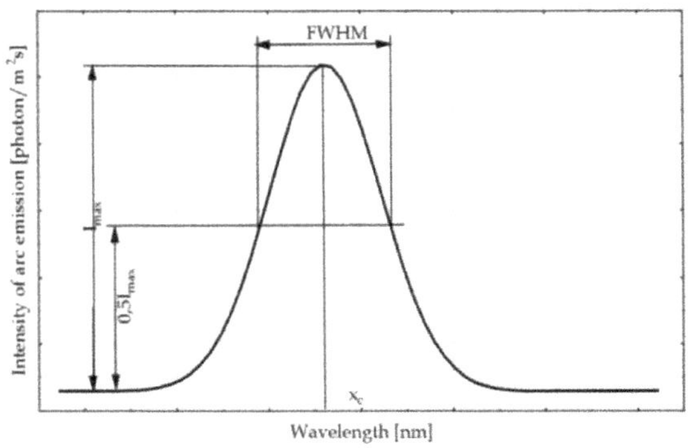

Figure 5. Typical spectral line profile

The second important factor is the Doppler broadening of the spectral lines, which is associated with the movement of the particles emitting the radiation. If the emitter has a velocity component of the direction consistent with the observation, the relative change in wavelength, involve a change in frequency is called Doppler effect. In the case of a thermal motion when emitting particle velocity distribution is Maxwell's distribution, the profile of the emitted spectral line is the Gaussian profile (Zielińska S., 2004).

Another kind of broadening, which can be encountered in the analysis of spectral lines, is a pressure broadening. This kind of broadening of the spectral line is the result of collisions with the other particles emitter. They can limit the lifetime of the excited atomic levels, and thus lead to an broadening of the line profile, which in this case is the Lorentz distribution. There are basically three types of the pressure broadening: the resonances, van der Waals and Stark (Zielińska S., 2004).

Components of the measurement system is a factor caused the further broadening of the spectral line. Apparatus profile in this case is the Gaussian profile. Theoretically, the spectrometer apparatus function should be linearly dependent on the wavelength. In reality, however, the profile apparatus is a convolution of functions associated with the matrix detector and functions of the optical elements of the spectrometer (Zielińska S., 2004).

The factors leading to the broadening of the spectral lines can be divided into given the Lorentz and Gaussian profiles. Their impact on the value of the broadening is different and may depend on conditions in the plasma. Resultant spectral line profile is a function which is a convolution of the Lorentz and Gauss functions (Huddlestone R.H., Leonard S.L., 1965) called the Voigt profile.

It should be noted that the photo-detector (CCD detector), in addition to the signal, measures also the background radiation. To eliminate the influence of background radiation, when analysing the intensity distribution of the welding arc radiation this radiation must be subtracted.

Taking into account the resolution of the transmitters, the recorded peaks are "cluster" of several spectral lines of a single element (Fig. 6), or even a few elements in the various degrees of ionization. So, matching the shape function is important in determining the exclusion of gravity for that group of peaks. The first element of the monitoring system of welding processes is to develop methods for the identification and measurement of the characteristic quantities of recorded spectral line, such as a peak width, position and amplitude of maximum. The examination which function better describes the profile of the peak ("cluster" of spectral lines) seems to be crucial for the detection of disturbances of the welding process. Profiles of the peaks can be matched using Gaussian, Lorentz and Voigt functions (Fig. 6).

Matching functions are carried out mostly using the least squares method (eg. Levenberg-Marquardt algorithm) and special software can be adapted for this purpose. On the basis of the matching function parameters the position of maximum spectral line (xc) and the spectral line width (FWHM) can be determined. In developing the experimental data even in so-called matching additive constant y_{00} must be considered. The constant is present due to the additional signals recorded by the measurement apparatus.

From the equation (7), a relationship between intensity of the welding arc radiation (energy radiated for a given spectral line xc) B_{iv}, arc length and welding current can be designated. Using a software, coefficients G_i and γ can be estimated basing on the collected data.

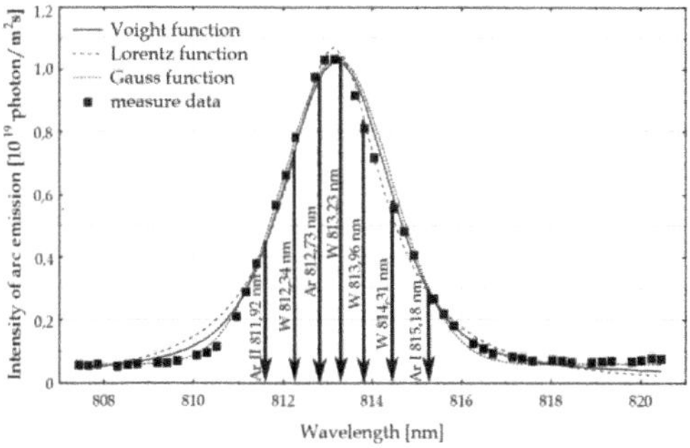

Figure 6. An example of peak matched Gaussian functions, Lorentz and Voigt, I=200 A, L=3 mm, 100 % Ar with marks of spectral lines .

The values of the parameters characterizing the welding process can be determined minimizing the sum of squares:

$$\chi^2 = \sum_l \sum_k \frac{1}{(\Delta B_{kl})^2} \left[B_{kl}(I_k, L_l, \lambda) - \tilde{B}_{kl}(I_k, L_l, \lambda, \{g_i\}) \right]^2 \tag{9}$$

where: I_k – welding current; L_l – arc length; $B_{kl}(I_k, L_l, \lambda)$ - intensity of light of wavelength λ recorded during the welding current I_k, at arc length L_l; ΔB_{kl} - uncertainty set of light intensity $B_{kl}(I_k, L_l, \lambda)$; $\tilde{B}_{kl}$ $(I_k, L_l, \lambda, \{g_i\})$ - defined by formula (7) the theoretical intensity of the light with a wavelength λ recorded during the welding current I_k, at arc length L_l; $\{g_i\}=\{G_1, \gamma, G_2, G_3, G_4\}$ - set of values of the parameters appearing in formula (7).

Uncertainty-determination and the parameter $g \in \{g\} = \{G_1, \gamma, G_2, G_3, G_4\}$ is determined by the method described in (Kończak S., Nowak M., 1981):

$$\varepsilon_i = \frac{\chi^2}{m_p - m} h_{ii}^{-1} \tag{10}$$

where: h_{ii}^{-1} - ii component of the inverse Hesse matrix; χ^2 - sum of squared deviations from the theoretical value of the experimental results; m_p- number of experimental results; m- number of parameters designated by the matching.

The components of the Hesse matrix model is defined as (Kończak S., Nowak M., 1981):

$$h_{ij} = \frac{\partial^2 \chi^2}{\partial g_i \partial g_j} =$$

$$= -2 \sum_l \sum_k \frac{1}{(\Delta B_{kl})^2} \left[B_{kl}(I_k, L_l, \lambda) \frac{\partial^2 B_{kl,teor}(I_k, L_l, \lambda, \{g_i\})}{\partial g_i \partial g_j} - \frac{\partial B_{kl,teor}(I_k, L_l, \lambda, \{g_i\})}{\partial g_i} \frac{\partial B_{kl,teor}(I_k, L_l, \lambda, \{g_i\})}{\partial g_j} \right] \tag{11}$$

4. Application of the arc light emission to monitor the welding process

In order to monitor the welding processes successfully, the optical sensing systems have been developed. Special procedures, models and modification of the monitoring devices have to be implemented with the sensing systems, some of the systems are discussed herein. Some typical application examples included sensing of the arc length in the TIG welding (Węglowski M.St., 2010), relationship between the welding conditions and intensity of the arc light emission in the MIG/MAG methods (Węglowski M.St., 2008; Węglowski M.St., Zhang Y. M., 2010) and the influence of the parameters and disturbance of the welding process on the shape of the spectrum of the arc light radiation (Węglowski M.St., 2009). In this part, their principles are being described.

4.1. Sensing of the arc length in the TIG welding method based on the arc light intensity

One of the main task of the monitoring systems in the robotized and automated welding stations is the measurement and control of the arc length. The main objective of the investigation was to study the possibilities of using of the visible radiation of the welding arc for stability monitoring of the TIG welding process, giving consideration to the changes of the intensity of visible light radiation with the changes of the welding current or welding arc length (reproducing the case of burn- through and arc migration). The arc length is one of the basic welding parameter in the TIG method, which directly influences the arc voltage. The arc length has an effect on the distribution of an arc energy, and as the consequence on the amount of heat put into the welded joint and on the width of the weld.

The tests have been performed on the stand for the automatic TIG welding. The measuring system consists of welding current and voltage transducers, an electrooptical converter, a measurement card and a PC computer. The analysed beam of the visible radiation is fed into the electrooptical converter by means of a standard optical wave guide. The electrical signal corresponding to the visible light intensity and signals from the welding circuit are recorded on the PC by the recording device, equipped with the NI DAQ 6036 measuring card. The recorded signals were then analysed. The intensity of the visible light radiation of the welding arc was measured in volts. The following experiment conditions were approved: the arc burns between the thoriated tungsten electrode (cathode) and a copper plate (anode), the welding torch is fixed, argon (Ar) as the shielding gas (gas flow rate q_g=10 dcm^3/min), welding current source: Kemppi Pro 5000 (DC current set in the range of 30÷300 A). It was assumed that the arc length is equal to the distance between the electrode tip and the welded metal surface. The range of the welding arc length L=2÷5 mm. In Figure 7 show the configuration of the optical system relative to the welding torch is shown.

The arc length was changed in the range from 2 to 5 mm during the experiments. Figure 8 shows the influence of the arc length L on the visible light intensity and arc voltage for the welding currents in the range of 50÷300 A. It can be seen that considerable changes of the arc length are followed by the substantial changes of radiation intensity of the welding arc (wave length 696 nm) and only by small changes of the arc voltage.

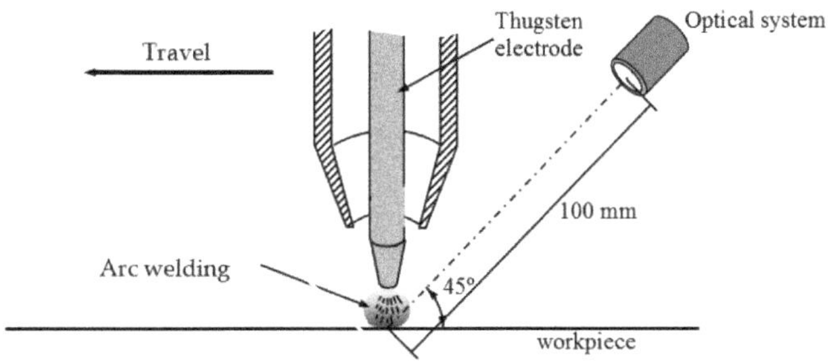

Figure 7. Configuration of the optical system relative to the welding torch

Three cases of transient states of the welding arc length have been also investigated: the abrupt change of the arc length (Fig. 9a), the abrupt change of the welding arc length, simulating the burn-through of the joint (Fig. 10a) and a smooth change of the welding arc length, simulating a bad preparation of welded elements or their distortion during welding (Fig. 11a). These are typical transient states in the welding practice.

The abrupt change of the welding arc was forced by a proper preparation of the 20 mm thick plate by milling (Fig. 9a). The height of the received steps was 1 and 2 mm, which resulted in the arc length of 1, 2 and 4 mm at a welding current of 100 A (DC). Results of the measurement performed at the welding speed of 60 cm/min are presented in Figure 9b. The moment of entering the step by the welding torch is shown by arrows.

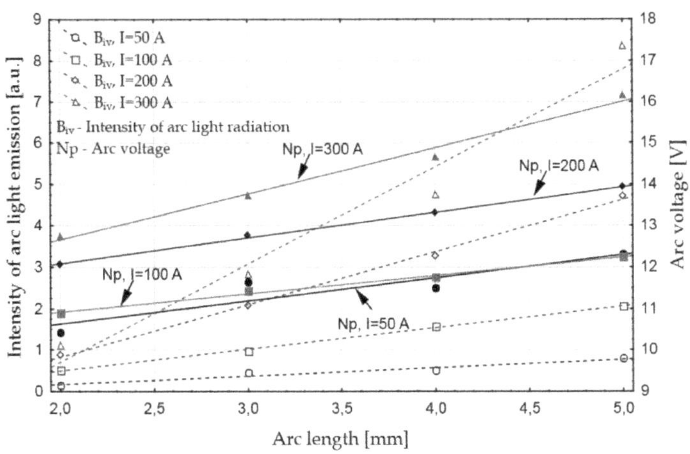

Figure 8. Influence of welding arc length on the light emission (B_{iv}) and the arc voltage (N_p) at the welding current in the range of 50÷300 A. Argon as shielding gas

The second tested transient state was the abrupt change of the arc length simulating the burn-through of the welded joint. A 20 mm thick plate were prepared by drilling holes with a diameter of 1,3÷6 mm (Fig. 10a). The arc length during the experiment was maintained at 3 mm at the

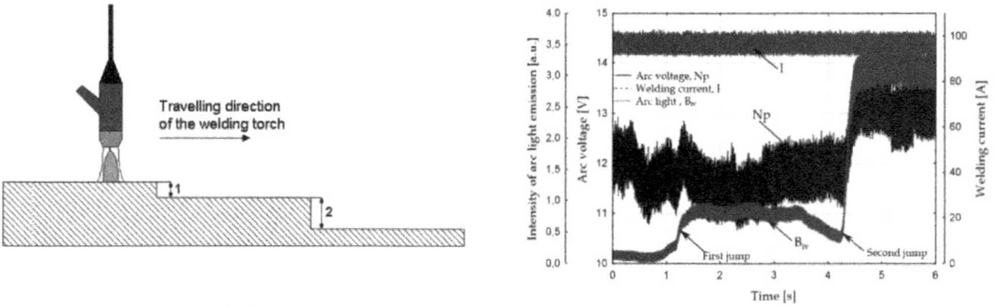

(a) (b)

Figure 9. (a) scheme of the experiment with the forced abrupt change of welding arc length, (b) measurement results of the welding current, arc voltage and intensity of arc radiation at abrupt changes of the welding arc length, welding speed 60 cm/min

welding current of 100 A (DC) and a welding speed of 60 cm/min. The test results are presented in Figure 10b. The moment of entering the holes by the welding torch is shown by arrows.

The third tested transient state was a smooth change of the arc length simulating the deformation of welded plates or improper preparation of the joint. Plates 20 mm in thickness were welded at the angle of 5° (Fig. 11a). The arc length during the experiment changed in the range of 1÷7 mm at the welding current of 100 A (DC) and a welding speed of 60 cm/min. The test results are presented in Figure 11b.

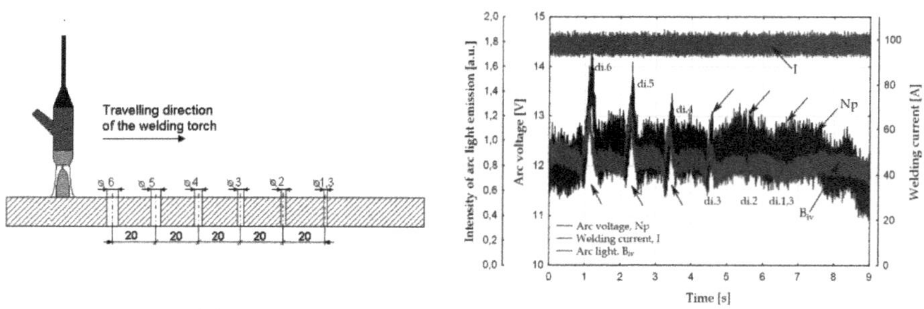

(a) (b)

Figure 10. (a) scheme of the experiment with the forced abrupt change of welding arc length simulating the burn-through of the welded joint, (b) measurement results of the welding current, arc voltage and intensity of arc radiation at abrupt changes of the welding arc length simulating the burn-through of the welded joint, welding speed 100 cm/min,

The one of the most important factor is the influence of the changes of welding current and the arc length during TIG welding on the intensity of the visible radiation. On the basis of the collected test data and equation 2 a relationship combining the intensity of the welding arc radiation (B_{iv}) with the arc length (L) and the welding current intensity (I) can be determined. The arc length is in the range of 2 ÷ 5 mm. Based on the formula (7) presented in the paragraph the relationship can be determined. Using a software, basis of the collected data and formulas (9-11), coefficients G_i and γ can be estimated.

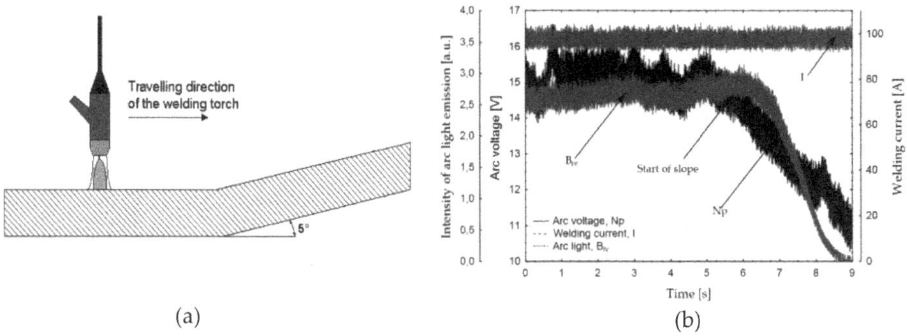

Figure 11. (a) scheme of the experiment with the smooth change of the welding arc length simulating improper preparation of the joint or deformation of welded plates, (b) measurement results of the welding current, arc voltage and intensity of the arc radiation at the smooth change of the welding arc length, welding speed 60 cm/min

Two cases were taken into account during calculation:

- theoretical Zhang model, in this model coefficient γ=2 and then it can be written (eq. 7):

$$B_{iv} = G_1 L I^2 \left(e^{\frac{G_2}{I}} - \frac{1}{2} \right) + G_3 I^2 + G_4 \tag{12}$$

- generalization model - coefficient γ is a parameter dependent on the measured data.

Basing on the formula (9) the calculation were carried out taking into account the following two cases:

- ΔB_{kl} - uncertainty set of the light intensity is constant for all data and it is not taken into account during calculations; in this case the fitting will be worse for smaller values,
- ΔB_{kl} - uncertainty set of the light intensity is not constant for all data and it is taken into account during calculations.

No	Coefficients	Theoretical model	Generalization model
1	G1	$3{,}8(2)\cdot10^{-5}$	$1{,}11(1)\cdot10^{-3}$
2	G2	56(3)	9(2)
3	G3	$-4{,}4(1)\cdot10^{-5}$	$-3{,}98(12)\cdot10^{-5}$
4	G4	$1(57)\cdot10^{-3}$	$-1{,}8(6)\cdot10^{-1}$
5	γ	2	1,455(4)
6	χ^2 sum of the least-squares of the deviations	7,33	3,6
7	correlation coefficient R_2	0,98	0,99

Table 2. Results of calculation of coefficients G_i for theoretical and generalization models at arc length in the range of 2 - 5 mm, ΔB_{kl} – constant

No	Coefficients	Theoretical model	Generalization model
1	G1	$4,6(2)\cdot 10^{-5}$	$1,7(1)\cdot 10^{-3}$
2	G2	34(2)	$2(65)\cdot 10^{-2}$
3	G3	$-4,06(10)\cdot 10^{-5}$	$-35,1(6)\cdot 10^{-6}$
4	G4	$-1,09(16)\cdot 10^{-1}$	$-11,4(9)\cdot 10^{-2}$
5	γ	2	1,364(2)
6	χ^2 sum of the least-squares of the deviations	5,75	1,07
7	correlation coefficient R^2	0,99	0,99

Table 3. Results of calculation of coefficients G_i for theoretical and generalization models at arc length in the range of 2 - 5 mm, ΔB_{kl} – is not constant

Taking into account the results given in Tables 2 and 3, the sum of the least-squares of the deviations is smaller for the generalization model and for case were weight ΔB_{kl} is not constant. Finally the formula 9 can be written as:

$$B_{iv} = 0,0017 L I^{1,364}\left(e^{\frac{0,02}{I}} - \frac{1}{2}\right) - 0,000035 I^2 - 0,114 \qquad (13)$$

This equation is satisfied for the wavelength 698 nm and the arc length in the range of 2 - 5 mm. The graphic presentation of this formula is shown in Figures 12a and 12b. The arc burns between the thoriated tungsten electrode (cathode) and a water cooled copper plate.

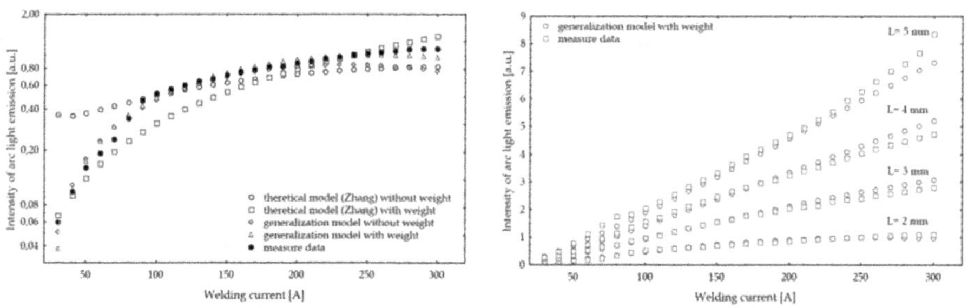

(a) (b)

Figure 12. (a) relationship between intensity of the welding arc radiation in the TIG method and welding current at arc length 2 mm, and the wavelength 698 nm, (b) relationship between an intensity of the welding arc radiation in the TIG method and the arc length, welding current at arc length in the range of 2-5 mm, and the wavelength 698 nm

The change of the TIG welding arc length causes also changes of the intensity of the visible arc radiation. An increase of the arc length results in the intensity increase of the selected spectral line (696 nm) of the TIG welding arc visual radiation. This increase depends on the welding current intensity. Larger increases of the visual arc radiation intensities are observed at the higher welding currents.

Three cases of the welding arc length transient states have been tested. It has been found that in all that cases the intensity of the arc radiation at the wave length of 696 nm is much more sensitive to abrupt changes of the welding arc length, than the arc voltage. In the first case (Fig. 9) the change of the arc length for 1 and 2 mm was followed by a 400% change of the radiation intensity and only by a 10% change of the arc voltage. In the second tested case, simulating the burn-through of the welded joint, considerable changes take place for both - the arc voltage and intensity of the arc radiation, but in the radiation intensity record the peaks corresponding to the consecutive holes (Fig. 10) can be more easily identified. By the arc voltage measurement a hole with a diameter of 4 mm can be identified, while the measurement of radiation intensity makes possible the identification of a 1,3 mm hole. The third tested transient state was a smooth change of the welding arc length, simulating the deformation of welded plates or incorrect setup for welding. Also in that case the changes of the welding arc length are followed by considerable changes of radiation intensity and smaller changes of the arc voltage.

4.2. Relationship between the welding conditions and intensity of the arc light emission in GMAW

This section describes the acquisition and analysis of the arc light emission and its correlation with the welding parameters and disturbances of the welding process. A spectrophotometer card PCI 2000 ISA-A in the visible spectral range of 340 nm to 860 nm was used in the study. The measurement system consisted of the welding current and voltage transducers, an electro-optical converter, a data acquisition card and a PC computer (Fig. 13). Signals from the welding circuit were recorded on the PC through the data acquisition card NI DAQ 6036. The measurements during bead-on-plate welding and joints welding were carried out. The signals were analyzed in time domain. During trials, a spectrophotometric card PCI 2000 ISA-A, which has been designed for the CCD Sony model ILX511 detector in the visible spectral range of 340 nm to 860 nm was used to image and record the arc light spectrum for the later analysis. The CCD detector was a line scan array of 2048 pixel.

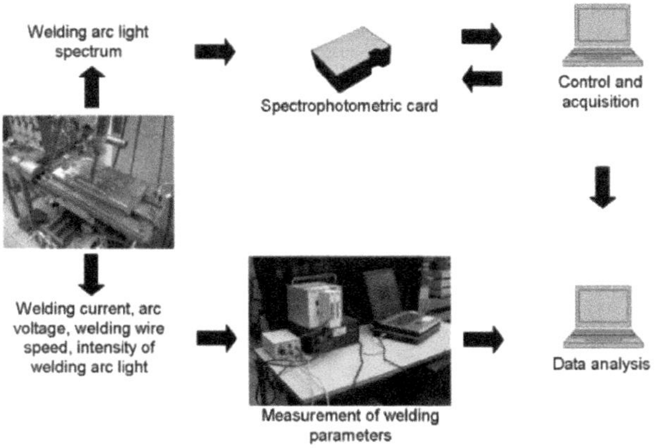

Figure 13. Experimental setup with the flow of data

The work-piece was moved while torch was in a fixed position such that the arc light sensor was stationary in relation to the work-piece. The sepctrophotometric card used a sampling time 3 ms. An optical system was used to focuse the welding arc light. The entire arc column has been analyzed as a single object. During welding the arc voltage, the welding current, the wire feed speed and the intensity of the arc light emission were continuously measured. The Hall effect current sensor Model PR 1001 was used to measure the welding current. This sensor provides electrical isolation between the current carrying conductor and the output of the sensor. The voltage was measured by a resistance bridge by a LV 25 P transducer in the output of the power supply. The wire feed speed was measured by E21 MPL10 transducer. The intensity of the arc light emission was measured by PIN BPW34 photodiode. Signals from the welding circuit were recorded on the PC through the IPP-2 measured system designed in the Instytut Spawalnictwa (Institute of Welding), witch based on SCXI data acquisition system National Instruments. This system consisting of the National Instruments SCXI-1125 is 8-channel isolated analog input modules and data acquisition board NI DAQ 6036 E. Whole system was placed in the SCXI-1000 chassis. The signals were recorded at a sampling rate of 20 kHz.The torch was moved at the travel speed 25 cm/min to make bead-on-plate welds and weld. Direct current levels between 104 A and 235 A were examined, all at an operating voltage in the range of 16.5 V - 25.5V. Figure 14a shows the arc spectrum obtained in the range of 360-860 nm at the welding current in the range of 104 – 235 A. The graph is presented in a logarithmic scale.

As shown in Figure 14a the increase of the welding current causes increase of the arc light intensity in the whole range. The shape of the spectrum was modeled by the three mathematic functions: Lorentz, Gausse and Voight (Fig. 14b). The fitting for both single wavelength and multiple wavelengths was carried out mathematically and the best result was achieved with the Lorentz function. The central wavelength, intensity and FWHM - Full With at Half Maximum were calculated. The main source of the arc light radiation in the GMAW is liquid metal. The lines from the shielding gases have not been found. The detailed analysis of influence of the welding current on the arc light spectrum was previously discussed (Węglowski M.St., 2008, 2009).

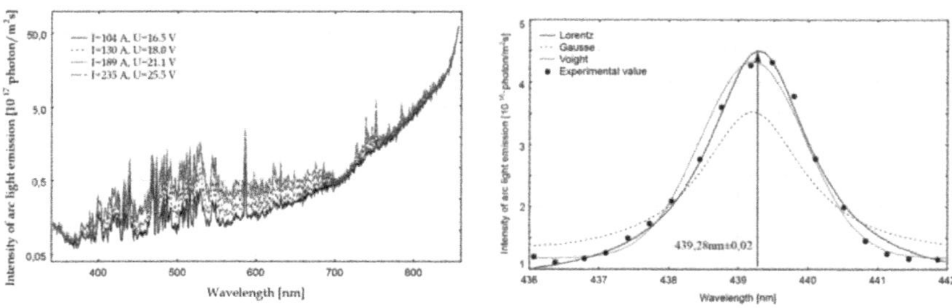

(a) (b)

Figure 14. (a) effect of the welding current on the arc light spectrum. The welding current in the range of 104-235 A, Ar + CO_2 as the shielding gas, wavelength in the range of 480-860 nm. Logarithmic scale, (b) calculated line profile of the wavelength 439,28 nm compared with measured values and the Gaussian, Lorentz and Voight functions

The purpose of these studies was also to check the influence of disturbances of the welding process on the arc light intensity. To this end, arc light intensity was measured during welding of real joint, 4 mm in thickness. The disturbances of the welding process was tha additional filler metal in the grove. The experiments were done under the following conditions: welding current I=160 A, arc voltage U=21.2 V, shielding gas M21 Ferromix C18, welding speed 25 cm/min), wire EN 440 G3Si1, base material S235, the grove was prepared for Y.

The weld produced is shown in Figure 15a. The scheme of the method of disturbance of the welding process is shown in Figure 15b. The macroscopic examination of the padding welds are shown in Figure 16. During welding the arc voltage, welding current, wire feed speed and the intensity of the arc light emission were continuously measured. The intensity of the arc light signal is shown in Figure 17.

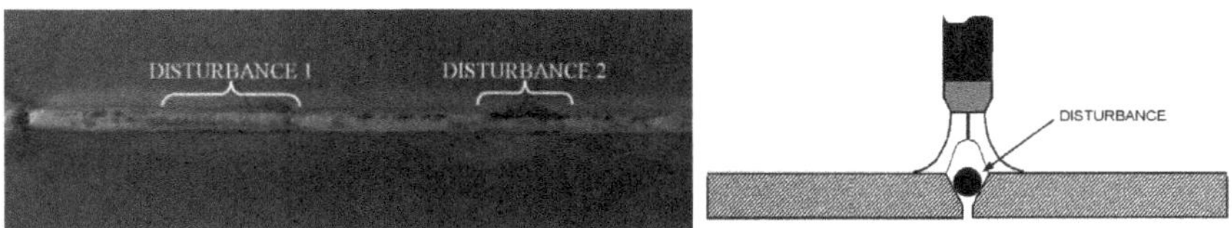

(a) (b)

Figure 15. (a) welding joint with marked areas of disturbances of the welding process, (b) scheme of disturbance of the welding process

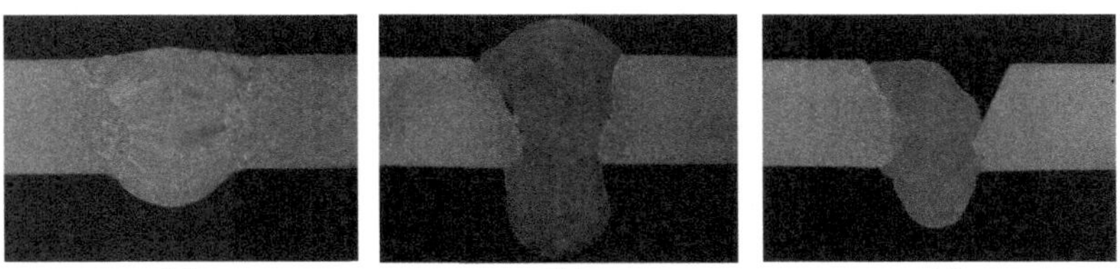

(a) without disturbances (b) disturbance 1 (c) disturbance 2

Figure 16. Macroscopic examination of the padding welds. Etching Adler, magnification x2

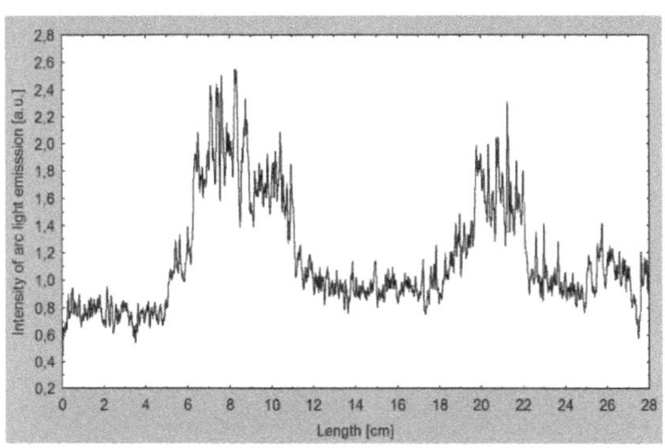

Figure 17. Intensity of the arc light signal recorded during welding of plate with disturbance

To estimate the stability of the welding process based on the arc light emission the least squares method was used. To model the arc light signal a cubic polynomial was used:

$$y = a_0 + a_1 x + a_2 x^2 + a_3 x^3 + \varepsilon \tag{14}$$

where: y – arc light emission, x – time, a_i – coefficients, ε – residual the differences between the observations and the model.

To calculate a residual the following formula should be solved:

$$Y = \begin{bmatrix} y(m-M \\ . \\ y(m) \\ . \\ y(m+M \end{bmatrix} \quad X = \begin{bmatrix} 1 & x(m-M & x^2(m-M) & x^3(m-M) \\ . & . & . & . \\ 1 & x(m) & x^2(m) & x^3(m-M \\ . & . & . & . \\ 1 & x(m+M) & x^2(m+M) & x^3(m+M \end{bmatrix} \quad \beta = \begin{bmatrix} a_0 \\ a_1 \\ a_2 \\ a_3 \end{bmatrix} \tag{15}$$

where: Y- intensity of the arc light signal matrix, X – time matrix, β – coefficients matrix, m – discrete time.

But this system is over determined. There are more equations than unknowns. So it cannot expect to solve the system exactly. Instead, it can be solved it in the least squares sense:

$$\min_{\beta} \|X\beta - Y\| \tag{16}$$

A theoretical approach to solve the over determined system begins by multiplying both sides by XT. This reduces the system to a square, *n*-by-*n* system known as the normal equations:

$$X^T X\beta = X^T Y \tag{17}$$

If there are thousands of observations and only a few parameters, the design matrix X is quite large, but the matrix X^TX is small. It has been projected Y into the space spanned by the columns of X. Continuing with this the theoretical approach, if the basis functions are independent, then X^TX is nonsingular and

$$\beta = \left(X^T X\right)^{-1} X^T Y \text{ and } \bar{Y} = \bar{X}\beta \tag{18}$$

then, the residual can be calculated as:

$$\varepsilon = \sqrt{\sigma^2} = \sqrt{\frac{\left(Y - \bar{Y}\right)^T \left(Y - \bar{Y}\right)}{2M+1}} \tag{19}$$

where: 2M+1 – the number of data used in the fitting.

To calculate formulas 15 and 19 the following parameters were established: m in the range of 10000 to 590000 at the step 100 and M=10. To estimate the points of unstability of the welding process, model by polynomial can be used. To estimate the best degree of the polynomial the F-test method can be used. Based on the previously least squares methodology, the residual of polynomials ε_P can be calculated as:

$$\varepsilon_P = \sqrt{\sigma^2} = \sqrt{\frac{\left(\varepsilon - \bar{Y}\right)^T \left(\varepsilon - \bar{Y}\right)}{n}} \tag{20}$$

where: ε – residual from eq. 7, Y – value of polynomial function, n – the number of data.

The calculation acc. to formula 20 were carried out for 50 different polynomials. The best results can be achieved for polynomial 28th degree (Figure 18).

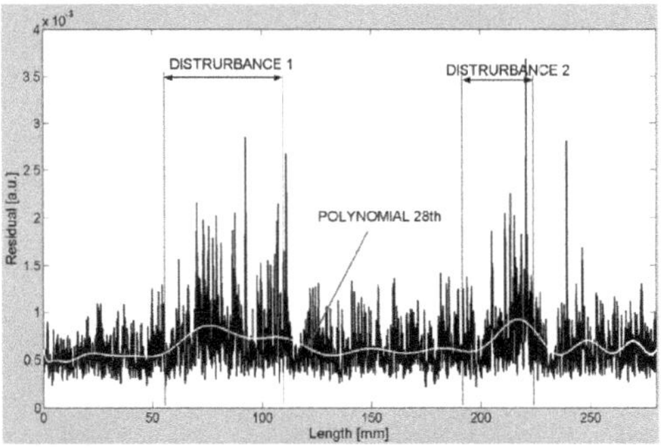

Figure 18. The results of model by polynomial 28th degree with marked the areas with disturbances.

It is shown that the arc light signal can be utilized to monitor the welding processes. This signal is sensitive to any changes in the welding area. The spectrophotometric card can be useful tool to investigate the properties of the welding arc.

4.3. The influence of the parameters and disturbance of the welding process on the shape of the spectrum of the arc light radiation

The tests were performed on the stand for automatic MAG welding operations by the control consol. The testing plate for welding was fixed while the welding was moved at a controlled speed. The torch was located perpendicularly to the welding surface. All experiments were performed by the bead-on-plate welding. The measuring system consisted of the welding current and the voltage transducers, a PC computer equipped with the CCD spectrophotometer card, and a speed wire measurement device. The electrical signals from the current and voltage transducers were recorded on the PC equipped with the NI DAQ 6036 measuring card. The analyzed radiation was fed into the CCD spectrophotometer by means of a standard fibre optics. A spectrophotometer card PCI 2000 ISA-A (Ocean Optics Inc.) was used in this research. The output of each pixel is converted to an electrical current which represents the amount of the energy that has fallen on each pixel in a relative manner.

Figure 19 shows the different influence of the welding current intensity on the amplitudes, additive constants and widths (FWHM) of the Lorentz functions that fit the best spectral peaks of the arc light. One can see that the influence manifests in different ways in the cases of the different spectral peaks. Generally, the welding current intensity strong influences the additive constants and amplitudes of the peaks in the spectral range from 400 nm to 500 nm. Figure 19d shows the dependence of the best fitted additive constants on intensity of the current in the welding process of the clean mild steel S 235 (I=104 A; U=16,5 V).

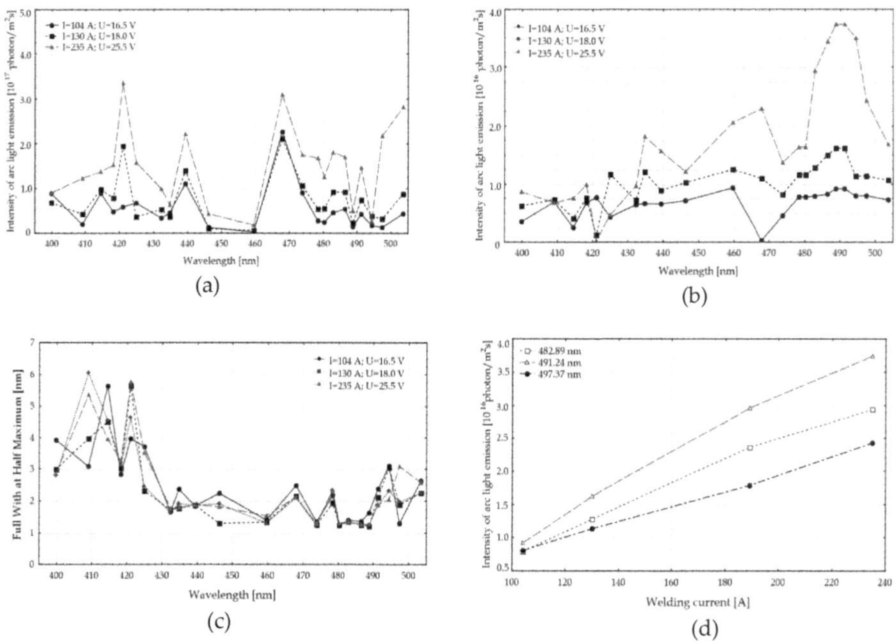

Figure 19. (a) influence of the welding current intensity on the amplitudes, (b) the values of additive constant in formula (3), (c) FWHM and (d) intensity of single emission lines,

Investigations on the effect of imposed disturbances, in the form of paint or grease layers on the plate surface, on the intensity of the MAG arc light radiation in the visible range, have been performed. The mild steel S 235 as the welding plate and the 1,2 mm diameter SG2 type welding wire were used. Experiments were performed with shielding gas 82% Ar and 18 % CO_2. The plate surface was clean, covered with oil paint or covered with a machine grease. It was found that both the paint and grease layer influence the recorded spectral characteristics of the MAG welding light radiation. Figure 20 presents the different influence of the existence of paint on the welded plate on the amplitudes, additive constants and the widths of the Lorentz functions that fit the best spectral peaks of the arc light. One can see that the influence manifests in different ways in the cases of the different spectral peaks.

The presented investigations show that the low resolution spectral characteristics of the arc light emission registered with CCD device can be applied for the purpose of monitoring of the welding process. The arranged measuring stand has made it possible to record the visible spectrum of the radiation of the welding arc within the range of wavelengths from 380 nm to 780 nm. The measuring stand comprised a spectrophotometer, a computer recording the results of the measurements and a device for mechanized welding.

It was found that the spectral distribution of a single peak in the low resolution spectral characteristics can be best fitted with the Lorentz function. In the recorded spectrum of the welding arc light emission, separation of the ionic or atomic lines is not possible. However, the correlation between the parameters of the fitted Lorentz function and welding parameters (i.e. welding current) was obtained. The Lorentz function parameters depend also on the disturbances in the MAG welding process, e.g. their values are different in the cases of clean and painted surface of the welded mild steel S 235 plate.

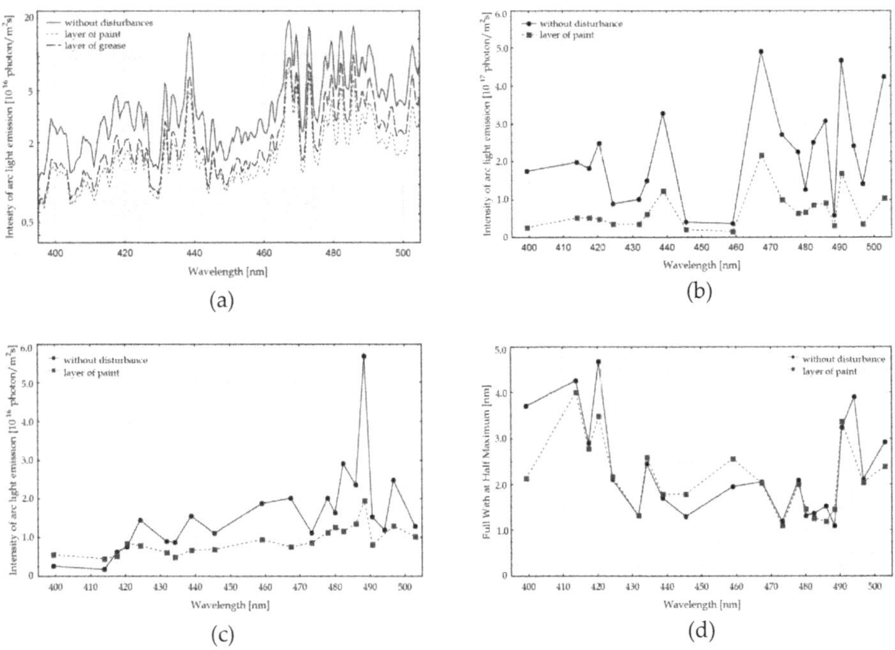

Figure 20. (a) the influence of disturbances on the MAG welding arc light spectrum (I=169 A, U=19,7 V) (b) influence of paint on the welded plate on the amplitudes, (c) the values of additive constant in formula (3) and (d) FWHM (Attention: curves presented in the figure cannot be interpolated)

5. Summary

Modern monitoring methods of the welding processes are inherent in each automatics and robotics production system. These systems detect very rapidly any incorrectly made weld joints during manufacturing and this way decreases the costs of production. That means possibility of detecting of any faulty parts without very costly nondestructive examinations. At present very popular conventional monitoring methods of the welding processes, based on measurements of the welding current and arc voltage in many cases are inefficient and are replaced or/and completed by nonconventional monitoring methods.

One of the most popular nonconventional monitoring method is sensing system based on the arc light emission. The main aim of these investigations was to check the possibilities of applying the visible radiation of the welding arc for the purpose of monitoring of the quality of the welding process.

The arranged measuring stand made it possible to record the visible spectrum of the radiation of the welding arc within the range of the wavelengths from 380 nm to 780 nm. The measuring stand comprised a spectrophotometer, a computer recording the results of measurements and a device for mechanized welding.

Results of the performed investigations in the field of measurement of light radiation intensity during TIG and MIG/MAG welding, shown in this chapter, indicate that this signal can be used for the monitoring of the welding process quality. The experience gained during these investigations allows for further research on the welding arc radiation phenomenon. The obtained knowledge increases the possibilities of using the signal for on-line monitoring of the welding process on the automated and robotized stands. The analysis of the spectrum of the welding arc radiation should help to develop the new vision sensor in the arc welding.

The investigations are continued in the many research centers, and cover the following issues:

- utilize the artificial intelligence method to estimate the stability of the welding process,
- develop the filtering method, and methodology to signal analysis in time, and frequency domain,
- laser diagnostic on the welding arc in the TIG and MIG/MAG methods, develop method of measurement of the arc light emission in many points simultaneously, and many others.

Author details

Marek Stanisław Węglowski
Instytut Spawalnictwa (Institute of Welding), Poland

6. References

Zhang, Y.M. (2008). *Real-time weld process monitoring*. Woodhead Publishing in Materials, ISBN 978-1-84569-268-1, Cambridge, England

Blakeley, P. J. (1990). Who should be using arc monitoring? *Welding and Metal Fabrication,* Vol. 58, No 6, pp. 268-272, ISSN 0043-2245

Siewert, T., Madigan, B. (1992). Through the arc sensing for measurement Gas Metal Arc weld quality in real time. *Materials Evaluation*, Vol. 50, No 11, pp. 1314-1318, ISSN 0025-5327

Chen, X. Q. (2002). *Advanced automation techniques in adaptive material processing*, Word Scientific Publishing, ISBN 981-02-4902-0, Rever Edge, New Jersey, USA.

Pan, J. (2003). *Arc welding control*. Woodhead Publisher Limited, ISBN 0-8493-1772-X, Cambridge, England.

Kim, J. W., Na, S. J. (1991). A study on an arc sensor for Gas Metal Arc welding of horizontal fillets. *Welding Journal*, Vol. 70, No 8, pp. 216s-221s, ISSN 0043-2296

Johnson, J. A., Carlson, N. M., Smartt H.B., et al (1991). Process control of GMAW: sensing of metal transfer mode. *Welding Journal*, Vol. 70, No 4, pp. 91s-99s, ISSN 0043-2296

Modenesi, P. J., Nixon, J. H. (1994). Arc instability phenomena in GMA welding. *Welding Journal*, Vol. 73, No 9, pp. 219s-224s, ISSN 0043-2296

Luksa, K. (2006). Influence of weld imperfection on short circuit GMA welding arc stability. *Journal of Materials Processing Technology*, Vol. 175, pp. 285-290, ISSN 0924-0136

Guu, A. C., Rokhlin, S. I. (1992). Arc welding process control using radiographic sensing. *Materials Evaluation*, Vol. 50, No 11, pp. 1344-1348, ISSN 0025-5327

Fan, H. Ravala, N. K., Wikle III, H. C., et al. (2003). Low-cost infrared sensing system for monitoring the welding process in the presence of plate inclination angle. *Journal of Materials Processing Technology*, Vol. 140, No 1-3, pp. 668-675, ISSN 0924-0136

Carlson, N. M., Johnson, J. A. (1992). Ultrasonic NDT methods for weld sensing. *Materials Evaluation*, Vol. 50, No 11, pp. 1338-1343, ISSN 0025-5327

Taylor-Burge, K. L., Harris, T. J. (1993). The real time analysis of acoustic weld emission using neural networks, *Proc. of the International Conf. on the Joining of Materials JOM-6* ed. Rasmus Knudsens Vej, pp. 60-67. Helsingor, Denmark

Saini, D., Floyd, S. (1998). An investigation of Gas Metal Arc welding sound signature for on-line quality control. *Welding Journal*, Vol. 77, No 4, pp. 172s-179s, ISSN 0043-2296

Luksa, K. (2003). Correspondence between sound emission generated in GMA [MIG/MAG] welding process and signals registered in the arc circuit. *Welding International*, Vol. 17, No 6, pp. 438-441, ISSN 0950-7116

Smith, J., Lucas, B. (1999). Putting intelligence into welding – rule based systems, fuzzy logic and neural networks. *Welding & Metal Fabrication*, Vol. 67, No 10, pp. 7-9, ISSN 0043-2245

Amin, M., Naseer, A. (1987). Synergic control in MIG welding – parametric relationship for steady DC open arc and short circuiting arc operation. *Metal Construction*, Vol. 19, No 1, pp. 22-28, ISSN 0026-0541

Kang, Y. H., Na, S. J. (2003). Characteristics of welding and arc signal n narrow groove GMAW using electromagnetic arc oscillation. *Welding Journal*, Vol. 82 No 5, pp. 93s-99s, ISSN 0043-2296

Lu, W., Zhang, Y. M. (2004). Sensing of weld pool surface using non-transferred plasma charge sensor. *Measurement Science and Technology*, Vol. 15, No 5, pp. 991-999, ISSN 0957-0233

Zheng, B., Wang, H. J. (2000). Control for weld penetration in VPPAW of aluminum alloys using the front weld pool image signal. *Welding Journal*. 70 (12): 363-s to 371-s, ISSN 0043-2296

Zhang, Y. M., Wu, L. (1993). Determination joint penetration in GTAW with vision sensing of weld face geometry. *Welding Journal*, Vol. 72, No 10, pp. 463s-469s, ISSN 0043-2296

Li, P. J., Zhang, Y. M. (2001). Precision Sensing of Arc Length in GTAW Based on Arc Light Spectrum. *Journal of Manufacturing Science and Engineering*, Vol. 123, No 1, pp. 62-65, ISSN 1087-1357

Wang, Q. L., Li, P. J. (1997). Arc light sensing of droplet transfer and ist analysis in pulsed GMAW processes. *Welding Journal*, Vol. 76, No 11, pp. 458s-469s, ISSN 0043-2296

Weglowski, M. St. (2007). Utilization of the arc light emission emitted during TIG welding to monitoring this process. *Archives of Mechanical Technology and Automation PAN*, Vol. 27, No 1, pp. 103-112. (in Polish), ISSN 1233-9709

Yoo, C. D., Yoo, Y. S., and Sunwoo H.K. (1997). Investigation on arc light intensity in gas metal arc welding. Part 1: relationship between arc light intensity and arc length. *Proceeding of the Institution of Mechanical Engineers. Part B: Journal of Engineering Manufacture*, Vol. 211 (B5), pp. 345-353, ISSN 0954-4054

Yoo, C. D., Yoo, Y. S., and Sunwoo H. K. (1997). Investigation on arc light intensity in gas metal arc welding. Part 2: application to weld seam tracking. *Proceeding of the "Institution of Mechanical Engineers. Part B: Journal of Engineering Manufacture*, Vol. 211 (B5), pp. 355-363, ISSN 0954-4054

Ancona, A., Lugara,P.M., Ottonelli,F., Catalano, I.M. (2004). A Sensing Torch for On-line monitoring of the Gas Tungsten Arc Welding Process of Steel Pipes. *Measurement Science and Technology*, Vol. 15, No 12, pp. 2412-2518, ISSN 1361-6501

Sadek, C.A., Diogo, D.S.M., Marcelo, S.M. (2006). Emission Spectrometry Evaluation in Arc Welding Monitoring System. *Journal of Materials Technology*, Vol. 179, No 1-2, pp. 219-224, ISSN 0924-0136

Li, Z. Y., Wang, B., Ding, J.B. (2009). Detection of GTA Welding Quality and Disturbance Factors with Spectral Signal of Arc Light. *Journal of Materials Processing Technology*, 2009, Vol. 209, No 10, pp. 4867-4873, ISSN 0924-0136

Mirapeix, J., Cobo, A., Fernandez, S., Cardoso, R., Lopez Higuera, J.M. (2008). Spectroscopic Analysis of the Plasma Continuum radiation for On-line Arc-welding Defect Detection. *Journal of Physics D: Applied Physics*, Vol. 41, No 3, pp. 135202–135210, ISSN 1361-6463

Johnson, C.A., Sciaky, A.M. (1966). *System for controlling length of welding arc*, Patent US, nr 3236997

Zhang, Y.M., Li, P.J. (2001). Precision sensing of arc length in GTAW based on arc light spectrum. *Journal of Manufacturing Science and Engineering*, Vol. 123, No 2, pp. 62-65, ISSN 1087-1357

Garcia-Allenda, P.B., Mirapeix, J., Conde, O.M., Cobo, A., Lopez-Higuera, J.M. (2008). Arc welding spectroscopic monitoring based on feature selection and neutral networks. *Sensors*, Vol. 8, No 10, 6496-6506, ISSN 1424-8220

Pattee, H.E, Myers L.B., Evans R.M., Monroe R.E. (1973). Effect of Arc Radiation and Heat on Welders. *Welding Journal*, Vol. 52, pp. 297s-308s. ISSN: 0043-2296

Hinrichs J.F., (1978). Radiation and arc welding: New data to enhance safety. *Welding and Metal Fabrication*, Vol. 51, pp. 102-103, ISSN 0043-224

Marzec, S., Janosik, E. (1995). Influence of arc radiation of health of welder Wpływ promieniowania łuku spawalniczego na organizm spawacza. *Builetyn of Welding Institute*, Vol. 39, No 6, 33-37, ISSN 0867-583X

Quigley, M. (1977). Physics of the welding arc. *Welding and Metal Fabrication*, Vol. 9, No 12, pp. 619-625, ISSN 0043-2245

Glickstein, S. (1976). Temperature measurements in a free burning arc. *Welding Journal*, Vol. 55, No 8, pp. 222s-229s, ISSN 0043-2296

Huddlestone, R.H., Leonard, S.L. (1965). *Plasma diagnostic techniques*. Academic Press, New York, USA

Szymański, A. (1991). *Spectroscopic and Computer Modeling Investigations of a Laser-Sustained Plasma*. Institute of Fundamental Technological Research Works, Polish Academy of Sciences, Warsaw, 30/1991

Petrie, T.W., Pfender, E. (1970). The influence of the cathode tip on temperature and velocity fields in a GTA. *Welding Journal*, Vol. 49, No 12, pp. 588s-596s, ISSN 0043-2296

Etemadi, K., Pfender, E. (1982). Computer – controlled plasma emission spectroscopy. *Review of Scientific Instruments*, No 2, pp. 255-257, ISSN 0034-6748

Lee, C.W., Na, S.J. (1996). A study on the influence of reflected arc light on vision sensors for welding automation. *Welding Journal*, Vol. 75, No 12, pp. 379s-387s, ISSN 0043-2296

Yu. J.Y., Kim. J.I., Na. S.J. (2003). Influence of reflected arc light on vision sensor for automatic GTAW systems. *Welding Journal*, Vol. 82, No 2, pp. 36s-42s, ISSN 0043-2296

Hinrichs, J.F. (1978). Radiation and arc welding: New data to enhance safety. Welding and Metal Fabrication, vol. 51, No. 2, pp. 102–103, ISSN 0043-2245

Farmer, A.J.D., Haddad, G.N. (1984). Local thermodynamic equilibrium in free-burning arcs in argon. *Applied Physics Letters*, Vol. 45, No 1, pp. 24-25, ISSN 0003-6951

Grove, L., Loseke, W.A., et al (1970). Development of portable direct reading spectrometer to monitor oxygen-hydrogen containing contaminants in gas tungsten-arc process shields. *Welding Journal*, Vol. 49, No 11, pp. 538s-545s, ISSN 0043-2296

Wang, Q.L., Li, P.J. (1997). Arc light sensing of droplet transfer and its analysis in pulsed GMAW processes. Welding Journal, Vol. 76, No 11, pp. 458-469, ISSN 0043-2296

Cho, Y.T., Na, S.J. (2005). Application of Abel inversion in real-time calculation for circularly and elliptically symmetric radiation sources. *Measurement Science and Technology*, Vol. 16, pp. 878-884, ISSN 0957-0233

Fan, H.G., Na, S.J., Shi, Y.W. (1997) Mathematical model of arc in pulsed current gas tungsten arc welding. *Journal of Physics D: Applied Physics*, Vol. 30, pp. 94-102, ISSN 0022-3727

Kang, Y.H., Na, S.J. (2002). A study on the modeling of magnetic arc deflection and dynamic analysis of arc sensor. *Welding Journal*, Vol. 81, No 1, pp. 8s-13s, ISSN 0043-2296

Ogawa, Y. (2004). Effect of active flux on anode reaction. Dokument MIS XII-1797-0

Zielińska S. (2004). *Physical properties of MIG/MAG plasma*. Doctoral thesis. Institute of Physics, Jagiellonian University, Krakow, Poland

Kończak, S., Nowak, M. (1981). The estimation of semiconductor parameters using least squares in photomagnetoelectric investigation. *Physica Status Solidi A - Application and Materials Science*, Vol. 63, pp. 305-311, ISSN 0031-8965

Węglowski, M. St. (2010), *Utilization of the arc light emission emitted by an electric arc to monitor the TIG process*. Doctoral thesis, AGH University of Science and Technology, Krakow, Poland

Węglowski, M.St. (2008). Modeling and analysis of arc light spectrum in GMAW. *Welding Journal*, Vol. 87, No 8, pp. 212-218, ISSN 0043-2296

Węglowski, M.St. (2009). Measurement of arc light spectrum in the MAG welding method. *Metrology and Measurement Systems PAN*, Vol. 16, No 1, 143-159. ISSN 0860–8229.

Węglowski, M.St., Zhang, Y.M. (2010). Relationship between welding conditions and intensity of arc light emission in GMAW. *Proceedings of the 5th International Conference on Advances in Production Engineering APE'2010*, Warsaw University of Technology, Warsaw, pp. 215-224, ISBN 978-83-7204-903-2

Chapter 14

Real-Time Measurement of Three Dimensional Weld Pool Surface in GTAW

Wei Jie Zhang, Yu Kang Liu and Yu Ming Zhang

Additional information is available at the end of the chapter

1. Introduction

Gas tungsten arc welding (GTAW) [1] is the primary process used by human welders for critical applications. In this process as shown in Fig. 1, an *arc* is established between the non-consumable tungsten electrode and *base metal*. The base metal is melted by the arc forming a *liquid weld pool* that joins the two pieces of base metal together after solidification. An optional filler metal (not shown in the figure) can be added if necessary but it is melted by the arc column, rather than directly by an arc spot as in gas metal arc welding (GMAW) where the anode can much more efficiently melt a continuously-fed wire than the arc column to increase the melting productivity. However, the detachment and impact of the associated droplets on the weld pool compromise the controllability of the process and limit its use in precision applications.

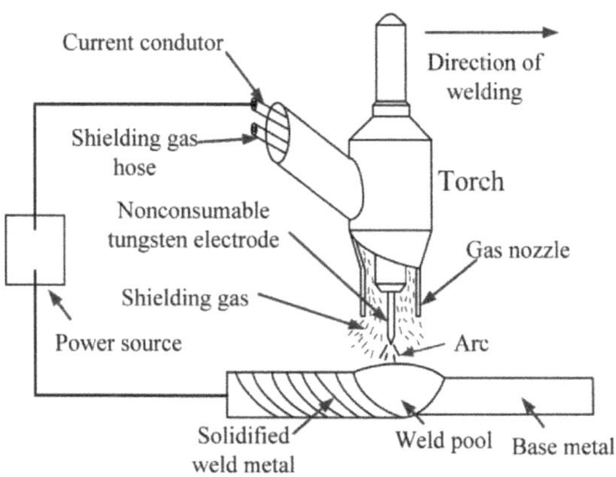

Figure 1. Illustration of GTAW

Because GTAW is primarily used in applications where appropriate *degree of full penetration* (if and how much the liquid metal has fully penetrated the entire thickness of the base metal) is critical for the service, the process should be mechanized or automated as long as it can be justified for production cycle, cost, and quality. However, there are a number of issues which adversely affect the automation significantly. The first one is the *accessibility*. That is, in many applications there is no sufficient space to allow a mechanized system's torch head to access. Second, mechanized systems require significant amount of time for on-site installation and joints be prepared with great precision. The *production cycle* in many applications is adversely affected substantially. The third issue is the *assurance* of the weld quality. In manual welding, welders who observe the weld pool can assure the desired full penetration is produced. However, in mechanized welding, no welder has the capability to interference with the system; they are not required or allowed in robotic welding to observe the welding process with the similar level of concentration as in manual operation. Mechanized/automated systems rely on precision control of joint fit-up and welding conditions and tedious programming of welding parameters to produce repeatable results. However, precision control of joints and welding conditions is very costly and not always guaranteed. Up to date, there are no satisfactory sensors/ways that can conveniently/ automatically monitor the *penetration depth* (how far the liquid metal penetrates along the thickness of the base metal) or the degree of the full penetration like a skilled welder.

The difficulty is primarily due to the *invisibility* of the liquid metal *bottom surface* underneath the weld pool and the extreme brightness of the arc and various methods have been studied, including pool oscillation, ultrasound, infrared sensor, and vision-based sensing method, etc. In the following subsections, each sensing method is briefly reviewed.

1.1. Pool oscillation method

Sensing the weld penetration by monitoring weld pool oscillation behavior is based on the fact that a weld pool can be brought into natural oscillation and the oscillation frequency of the weld pool is related to the weld pool geometry. This phenomenon may be used to monitor the weld pool in a feedback control system. The pioneering work in pool oscillation was conducted by Kotecki [2], and Richardson [3].

Hardt [4] and their co-workers proposed a method to determine the back-side bead width by measuring the natural frequency of pool motion when driven by a time varying arc plasma force. The method was developed analytically and verified by experiments. However, the results were obtained for stationary weld pools, and it was unclear if similar results occur when the welding torch was moving. G. den Ouden found an abrupt change in the oscillation frequency of the pool during the transition from partial to full penetration [5, 6].

Andersen [7] developed a synchronous weld pool oscillation method for controlling the weld pool dimensions and state of penetration. The approach used to induce pool oscillations was to excite the weld pool with current pulses synchronized to the natural oscillations of the pool. An optical sensor was utilized to detect the pool oscillations. A model of the weld pool was also developed using a fluid droplet formulation for the relation of weld pool geometry and other physical parameters to the natural frequencies of the weld pool. Comparison of the weld pool's natural frequency as predicted by the developed weld pool geometry models and measurements of the pool width thus allowed the assessment of the penetration state. Hartman [8] further evaluated this synchronous excitation method and developed a control system that regulated the total heat input to maintain constant fusion zone geometry by monitoring the arc light reflection from the oscillation of the molten metal surface.

Ju [9] proposed a new vibration method: the Pulse Shielding Gas (PSG) oscillating method. A control system was constructed by controlling the welding current based on the natural vibration frequency measurements from an arc sensor. It was found that spectrum analysis using the Fast Fourier Transformation (FFT) was effective for detecting peculiar frequency of the molten pool.

Yudodibroto [10] implemented the weld penetration control based on weld pool oscillation sensing method during GTAW process with cold filler wire addition. The frequency of the weld pool oscillation was obtained from the arc voltage variation via analysis. It was found that the weld pool oscillation approach is suitable for penetration control during cold wire GTAW when the metal transfer occurs in an uninterrupted bridging manner.

1.2. Ultrasonic sensing method

Ultrasonic sensors [11-17] are widely used to determine the boundaries of the liquid and metal in the weld pool.

In [15] the developed ultrasonic sensing system could locate and track the welding seam ensuring correct positioning of the welding head relatively to the joint preparation. The system was able to monitor the joint profile of the molten weld pool and modified the relevant heat input parameters ensuring consistent penetration, joint filling and acceptable weld bead shape. It also made use of both the above information to reconstruct 3D images of the weld pool silhouettes providing in-process inspection capabilities of the welded joints.

At Georgia Institute of Technology, Ume leaded the development of non-contact ultrasonic penetration sensors based on laser-phased array techniques [13, 14]. Recently, in order to overcome the contact requirement of the ultrasonic sensing method, various non-contact ultrasonic sensing methods have been developed, such as laser ultrasonic sensing [11, 17], electromagnetic acoustic transducer (EMAT) ultrasonic sensing [12], and laser-EMAT ultrasonic sensing [16], etc.

Mi [17] developed a ultransonic sensing system to monitor the weld penetration. The sensing system was based on using a laser phased array technique to generate focused and steered ultrasound, and an EMAT as a receiver. Both the ultrasound generation by the laser phased array and the reception by the EMAT were non-contact, which could thus eliminate the need for a couplant medium. This made the system capable of operating at high temperatures involved in the welding process. A signal processing algorithm based on a cross-correlation technique was further developed to estimate the time-of-flight (TOF) of the ultrasound.

1.3. Infrared sensing method

Infrared sensing is a type of non-contact thermal measurement technique which has been widely used in various applications. Because the temperature distribution in the weld zone contains abundant information about the welding process, infrared sensing of welding processes has drawn considerable attentions from various research institutions.

Chin at Auburn University [18-21] developed a thermal imaging system to measure the variations in weld process parameters such as bead width, penetration depth, and torch offset. The penetration depth has been correlated with the infrared characteristics of the infrared image. The interference of arc radiation was reduced by selecting scanner with specific wavelength region.

At MIT, Hardt used an infrared camera to view the temperature field from the back-side [22]. The penetration depth was precisely estimated from the measured temperature distribution and then controlled [23]. In particualr, a discrete time transfer function matrix empirical model for gas metal arc welding process was proposed, which took the common dynamics for each output and inherent process and measurement delays into account. The adaptation mechanism employed in the control system rendered this model useful over a wide operating range.

In [24] infrared sensor was used to monitor weld process parameters including the weld bead width, penetration depth, and torch position. Analysis of the computed ellipse showed that the temperature gradient or heat energy distribution (minor axis of the ellipse) and the heat input (volume under the temperature profiles) varied with the penetration depth.

1.4. Vision-based sensing method

Based on the observation of the weld pool, a skill welder can assure the desired full penetration. The weld pool thus should contain abundant information of weld penetration. To this end, vision-based systems have been applied to monitor the weld poo by emulating human welders' visual sensory ability. Continued advances in computational capabilities and reduction in cost have recently led to an increase in researches and applications of vision-based systems for the weld pool measurement and welding process control. In the following subsections, vision-based sensing methods are extensively reviewed, including 2D weld pool sensing, and 3D weld pool sensing methods.

1.4.1. 2D weld pool sensing

2D weld pool geometry contains certain information of the welding process, and has been used to monitor the welding process and control the weld penetration [25-27].

Fan et.al [26] studied 2D visual sensing and penetration control in aluminum (Al) alloy pulse GTAW process. A three-optical-route visual sensor was designed. The sensor could capture the weld pool from three directions at the same time. The authors finally used PID and a multiplex controller to control the penetration.

Ma [27] used two normal CCD cameras for capturing clear images from two directions: one of them was used to measure the root gap and another one was used to measure the geometric parameters of the weld pool. Seam tracking and penetration control of robot welding process was simultaneously established based on the proposed binocular vision sensor.

1.4.2. 3D weld pool sensing

Although 2D *weld pool geometry* has been obtained with above different techniques, the convexity/deformation of the weld pool is not yet fully explored. Early researchers have found that important information such as weld defects and penetration are contained in the surface deformation of the weld pool [28, 29]. A recent study suggests that compared with the 2D weld pool geometry, the 3D geometry can better predict the weld penetration which is measured by the *backside weld bead width* [30]. Therefore, numerous methods have been developed to reconstruct the 3D weld pool surface.

The measurement of 3D surface has been recently studied extensively with techniques which can be roughly categorized into three branches: 1) reflectometry/deflectometry with fringe reflection technique [31-33]; 2) phase shifted digital fringe projection technique for diffuse objects [34, 35]; 3) shape from shading technique [36]. Unfortunately, the dynamic and specular nature of the weld pool and the interference from the strong arc radiation complicate the observation and deteriorate the effectiveness of most of those methods.

The most popular techniques currently being studied for 3D weld pool measurement can be divided into four categories:

1. **Model-based reconstruction**

   The 3D weld pool surface was partially reconstructed based on a simple model proposed in [37]. The 2D weld pool images were captured under the base-current period in GMAW. The proposed model then used the capturing angle of the camera and the 2D weld pool profile to calculate the weld pool width, the length of the pool tail, the height of the rear of the pool, etc. The reconstruction algorithm was further applied in [38] for the control of weld pool shape. A fuzzy logic controller was constructed to control weld penetration. It was found that the correlation was nonlinear and thus suitable to employ the proposed fuzzy controller. Simulation and control experiments were carried out to verify the effectiveness of the proposed control algorithm.

   Although this model-based reconstruction algorithm is simple and fast, it can only measure the height of the weld bead that is solidifying or have already solidified at the rear of the weld pool. The 3D geometry of head of the weld pool cannot be acquired using this method. Further, the model-based reconstruction algorithm only suits for thin work piece welding application.

2. **Stereovision measurement**

In [39], two cameras were synchronized to capture the two images of weld pool surface simultaneously in the short circuit period during the Surface Tension Transfer (STT) process and external illumination was used. The paired images were rectified using calibration parameters obtained through the stereo calibration procedure. As the weld pool surface was highly patterned in the experiment, an image correlation-type measure was used to match points between the two rectified images. Then by using stereo image processing algorithms the weld pool shape was rendered in 3D. A closed-loop control system was further developed using the technique for robot welding process [40]. However, the shape of the bright part in the head of the weld pool cannot be acquired by using this method. Further, the accurate reconstruction of the weld pool requires both precise synchronization of the two cameras and high quality of the captured images.

To avoid the synchronization problem, the biprism stereo vision sensing was proposed in which one camera was used with a biprism attached on its head [41]. However, only the height of the weld pool boundary was extracted in real-time, the 3D geometry inside the weld pool was missed. Furthermore, the reconstruction accuracy might be an issue since the visual differences are comparatively small between the two simultaneously captured images. A similar reconstruction algorithm has been utilized in a stereo sensing system using single camera with a stereo adapter developed to reconstruct the 3D weld pool for tracking particle flow on the weld pool surface [42].

3. **Shape from shading (SFS) reconstruction**

3D weld pool reconstruction algorithms have also been proposed based on shape from shading method [43-47]. Zhao et al. [46] use SFS algorithm to reconstruct the surface from one single weld pool image. Two-dimensional shape parameters were extracted from a 2D image processing algorithm. Finally, a SFS algorithm on a single image was used to recover the surface height from a single weld pool image. The extracted three-dimensional parameters for the weld pool surface were verified and used for double-sided shape control.

However, SFS algorithms are usually complex and thus used for off-line reconstruction of the 3D weld pool surface. Furthermore, the reconstruction algorithms are based on two assumptions: 1) The object surface is a Lambertian surface which reflects light with equal intensity in all directions; 2) The camera and the light source are at the infinite far distance from the object surface. The weld pool, on the other hand, is a specular surface which is not a Lambertian surface. The camera and light source in the experiment systems are not far enough from the weld pool such that the infinite far position assumption is invalid. Therefore, the 3D weld pool reconstruction using SFS might not be an ideal solution.

4. Structured-light based sensing

A structured-light vision system was developed in [48] projecting a pulsed laser on the weld pool surface through a special grid. A high shutter-speed camera was used to capture the laser stripe pattern reflected by the weld pool surface. First, to eliminate the influence of the bright arc light, a short duration pulsed laser was projected onto the weld pool surface. The camera shutter was synchronized with the pulse duration. Second, a frosted glass was used to allow each laser ray as a new point light source which disperses light with a certain diffuse angle. The camera viewed the grid openings through their reflection from the weld pool surface and obtained image consisted of bright strips deformed by the weld pool surface deformation. The proposed method could obtain specular reflection from the weld pool under the presence of the bright arc. An iterative algorithm was used to calculate the surface of the weld pool. The time cost of the reconstruction was about 1s. However, the synchronization of the laser and high-speed shutter required specific, high-costly, and sophisticated equipment. The boundary of the weld pool was also hard to extract using this sensing method.

Follow-up study [49] provided a measurement system based on a mathematical model of weld pool surface. The captured image from [48] was applied as an example in the study. Although this work did not propose a new reconstruction algorithm, it provided some novel insights of 3D weld pool surface measurement.

A laser grating sensing technique was proposed in [50]. The reflected grating was captured by a two-lens system. The depth of weld pool was determined based on the phase changes of the deformed grating image [51]. However, using this method the boundary of the weld pool was hard to be determined. Further, it was only a primarily study since there is no detailed quantitative analysis of the reconstruction.

A novel reconstruction algorithm using the slope field and point tracking of the dot matrix was proposed in [52]. A single laser line was projected onto the weld pool surface from a known position with a certain angle. The reflected laser beam from the weld pool surface was captured by a calibrated compact CCD sensor. From the acquired images, the profile of the weld pool surface can be extracted according to ray-tracing technique and the parameters of the CCD sensor. If the line was projected onto the center of the weld pool, the depth of weld pool could also be extracted.

In this technique incorporated use of a calibrated CCD sensor and structured light made it possible to extract the depth of pool from captured images. Although the height reconstruction error was small, the point tracking procedure was complex such that the point matching for each frame requires to process three consecutive frames. It was thus only suitable for off-line reconstruction of 3D weld pool surface. Also, the boundary information of the weld pool in the reconstruction was not addressed.

A laser pattern reflected from the weld pool surface has been intercepted/imaged by/on a diffusive imaging plane placed with a distance from the weld pool [53]. The camera aimed at the imaging plane (rather than the weld pool illuminated by the extremely strong arc) to acquire the reflected laser pattern. Its uniqueness lied in its simultaneous use of the distance and specular nature of the weld pool surface to significantly decay the arc radiation but not the intensity of the laser reflection from the specular weld pool surface despite the distance. To compute the weld pool surface from the reflected patterns, an iterative algorithm has been proposed using the slope field of the projected dot matrix. The slope differences between the neighborhood laser dots were used to find the estimated height of the weld pool surface.

However, this slope error based algorithm requires numerous iterative loops till the estimated surface approaches the actual weld pool surface resulting in relatively large reconstruction errors. Similarly, this imaging method and reconstruction algorithm have been used to image and reconstruct the weld pool surface in gas metal arc welding (GMAW) using a five line laser pattern [54].

This chapter focuses on the development of a procedure of *image processing algorithms* and an *analytical solution* that allows the 3D weld pool surface in GTAW be reconstructed in real-time using the aforementioned innovative imaging principle [53]. The *effectiveness, time cost, accuracy* and *robustness* of the proposed algorithm are quantitatively studied. The accuracy and speed are tested using objects with known geometry and compared with those from previous studies. In particular, the chapter is organized as follows: Section 2 details the vision-based monitoring system. The proposed image processing algorithm procedure is presented in Section 3. The proposed analytic reconstruction algorithm is detailed in Section 4. In Section 5 one object with known 3D geometry is used to emulate the weld pool surface. By comparing the reconstruction surface of the object with its actual surface, the effectiveness and accuracy of the proposed algorithms are verified. The time cost of the reconstruction algorithm is then analyzed. Section 6 presents the summaries of this chapter.

2. Vision-based monitoring system

2.1. Monitoring system

The configuration of the sensing system and the 3D rectangular coordinate systems *oxyz* are shown in Fig. 2.

A 20 mW illumination laser generator at a wavelength of 685 nm with variable focus is used to generate a structure light pattern, i.e., a 19×19 dot matrix pattern (Lasiris SNF-519×0.77-685-20). The laser pattern is projected onto the area under the torch electrode and covers the whole possible weld pool region. During the welding process, the base metal is melted by the arc forming a *liquid weld pool* which has a mirror-like specular surface. It can reflect the majority of the incident laser rays. Therefore, only the dots projected on the weld pool are reflected by its specular surface [49].

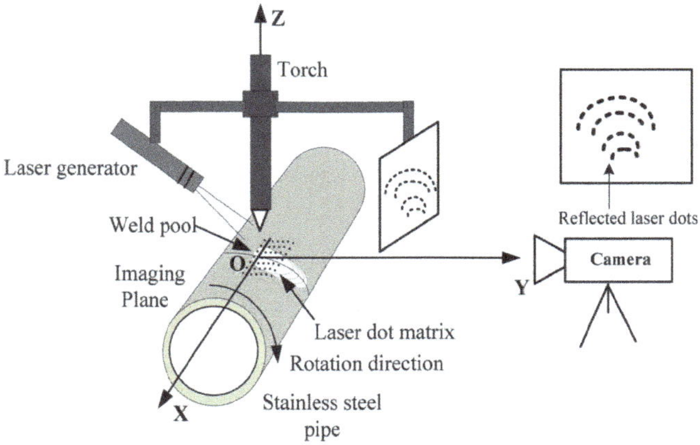

Figure 2. Monitoring system

In order to capture the reflected dot matrix, an imaging plane made by a sheet of glass attached with a piece of paper is installed with a distance of 100 mm approximately from the electrode. A camera (Pointgrey Flea 3 FL3-FW-03S1C-C) is located behind the imaging plane directly aiming at it. The camera captures the images of the reflected pattern from the imaging plane. The captured image is 8-bit monochrome with a resolution of 640×480 or 480×640. A band-pass filter of 20 nm band-width centered at a wavelength of 685 nm is attached to the camera to block the majority of the arc radiation. A computer connects with the camera using a 9-pin 1394b interface. With a maximal frame rate of 200 fps (frame per second), the high transfer rate from the camera to PC (maximum rate 800Mbit/s) makes possible the real-time monitoring and measurement of the 3D weld pool surface in GTAW.

The projection pattern, the dot matrix, is shown in Fig. 3a. The reflection patterns at the solution 480×640 and 640×480 are presented in Fig. 3b and 3c, respectively. The reference dot, i.e., the center ray of the dot matrix is intentionally missed. Please note that the brightness of reflected patterns in the captured images is intentionally enhanced for readability. The original images captured from welding process are much darker than the presented ones.

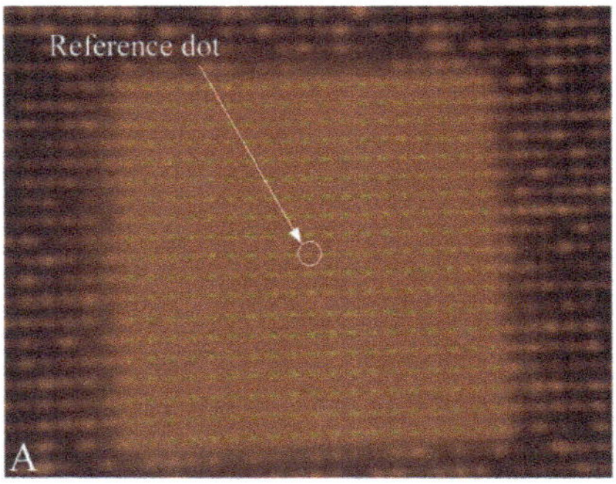

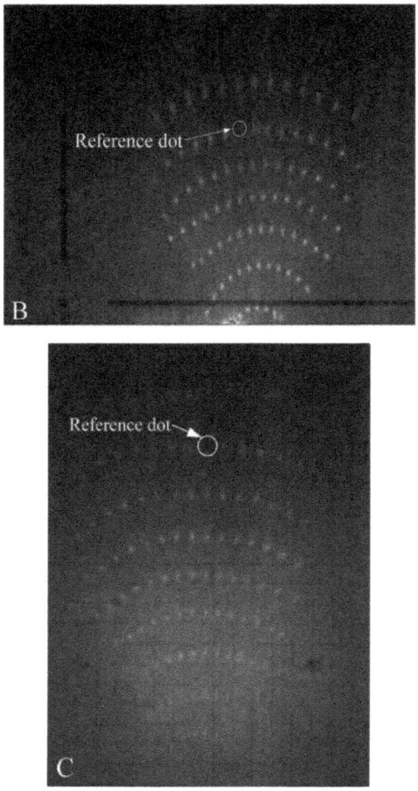

Figure 3. Projection and reflection patterns. A: Dot matrix laser pattern; B: Reflection pattern at resolution 480×640; C: Reflection pattern at resolution 640×480

2.2. Experiment conditions

The welding process used is direct-current electrode-negative (DCEN) GTAW. The material of the pipe is stainless steel (4 inch normal, stainless T-304/304L, schedule 5). The pipe rotates during welding while the torch's orientation, imaging plane, laser projector, and camera are stationary. The rotation speed and the distance from the tungsten tip to the pipe surface are controlled by a computer to weld at required welding speed and arc length.

Ranges of parameters selected to conduct the welding experiments and acquire images in this chapter are shown in Table 1. The full penetration, i.e., the liquid weld pool extends from the front to the back face of the work piece, can be produced on the work piece with those welding parameters. Shielding gas is pure argon. The 2% ceriated ground tungsten electrode (3/32 ×7") grinding to 30º is used.

Welding Parameters			
Current/A	Welding speed/mm/s	Arc length/mm	Argon flow rate/L/min
65	1.5	4.5	11.8
Monitoring Parameters			
Project angle/°	Laser to weld pool distance/mm	Imaging plane to weld pool distance/mm	
35.5	24.7	101	
Camera Parameters			
Shutter speed /ms	Frame rate/ fps	Camera to imaging plane distance/mm	
4	30	57.8	

Table 1. Major experiment parameters

3. Image processing scheme

Using the vision-based monitoring system, images of laser reflection pattern, as shown in Fig. 3b and 3c, can be captured during welding process. However, those two images are deliberately enhanced in brightness such that clear reflection patterns can be seen. The original captured image corresponding to Fig. 3b is shown in Fig. 4a. Fig. 3b is also presented in Fig. 4b for a comparison.

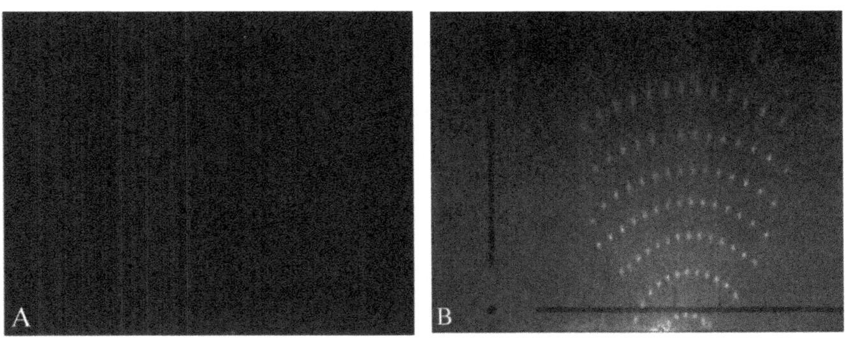

Figure 4. Captured image of reflection pattern. A) Original image captured during welding process; B) Brightness-enhanced image

A human with naked eyes can identify the distorted reflection pattern from captured images as shown in Fig. 4b, while the dots in Fig. 4a can hardly be seen. Images captured during the experiments with different conditions (see Table 1) might obtain even lower brightness and contrast. Furthermore, even in the same captured image, Fig. 4 for example, the gray levels of reflected dots are of great difference. The dots located in the lower part of the image are brighter than those in the upper part of the image. The reflected laser dots are highly coupled with the background in gray scale, especially those in the upper part of the image. In addition, the background of the image, i.e., the part other than the reflection pattern in the captured image, has a severely unbalanced brightness. It can be clearly observed that the brightness of the lower part of the background is much stronger than that of the upper part.

In order to reconstruct the 3D weld pool surface based on the reflection pattern, the reflected dots in the pattern should be extracted from the captured image first. Based on those features of the captured image, noise reduction operation should be conducted first to smooth the image before unifying the brightness. To this end, the flowchart for the proposed image processing scheme is shown in Fig. 5. In particular, a Wavelet-based method is used here for noise reduction which is a pre-processing step to remove certain noises from the image to assure the effectiveness of subsequent processing steps shown in the chart. A Top-hat operation is then performed to unify the background brightness of the captured image while enhancing the reflected dots in the meantime. After a binary thresholding the reflected dots are extracted from the image along with certain noises that are considered as ``fake dots". An adaptive identification algorithm is thus proposed to distinguish the reflected laser dots from the "fake dots". In order to calculate the weld pool surface, each reflected dot is matched with its corresponding incident ray as defined by its row and column numbers in the projected laser matrix. This is done through the row and column pattern recognition.

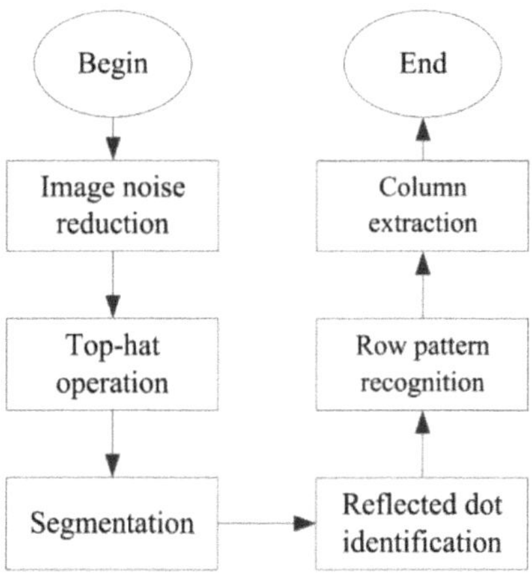

Figure 5. Flowchart for the proposed image processing scheme

In the following sections, the image Fig. 4a is taken as an example to demonstrate the proposed image processing scheme. In particular, each block of image processing operation/algorithm in Fig. 5 is detailed. The effectiveness and robustness of the proposed procedure will be verified in Section 5.

3.1. Noise reduction

Effective noise reduction is a prerequisite for high quality image segmentation. To this end, a wavelet thresholding method is employed to reduce noises in the image. Wavelet noise reduction procedures rely on the recurrent fast wavelet transform (FWT) algorithm proposed by Mallet [55]. The principle of a wavelet-based noise reduction can be described as

$$\hat{x} = FW^{-1}\{T_{\lambda}FW(\tilde{x})\} \tag{1}$$

where $\hat{x}$ is an estimation of x, $T_{\lambda}(\cdot)$ is the thresholding operation, $FW(\cdot)$ and $FW^{-1}(\cdot)$ are the forward and inverse FWT respectively. In order to avoid loss of useful information in the captured image, a soft thresholding is applied [56] with a 8th Symlet Wavelet with 3 levels used for the FWT [57]. After performing noise reduction as shown in Eq. 1 to Fig. 4a, the resultant image is shown in Fig. 6.

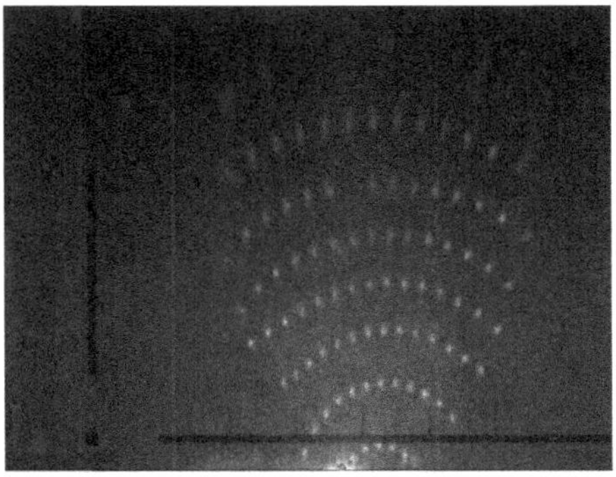

Figure 6. Image after wavelet noise reduction, brightness of the image is enhanced for readability

It can be observed that the lower part image is much brighter than that of the upper half as shown in Fig. 6. To balance the uneven brightness distribution of captured images a top-hat operation is performed. The resultant image is shown in Fig. 7. It can be observed that the unbalance of background in gray scale is much less after the top-hat operation. The reflected dots are clearly seen, and the gray level of background is low enough such that the reflected dots are in good contrast with their local areas.

Figure 7. Result image for top-hat operation

3.2. Adaptive segmentation

A binary thresholding is performed to Fig. 7, and the resultant image is shown in Fig. 8. From the figure, the sizes of the identified dots (including both the reflected dots and the fake dots/noise) can be calculated; the size histogram for all the dots thus can be obtained, as shown in Fig. 9. It can observed that most reflected laser dots' sizes are roughly similar while fake dots' sizes are much smaller than those of the laser dots. Therefore, a bimodal histogram is obtained in Fig. 9 in which most fake dots are in area F, while the majority of the reflected laser dots are concentrated in area R.

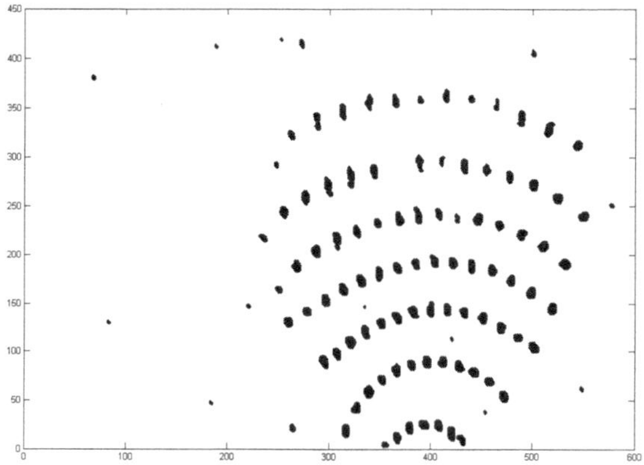

Figure 8. Resultant image after the binary thresholding

To distinguish the reflected laser dots from fake dots an adaptive threshold is required. To this end, Otsu adaptive thresholding method is applied to the size histogram to find the optimal threshold such that the reflected laser dots can be identified from the fake dots [58].

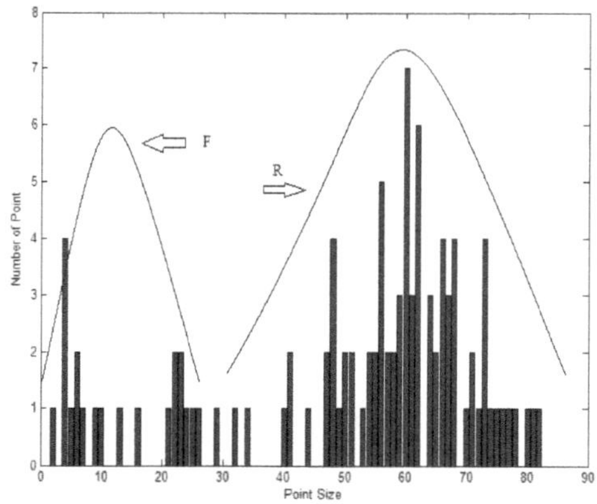

Figure 9. Histogram of dots' sizes, fake dots' size is small, thus they are concentrated in F. The laser dots' size is comparatively large; therefore they are focused in R

Taking the dots in Fig. 8 into calculation, the resultant threshold is 23. Using this threshold, the reflected laser dots are identified, and their positions are shown in Fig. 10.

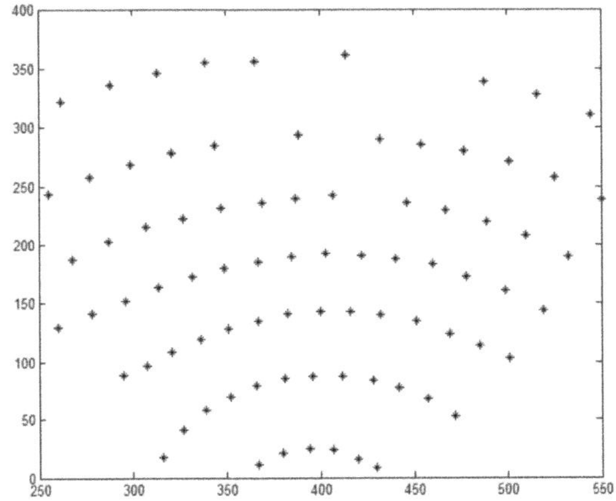

Figure 10. Positions of the identified laser dots

One can find all the fake dots are filtered out after the thresholding, while majority of the reflected dots is preserved. However, it can also be observed that a few reflected laser dots, 5 out of 86 dots in this case, are misjudged as fake dots. This is understandable the adaptive thresholding might not be "intelligent" enough to be able to identify all the reflected dots in each image captured during the welding process with different experiment conditions. However, although a small portion of reflected dots are temporally lost, they can be retrieved in the next section.

3.3. Row/column recognition of the reflection pattern

In order to apply the reflection pattern of laser dots to reconstruct the 3D weld pool surface, each reflected dot should be matched to its corresponding incident ray from the dot matrix first. This subsection first develops the recognition process to identify the row number of the corresponding of each laser dot in the reflection pattern. The column numbers are extracted latter in this subsection.

One can observed that the laser dots in the reflection pattern are well distributed in several smooth top-convex curves which can be roughly considered as quadratics. Fig. 11 is the illustration of the second-order polynomial fitting for the rows of the reflected dots. It can be observed that all rows of reflected dots can be modeled as the following second-order polynomial, as shown in Eq. 2, where a pixel location is presented by coordinate (x, y), and a, b, c>0. Row r_α can be denoted by $r_\alpha(a_\alpha, b_\alpha, c_\alpha)$, where variables a_α, b_α, and c_α are what represent the row using Eq. 2, where α=1, …,R, and R is the number of the rows in the reflection pattern, in the case shown in Fig. 11, R=7.

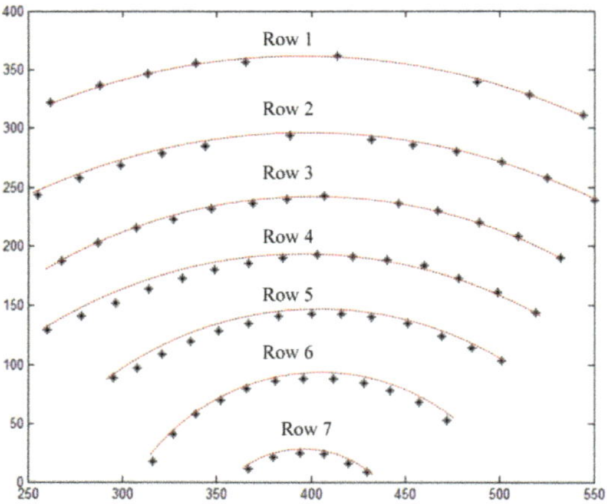

Figure 11. Illustration of curve fitting for the reflected laser dots

$$y = a(x - b)^2 + c \tag{2}$$

Since the correspondence of the mapping is known, the recognition process is to first assign all the reflected dots to different rows, second match to the rows to the corresponding rows in dot matrix. In particular, the first step is to define the 7 rows using Eq. 2; Second is to find the a row for each reflected dot with a pre-defined offset. Then the 7 rows are thus formed; Last step is use the reference dot (in 10th row and 10th column shown in Fig. 3) to map the rows into the dot matrix.

Using the Hugh transform method, all the rows in the reflection pattern in Fig. 10 can be identified, as shown in Fig. 12. The curves present the identified row. The reflected dots are distributed around those rows.

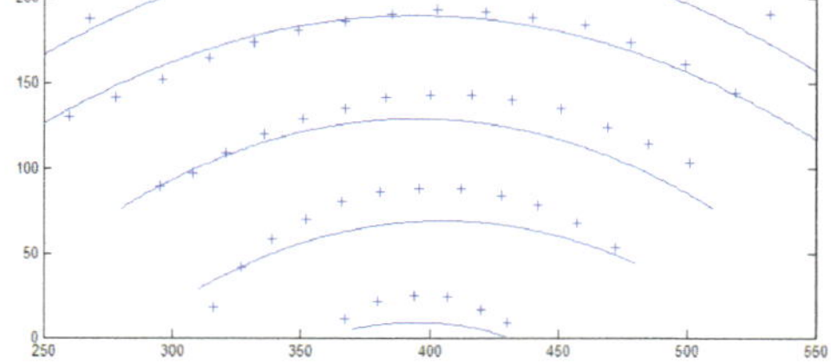

Figure 12. Results of row identification

It can be found some of the reflected dots are missed in some of the rows. That is because those reflected dots are mistakenly filtered out in Section 3.2. Since all the rows are extracted, those reflected dots can be retrieved. The process is, for one specific row, 1) the median distance between two neighbor dots in the row is first calculated; 2) all the dots are visited to find out the abnormal large distance between neighbor dots. It is considered that there is for a reflected dot missed; 3) starting from the largest size, all the filtered out dots (including the fake dots, i.e., the noise) are visited to find the fittest to fill in the position of the missed reflected dot. The process goes through recursively for every row, and the filtered out reflected dots can be retrieved. The result is shown in Fig. 13. Reflected dots are marked with different shapes in different rows.

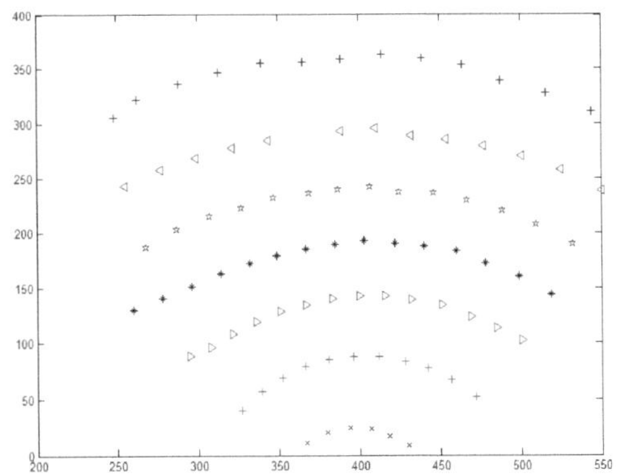

Figure 13. Reflected dots in different rows

It can be found in Fig. 13 that there is one point missed in the second row of reflected dots from the top of the reflection image. That intentionally absent dot is the center dot (in 10th row and 10th column of the 19×19 dot matrix). It serves as the reference dot (see Fig. 3) to facilitate the row/column identification and the corresponding match between reflected dots and incident rays. To this end, the row pattern recognition can be accomplished. The result is shown in Fig. 14. The numbers in the image indicate the corresponding row match between the row in the reflection image and the incident rows in the dot matrix.

After the reference dot found in one row in the reflection image, the column number for each dot in the row can be identified, shown in Fig. 15. The distortion of the laser pattern in the vertical direction is much less severe than that in the horizontal direction. Therefore, a center line $y=kx+b$ is fitted using reference dot together with its nearest adjacent dots in the neighbor rows.

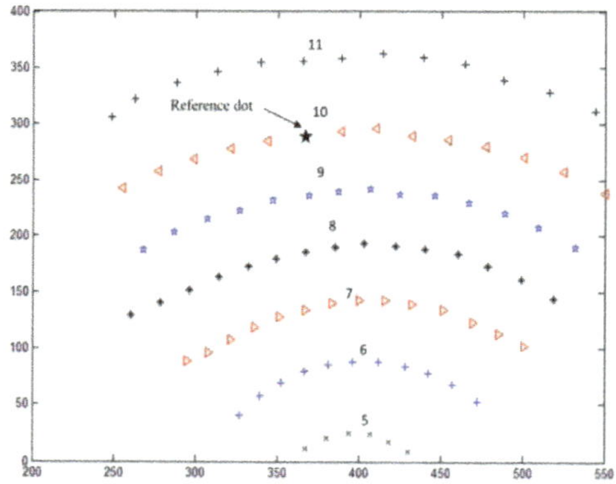

Figure 14. Reflected rows matching with the corresponding incident rows in the dot matrix. The numbers represent the incident rows in the dot matrix that the reflected rows match respectively.

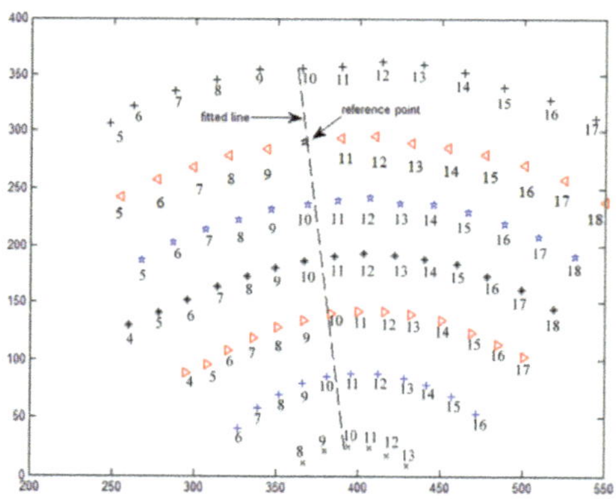

Figure 15. Column extraction with fitted line

4. Weld pool reconstruction

This section focuses on the development of an analytical solution that allows the 3D weld pool surface in GTAW be reconstructed in real-time. In particular, the boundary of the weld pool is extracted in section 4.1 ; the reconstruction of the 3D weld pool surface is detailed in section 4.2; An reconstruction example is given in section 4.3.

4.1. Extraction of weld pool boundary

Before the 3D surface is reconstructed, the boundary of the weld pool should be determined first. The model used to fit the boundary of the weld pool is from literature [59] and demonstrated in Fig. 16:

$$y_r = \pm a x_r^b (1\text{-}x_r), \quad (a > 0, 1 \geq b > 0) \tag{3}$$

where $x_r = x / L$, $y_r = y / L$, and L is the length of the weld pool, which is the distance from the head to the tail of the weld pool. The width of the weld pool can be calculated using parameter a, b and L.

$$w = w_r \times L = 2aL\left[\frac{b}{1+b}\right]^b \left(\frac{b}{1+b}\right) \tag{4}$$

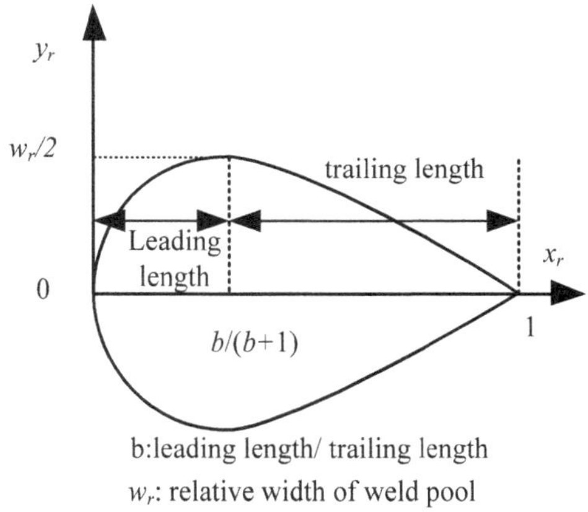

Figure 16. Illustration of the weld pool boundary model.

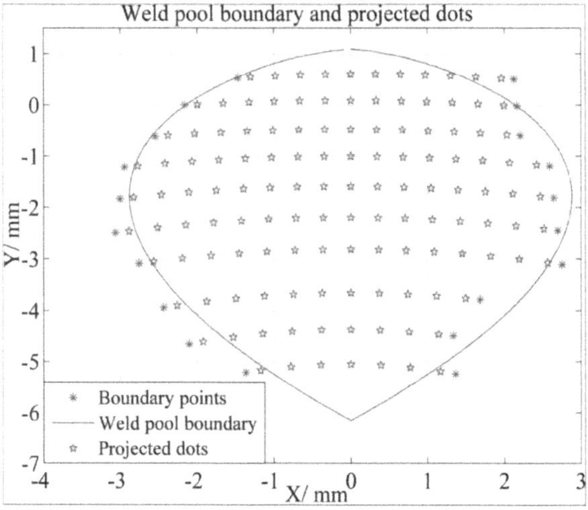

Figure 17. Modeling of the weld pool boundary

Fig. 3c is taken as an example to demonstrate the boundary fitting. The resultant boundary of the weld pool is shown in Fig. 17. Red stars and blue crosses show the laser dots projected on the specular weld pool surface and the raw estimates of the boundary points respectively. The rays and boundary shown in the figure are the projection of them on the *oxy* plane. The red stars in each row are curved because the projected rays intercept the pipe surface rather than a plate. The blue crosses are the raw estimates of the boundary points as defined above and the blue lines are the fitted boundary using Eq. 3. It can be seen that the weld pool is about 7 mm long and 6 mm wide.

4.2. Reconstruction of 3D weld pool surface

The formation of the image on the imaging plane is governed by the law of specular reflection. A reconstruction scheme thus is required to extract the 3D geometry of the weld pool surface by solving an inverse problem of the reflection law. The dot matrix reflection from the weld pool surface is demonstrated in Fig. 18.

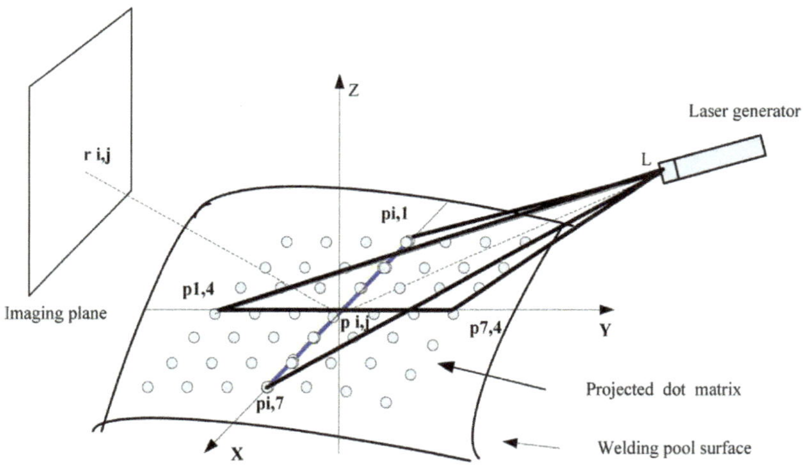

Figure 18. Demonstration of dot matrix reflection

The equations for all incident rays projected from laser to the weld pool surface are known. For ray $Lp_{i,j}$, the ith row j^{th} column in the pattern, the position of its reflection image $r_{i,j}$ in *oxyz* coordinate in Fig. 18 can be obtained using its position in the imaging plane (which has been extracted by the image processing algorithm) and the equation of the imaging plane in *oxyz* coordinate. The goal of the surface reconstruction algorithm is to calculate the coordinates of each $p_{i,j}$ in *oxyz* coordinate. Then the 3D weld pool surface can be interpolated using the coordinates from all $p_{i,j}$'s.

However, without further constraints or assumptions, $p_{i,j}$'s positions cannot be directly determined. Fortunately, the weld pool surface in GTAW is smooth such that two reasonable assumptions can be made:

1. For the most left and right dots in a row, i.e., edge dots, reflected from the weld pool surface that are close to the boundary of the weld pool surface, their deviations from the original pipe surface as measured by their z-coordinates are approximately zero.
2. $z_{i,j} = z_i(x_{i,j}(j))$ can be modeled as a polynomial of the column number j where $z_{i,j}$ and $x_{i,j}$ are the z and x coordinate respectively for $p_{i,j}$. Based on the observation of the extracted pattern, a second order polynomial should be sufficient to meet the reconstruction accuracy requirement defined later in the paper.

A procedure can thus be proposed to reconstruct the 3D weld pool surface from the extracted image, i.e., the extracted laser reflection pattern, obtained from the image processing:

Step 1. Determination of Intersection Points: The surface of work piece or a previous estimate if available is used as the "previously estimated weld pool surface". With this known surface together with known knowledge for the origin, projection angle, and internal angle of the laser pattern, the positions of the intersections of the projected rays with it can be easily acquired. That is, the coordinates for all $p_{i,j}$'s in the $oxyz$ coordinate system shown in Fig. 18 are obtained.

Step 2. Updating of Slope Field at Intersection Points: With the known equations of the incident rays and equations of the estimated reflection rays $p_{i,j}r_{i,j}$'s , slopes and components $\frac{\partial z}{\partial x}$'s and $\frac{\partial z}{\partial y}$'s at all $p_{i,j}$'s, i.e., the row and column slopes are obtained/updated.

Step 3. Updating of Intersection Points: The projection of the ith row of the incident rays in the dot matrix on the oxz plane is illustrated in Fig. 19. Updating from $p_{i,j}^{(0)'}s$ (on the previously estimated or initial surface obtained in Step 1) requires to find a new set of points $p_{i,j}$'s along with this row of incident rays to satisfy the row slopes $\frac{\partial z}{\partial x}$'s obtained in Step 2 from the previously estimated or initial surface. Fig. 19 shows the case when the previously estimated surface is the initial surface (the surface of the work piece).

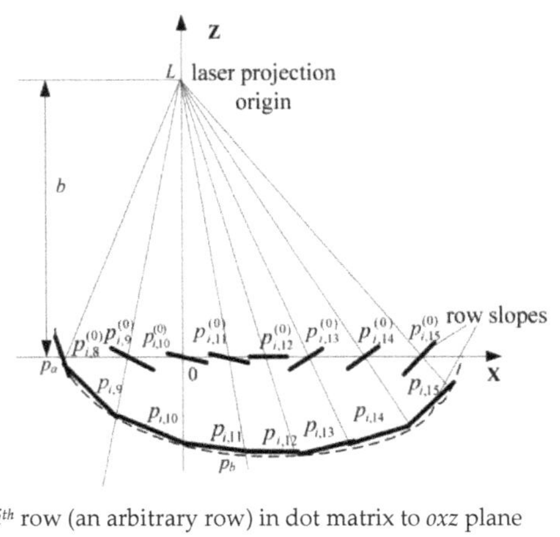

Figure 19. Projection of ith row (an arbitrary row) in dot matrix to oxz plane

To analytically solve for $p_{i,j}$'s, assumption (2) is used such that $p_{i,j}$'s are constrained in a 2nd order polynomial:

$$z = a_2 x^2 + a_1 x + a_0 \tag{5}$$

where a_0, a_1 and a_2 are the parameters. (Higher order polynomials may be used if a 2nd order one is not sufficient.) To be convenient, denote the edge point by $p_a(x_a, z_a)$ and inner point by $p_b(x_b, z_b)$, where x_m and z_m is the x coordinate and z coordinate of p_m. Denote their row slopes as S_a and S_b, respectively. The slopes of the two incident rays Lp_a and Lp_b in Fig. 19 are denoted as k_a and k_b. The distance of the laser projection origin to the x-axis is denoted as b. Since the internal angle of the laser pattern is fixed (0.77°), the slope for each incident ray can be easily determined.

At point $p_m(x_m, z_m)$ where m=a, b, the geometric relationships between incident rays and the row constraint polynomial Eq. (5) are:

$$\begin{bmatrix} z_m \\ z_m \\ dz_m / dx\big|_{x=x_m} \end{bmatrix} = \begin{bmatrix} a_2 & a_1 & a_0 \\ 0 & k_m & b \\ 0 & 2a_2 & a_1 \end{bmatrix} \begin{bmatrix} x_m^2 \\ x_m \\ 1 \end{bmatrix} \tag{6}$$

$$dz_m / dx\big|_{x=x_m} = S_m \tag{7}$$

Solving Eq. (6) and (7), the parameters for the row constraint polynomial Eq. (3) are obtained:

$$a_0 = z_a - (a_2 x_a^2 + a_1 x_a) \tag{8}$$

$$a_1 = S_a - 2a_2 x_a \tag{9}$$

$$a_2 = \left[(S_a - S_b)^2 / 4 - (k_b - S_b)(S_a - S_b)/2 \right] / \left[(k_a - k_b) x_a \right] \tag{10}$$

when $x_a \neq 0$.

$$a_0 = z_a \tag{11}$$

$$a_1 = S_a \tag{12}$$

$$a_2 = \left[4(a_1 - k_b)(S_b - a_1) + (S_b - a_1)^2 \right] / (b - a_0) \tag{13}$$

when $x_a = 0$.

From Eq. (8)-(13), the parameters of the row constraint polynomial are completely solved. However, this requires the position of the dot p_a. According to the assumption (1), the z coordinate of each edge dot on the weld pool surface can be considered zero, that is z_a=0. Its x_a can be calculated by intersecting the corresponding incident ray with z=0. The coordinates for p_a is thus considered known.

With this constraint polynomial, the intersections of the incident rays (projections on the *oxz* planes shown in Fig. 19) with it are the projection of the updated $p_{i,j}$'s. The updated $p_{i,j}$'s in the *oxyz* coordinate system as shown in Fig. 18 can thus be obtained through the inverse projection.

Step 4. Interpolating Surface: With all the coordinates of the $p_{i,j}$'s, a triangle-based cubic interpolation [60] method is applied to interpolate the 3D weld pool surface. With the obtained surface, the slopes of the surface at the $p_{i,j}$'s can be easily acquired. Therefore, the resultant reflection pattern using the reconstructed surface is obtained.

4.3. Reconstruction example

Taking Fig. 3c as an example, the results of the weld pool reconstruction using the proposed reconstruction scheme are shown in Fig. 20.

a

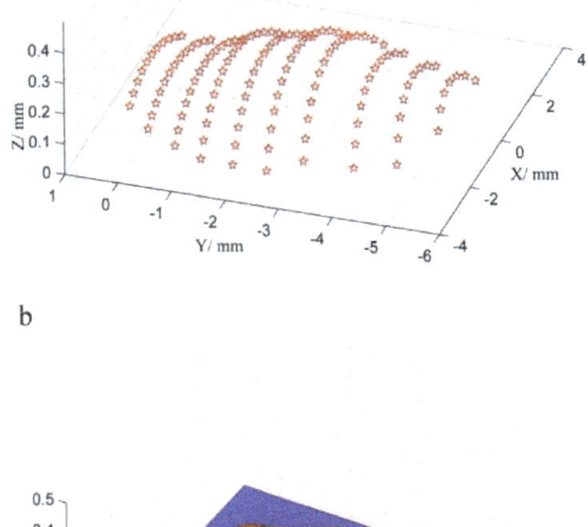

b

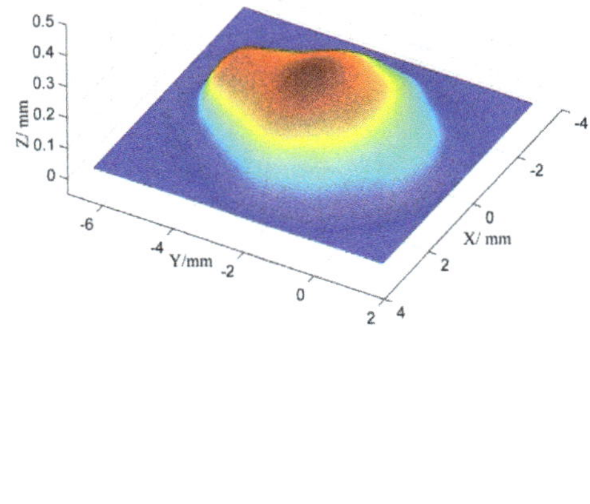

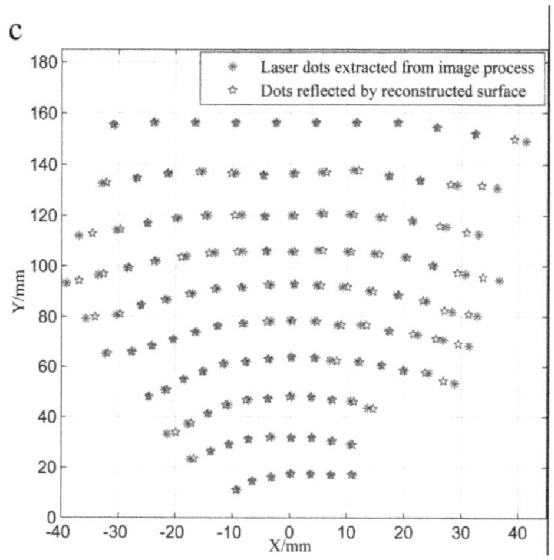

Figure 20. Results of 3D reconstruction of the weld pool. (a) shows the 3D coordinates of all the projected laser dots on the weld pool surface, (b) is the weld pool surface interpolated using the laser dots in (a), (c) shows the reflected pattern from imaging process and that from the reconstructed weld pool surface.

Fig. 20a shows the position for each projected dot on the weld pool surface. The interpolation result for the 3D weld pool in Fig. 20b gives a better view of the weld pool surface. The comparison between the two reflection patterns (calculated from the reconstructed surface and extracted from the acquired image) is shown in Fig. 20c. An acceptable match of the reflection patterns is obtained using the proposed reconstruction algorithm.

In order to verify the accuracy of the proposed reconstruction algorithm, the results in this study are compared with the previous work in [52, 53]. The reason the two studies are selected for comparison is that high reconstruction accuracies are acquired through detailed quantitative analysis in these two studies. In order to compare the reconstruction accuracy with the previous work [53] which uses the similar sensing system, the same reconstruction error measurement parameters are adopted here, i.e., the average reflection error (ARE) and maximum reflection error (MRE):

$$ARE = \frac{1}{N}\sum_{k=1}^{N} E_k \tag{14}$$

$$MRE = \max\left\{E_k, k = 1, \ldots, N\right\} \tag{15}$$

$$E_k = \sqrt{\left[\left(x_k^{(e)} - x_k^{(c)}\right)\frac{W_p}{W_r}\right]^2 + \left[\left(y_k^{(e)} - y_k^{(c)}\right)\frac{L_p}{L_r}\right]^2} \tag{16}$$

where W_r and L_r are the horizontal and vertical ranges of the actual reflected dots extracted from image processing, respectively and W_p and L_p represent the horizontal (X axis) and vertical (Y axis) ranges of the corresponding projected dots on the work piece, which are the width and length of the weld pool boundary.

Using the same example (Fig. 3c), E_k can be calculated using the data shown in Fig. 4 and Fig. 7. The ARE and MRE of the reconstruction are thus obtained: $ARE = 0.03mm$, $MRE = 0.20mm$. The minimal ARE and MRE obtained is [53] are 0.08 mm and 0.22 mm, respectively. Therefore, using the proposed algorithm, the error measurement parameters are 0.05 mm smaller in ARE and 0.02 mm smaller in MRE than those parameters obtained in [53]. The reconstruction accuracy increases 62.5 % in ARE and 9% in MRE.

5. Experiments results and analysis

In Section 3, the reflected laser dots have been identified from the reflection pattern. Then they have been used to reconstruct the profile of the 3D weld pool surface in Section 4. To verify the accuracy with respect to ground truth, the proposed algorithm needs to be tested by reconstructing objects with known geometry. Also, the real-time performance of the proposed scheme is not validated. In this section, using a spherical bench mark with a known geometry having a specular surface to emulate a weld pool, a simulation has been conducted to evaluate the effectiveness and accuracy of the proposed algorithm; the time cost of the algorithm was evaluated to see if it is suitable for real-time welding process; During simulation, the error measurement parameters ARE and MRE are used to evaluate the reconstruction accuracy.

To evaluate the accuracy and effectiveness of the proposed algorithm, the 3D weld pool is characterized using the three geometric parameters, i.e., the length, the width and the height, in which the length is the distance from the head to the tail of the weld pool; the width can be obtained using Eq. 4; the height is defined as the maximum height of the weld pool.

5.1. Simulation and results

The verification of the proposed reconstruction scheme starts with a bench mark with a known geometry to imitate the weld pool surface which is shown in Fig. 21. The top cap of the sphere is obtained by using a plane to intercept a sphere. It is used to imitate the weld pool surface in the experiment. Its height (d) is 0.5 mm and the diameter (r) is 5 mm.

In Fig. 22b the boundaries of the reconstructed bench mark fitted by the weld pool boundary model illustrated in Fig. 16 and a circle are presented. For the dimension of the boundary fitted by the weld pool boundary model, its length is 10.024 mm and its width is 10.32 mm which lead to 0.024 mm error in length and 0.32 mm error in width. Based on the dimension of the actual bench mark, the reconstruction errors are 0.48% of the length and 3.2% of the width respectively. Having a circle to fit the reconstructed bench mark boundary using the least squared method, the boundary diameter is 9.989 mm which is only 0.11% of reconstruction error. It can be seen that the circle can fit the boundary of the reconstructed bench mark better than the weld pool boundary model. This is understandable since the model only can fit the moving weld pool which rarely has a circular boundary.

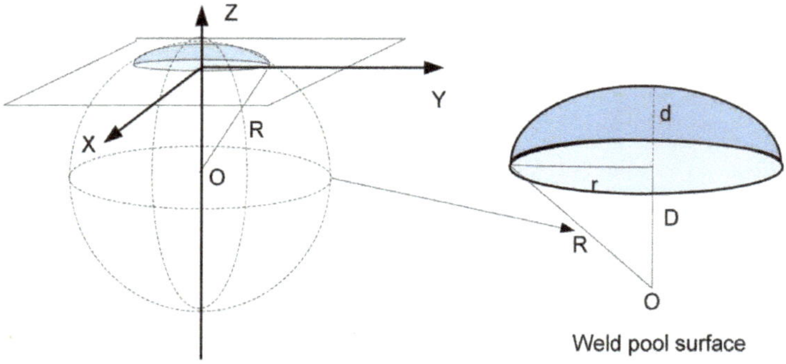

Figure 21. The illustration of the imitation of the weld pool

A simulation was designed to project dot matrix on the bench mark. The projected dots on the spherical surface were reflected and intercepted by an imaging plane. The results of the simulation for the reconstruction of the bench mark are shown in Fig. 22.

a

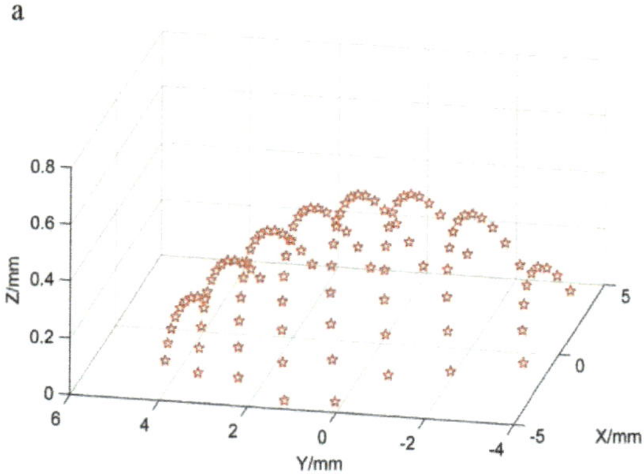

b

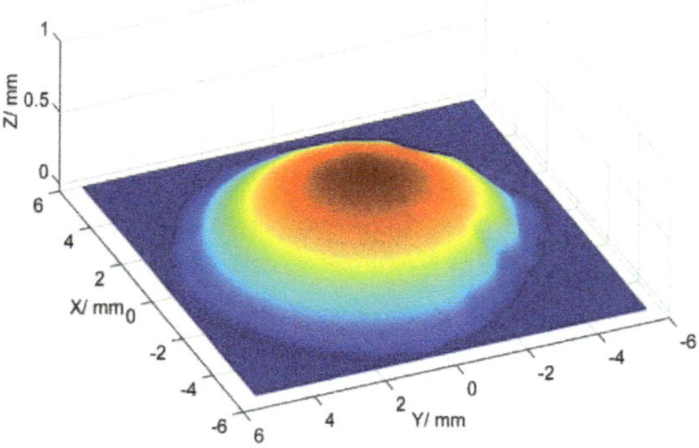

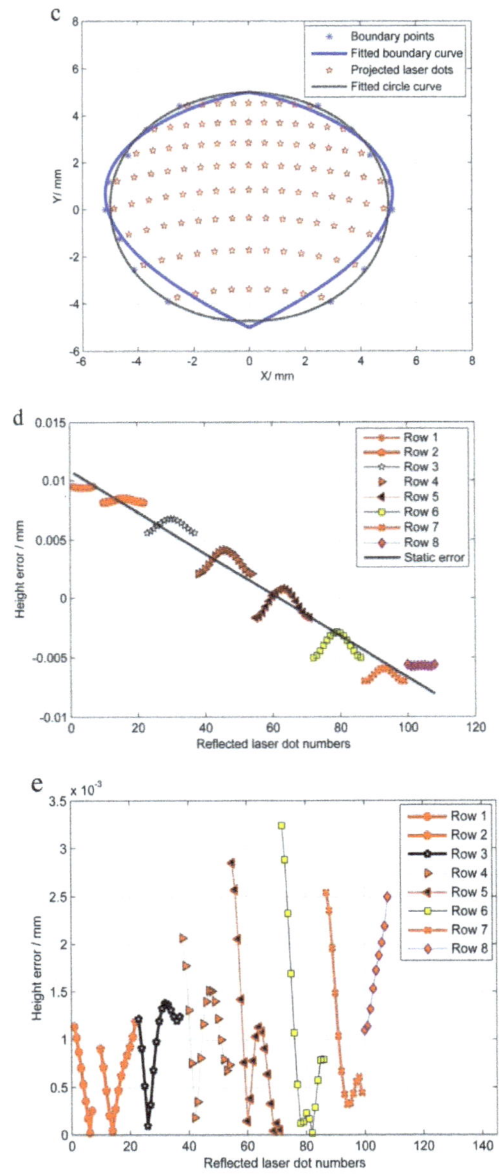

Figure 22. Simulation results of the proposed reconstruction scheme. (a) The 3D coordinates of all the projected laser dots; b) The reconstructed surface of the bench mark; (c) The boundary of the reconstructed bench mark; (d) The height errors between the actual and reconstructed bench marks row by row and the static height error. Row 1 is defined as the most left reflected dots row in (a). The absolute errors between height errors and the corresponding static errors are shown in (e).

Fig. 22d show the height errors of the reconstruction at the all the x, y coordinates of the reflected dots. The sphere function for the bench mark is:

$$x^2 + y^2 + \left[z - \left(d - R\right)\right]^2 = R^2 \tag{17}$$

The definitions of d and R can be seen in Fig. 21. The heights (z-coordinates) of bench mark at the same (x, y) coordinates of the reflected dots thus can be easily obtained:

$$z_k^{(b)} = \sqrt{R^2 - x_k^2 - y_k^2} + (d - R) \tag{18}$$

where $z_k^{(b)}$ is the height of bench mark at position (x_k, y_k) which is the x, y coordinate of corresponding projected laser dot $p_k(x_k, y_k, z_k)$. It should be noted that there is another possible solution for the height from Eq. 17, which is $-\sqrt{R^2 - x_k^2 - y_k^2} + (d - R)$. However, according to the definition in Fig. 21, the height in Eq. 18 is the right solution. Therefore, the height error at position (x_k, y_k) is:

$$z_k^{error} = z_k^{(b)} - z_k \tag{19}$$

The height errors of the reconstruction obtained for all the projected laser dots are shown in Fig. 22(d). From Row 1 to Row 8, the errors can be observed turning from positives to negatives. That indicates static errors in the reconstruction. The static errors are obtained by fitting the height errors into a straight line as shown in Fig. 22(d). In general, static errors might be caused (in real system) by the noises from the image processing, welding machine, the data acquisition process and calibration errors of experiment set-up. Static error also occurs when the laser projection direction does not aim at the center of the weld pool. (This may also occur in simulation). Hence, the calibration is needed such that the point where the tungsten axis intersects with the surface of the work piece is also the point where it interests with the reference incident ray, i.e., the central ray or 10th row 10th column ray, in order to minimize this possible static error.

In this simulation, the experiment condition is considered ideal. The static error shown in Fig. 22(d) is thus comparatively small. Having the height errors subtracted by the corresponding static errors, the resultant absolute errors are shown in Fig. 22(e).

It can be seen in Fig. 22(e) that maximum height error is 3.20×10^{-3} mm at 1st point of Row 6, and the minimum height error is 1.93×10^{-5} mm at 11th point of Row 6. The heights of the bench mark at the corresponding positions are 0.104 mm and 0.468 mm. That is that the height error is 0.3% and 0.004% of the two heights respectively. The height error at the maximum height (8th point of Row 6) is 1.2×10^{-4} mm which is 0.024% of the maximum height (0.5 mm). In comparison with the best result in study [52] which is 0.16%, the proposed algorithm achieves higher reconstruction accuracy in addition to being analytic.

The obtained results are then used to calculate the ARE and MRE (see Eq. 14 and Eq.15) to evaluate the reconstruction accuracy. The resultant ARE and MRE for the reconstruction of the bench mark are 0.06 mm and 0.21 mm, respectively. Compared with the best results in [53] which are 0.08 mm of ARE and 0.22 mm of MRE, the accuracy of the proposed reconstruction algorithm in this study is improved by 25% in ARE and 4.55% in MRE.

5.2. Time cost of the reconstruction algorithm

Since there is no difference for the reconstruction scheme for a bench mark or an actual weld pool in a GTAW experiment, the bench mark reconstruction thus was used to verify the computation time cost for the proposed reconstruction scheme. To this end, the reconstruction process of the bench mark is performed 2000 times in this study to evaluate the time cost for each reconstruction process.

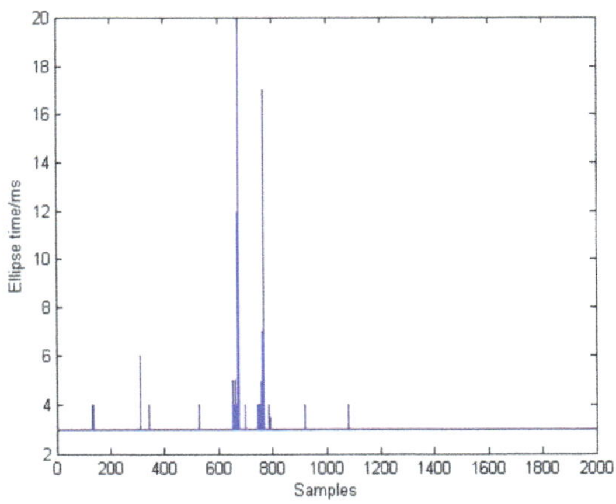

Figure 23. Computation time obtained in the reconstruction for the bench mark in 2000 attempts

The scheme is programmed by C++ in Visual Studio 2008. The computer used is a desktop with Intel i7, 2.8 GHz processor with 4 GB RAM. The computation time is recorded and shown in Fig. 23. The minimum time of one loop is 3 ms, the maximum is 20 ms, and the average of time cost is 3.04 ms. It is believed that an average time cost of 3.04 ms for the reconstruction of 3D weld pool surface is fast enough for the real-time monitoring a GTAW process, and for the future control of GTAW process.

6. Summary

The monitoring and measurement of three-dimensional weld pool surface provides the foundation for a possible advanced control for the welding process. An analytic reconstruction algorithm based on the slope field of the reflected laser pattern is proposed in this chapter to measure the 3D weld pool surface in real-time. The reconstruction is mathematically formulated. The virtual spherical convex surface, considered as similar in shape of a weld pool surface which cannot be precisely measured by any existing methods, were used to test the effectiveness and accuracy of the proposed algorithm. The following are concluded:

- The proposed algorithm can effectively reconstruct the weld pool surface.
- Reconstruction accuracy is improved. The boundary reconstruction error in the simulation of the bench mark is 0.11%; the height reconstruction error at the maximum height is 0.024%. The two reconstruction error parameters ARE (average reconstruction error) and MRE (average reconstruction error) are 0.06 mm and 0.21 mm. Compared with the previous works, the reconstruction accuracy is significantly improved.
- The time cost for the reconstruction algorithm is 3.04 ms on average in the simulation and 3.22 ms on average obtained in the real-time welding experiment. The proposed algorithm is thus considered to be real-time for GTAW application.

Author details

Wei Jie Zhang, Yu Kang Liu and Yu Ming Zhang
Department of Electrical and Computer Engineering Institute for Sustainable Manufacturing, University of Kentucky, Lexington, USA

7. References

[1] R. O'Brien, Ed., Welding Handbook, 8th Edition VOL. 2 - Welding Processes. AWS, 1998.

[2] D. J. Kotecki, D. Cheever, and D. Howden, "Mechanism of ripple formation during weld solidification," Welding Journal, vol. 51(8), pp. 386s- 391s, 1972.

[3] R. Renwick and R. Richardson, "Experimental investigation of gta weld pool oscillations," Welding Journal, vol. 62(2), pp. 29s-35s, 1983.

[4] M. Zacksenhouse and D. Hardt, "Weld pool impedance identification for size measurement and control," ASME Journal of Dynamic Systems, Measurement, and Control, vol. 105(3), pp. 179-184, 1984.

[5] X. Y.H. and G. den Ouden, "Weld pool oscillation during gta welding of mild steel," Welding Journal, vol. 72(8), pp. 428s-434s, 1993.

[6] A. J. R. Anedenroomer and G. den Ouden, "Weld pool oscillation as a tool for penetration sensing during pulsed gta welding," Welding Journal, vol. 77(5), pp. 181-187, 1998.

[7] G. B. R. e. a. Andersen, K; Cook, "Synchronous weld pool oscillation for monitoring and control," IEEE TRANSACTIONS ON INDUSTRY APPLICATIONS, vol. 33(2), pp. 464-471, 1997.

[8] D. D. C. G. Hartman, D.A. and R. Barnett, "Intelligent fusion control throughout varying thermal regions," in Proceedings of the IEEE Industry Applications Conference, 1999.

[9] Y. O. K. Ju, J; Suga, "Penetration control by monitoring molten pool oscillation in tig arc welding," INTERNATIONAL JOURNAL OF OFFSHORE AND POLAR ENGINEERING, vol. 14(2), pp. 145-149, 2002.
[10] B. Yudodibroto, M. Hermans, Y. Hirata, and G. den Ouden G, "Influence of filler wire addition on weld pool oscillation during gas tungsten arc welding," Science and Technology of Welding and Joining, vol. 9(2), pp. 163-168, 2004.
[11] J. B. J. Aussel, B. Le, "Generation acoustic wave by laser: Theoretical and experimental study of the emission source," Ultrasonics, vol. 24, pp. 246-255, 1988.
[12] C. M. F. A. V. Clark, S. R. Schaps, "Well-shielded emat for on-line ultrasonic monitoring of gma welding," in Proceedings of Institute of Electrical and Electronics Engineers Ultrasonic Symposium, Orlando, FL, 1992.
[13] G. Graham and I. Ume, "Automated system for laser ultrasonic sensing of weld penetration," Mechatronics, vol. 7(8), pp. 711-721, 1997.
[14] S. Hopko and I. Ume, "Laser generated ultrasound by material ablation using fiber optic delivery," Ultrasonics, vol. 37(1), pp. 1-7, 1999.
[15] E. Siores, "Development of a realtime ultrasonic sensing system for automated and robotic welding," Ph.D. dissertation, Brunel University, 1988.
[16] A. Kita, "Measurement of weld penetration depth using non-contract ultrasound method," Ph.D. dissertation, Georgia Institute of Technology, 2005.
[17] B. Mi and C. Ume., "Real time weld penetration depth monitoring with laser ultrasonic sensing system," Transactions of ASME: Journal of Manufacturing Science and Engineering, vol. 128, pp. 280-286, 2006.
[18] W. Chen and B. Chin, "Monitoring joint penetration using infrared sensing techniques," Welding Journal, vol. 69(4), pp. 181s-185s, 1990.
[19] S. Nagarajan, P. Banerjee, W. Chen, and B. Chin, "Control of the welding process using infrared sensors," IEEE Transactions on Robotics and Automation, vol. 8(1), pp. 86-93, 1992.
[20] P. Banerjee and et al., "Infrared sensing for on-line weld shape monitoring and control," ASME Journal of Engineering for Industry, vol. 117, pp. 323-330, 1995.
[21] H. C.Wikle, R. H. Zee, and B. Chin, "Sensing system for weld process control," Journal of Materials Processing Technology, vol. 89-90, pp. 254-259, 1999.
[22] J. Song and D. Hardt, "Closed-loop control of weld pool depth using a thermally based depth estimator," Welding Journal, vol. 72(10), pp. 471s-478s, 1993.
[23] J. Song and D. Hardt, "Dynamic modeling and adaptive control of the gas metal arc welding process," ASME Journal of Dynamic Systems, Measurement, and Control, vol. 116(3), pp. 405-413, 1994.
[24] W. C. S. Nagarajan, B. A. Chin, "Control of the welding process using infrared sensors," IEEE Transactions on Robotics and Automation, vol. 8(1), pp. 86-93, 1992.
[25] C. Balfour, J. Smith, and A. Al-Shamma, "A novel edge feature correlation algorithm for real-time computer vision-based molten weld pool measurements," Welding Journal, vol. 85(1), pp. 1s-8s., 2006.

[26] C. Fan, F. Lv, and S. Chen, "Visual sensing and penetration control in aluminum alloy pulsed gta welding," The International Journal of Advanced Manufacturing Technology, vol. 42(1), pp. 126-137, 2009.
[27] S. C. W. H. B. Ma, "Binocular vision system for both weld pool and root gap in robot welding process," Sensor Review, vol. 30(2), pp. 116-123, 2010.
[28] M. Lin and T. Eagar, "Influence of arc pressure on weld pool geometry," Welding Journal, vol. 64(6), pp. 163-169, 1985.
[29] A. G. S.I. Rokhlin, "A study of arc force, pool depression, and weld penetration during gas tungsten arc welding," Welding Journal, vol. 72(8), pp. 381-390, 1993.
[30] W. Zhang, Y. Liu, and Y. M. Zhang, "Characterization of three-dimensional weld pool surface in gtaw," Welding Journal, vol. 91(7), pp. 195s-203s, 2012.
[31] Y. Tang, X. Su, F. Wu, and Y. Liu, "A novel phase measuring deflectometry for aspheric mirror test," Optic Express, vol. 17(22), pp. 19778-19784, 2009.
[32] W. Zhao, X. Su, Y. Liu, and Q. Zhang, "Testing an aspheric mirror based on phase measuring deflectometry," Optic Engineering, vol. 48(10): 103603, 2009.
[33] L. Huang, C. S. Ng, and A. K. Asundi, "Dynamic three-dimensional sensing for specular surface with monoscopic fringe reflectometry," Optics Express, vol. 19(13), pp. 12 809-12 814, 2011.
[34] S. Zhang and P. S. Huang, "High-resolution, real-time three-dimensional shape measurement," Optic Engineering, vol. 45(12):123601 2006.
[35] S. Zhang, "Recent progresses on real-time 3d shape measurement using digital fringe projection techniques," Optic Lasers Engineering, vol. 48(2), pp. 149-158, 2010.
[36] J. Balzer and S. Werling, "Principles of shape from specular reflection," Measurement, vol. 43(10), pp. 1305-1317, 2010.
[37] G. J. Zhang, Z. H. Yan, and L. Lin, "Reconstructing a three- dimensional p-gmaw weld pool shape from a two-dimensional visual image," Measurement Science and Technology, vol. 17(7), pp. 1877-1882, 2006.
[38] Z. H. Yan, G. J. Zhang, and L. Wu, "Simulation and controlling for weld shape process in p-gmaw based on fuzzy logic," Proceedings of the 2011 IEEE International Conference on Mechatronics and Automation, pp. 2078-2082, 2011.
[39] C. Mnich, F. Al-Bayat, C. Debrunner, and et al., "In situ weld pool measurement using stereovision," Proceedings of 2004 Japan-USA Symposium on Flexible Automation, ASME, pp. 1-2, 2004.
[40] J. Steele, C. Mnich, C. Debrunner, T. Vincent, and S. Liu, "Development of closed-loop control of robotic welding processes," Industrial Robot: An International Journal, vol. 32(4), pp. 35-{355, 2005.
[41] D. Y. Choong and J. Lee., "3-d measurement of weld pool using biprism stereo vision sensor," Http://joining1.kaist.ac.kr/research/vision.htm, Seoul National University, 2004.

[42] C. X. Zhao, I. M. Richardson, S. Kenjeres, C. R. Kleijn, and Z. Saldi, "A stereo vision method for tracking particle ow on the weld pool surface," Journal of Applied Physics, vol. 105(12): 123104, 2009.
[43] Q. Y. Du, S. B. Chen, and T. Lin, "Reconstruction of weld pool surface based on shape from shading," Journal of Mechanical Engineering, vol. 19(2), pp. 168-171,2006.
[44] J. F. Wang, W. Y. Wang, and S. B. Chen, "Extraction of welding pool shape using linear approximation." Transactions of the China Welding Institution, vol. 28(8), pp. 54-56, 2007.
[45] J. Wang, W. Wang, and S. Chen, "Inspection of welding pool height from shading in pulsed gtaw with wire _ller," Industrial Robot: An International Journal, vol. 36 (3), pp. 270-276, 2009.
[46] D. Zhao, J. Q. Yi, S. Chen, L. Wu, and Q. Chen, "Extraction of three-dimensional parameters for weld pool surface in pulsed gtaw with wire filler," Journal of Manufacturing Science and Engineering, vol. 125, pp. 493-503, 2003.
[47] L. P. Li, X. Q. Yang, F. Y. Zhang, and T. Lin, "Research on surface recover of aluminum alloy p-gtaw pool based on sfs," Robotic Welding, Intelligence and Automation, Lecture Notes in Electrical Engineering., vol. 88, pp. 307-314, 2011.
[48] R. Kovacevic and Y. M. Zhang, "Sensing free surface of arc weld pool using specular reflection: principle and analysis," Proceedings of the Institution of Mechanical Engineers, Part B, Journal of Engineering Manufacturing, vol. 210(6), pp. 553-564, 1996.
[49] G. Saeed and Y. M. Zhang, "Mathematical formulation and simulation of specular reection based measurement system for gas tungsten arc weld pool surface," Measurement Science and Technology, vol. 14(8), pp. 1671-1682, 2003.
[50] X. P. Ai, N. S. Liu, Y. Q. Wei, X. Hu, and X. R. L. S. Wei, "Study on image acquisition in 3-d sensor system of arc welding pool surface shape using grating projection," Proceedings of SPIE, vol. 7506-750628, 2009.
[51] Y. Q. Wei, N. S. Liu, X. Hu, and X. Ai, "Phase-correction algorithm of deformed grating images in the depth measurement of weld pool surface in gas tungsten arc welding," Optical Engineering, vol. 50:5, 2011.
[52] G. Saeed, M. J. Lou, and Y. M. Zhang, "Computation of 3d weld pool surface from the slope field and point tracking of laser beams," Measurement Science and Technology, vol. 15(2), pp. 389-403, 2004.
[53] H. S. Song and Y. M. Zhang, "Image processing for measurement of three-dimensional gta weld pool surface," Welding Journal, vol. 86(10), pp. 323s-330s, 2007.
[54] X. J. Ma and Y. M. Zhang., "Gas metal arc weld pool surface imaging: Modeling and processing," Welding Journal, vol. 90(5), pp. 85s-94s, 2011.
[55] S. Mallat, A Wavelet Tour of Signal Processing. Academic Press, 1999.
[56] D. L. Donoho and I. M. Johnstone, "Ideal spatial adaptation by wavelet shrinkage," Biometrika, vol. 81, pp. 425-455, 1994.

[57] I. Daubechies, Ten lectures on Wavelets. Philadelphia: Society for Industrial and applied Mathematics, 1992.

[58] K. Fukunage, Introduction to Statistical Pattern Recognition. New York Academic, 1972.

[59] Y.M. Zhang and L. Li., "Dynamic estimation of full penetration using geometry of adjacent weld pools," Journal of Manufacturing Science and Engineering, vol. 19, pp. 631-644, 1997.

[60] C. B. Barber, D. D P, and etc, "The quick hull algorithm for convex hulls," ACM Transaction of Mathematic Software, vol. 22, pp. 469-483, 1996.

Section 4

General Topics in Welding

Chapter 15

The Physics of Weld Bead Defects

P. S. Wei

Additional information is available at the end of the chapter

1. Introduction

Productivity in different welding techniques can be improved by increasing welding speed and current. This strategy, however, is limited by the appearance of surface defects such as rippling, humping, undercutting, etc [1]. Weld ripples exhibit rather regular, arc-shaped topographic features on a solidified surface, for example, as shown in Figure 1 in EBW of Al 6061 [2]. The ripples slightly elevate above the surface. Figure 2 shows a rippling structure on silicon surface irradiated by a p-polarized laser beam, provided by Pedraza et al. [3]. Notice tiny little "fingers" in lower rim of fringes and asymmetry in fringe profile taken in downward direction.

Figure 1. Rippling in welding Al 6061 [2]

On the other hand, more complicated humping shows an irregular and unpredictable surface contour consisting of a series of swelled protuberance, as can be seen in Figures 3(a)-(d). Morphologies of humped welds are quite complicated, which were roughly categorized into the gouging region and beaded cylinder morphologies [4]. Typical gouging region morphology defects in GTAW at high currents and high travel speeds are shown in Figure 3(a) [5]. The front of the weld pool exhibits a large depression known as the gouging region. Open, unfilled dry spots in between the humped beads can also be seen. In some cases, two small channels appear at the walls of the gouging region. The weld bead having parallel grooves at the side is the undercutting defect.

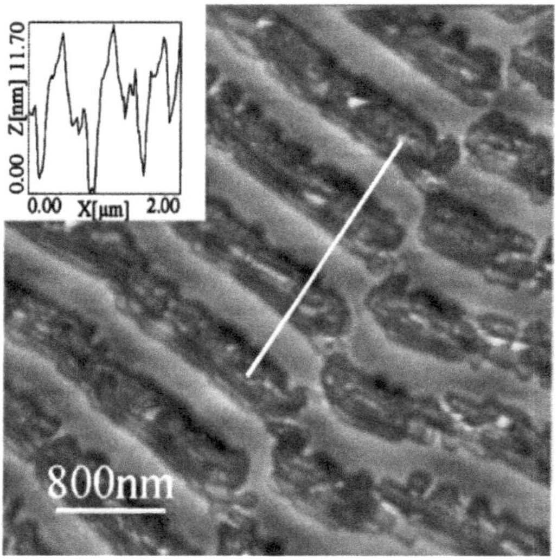

Figure 2. Rippled structure irradiated by 50 pulses at a laser fluence of 0.7 J/cm^2, using a p-polarized beam and incident angle of 38.5°. Notice "fingers" in lower rim of fringes [3].

Figure 3(b) shows a parallel humping or a split bead, separated by an empty channel [5]. Figure 3(c) shows beaded cylinder morphology defects, which are quite different from the gouging region morphology [6]. The beaded cylinder morphology includes beadlike protuberances that sit above the workpiece surface and are connected by a narrow central channel. In some cases of disconnected protuberances, traces of a central channel can still be seen. It is interesting to find that the gouged region and bead cylinder morphology are inverse phenomenon. Different morphologies therefore can be revealed by simply interchanging the liquid and gas phases. Figure 3(d) shows humping in EBW of Al 5083 [2]. Figure 4 also shows the a "star-like" or finger structure develops as well as the number of rays increases with laser power in etching of regular holes in Mo films with an Ar^+ laser beam immersed in Cl_2 atmosphere [7]. Surface patterns of weld beads therefore are very complicated.

Even though formation of the defects have been extensively proposed in the past, more systematical understanding of pattern formations is still limited. The aims of this work are to provide rigorous and pictorial interpretation based on thermal science concepts, and clarify and propose some physics involved in the mechanisms of weld bead formation. Clear physical concepts associated with quantitative scale analysis are important and beneficial for predicting, controlling and avoiding the occurrence of surface defects [1].

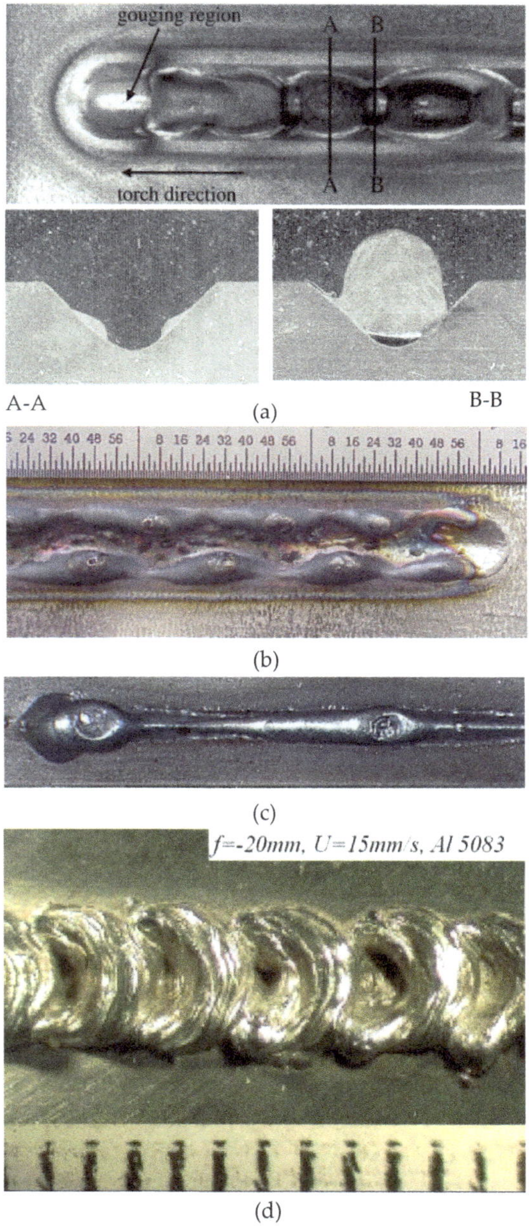

Figure 3. (a) Humps with gouged region and (b) parallel hump in GTAW [5], (c) cylinder beads in GMAW [6], and (d) humps in EBW [2].

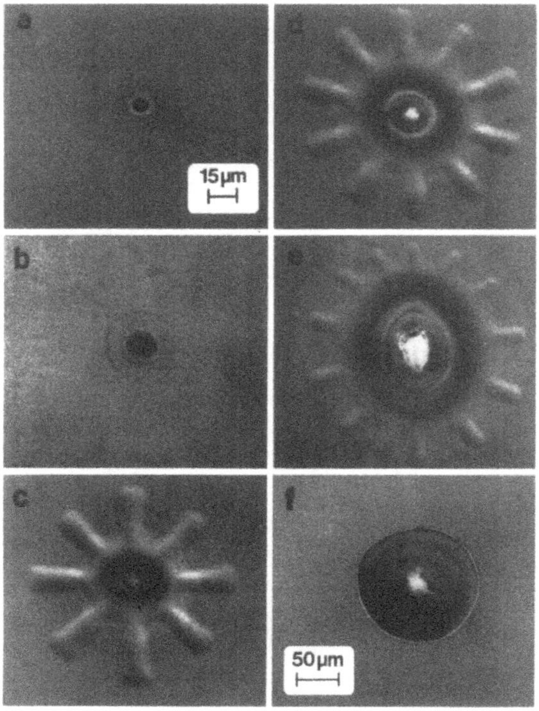

Figure 4. A "star-like hole" in etching Mo films with an Ar^{+}-laser in Cl_2 atmosphere with powers (a) 10 mW, (b) 20 mW, (c) 50 mW, (d) 100 mW, (e) 500 mW, and (f) 150 mW [7].

2. Mechanisms of surface patterns

Different mechanisms of rippling have been proposed and summarized in Table 1.

Solidification rate fluctuations	Cheever and Howden [8], D'annessa [9]
Power source effects	Garland and Davies [10], Ecer et al. [11]
Thermocapillary instability	Fujimura et al. [12]
Kelvin-Helmholtz instability	Ang et al. [13]
Rayleigh-Taylor instability	Bennett et al. [14], Lugomer [15]
Instability due to evaporation	Emel'yanov et al. [16]
Morphological instability	Weizman et al. [17], Style and Wettlaufer [18]
Thermocapillary edge flow	Anthony and Cline [19], Wei et al. [20, 21]
Laser interactions	Birnbaum [22], Siegman and Fauchet [23]

Table 1. Mechanisms proposed for rippling

On the other hand, different mechanisms responsible for humping are presented in Table 2.

Rayleigh's capillary instability	Bradstreet [24], Gratzke et al. [25], Albright and Chiang [26]
Kelvin-Helmholtz instability	Kumar and DebRoy [27], Tytkin et al. [28]
Hydraulic jump	Shimada and Hoshinouchi [29]
Thermocapillary edge flow	Wei [1]

Table 2. Mechanisms proposed for humping

Kelvin-Helmholtz instability	Dynamic pressure difference
Rayleigh-Taylor instability	Density difference
Rayleigh capillary instability	Capillary pressure difference
Morphological instability	Solute supersaturation
Thermocapillary instability	Surface tension gradient
Evaporation instability	Evaporation pressure difference
Hydraulic jump	Hydraulic pressure difference
Laser interaction	Polarizations
Gravitational-electromagnetic instability	Interactions between gravitational and electromagnetic forces

Table 3. Mechanisms of instabilities

Surface roughness including rippling, gouging, humping, fingers, etc., therefore can be affected by mechanisms shown in Table 3. Understanding their physical meanings is described as follows.

3. Thermal science analysis of surface patterns

Rippling or humping is determined by the formation of capillary wave on the free surface. Under surface tension, the pressure differences created at a curve interface support deformation of the interface, as sketched in figure 5.

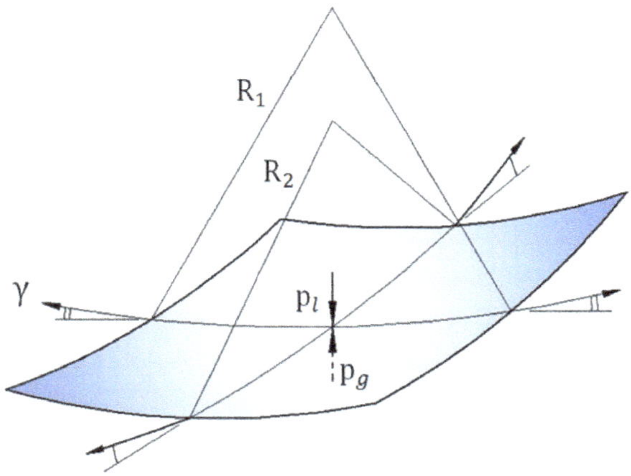

Figure 5. Normal pressure balance on an interface (Young-Laplace equation)

Mathematically speaking, it is governed by Young-Laplace equation

$$p_\ell - p_g = \gamma\left(\frac{1}{R_1} + \frac{1}{R_2}\right) \tag{1}$$

where p_ℓ, p_g are liquid and gas pressures, γ, the surface tension, R_1 and R_2 two principal radii of a surface, respectively. Equation (1) can be simply derived from a normal stress balance on an interface [30]. Any physical or chemical variables affecting pressures at the interface are responsible for different surface roughness patterns. The onset and mechanisms of instability are determined from perturbed deformation governed by equation (1) by substituting perturbed liquid and gas pressures. In this work, free surface instability can also be relevantly revealed from the concept of mass conservation, as shown in figures 6(a) and (b). Provided that velocity profiles maintain the same, the free surface is flat. However, the free surface deforms downward, if the mass of the outflow is greater than that of the incoming flow. Since surface deformation is enhanced, the free surface is suffered from instability.

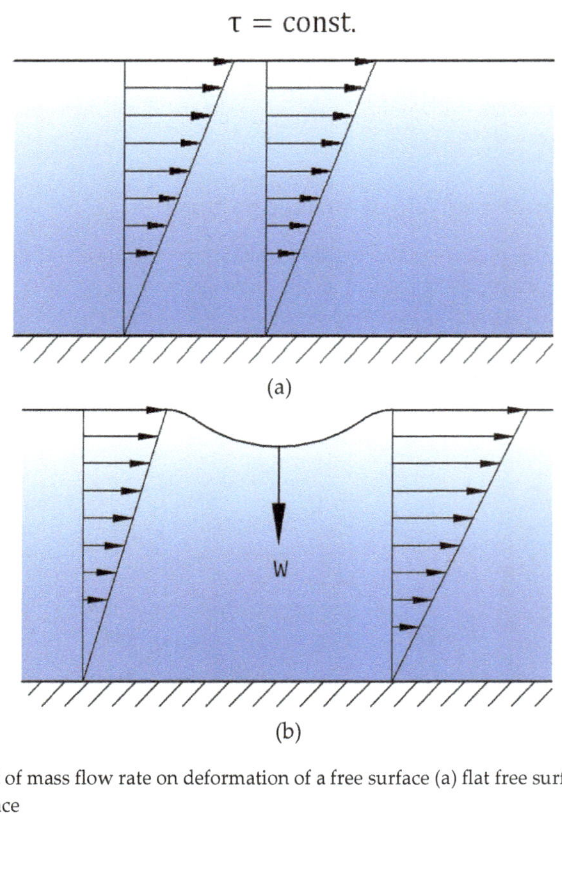

Figure 6. The effect of mass flow rate on deformation of a free surface (a) flat free surface, and (b) deformed free surface

The factors affecting surface patterns listed in Table 3 are rigorously described as follows:

4. Kelvin-Helmholtz instability

KH instability arises due to difference in velocities between gas and liquid. KH instability is the simplest and widely encountered instability, derived from Young-Laplace equation, where liquid and gas pressures are determined from conservation of mechanical energy (namely, the Bernoulli's equation) with specified constant liquid and gas velocities $U_{\ell 0}$ and U_{g0} far away the location considered. Bernoulli's equation indicates that the lower the velocity is, the higher the pressure is. Provided that deformation is toward the gas, the decrease in perturbed velocity results in an increase in perturbed pressure in the liquid, as illustrated in Figure 7. Opposite phenomenon occurs in gas phase. This results in a further increase in deformation under the action of surface tension. Wavelength for surface deformation can be scaled from equation (1) [1] and given by

$$\lambda_{KH} \sim \frac{\gamma}{\rho_\ell (U_{\ell 0} - U_{g0})^2} \tag{2}$$

Wavelength of surface deformation due to KH instability therefore reduces if the difference in velocities between gas and liquid increases. Provided that gas velocity is 10 m/s, the length for roughness in liquid metals is around 10 μm within a relevant range of rippling spacing.

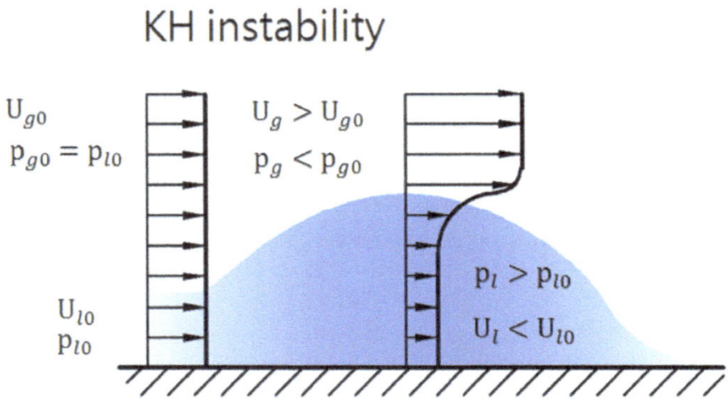

Figure 7. Kelvin-Helmholtz instability

5. Rayleigh-Taylor instability

RT instability occurs when a heavier liquid overlies a lighter liquid. The pressure involved is hydrostatic pressure, which is a function of gravitational acceleration and the depth location considered. As illustrated in figure 8, the difference in hydrostatic pressures of liquid and gas across a surface with a deformation η simply yields

$$p_\ell - p_g = -(\rho_\ell - \rho_g) g \eta \tag{3}$$

Scaling Young-Laplace equation (1) by substituting equation (3) leads to RT instability with the critical wavelength

$$\lambda_{RT} = \sqrt{\frac{\gamma}{g(\rho_\ell - \rho_g)}} \tag{4}$$

Provided that deformation is toward the lighter fluid, a positive perturbed pressure deviated from the base state results. A negative perturbed pressure simultaneously occurs with deformation toward the heavier liquid. Deformation further increases. At later times, initial perturbations grow into spikes of heavier fluid "falling" into lighter fluid and bubble of the lighter fluid "rising" into the heavier fluid. Hence, RT instability has also been stated to occur when the pressure and density gradients are in opposite directions, or a lighter fluid pushes or accelerates a heavier fluid. Equation (4) indicates that an increase in surface tension or decrease in difference in densities across a free surface increases wavelength of ripples.

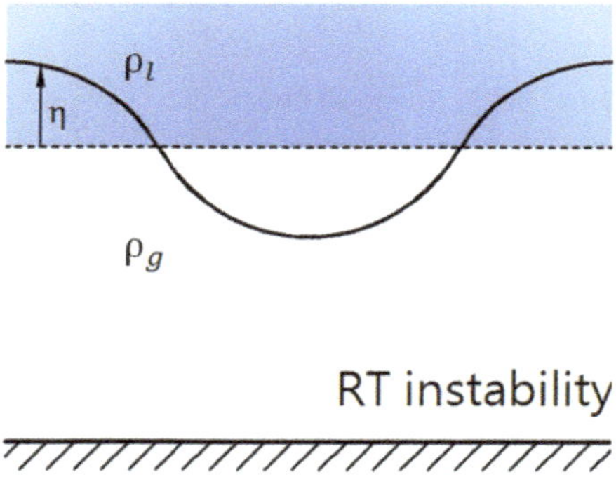

Figure 8. Rayleigh-Taylor instability

Acceleration in equation (4) may not be the earth's gravity. RT instability can also occur at an interface separating fluids through which a blast wave has been transmitted from a heavier to a lighter fluid. This instability is Richtmyer-Meshkov instability, often called impulsive or shock-induced RT instability. In high intensity beam welding, the produced shock waves propagate with discontinuities of densities, pressures and velocities in different magnitudes and directions across the free surface. Provided that the interface subject to an oblique shock, it will give rise to complicated instabilities or patterns at the interface. The normal component of the shock generates RM instability, and the parallel component generates KH instability. If a normal acceleration is also present, RT instability occurs.

6. Rayleigh's capillary instability

Rayleigh's capillary instability is a crucial factor to understand a bulged or gouging region of surface roughness. The gouged region exhibits an inverse feature of the bulged region.

Rayleigh's capillary instability can be revealed from figure 9. Radius r_B is at a location B near the minimum radius r_A. Provided that wavelength of surface deformation is long, curvature $1/R_2 >> 1/R_1$ at location A, as illustrated in figure 9(a). Since liquid pressure is primarily balanced by capillary pressure due to curvature $1/R_2$, a greater liquid pressure is induced by a smaller radius of the cylinder. The positive difference in the pressures thus induces a perturbed flow from locations A to B. The system therefore is unstable and breakups the cylinder into droplets. However, figure 9(b) shows that for deformation with small wavelength, curvature $1/R_1 >> 1/R_2$ at location A. The perturbed pressures at location A therefore decrease and it can be smaller than perturbed pressure at location B, leading to the flow from points B to A, and stabilizing the system.

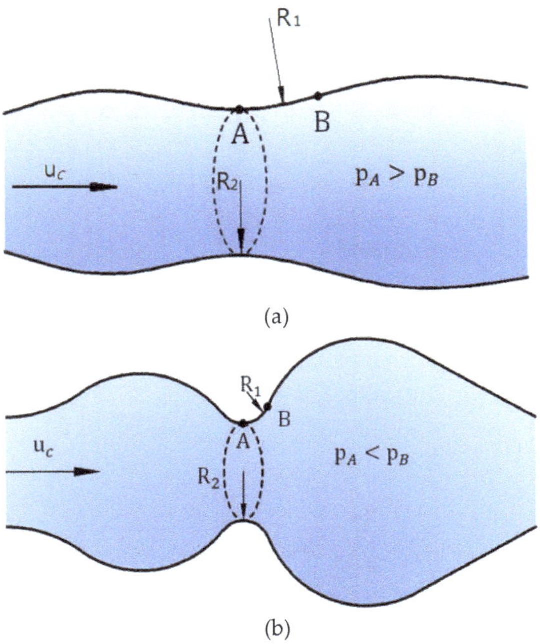

Figure 9. Cylinder is unstable for (a) $1/R_2 >> 1/R_1$, and stable for (b) $1/R_1 >> 1/R_2$ at location A

Since the difference in the perturbed pressures at two locations affects instability, the minimum wavelength for onset of instability is the balance between two components of capillary pressure. The critical wavelength thus is of the same order of the radius of the liquid cylinder [31]. Stability of a bead should depend on the boundary conditions at its contact lines on the surface. Gau et al. [32] experimentally found that cylindrical segments for water on hydrophilic stripes with the apparent contact angle less than 90° did not break up into droplets, as would be expected. It displayed long-wavelength instability where all excess fluid gathered into a single bulge on a hydrophilic stripe. Speth and Lauga [33] theoretically confirmed the most unstable wavenumber for the instability —the one which was observed in an experimental setting — decreases to zero when the apparent fluid contact angle reached 90° . The creation of bulges in the experiment corresponded with a zero-wavenumber capillary instability.

7. Morphological instability

Morphological instability is a consequence of thermal and metallurgical processes. Surface morphology becomes unstable by decreasing surface tension and increasing constitutional supercooling (namely, $mG_C - G_T > 0$) [34,35], where m is the negative slope of the liquidus line in the phase diagram of a dilute solution, G_C and G_T are negative concentration and positive temperature gradients ahead of the solidification front, respectively. The decreased solute concentration increases liquidus temperature near the solidification front, as illustrated in figures 10(a) and (b), respectively, and revealed from the phase diagram of figure 10(c).

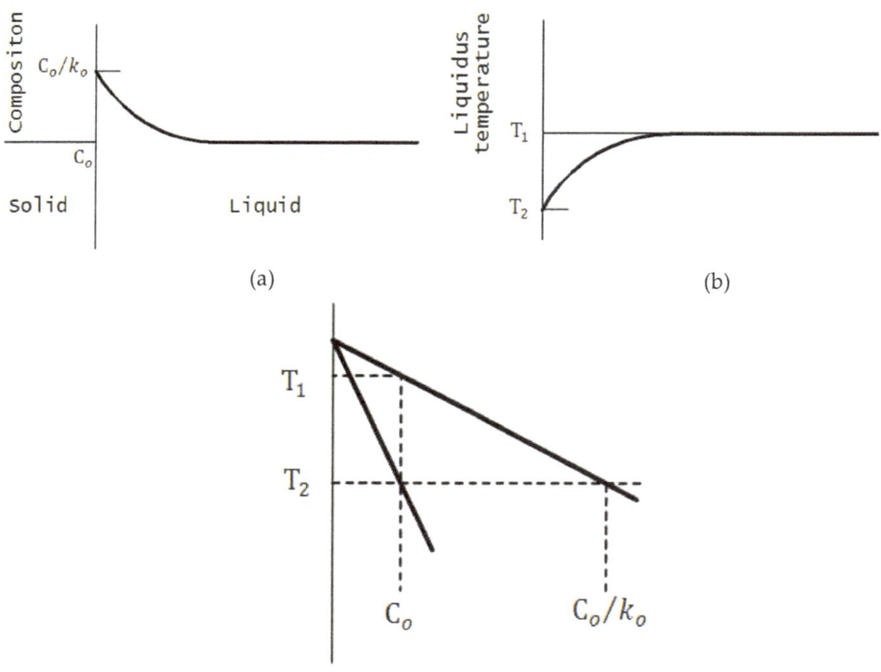

Figure 10. (a) Concentration and (b) liquidus temperature profiles ahead of a solidification front, and (c) the corresponding phase diagram

In view of the rapid drop of solute concentration ahead of the freezing front, the actual temperature may be below the liquidus temperature and result in constitutional supercooling or morphological instability, as shown in figure 11. Mathematically speaking, morphological instability can be simply found by considering interfacial temperature governed by Gibbs-Thomson equation [35]

$$T_s - T_m = \frac{\gamma T_m}{\rho h_{s\ell}} \frac{\partial^2 \eta}{\partial x^2} + mC' \tag{5}$$

where T_S, T_m, $h_{s\ell}$ are, respectively, interfacial temperature, melting temperature and latent heat for solid-liquid transition. The variations of temperature and concentration at the interface subject to a forward deformation can be expressed by their spatial gradients, $T_s - T_m - mC' \sim (G_T - mG_c)\eta_c$. Scaling equation (5) therefore leads to

$$\lambda_M \sim \sqrt{\frac{\gamma T_m}{\rho h_{s\ell}(mG_c - G_T)}} \tag{6}$$

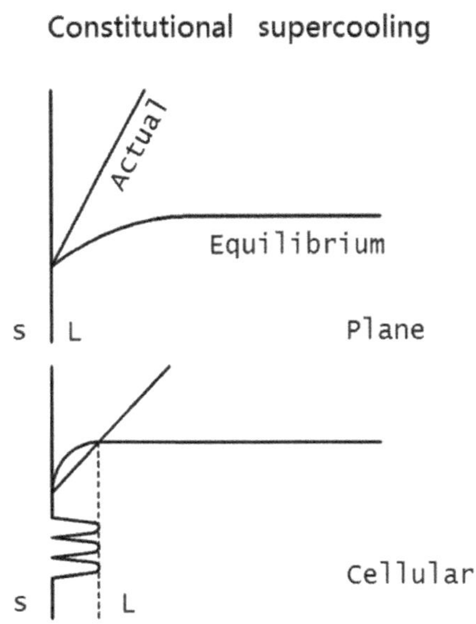

Figure 11. Morphological instability due to constitutional supercooling

8. Instability due to evaporation

It is well-known that a semi-infinite liquid heated below is susceptible to evaporative instability, as illustrated in figure 12(a). As surface deformation is closer to the bottom surface than the flat surface, temperature at the trough is greater than the equilibrium

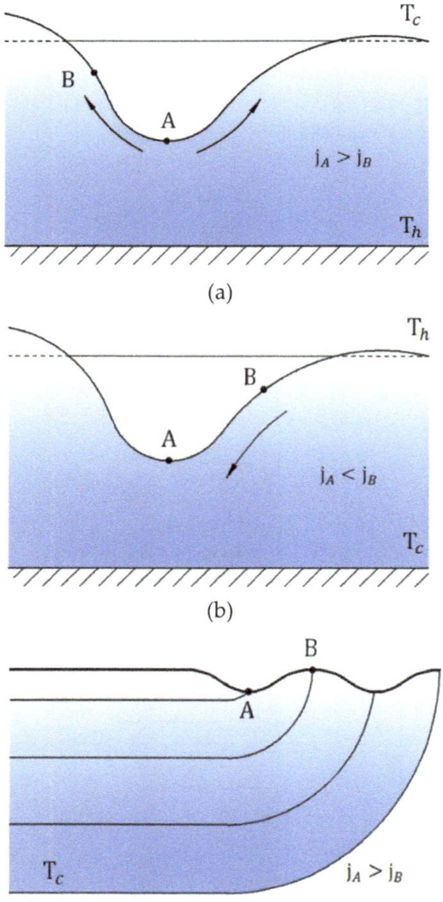

Figure 12. Instability due to evaporation for (a) unstable system subject to heating from below, and (b) stable system and (c) unstable system subject to heating from above

temperature, meaning that higher evaporation takes place at the trough than the base state. On the other hand, a deformation away from the bottom surface leads to lower evaporation rate than that at the base state. Liquid pressure therefore increases from the surface trough to crest. The induced flow from the trough to crest thus enhances surface deformation, leading to evaporative instability (referring to figure 6(b)). However, in welding and manufacturing processes, workpieces are irradiated by incident flux at the top surfaces. A liquid heated from above, which is the case opposite to previous figure 12(a), is stable, as illustrated in figure 12(b). In this work, it is found that evaporation may also induce instability subject to base temperature decreasing in radial directions, as illustrated in figure

12(c) [1]. For a typical welding process the isothermal field is in a roughly spherical shape. Provided that surface deforms toward the bottom, temperature at the crest, which is close to the pool edge, can be lower than that at the trough. The decrease in pressures therefore pushes the liquid from the trough to crest and gives rise to instability.

9. Thermocapillary instability

All pure liquid metals and alloys containing minor surface active solutes such as O, S, Se, Te, et al. have negative surface tension coefficient ($d\gamma / dT < 0$). As illustrated in figure 13, thermocapillary force balanced by viscous shear stress at the interface is given by

$$\tau = \mu \frac{\partial u_t}{\partial n} = \frac{d\gamma}{dT} \frac{\partial T}{\partial s} \tag{7}$$

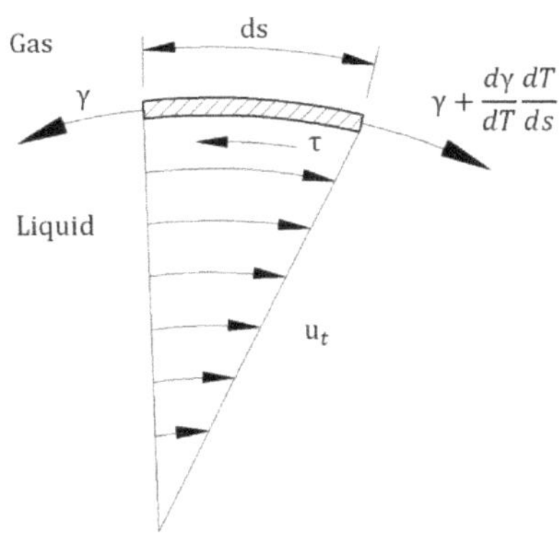

Figure 13. Tangential stresses balance between thermocapillary force and shear stress.

A negative surface tension coefficient induces an outward surface flow, provided that surface temperature decreases in the outward direction. Consider an interface to be displaced toward the hot surface at the bottom, as illustrated in figure 14(a). Temperature at the trough thus is hotter than other points on the deformed surface. This results in an outward lateral flow from the trough to crest along the interface. To conserve mass, the perturbed liquid flows downwards and further deforms the interface with speed w (referring to figure 6(b)). The system thus is unstable. The well-accepted interpretation is incorrect. Rather than perturbed downward flow, it is usually interpreted as that instability results from incessant and amplified upward flow accompanying with enhanced energy at the trough due to increased thermocapillary force from the trough to crest [31].

In this work, it is proposed that a free surface heated from above and a negative surface tension coefficient may still cause instability, as illustrated in figure 14(b) [1]. Provided that the surface is strongly deformed toward the bottom, significant heat conduction is transport from the interior to free surface. Enhanced downward thermocapillary surface flow therefore is required to balance horizontal conduction, and results in further deformation. This work also proposes that thermocapillary instability takes place near the edge of the molten pool, where $d\gamma / dT > 0$ in the presence of surface active solutes in the case similar to the mechanism (see figure 14(a)) provided by Pearson [36], as illustrated in figure 14(c). This is another reason responsible for serious roughness encountered in alloys having surface active solutes [21].

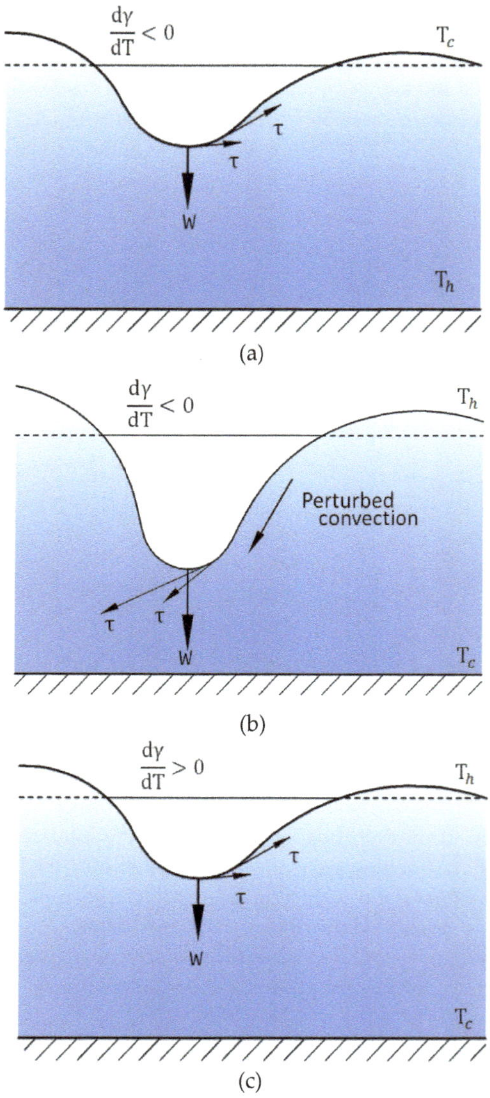

Figure 14. Thermocapillary instabilities: (a) heated from below and (b) heated from above with negative surface tension coefficient, and (c) heated from above with positive surface tension coefficient.

10. Thermocapillary edge flow

Deformation of the free surface near the solidification front is responsible for rippling or humping. As illustrated in figure 15, an increase in liquid pressure due to a decreased surface flow from the central to rear edge of the pool give rise to deformation of the free surface near the solidification front. Subtracting Young-Laplace equation at two locations which are away and near the edge of the pool surface, and introducing the pressure difference between two locations from Bernoulli's equation,, the amplitude of ripples can be found to be [1,20]

$$\frac{a\gamma}{\rho\alpha^2} = (1 - K_{loss})(\frac{u_c r_m}{\alpha})^2 \tag{8}$$

where the loss coefficient K_{loss} is introduced to account for the energy loss near the pool edge, a, α and r_m are, respectively, roughness amplitude, thermal diffusivity and distance of the rear edge measured from the pool center. Surface speed can be scaled and related to thermocapillary force [37,38]. Based on a typical surface velocity of 1 m/s, pool width of 1 mm and loss coefficient of 0.99, rippling amplitude predicted from equation (8) is around a reasonable value of 10 μm .

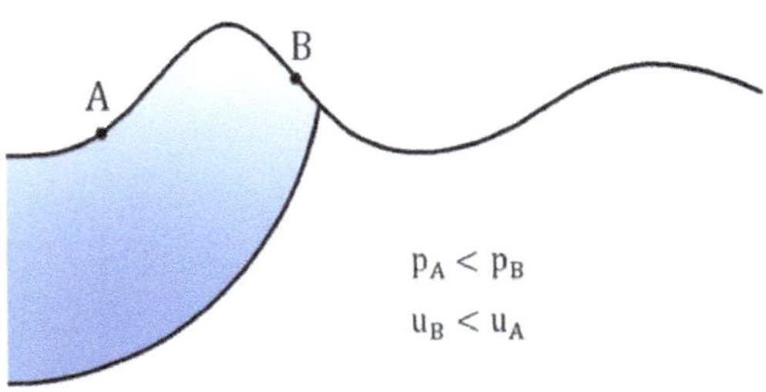

Figure 15. Thermocapillary edge flow

11. Hydraulic jump

Studying hydraulic jump usually assumes liquid pressure to be hydrostatic pressure. Hydraulic jump occurs if the pressure gradient becomes increasingly adverse as the flow proceeds downstream. As sketched in figure 16(a), an increased liquid height ($h_1 > h_0$) increases hydrostatic force (= pressure x height) against the downward flow, and decreases downward velocity to satisfy conservation of momentum and mass. However, liquid height at the downstream location can also be less than that at the upstream location, as illustrated in figure 16(b). Hydrostatic force decreases whereas velocity increases. Provided that the Froude number, $U_0^2 / gh_0 > 1$, hydraulic jump occurs because of $h_1 / h_0 > 1$ [30]. As Froude number $U_0^2 / gh_0 < 1$, the height ratio $h_1 / h_0 < 1$. Surface tension can also play a role in hydraulic jump [39], as observed the occurrence of polygonal patterns from breaking axial symmetry of a circular hydraulic jump.

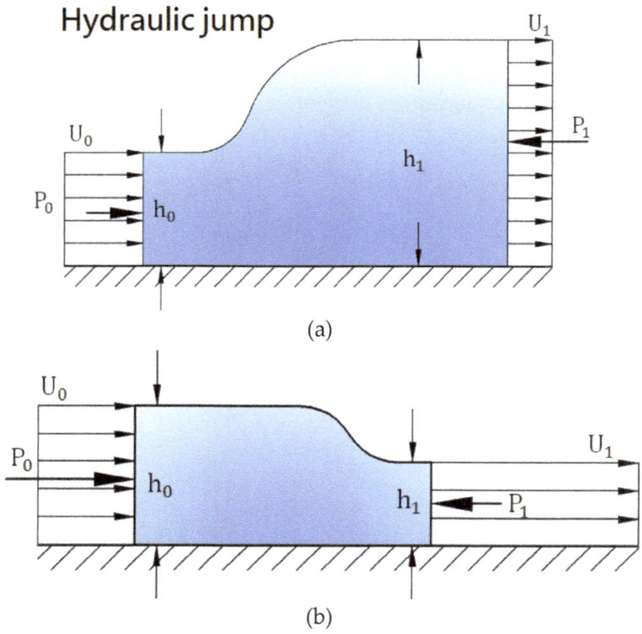

Figure 16. Hydraulic jumps with (a) increase and (b) decrease of height at downstream location.

12. Fingers

Fingers can be accompanied with hydraulic jump (see figure 4). This is because pressure and density gradients are in opposite directions in the course of splashing, leading to RT instability. Allen [40] was the first to propose that the splashing of a droplet impact on a surface is an example of RT instability, caused by a rapidly decelerating interface. Replacing gravitational acceleration by u_c / t_c in equation (4) and substituting $r_c = n\lambda_{RT}$ from geometrical consideration lead to

$$r_c \sim n\sqrt{\frac{\gamma t_c}{u_c \rho_\ell}} \tag{9}$$

Provided that time is short, equation (9) indicates that radius of the ring-shaped spread is less than wavelength of fingers. The ring-shaped spread thus is free from fingers. On the other hand, larger time results in the ring-shaped spread covered with n fingers.

13. Gravitational-electromagnetic instability

In plasma, Rayleigh-Taylor instability can occur because the magnetic field acts as a light fluid supporting a heavy fluid (plasma), [41]. This is because the ion drift velocity $\mathbf{U_0}$ is in the direction of $\mathbf{g} \times \mathbf{B_0}$. If a ripple in plasma develops on the interface as the result of random thermal fluctuations, the drift velocity $\mathbf{U_0}$ will cause the ripple to grow, as sketched in figure 17. Gravitational acceleration $\mathbf{g}$ is in the downward direction. The drift of ions causes a charge to build up on the sides of the ripple, and an electric field develops which changes sign as one goes from crest to trough in the perturbation. The drift $\mathbf{E}' \times \mathbf{B_0}$ is thus always upward in those regions where the surface has moved upward, and downward where it has moved downward. The ripple therefore grows as a result of these properly phased $\mathbf{E}' \times \mathbf{B_0}$ drifts.

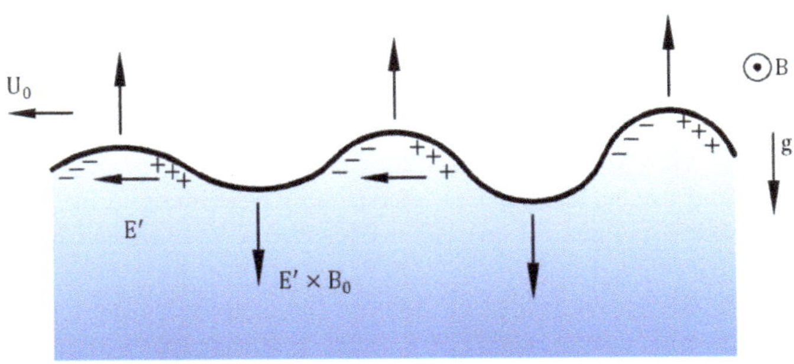

Figure 17. Gravitational-electromagnetic instability

14. Conclusions

Physical interpretation of bead defects is important and beneficial for controlling quality of welding joint. It involves inter-discipline among different sciences of thermal physics, aerodynamics, electromagnetism, optics and metallurgy, morphology, pattern selection, instabilities, and contact line dynamics. Phase transitions between liquid and gas, and solid and liquid are also included. Spaces and amplitudes of rippling and humping can be effectively revealed from scaling of a force balance between perturbed liquid and gas pressures and surface tension. Any factor which can induce pressure differences or influence surface tension is responsible for specific surface patterns. This study provides a general, relevant and rigorous interpretation of physical mechanisms involved in surface roughness.

Author details

P. S. Wei

Department of Mechanical and Electro-Mechanical Engineering,
National Sun Yat-Sen University, Kaohsiung, Taiwan, China

Acknowledgement

The author is grateful for Mr. Sheng-You Tsai drawing the pictures

15. References

[1] Wei, P. S., 2011. Thermal science of weld bead defects- a review, *ASME Journal of Heat Transfer*, 133: 031005

[2] Wei, P. S., Chuang, K.C., Ku, J. S., and DebRoy, T., 2012, Mechanisms of spiking and humping in keyhole welding. *IEEE Transactions on Components, Packaging and Manufacturing Technology* 2: 383-394.

[3] Pedraza, A. J., Guan, Y. F., Fowlkes, J. D., and Smith, D. A., 2004. Nanostructures produced by ultraviolet laser irradiation of silicon. I. rippled structures. *Journal of Vacuum Science and Technology* B 22: 2823-2835.

[4] Soderstrom, E., and Mendez, P. 2006. Humping mechanisms present in high speed welding. *Science and Technology of Welding and Joining* 11: 572-579.

[5] Mendez, P. F., and Eagar, T. W., 2003. Penetration and defect formation in high-current arc welding. *Welding Journal* 82: 296-s to 306-s.

[6] Cho, M. H., and Farson, D. F., 2007. Understanding bead hump formation in gas metal arc welding using a numerical simulation. *Metallurgical and Materials Transactions B* 38: 305-319.

[7] Bäuerle, D., 1996, *Laser Processing and Chemistry*, Springer-Verlag, New York.

[8] Cheever, D. L., and Howden, D. G., 1969. Technical note: nature of weld surface ripples. *Welding Journal* 48: 179-s to 180-s.

[9] D'annessa, A. T., 1970. Sources and effects of growth rate fluctuations during weld metal solidification. *Welding Journal* 49: 41-s to 45-s.

[10] Garland, J. G., and Davies, G. J., 1970. Surface rippling and growth perturbations during weld pool solidification. *Metal Construction and British Welding Journal* 2: 171-175.

[11] Ecer, G. M., Tzavaras, A., Gokhale, A., and Brody, H. D., 1982. Weld pool fluid motion and ripple Formation in pulsed-current GTAW. *Trends in Welding Research in the United States*, (editor David, S. A.), Proceedings of a conference sponsored by the Joining Division of American Society for Metals, New Orleans, Nov. 16-18, 1981, pp.419-442.

[12] Fujimura, K., Ogawa, M., and Seki, M., 1992. Possible mechanism of the roughness formation on a liquid layer caused by a high heat flux. *Fusion Engineering and Design* 19: 183-191.

[13] Ang, L. K., Lau, Y. Y., Gilgenbach, R. M., Spindler, H. L., Lash, J. S., and Kovaleski, S. D., 1998. Surface instability of multipulse laser ablation on a metallic target. *Journal of Applied Physics* 83: 4466-4471.

[14] Bennett, T. D., Grigoropoulos, C. P., and Krajnovich, D. J., 1995. Near-threshold laser sputtering of gold. *Journal of Applied Physics* 77: 849-864.

[15] Lugomer, S. 2007. Micro-fluid dynamics via laser-matter interaction: vortex filament structures, helical instability, reconnection, merging, and undulation. *Physics Letters A* 361: 87-97.

[16] Emel'yanov, V. I., Konov, V. I., Tokarev, V. N., and Seminogov, V. N., 1989. Formation of periodic surface ripples under the action of pulsed carbon dioxide laser radiation on fused silica. *Journal of Optical Society of America* B 6: 104-114.
[17] Weizman, M., Nickel, N. H., Sieber, I, and Yan, B., 2008. Laser-induced self-organization in silicon-germanium thin films. *Journal of Applied Physics* 103: 093536.
[18] Style, R. W., and Wettlaufer, J. S., 2007. Evaporatively driven morphological instability. *Physical Review E* 76: 011602.
[19] Anthony, T. R., and Cline, H. E., 1977. Surface rippling induced by surface-tension gradients during laser surface melting and alloying. *Journal of Applied Physics* 48: 3888-3894.
[20] Wei, P. S., Chang, C. Y., and Chen, C. T., 1996.Surface ripple in electron-beam welding solidification. *ASME Journal of Heat Transfer* 118: 960-969.
[21] Wei, P. S., Chen, Y. H., Ku, J. S., and Ho, C. Y., 2003. Active solute effects on surface ripples in electron-beam welding solidification. *Metallurgical and Materials Transactions B* 34: 421-432.
[22] Birnbaum, M., 1965, Semiconductor Surface Damage Produced by Ruby Lasers, *Journal of Applied Physics*, 36: 3688-3689.
[23] Siegman, A. E., and Fauchet, P. M., 1986. Stimulated Wood's anomalies on laser-illuminated surfaces. *IEEE Journal of Quantum Electronics* QE-22: 1384-1403.
[24] Bradstreet, B. J., 1968. Effect of surface tension and metal flow on weld bead formation. *Welding Journal*, 47: 314-s to 322-s.
[25] Gratzke, U., Kapadia, P. D., Dowden, J., Kroos, J., and Simon, G., 1992. Theoretical approach to the humping phenomenon in welding processes. *Journal of Physics D: Applied Physics* 25: 1640-1647.
[26] Albright, C. E., and Chiang, S., 1988. High-speed laser welding discontinuities. *Journal of Laser Applications* 1: 18-24.
[27] Kumar, A., and DebRoy, T., 2006. Toward a unified model to prevent humping defects in gas tungsten arc welding. *Welding Journal* 85: 292-s to 304-s.
[28] Tytkin, Yu. M., Ryazantsev, O. V., and Chuvilo, V. N., 1981. The mechanism of formation of a coarse flaky surface of weld metal in welding under high-power conditions. *Svar. Proiz.* 2: 4-5 (pp.3-5).
[29] Shimada, W., and Hoshinouchi, S., 1982. A study on bead formation by low pressure TIG arc and prevention of under-cut bead. *Transactions of the Japan Welding Society* 51: 280-286.
[30] Li, W. H., and Lam, S. H., 1964, *Principles of Fluid Mechanics*, Addison-Wesley, Reading, Mass., 1964.
[31] Miller, C. A., and Neogi, P., 1985, *Interfacial Phenomena, Equilibrium and Dynamic Effects*, Marcel Dekker, New York.
[32] Gau, H., Herminghaus, S., Lenz, P., and Lipowsky, R., 1999. Liquid morphologies on structured surfaces: from microchannels to microchips. *Science* 283: 46-49.
[33] Speth, R. L., and Lauga, E., 2009. Capillary instability on a hydrophilic stripe. *New Journal of Physics* 11: 075024.

[34] Rutter, J. W., and Chalmers, B., 1953. A prismatic substructure formed during solidification of metals. *Canadian Journal of Physics* 31: 15-39.

[35] Mullins, W. W., and Sekerka, R. F., 1964. Stability of a planar interface during solidification of a dilute binary alloy. *Journal of Applied Physics* 35: 444-451.

[36] Pearson, J. R. A., 1958. On convection cells induced by surface tension. *Journal of Fluid Mechanics* 4: 489-500.

[37] Ostrach, S., 1982, Low-Gravity Fluid Flows, *Annual Review of Fluid Mechanics* 14: 313-345.

[38] Wei, P. S., Liu, H. J., and Lin, C. L., 2012, Scaling weld or melt pool shape induced by thermocapillary convection, *International Journal of Heat and Mass Transfer* 55: 2328-2337.

[39] Ellegaard, C., Hansen, A. E., Haaning, A., Hansen, K., Marcussen, A., Bohr, T., Hansen, J. L., and Watanabe, S., 1998. Creating corners in kitchen sinks. *Nature* 392: 767-768.

[40] Allen, R. F., 1975. The role of surface tension in splashing. *Journal of Colloid and Interface Science* 51: 350-351.

[41] Chen, F. F., 1974, *Introduction to Plasma Physics,* Plenum Press, New York

Chapter 16

Welding Techniques in Dentistry

Clébio Domingues da Silveira-Júnior, Morgana Guilherme de Castro,
Letícia Resende Davi, Flávio Domingues das Neves,
Verediana Resende Novais and Paulo Cézar Simamoto-Júnior

Additional information is available at the end of the chapter

1. Introduction

Welding involves a metallurgical union process that relies on base metal fusion, i.e. the constituent metal of the structure, with or without filler metal, to form the soldered joint. Thus, a definition of welding proposed by the AWS is an "operation which aims to get localized coalescence produced by heating to a suitable temperature, with or without applying pressure and filler metal, producing parts with strong welded union, nonporous and free of corrosion. According to some authors, when the welding conditions are suitable, the mechanical strength of a welded union is equal to or greater than the base metal intact.

Currently, over 50 different welding techniques have some industrial use and are the most important methods for permanent union of metals. This importance is further evidenced by the presence of welding processes in different areas, such as: aerospace, aviation, automotive petrochemical, nuclear, and medicine.

In dentistry, welding between abutment elements, during construction of the metal framework or even after ceramics application, has been used by the vast majority of dentists to solve problems related to laboratory distortions that are reflected in prostheses' misfit in the marginal area. The welding technique has the advantage of working with segments of the prosthesis to generate prosthetic framework with lower distortion, which enables better fit to the abutments. It promotes uniform stress distribution and minimizes trauma and failures in the bone, implant, and prostheses.

In addition, the recent and promising use of titanium (Ti) alloys in dentistry has been driven mainly by the desire to produce structures of low weight and high resistance to chewing efforts, since titanium presents favorable biological and mechanical properties, including: low density, excellent biocompatibility, good resistance/weight ratio, and low modulus of elasticity.

However, its high melting temperature, which is close to 1,700°C, and high reactivity with the nitrogen, hydrogen, and oxygen from air, when subjected to high temperatures, end up making it more fragile. Titanium, therefore, requires argon gas for inert gas protection during its manufacturing process.

Thus, the achievement of welded joints in dentistry has been enhanced by the development and incorporation of knowledge of other areas, such as engineering; this allowed the emergence of new techniques and equipment to improve welding, such as Laser Welding, Tungsten Inert Gas (TIG) Welding, and Plasma Arc Welding (PAW). These new techniques are alternatives to conventional brazing, which use a gas torch.

This chapter seeks to present the welding techniques applied to dentistry, as well as scientific studies related to welding techniques.

2. Welding processes in dentistry

Although it is relatively common practice in dentistry to obtain one-piece cast frameworks in order to avoid the soldering step, it is a process that incorporates numerous errors and can lead to an unsatisfying result.

One-piece cast metal frameworks over teeth should be avoided, because their apparent fit is at the expense of tooth movement, developing areas of pressure and traction in the periodontal ligament. The same is not true of the frameworks over implants. Osseointegrated implants are rigidly connected to the surrounding bone, and this connection lacks the inherent resilience of the periodontal ligament. It has been documented that movement of implants within bone is limited to 50–150 μm. In case of a misfit between implant and abutment, or between abutment and prosthesis, it is bone deformation that causes implant movement. Thus, compressive and tensile loads could be directed to the restoration, which could result in a loosening of the prosthesis and abutment screws, fracture of the restoration, bone micro-fractures surrounding the implants, and even fracture of the implant.

The fabrication of a metal framework of multiple elements cast in one piece is no more suitable for dental work, because it is potentially prejudicial to framework fit. Casting separated fragments for subsequent welding is preferred. Barbosa et al. (2010) comparatively analyzed, by SEM, the vertical and horizontal fit between UCLA abutments and implants used in frameworks of five elements that were cast in one piece after laser welding. Three different materials were used: titanium CP (grade 1), Co-Cr alloys, and Ni-Cr-Ti alloys. The passive fit of the frameworks was evaluated by testing the single screw and the stresses generated around the implants, by means of photoelasticity. There was a statistically significant improvement in the frameworks fit for all materials after sectioning and laser welding.

The welding processes are advised, regardless of the extension of the metal structure, on prosthesis over teeth or implants. In dentistry, different welding techniques are used, among which we highlight the brazing, TIG and PAW welding, and laser welding.

2.1. Brazing

2.1.1. Definition

The brazing process, also called oxygas or welding with direct flame, produces a coalescence of metals by heating the parts to be welded with a flame. The process needs another type of alloy, called a solder alloy, which is used to join two or more metal parts, both with or without the same metal, at a temperature greater than 450ºC and less than the melting point of the metal base. To execute this type of welding, an oxygen-propane torch is used with circular movements of the flame over the joint. The parts to be joined are heated until they are red hot, and then the reducing zone of the flame is directed obliquely to the weld area. The investment is left on the bench until it has fully cooled.

2.1.2. Technical description

2.1.2.1. Soldering space

A minimum space of 0.2–0.3 up to 0.5 mm is obtained (Figure 1) for welding. This can be accomplished with aluminum oxide disks or stones or fine diamond burs. To assess whether there is enough space for the weld, a radiographic film or paper card may be placed in the gap. The surface should be finished and polished properly, leaving it clean and without irregularities.

To obtain a uniform solder thickness, it is important to obtain a homogenous space throughout the extension of the area to be welded. Irregular spaces with thick, sharp discrepancies can cause contraction of the solder and result in traction of retainers, which are displaced from their original positions on the investment.

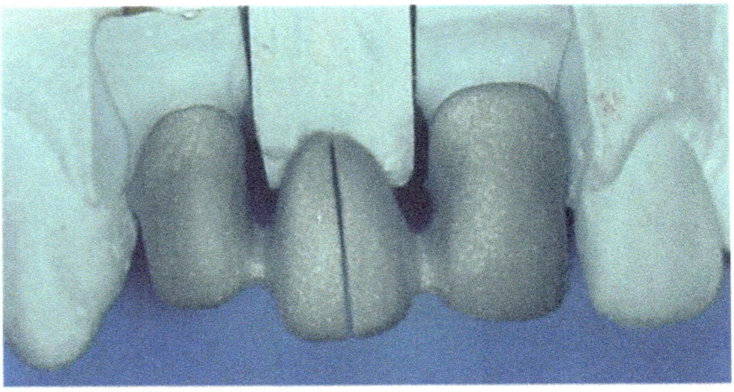

Figure 1. Sectioned framework cast in Ni-Cr alloy with correct space for soldering with the brazing technique.

2.1.2.2. Inclusion and soldering

The various stages below should be followed:

1. Inclusion of frameworks at investment suitable for welding or for casting, forming a block approximately 1.5–2.0 cm in height (Figure 2);

2. Take the investment to the furnace to eliminate moisture and dehydration;
3. Remove the investment from the furnace and wait until it has cooled completely;
4. Clean the areas to be welded with aluminum oxide jet;
5. Apply flux in the joint area and start heating. When heated to red hot, position the welding rod, which is held with clamps, in the area to be welded. The solder melts and flows into the joint under influence of heat and flux (Figure 3);
6. The investment is cooled slowly after the filler metal has completely covered the surface of the joint;
7. Divest and clean with instruments and aluminum oxide jets.

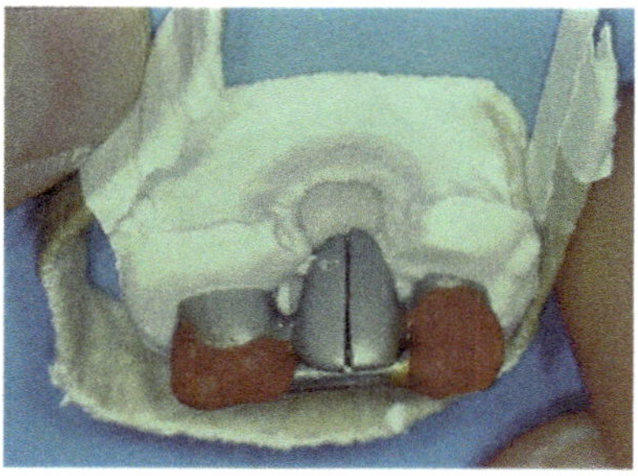

Figure 2. Framework inclusion in soldering investment for high temperatures.

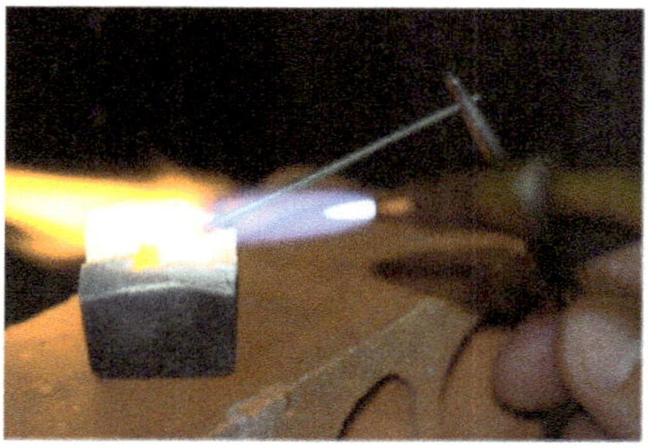

Figure 3. Brazing process using filler metal.

2.1.3. *Advantages*

- It has been used for years, therefore well known;
- Low cost;
- Relative effectiveness.

2.1.4. *Disadvantages*

- Problems such as oxidation of the parts joined by weld;
- Joint porosity and overheating of the union during the welding process can promote small structural defects and failure of the rehabilitation treatment.

2.2. Tungsten Inert Gas (TIG) Welding and Plasma Arc Welding (PAW)

2.2.1. Definition

TIG and PAW welding are techniques in which a union is obtained by heating materials by an arc established between a non-consumable tungsten electrode and the part to be welded (Figure 4).

The electrode and the area to be welded are protected by using an inert gas, usually argon or a mixture of inert gases (argon and helium). The basic equipment consists of a power supply, a torch with a tungsten electrode, a shielding gas source, and an opening system for the arc. The main difference between TIG and plasma welding is the use of a constrictor torch that concentrates the electric arc in plasma welding (Figure 5). Filler metal can be used or not.

Figure 4. Ceramic torch, with tungsten electrode, positioned over the sample.

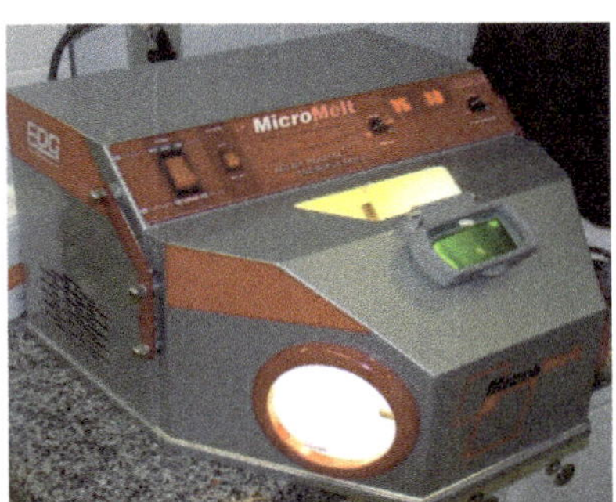

Figure 5. Plasma welding machine for Dentistry.

2.2.2. Technical description

To accomplish this welding technique, it is necessary for the equipment to be regulated for the purpose of welding. The equipment allows for the adjustment of both the pulse and current. After adjusting the machine, screw into one of the claws a structure without use and position the parts to be welded with hands or through specific tables of equipment, which position the parts to be joined. The argon activation is done by a foot pedal, so to start the welding process, press the foot pedal until the argon flows, and then pull off the structure in the electrode without pressing. Maintain steady hands; the buzzer will indicate when contact is made. Quickly release the pedal. The weld will be made, and the flow of argon will continue for a few seconds. It is possible that in the first few attempts the electrode will stick to the piece being necessary to regrind the same.

2.2.3. Advantages

- This allows execution of welds of high quality and excellent finishing, particularly in small joints;
- The thickness of the joint allows for welding in any position, e.g. repairing removable partial prosthesis;
- Excellent control of the weld pool, i.e. the region being welded;
- Expending less time;
- It can be executed directly in the working model;
- The equipment is affordable compared to that of laser welding;
- Allows welding in regions near the resins and porcelains;
- Allows welding with the frameworks in close contact or with minimal space for welding, using filler metal.

2.2.4. Disadvantages

- The electric arc welding processes, such as TIG and plasma welding, are characterized by the imposition of a large amount of heat to achieve fusion of the base and filler material, which causes important microstructure transformations. These transformations occur in a region called the HAZ (Heat Affected Zone), which is a base metal region whose structure or properties were changed by temperature variation during welding. These changes generate a complex region of stresses and deformations, leading to results that are not always desired, including material distortion, residual stresses, generation of fragile microstructures, grain grow, cracks, fissures, and changes in mechanical, physical, and chemical properties, among others;
- Insufficient weld penetration in butt type joints (Figure 6A);
- The presence of porosities in the region of the union (Figure 6B) that is due to the inclusion of argon gas, which is necessary to maintain the inert atmosphere during the welding procedure and thus minimize interaction with air elements. These bubbles and crashes act as initiators of fractures and points of stress concentration, and can lead to the failure of welded structures in a short period of time.

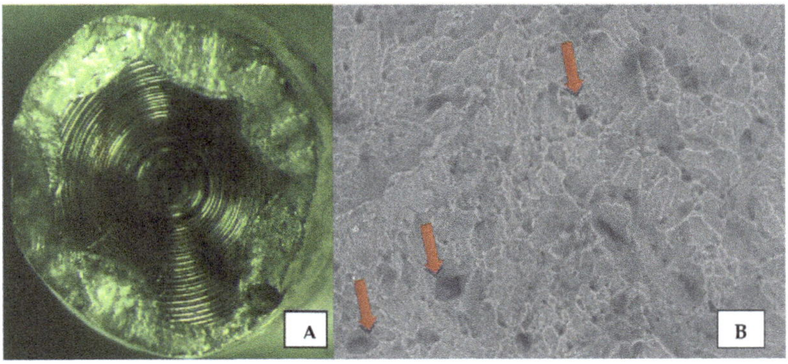

Figure 6. A) Insufficient weld penetration; B) Presence of bubbles and porosities.

2.3. Laser welding

2.3.1. Definition

It is a union process based on localized fusion in the joint, through bombardment from a high-intensity, concentrated, monochromatic and coherent light beam. The area to be welded is protected by using an inert gas, usually argon or a mixture of inert gases.

When the light beam reaches the surface of the metal, the metal absorbs its energy, converting it into heat that penetrates into the interior of the metal by conduction. Owing to a high concentration of heat, the metal is taken to its melting point, and a series of events culminates in the formation of a keyhole or spots that will be filled with the melted metal.

Figure 7. Laser welding machine (output of the light beam and the shielding gas).

2.3.2 Technical description

Looking through the eyepiece of the working chamber, the technician controls with his feet the number of pulses issued for welding. There are rubber gloves inside the working chamber to manipulate the structure to be welded.

2.3.3. *Advantages*

- it produces a keyhole that concentrates the energy absorbed in a small region, resulting in high penetration and formation of a narrow heat affected zone (HAZ) that results in less distortion compared to conventional welding methods;
- It can be executed directly in the working model;
- Expending less time;
- Both of which are optimizing steps needed for the brazing technique;
- Allows welding in regions near the resins and porcelain;
- Allows welding with the structures in close contact or with minimal space for welding, using filler metal.

2.3.4. *Disadvantages*

- Unions soldered by laser welding suffer from resulting defects in, among other things, residual stress. Typically, the residual stress introduced into welding joints is a consequence of thermal stress caused by the heating and cooling cycles of the welding process—this affects the mechanical behavior of laser-welded structures;
- The presence of porosities in the region of the union (figure 8A) that is due to the inclusion of argon gas, which is necessary to maintain the inert atmosphere during the welding procedure and thus minimize interaction with air elements. These bubbles and crashes act as initiators of fractures and points of stress concentration, and can lead to the failure of welded structures in a short period of time;
- Insufficient penetration of the laser beam, causes a big bubble or internal failure (figure 8B). According to some authors, the depth of penetration of the weld is the main factor that affects the values of resistance for laser-welded frameworks. Therefore, for better results, the adjustment of equipment is a key point, especially for larger diameters;
- High cost of the equipment.

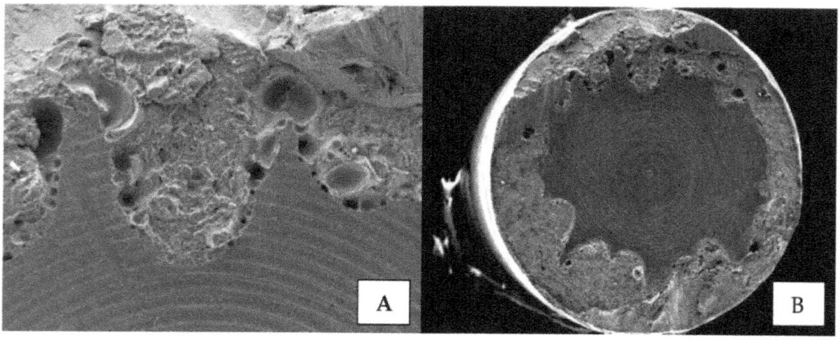

Figure 8. A) Presence of bubbles and porosities, B) Insufficient laser beam penetration.

Type of Welding	Advantages	Disadvantages
Brazing	- It has been used for years - Low cost - Relative effectiveness	- Oxidation of the parts joined by the weld - Joint porosity and overheating of the union during the welding process
Laser	- High penetration and formation of a narrow heat affected zone (HAZ) that results in less distortion compared to conventional welding methods - Procedures can be done directly in the working model - Expending less time - Optimizes steps needed for the brazing technique - Allows welding in regions near resins and porcelains - Allows welding of the structures in close contact - Filler metal may or may not be used	- Residual stress that affects the mechanical behavior of laser-welded structures - The presence of porosities in the region of the union - Insufficient penetration of the laser beam - High cost of the equipment
TIG/PAW	- High quality and excellent finishing of the weld - Allows for welding in any position - Excellent control of the weld pool - Expending less time - Procedures can be done directly in the working model - Accessible cost of the equipment - Allows welding in regions near resins and porcelains - Allows welding of the structures in close contact - Filler metal may or may not be used	- Generates a complex region of stresses and deformations, leading sometimes to undesirable results, such as material distortion, residual stresses, generation of fragile microstructures, grain grow, cracks, fissures, and changes in mechanical, physical, and chemical properties, among others - Insufficient penetration of the laser beam - Presence of porosities in the region of the union

Table 1. Shows a summary of the advantages and disadvantages of different welding techniques.

3. Employment of titanium alloys in dentistry

Until the 1970s, the main material used in fixed and removable prostheses frameworks was gold; however, with the increase in the price of gold, lower-cost alloys were introduced, which were comprised of nickel, beryllium, and cobalt. Over the years, important issues have been reported in relation to the allergenic capacity of nickel and carcinogenic power of beryllium. This has led to a constant search for biocompatible materials that meet dental requirements in chemical, physical, aesthetic, and economic aspects.

Thus, the use of titanium (Ti) in dentistry has been increasing in recent decades due to its favorable physical, mechanical, and biological properties, such as: low density, excellent biocompatibility, corrosion resistance, good resistance/weight ratio, low thermal conductivity, low thermal expansion coefficient, low modulus of elasticity, and relatively low cost. Figure 9 is an example of the metallic framework for the Brånemark Protocol made with titanium.

However, many practical problems are associated with the use of Ti and its alloys because of its high melting point of nearly 1,700ºC, which necessitates high processing temperatures. Also, its high chemical reactivity with oxygen, nitrogen, and hydrogen elements, especially at high temperatures, make Ti fragile, since significant concentrations of these elements are introduced into its surface layer. Contamination with these elements during the process of Ti union and its alloys can result in modification of the microstructure, which causes profound effects in its mechanical properties such as lower ductility and lower tensile strength values, even when soldered in welding machines with inert gas protection.

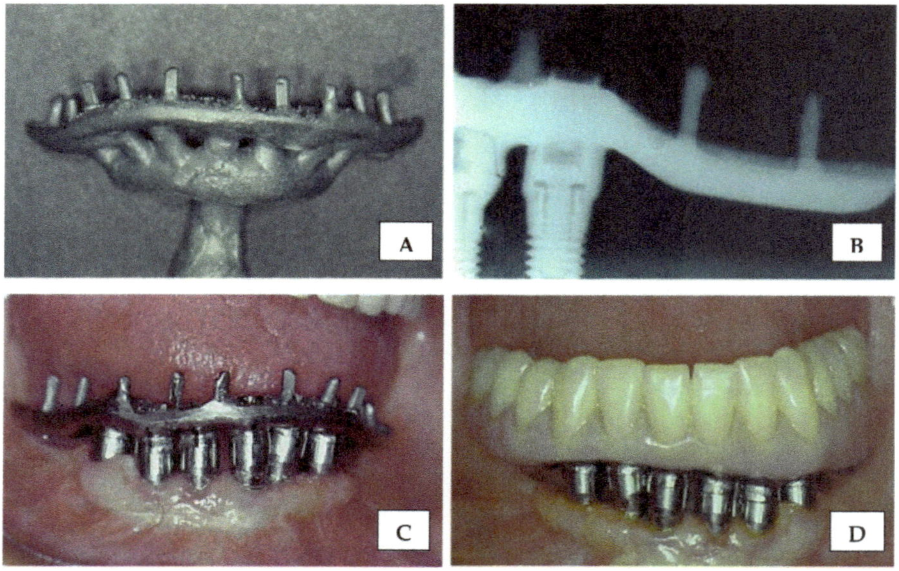

Figure 9. A) Titanium framework; B) X-ray to verify the quality of the laser welding; C) Clinical fit over the implant abutments; D) Installed prosthesis.

Thus, conventional methods that use oxygen flame welding are unsuitable to be used for Ti welding. TIG welding, laser welding, and brazing with infrared radiation are techniques that have been used for metals with gaseous protection, minimizing contamination by oxygen during the welding process and preserving the unique properties of the metal.

4. Use of welding technique to obtain prefabricated frameworks

The use of dental implants in the rehabilitation of edentulous patients is a consolidated treatment modality with high success rates. However, a large proportion of people in the world lack access to this treatment, due to high costs. Research in optimization, simplification of the original Brånemark protocol, such as the use of new technologies, alternative alloys to noble alloys, and new welding methods are presented as viable alternatives to the popularization of the implant in the rehabilitation of edentulous patients.

Therefore, studies using alternative frameworks to the Brånemark protocol have been done, using a titanium and titanium-aluminum-vanadium (Ti-6Al-4V) alloy, in the form of prefabricated bars welded in titanium abutments for construction of simpler frameworks. This reduces costs and the steps required to produce prosthesis, while increasing the speed of treatment; besides this, the alloy offers a great fit and passivity.

Thus, a series of studies involving prefabricated bar welding has been developed at the Federal University of Uberlândia, in partnership with the School of Dentistry, Mechanical Engineering and Technical College of Health. These studies seek to evaluate the use of prefabricated welded bars with different welding techniques. Despite being a line of research that is still unfinished, some of them are shown here.

One of the precursor studies compared the fit of metallic infrastructures welded by laser and brazing under different conformations (Simamoto-Júnior et al. 2008). In this study it was found that the laser-welded structures did not have the best fit as expected, since these have a smaller ZAC. Namely, other technical factors were directly related to the quality of the welded infrastructures' settlement and not just the heating zone factor.

Thus Simamoto-Júnior et al. (2008) evaluated the effect of the type of welding at the interface of three elements of fixed prostheses, which were processed from two master models with implants that were positioned aligned (straight) and misaligned (arc). Twelve models were divided into four groups (n=3) to compare the fit quality of the processes between laser welding and brazing: LA= laser welding/arc, BA= brazing/arc, SR= laser welding/straight, and BR= brazing/straight. At the end of each laboratory stage, casting/grinding, and welding, the structures were placed on the master model for evaluation of the abutment/implant interface, and the quality of both the horizontal and vertical fit were checked (Figure 10). It was expected that laser-welded frameworks would result in better fit, since these have a lower HAZ and, therefore, would cause more minor distortions. As mentioned previously, this did not happen.

This result motivated the study of Silveira-Júnior et al. (2009), which explored the influence of abutment screw tightening force before laser-welding procedures on the vertical fit of metal frameworks over four implants (Figure 11). The hypothesis was that laser welding would result in better fit for the frameworks of implant prostheses, if the tightening force applied to the abutment screws was controlled before the welding procedure. To construct the frameworks, prefabricated titanium abutments and cylindrical titanium bars were joined to compose three groups: GMT, GT10 and GT20. Before welding, manual torque that simulated routine laboratory procedures was applied to the GTM group. In GT10 and GT20, the abutment screw received 10 and 20 Ncm torque, respectively. Although the statistical results have not demonstrated differences in the three groups, it is not known whether this experience technique could influence the results. Therefore, the authors recommended a torque controller device to guarantee standardized framework tightening before welding, particularly by inexperienced technicians.

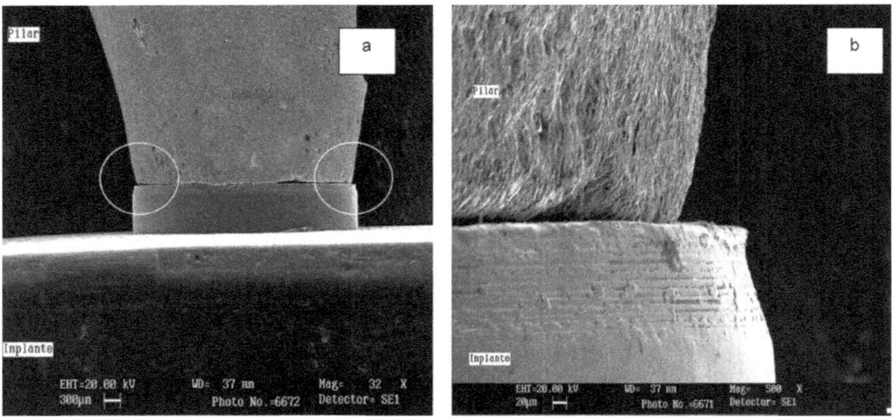

Figure 10. A) UCLA abutment on the implant using a lower zoom, detailing the areas to be analyzed; B) Image increased by 500x of the area to be evaluated.

Figure 11. A) The prosthetic framework with Ti abutments, tightened directly on the implants, after the abutments received varied torques and were welded using the laser technique. B) Evaluation of horizontal and vertical fit of the implant/abutment interface by Scanning Electron Microscopy (SEM).

The distal extension of pre-fabricated steel infrastructures was studied by Oliveira et al. (2010). They evaluated the maximum force required to fracture or bend cantilevers, using three different configurations of cylindrical prefabricated titanium bars, grade 5 (Ti-6Al-4V), welded by the TIG method (Tungsten Inert Gas); the control group consisted of frameworks welded by the laser method. They were divided into four groups (n=6). These groups included a control group (GC) comprised of simple distal bars with 3.18mm diameter that were welded using the laser method, and three experimental groups, all welded by TIG method: with simple distal bars of 3.18mm diameter (GDS); with 2.5mm diameter double distal bars welded together (GDD), and with double distal bars with mixed diameters of 2.5 mm and 3.18 mm welded together (GDDM). The results showed that the control group presented statistically significant differences with the GDS and GDD groups, with higher values of strength; when compared to GDDM, there were no statistically significant differences. Therefore, it is concluded that GDDM, in relation to the other groups, is the most promising method, since its performance is similar to that of titanium frameworks welded using the laser method (Figure 12).

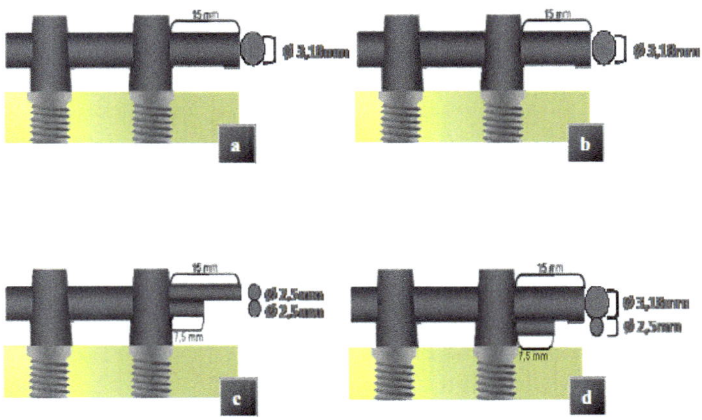

Figure 12. Schematic illustration of the different configurations of the distal bars to be evaluated: (a) GC (b) GDS (c) GDD (d) GDDM.

Silva et al. (2011) evaluated the effect of different plasma arc welding parameters on the flexural strength of titanium alloy beams (Ti-6Al-4V). Forty Ti-6Al-4V and ten Ni-Cr alloy beam specimens were prepared that were 40 mm in length and 3.18 mm in diameter, and were divided into 5 groups (n=10). The titanium beams for the control group were not sectioned or subjected to welding. Groups PL 3-10, PL 3-12, and PL 3-14 contained titanium beams sectioned and welded at a current of 3 A for periods of 10, 12, and 14 months, respectively. Group Ni-Cr-Be consisted of Ni-Cr beams welded using conventional torch brazing. Torch-brazed Ni-Cr alloy beams and non-welded titanium bars served as negative and positive controls, respectively. After the beams were subjected to a three-point bending test, the values obtained were analyzed to find the flexural strength (MPa). No significant differences were observed between the plasma welded groups ($p>0.05$). The Ni-Cr-Be group presented lower flexural strength results, although they were statistically similar to the PL 3-14 group. The weld penetrations were not significantly different between the plasma welded groups ($p=0.05$). This study provides an initial set of parameters for use of plasma welding during the fabrication of titanium alloy dental components. The plasma arc welding technique used with Ti-6Al-4V alloy showed improved performance over conventional torch brazing with Ni-Cr alloy.

Studies comparing plasma and laser welding were made by Castro et al. (2012) that evaluated the mechanical resistance of TI-6AL-4V alloy submitted to the processes of Plasma and Laser welding by means of tensile test and Finite Element Models (FEM). Forty-five dumbbell-shaped rods (n=5) with different central diameters were created from Ti-6Al-4V bars: CG (Control group) with 3 mm diameter and intact bars and PL2.5, PL3, PL4 and PL5 groups with 2.5, 3, 4 and 5 mm diameters submitted to Plasma welding process and L2.5, L3, L4 and L5 groups with 2.5, 3, 4 and 5 mm diameters submitted to Laser welding process. The results demonstrated that the control group showed higher values to tensile strength than test groups. There was statistical difference between control group and test groups but not among test groups to percentage of elongation. There was a positive correlation between welded area percentage and tensile strength in all the specimens in the test groups and a negative correlation between these parameters and the diameters of the specimens. There were no statistical differences between welding processes. The authors conclude that the diameter of 2.5 mm and 3.0 showed the highest values of tensile strength and percentage of welded area and appears to be the best option for the union of prefabricated bars for the use in prosthetics frameworks for both Plasma and Laser welding.

5. Use of welding techniques for repair of fracture prosthesis

In cases where there is a fracture of the metal framework or even where the metal structure is in need of some repair after loss of an implant, for example, the laser welding process known as TIG and PAW is indicated.

Laser welding is a safe process and can be accomplished around regions of ceramic and resin without risk of damage, because of its reduced HAZ, bond strength of the weld that is compatible with the metal source, and preservation of the metallic framework anatomy. Like laser welding, TIG and PAW welds exhibit reduced ZAC compared to brazing, they allow thickness joint to be welding in any position and allows welding in regions near the resins and porcelain. Moreover, it allows welding with the frameworks in close contact or with minimal space for welding using filler metals.

Two cases of prosthesis repair are presented in sequence. The first one shows the welding of fractured structure porcelain fused to metal, and the second shows the reconstruction of a structure after the loss of an implant.

5.1. Clinical case 1

Figures 13–16 display the employment of laser welding in the repair of a fractured metal ceramic fixed implant-supported prosthesis.

In this case, the patient presented the dentist with a fractured metal ceramic fixed implant-supported prosthesis (Figures 13A and 13B).

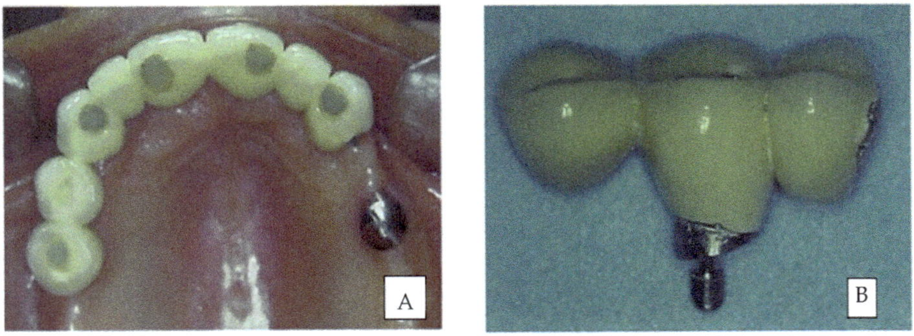

Figure 13. A) Occlusal view of a fractured metal ceramic fixed implant-supported prosthesis; B) Fractured part of the prosthesis.

Then the prosthesis was reseated intraorally and was removed for welding through drills and acrylic resin in order to keep the piece in the correct position (Figures 14A and 14B).

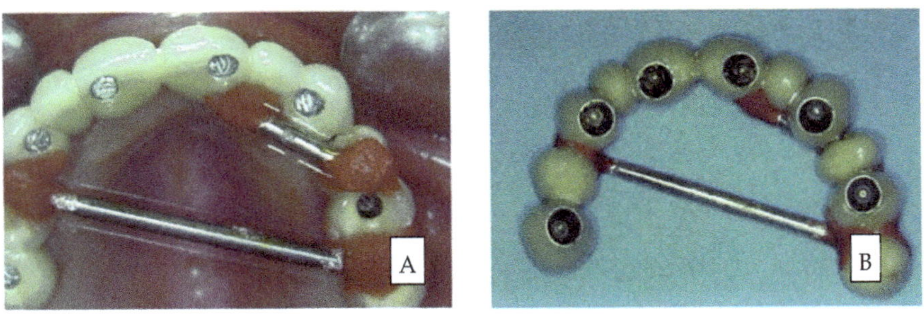

Figure 14. A) Prosthesis positioned intraorally and recorded for welding; B) Prepared structure for inclusion.

After the piece was removed from the patient's mouth, a plaster model index was made in the correct position (Figure 15A) and the piece was welded with laser welding (Figure 15B).

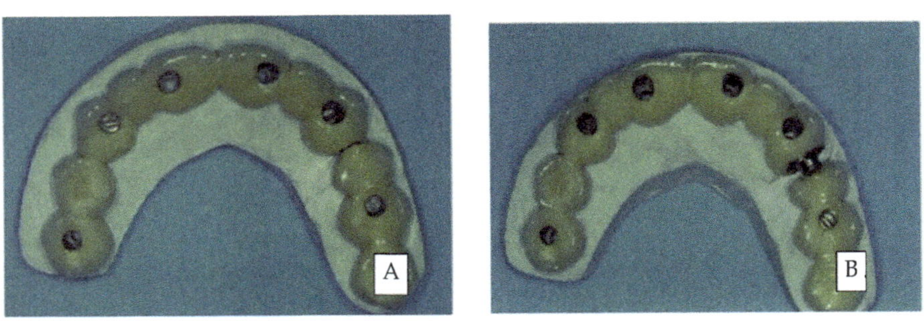

Figure 15. A) Plaster model index for welding; B) Union of the fractured region by means of laser welding.

After the welding composite resin has been applied over the welded region and again installed in the patient's mouth (Figures 16A and 16B).

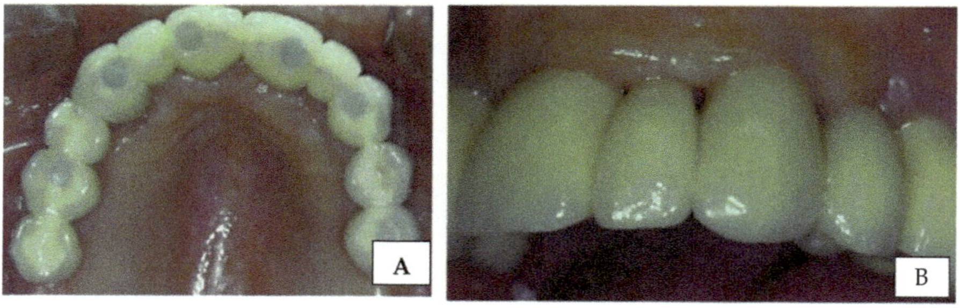

Figure 16. A) Repair terminated with composite resin; B) Details of finishing repair in the welded region.

5.2. Clinical case 2

In this case, the patient was already using a mandibular fixed implant-supported prosthesis over three implants when one of them was lost. Procedures were scheduled to replace the implant and readjust the prosthesis that the patient already used (Figure 17).

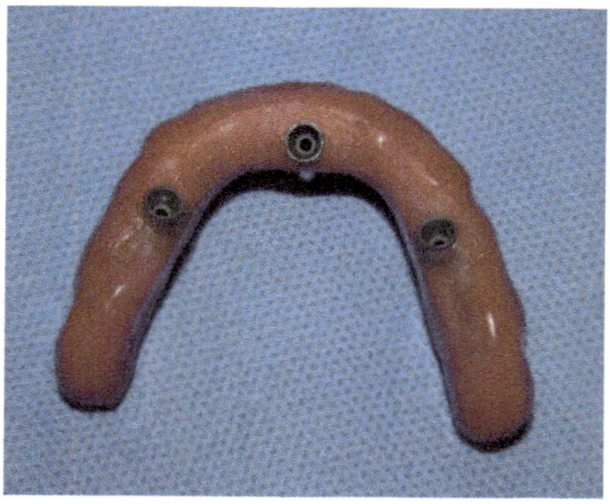

Figure 17. Prosthesis that the patient already used.

This required the assembly of top and bottom templates in the articulator (Figure 18A) and the subsequent construction of a silicone wall (Figure 18B). This device has the function of registering the resin teeth positioning with respect to antagonistic teeth, since the metallic framework of the prosthesis should be sectioned.

Before the welding procedure, the teeth were removed and also all excess acrylic resin around the metal (Figures 19A and 19B). Already at the clinical stage, the lost implant was removed and immediately replaced with another in virtually the same position. For this a surgical guide was used.

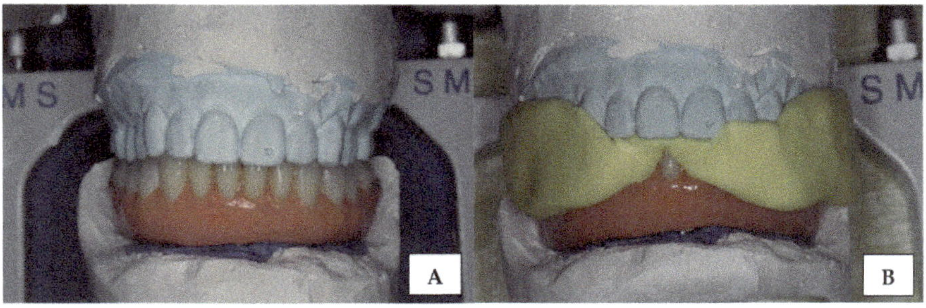

Figure 18. A) Articulating the prosthesis in a semi-adjustable articulator; B) Registration in silicone.

After installing the new implant, the prosthesis and its segmented fragment were screwed into the mouth over the implants. There was a small difference in leveling between these two parts (Figure 20), which was expected because it would be impossible to reposition the implant in an identical position to the previous one.

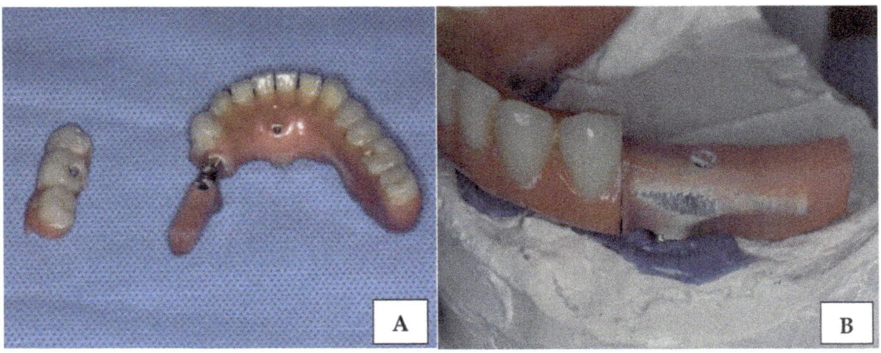

Figure 19. A) Teeth removal; B) Removal of excess resin.

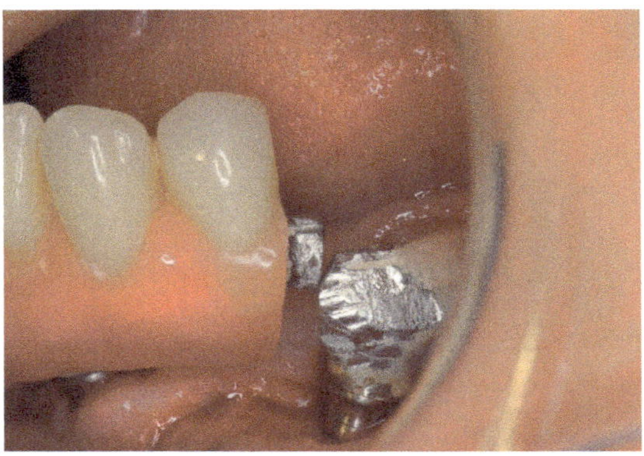

Figure 20. Metal framework cut and repositioned on the new implant.

The fragments were reattached in the mouth with chemically activated acrylic resin for subsequent welding in the laboratory (Figure 21).

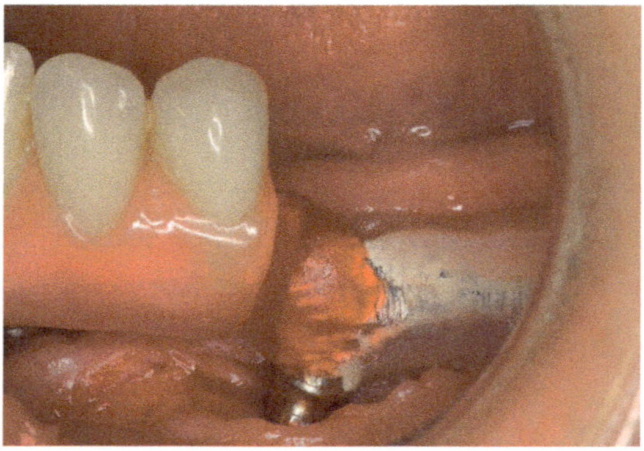

Figure 21. Reattachment of fragments with chemically activated acrylic resin.

In the laboratory it was necessary to adjust the working model. The analogue component of abutment concerning the lost implant was removed and replaced. The analogue was screwed under the sectional fragment and the whole set was screwed over the other two remaining analogues (Figure 22).

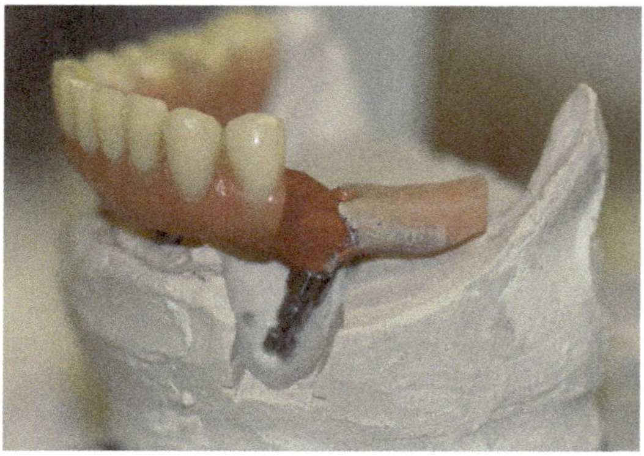

Figure 22. Analogue positioning according to the implant position in the patient's mouth.

New plaster was poured around the new analogue to secure it in order to implement the next steps (Figure 23).

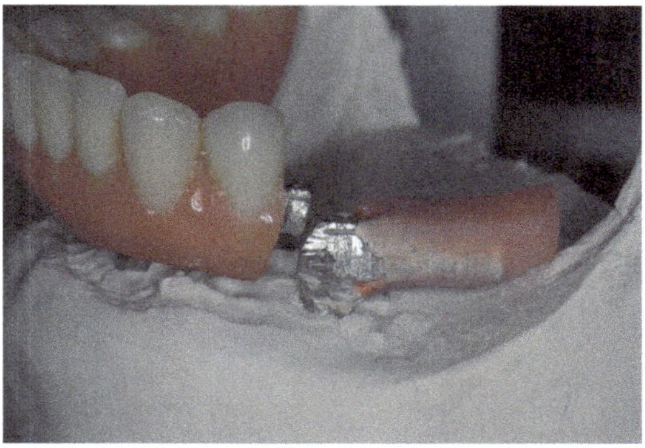

Figure 23. Plaster to fasten the new analogue.

The welding of the segments was done with TIG welding (Kernit IND. e Comércio Ltda, Indaiatuba – SP-Brazil) (Figures 24A and 24B) without the need of filler metal. Note that the covering material of the rest of the prosthesis was not damaged, even in the areas closest to the welding.

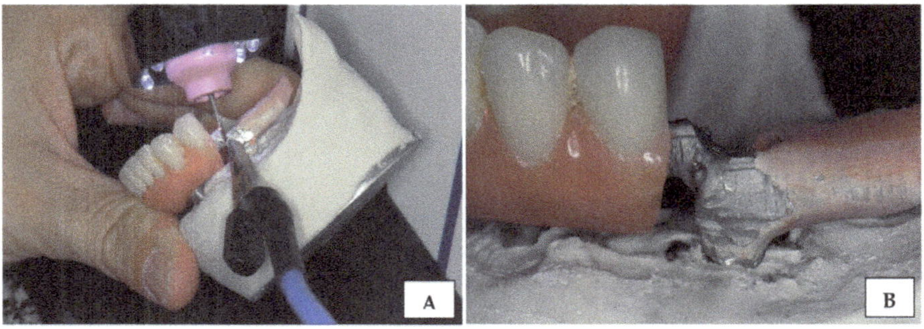

Figure 24. A) TIG welding of the metal bar; B) Structure after welding.

After the welding the teeth were repositioned on the bar using the silicone registration that had already been cut and fixed onto the top model (Figure 25A). The resin teeth were already embedded in the registration, and between these and the bar there was a space that was filled with acrylic resin to fasten them (Figure 25B).

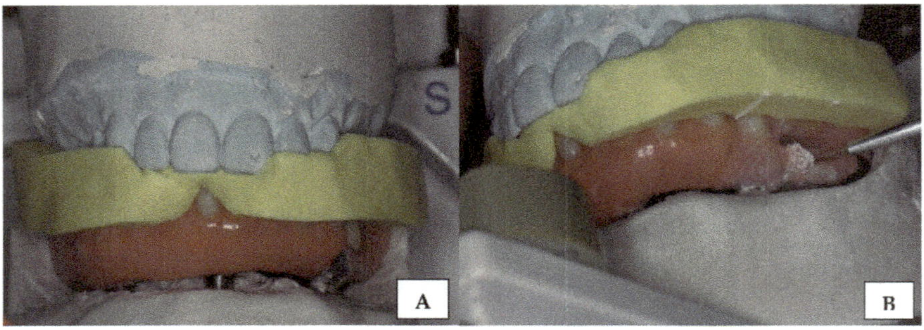

Figure 25. A) Silicone registration positioned; B) Fixation of teeth with acrylic resin.

After insertion of chemically activated acrylic resin in the same color as the rest of the prosthesis, finishing and polishing were performed (Figure 26).

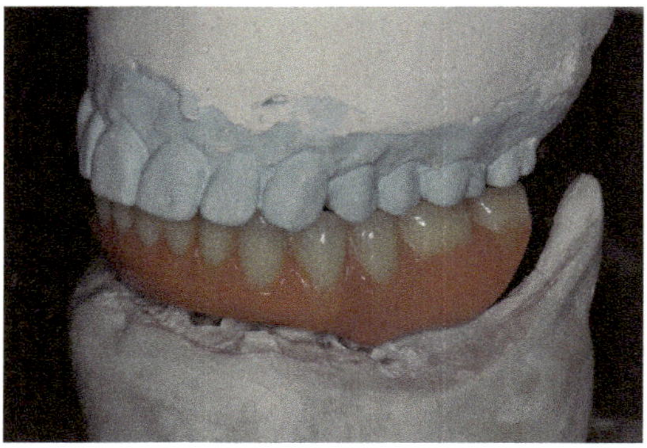

Figure 26. Final appearance of the piece.

6. Conclusion

The information available in the literature and the research shows that the mechanical behaviors of the plasma-, TIG-, and laser-welding techniques do not show major differences in their behavior; laser welding had an increased availability of studies and follow-ups, compared to the other methods. In addition, all feature similar advantages, such as: the ability to be performed on a plaster model; allowing welding in areas close to the resin and ceramic; allowing welding in any position, and requiring less chair time.

Author details

Clébio Domingues da Silveira-Júnior, Morgana Guilherme de Castro, Letícia Resende Davi1, Flávio Domingues das Neves, Verediana Resende Novais and Paulo Cézar Simamoto-Júnior
Federal University of Uberlândia, Brazil

Acknowledgement

The authors would like to thanks the financial support granted by FAPEMIG and CNPq. The authors also thanks for the use of laboratories and equipments of the institutions: Technical Health School of Federal University of Uberlândia (UFU) and School of Dentistry (UFU). Thanks to Professor Cleudmar Amaral de Araújo of Mechanical Engineering School (UFU) and to technicians Takeo Endo and Marco Aurélio Dias Galbiatti, for the technical support.

7. References

American Welding Society - AWS, Rules for welding in structural steel. 1992.

Aparicio C, Lang NP, Rangert B. Validity and clinical significance of biomechanical testing of implant/bone interface. Clin Oral Imp Res. 2006; 17(Suppl 2):2-7.

Baba N, Watanabe I. Penetration depth into dental casting alloys by Nd:YAG laser. J Biomed Mater Res B Appl Biomater. 2005;72(1):64-8.

Barbosa GAS, Simamoto Júnior PC., Fernandes Neto AJ, Mattos MGC, &Neves, FD. Effect of different torque levels on the implant/abutment interface using castable "UCLA" abutments. Braz J Oral Sci. 2005;14:912-92.

Barbosa GA, das Neves FD, de Mattos M da G, Rodrigues RC, Ribeiro RF. Implant/abutment vertical misfit of one-piece cast frameworks made with different materials. Braz Dent J. 2010;21(6):515-9.

Blackman R, Barghi N, Tran C. Dimensional changes in casting titanium removable partial denture frameworks. J Prosthet Dent. 199;65(2):309-15.

Berg E, Wagner WC; Davik G, Dootz ER. Mechanical properties of laser-welded cast and wrought titanium. J Prosthet Dent 1995;74:250-7.

Bergman M. Cast titanium in dental constructions. DPNOVA AB Malmo: Sweden, Aug. 1990. [Manual].

Byrne D, Houston F, Cleary R, Claffey N. The fit of cast and premachined implant abutments. J Prosthet Dent. 1998;80(2):184-192.

Byrne G. Soldering in Prosthodontics—An Overview, Part I. J Prosthodont. 2011 Apr;20(3):233-43.

Castro MG, Silva JPL, Júnior UG, Martins PO, Cavalcante LAL, Menegaz GL, Neto AJF, Novais VR, Araújo CA, Simamoto-Júnior PC. (2012). International Association of Dental Research, Proceeding of Journal of Dental Research, Foz do Iguaçu, June 2012.

Chai T, Chou CK. Mechanical properties of laser-welded cast titanium joints under different conditions. J Prosthet Dent. 1998;79 (4): 477-483.

Cho SK, Yang YS, Son KJ, Kim JY. Fatigue strength in laser welding of the lap joint. Finite Elements in analysis and design. 2004;40:1059-1070.

Craig, R.G. Restorative dental materials. 10 ed. Saint Louis: Mosby, 1997.

Degidi M, Nardi D, Piattelli A. Immediate Rehabilitation of the Edentulous Mandible with a Definitive Prosthesis Supported by an Intraorally Welded Titanium Bar. Int J Oral Maxillofac Implants 2009;24:342–347.

Gordon TE, Smith DL. Laser welding of prostheses - an initial report. J Prosthet Dent. 1970;24 (4): 472-476.

Hart CN, Wilson PR. Evaluation of welded titanium joint used with cantilevered implant-supported prostheses. J Prosthet Dent. 2006;96:25-36.

Liu J, Watanabe I, Yoshida K, Atsuta M. Joint strength of laser-welded titanium. Dent Mater. 2002;18:143–148.

Meadows C & Fritz JD (2005). Understanding Stainless Steel Heat-Affected Zones. Welding Journal. 26-30.

Neo TK, Chai J, Gilbert JL, Wozniak WT, Engelman MJ. Mechanical properties of titanium connectors. Int J Prosthodont. 1996;9(4):379-93.

Nuñez-Pantoja JM, Takahashi JMFK, Nóbilo MA, Consani RLX, Mesquita MF. Radiographic inspection of porosity in Ti-6Al-4V laser-welded joints. Braz Oral Res. 2011;25(2):103-8

Nuñez-Pantoja JM, Vaz LG, Nóbilo MA, Henriques GE, Mesquita MF. Effects of laser-weld joint opening size on fatigue strength of Ti-6Al-4V structures with several diameters. J Oral Rehabil. 2011;38: 196-201.

Paiva BV, Oliveira NCM, Viadanna APO, Silva JPL., Novais VR, Neves FD, Araújo CA, Simamoto-Júnior, PC. Avaliação da força máxima necessária para fratura ou flexão de "cantilevers"em protocolos totais fixos sobre implantes. Braz Oral Res. 2010;24(Suppl.1):106-27.

Ramirez JE, Mishael S, Shockley R (2005). Properties and Sulfide Stress Cracking Resistance of Coarse-Grained Heat-Affected Zones in V-Microalloyed X60 Steel Pipe. Welding Journal. 113-123.

Rocha R, Pinheiro AL, Villaverde AB. Flexural strength of pure Ti, Ni-Cr and Co-Cr alloys submitted to Nd:YAG laser or TIG welding. Braz Dent J. 2006;17(1):20-3.

Roggensack M, Walter MH, Boning KW. Studies on laser- and plasma-welded titanium. Dent Mater. 1993;9(2):104-7.

Sadoswsky SJ, Caputo A. Stress transfer of four mandibular implant overdenture cantilever designs. J Prosthet Dent. 2004;92: 328-36.

Sahin S, Çehreli MC, Yalçin E. The influence of functional force on biomechanics of implant-supported prostheses – a review. Journal of Dentistry. 2002;30: 271- 282.

Silva JPL, Cavalcante LAL, Novais VR, Araújo CA, Correr-Sobrinho L, Fernandes-Neto AJ. Avaliação da resistência à flexão em liga de titânio soldadas com diferentes configurações pelo processo plasma. Braz Oral Res. 2011;25(Suppl.1):227-58.

Silveira-Júnior CD; Neves FD; Fernandes-Neto AJ; Prado CJ; Simamoto-Júnior PC. Influence of differente tightening forces before laser welding to the implant/framework fit. Journal of Prosthodontics. 2009;18:337-341.

Simamoto-Júnior PC, Novais VR, Nóbilo MAA, Barbosa GAS, Soares CJ, Neves FD. Influence of the soldering type in fitting prostheses on implants: a scanning electronic microscopy analysis. Arquivos em Odontologia. 2008;44(1):23-29.

Sjögren G, Andersson M, Bergman M. Laser welding of titanium in dentistry. Acta Odontol Scand. 1988;46: 247-253.

Shimakura M, Yamada S, Takeuchi M, Miura K, Ikeyama J. Influence of irradiation conditions on the deformation of pure titanium frames in laser welding. Dental Materials Journal 2009;28(2): 243－247.

Srimaneeponga V, Yoneyamab T, Kobayashic E, Doid H, Hanawad T. Comparative study on torsional strength, ductility and fracture characteristics of laser-welded Ti–6Al–7Nb alloy, CP Titanium and Co–Cr alloy dental castings. Dental Materials. 2008;24:839–845.

Tahmaseb A, van de Weijden JJ, Mercelis P, De Clerck R, Wismeijer D. Parameters of passive fit using a new technique to mill implant-supported superstructures: an in vitro study of a novel three-dimensional force measurement-misfit method. Int J Oral Maxillofac Implants. 2010;25(2):247-57.

Tambasco J, Anthony T, Sandven O. Laser welding in the dental laboratory: an alternative to soldering. J Dent Technol. 1996;13(4):23-31.

Taylor JC, Hondrum SO, Prasad A, Brodersen CA. Effects of joint configuration for the arc welding of cast Ti6Al4V alloy rods in argon. J Prosthet Dent. 1998;79:291-297.

Tehini GE, Stein RS. Comparative Analysis of Two Techiniques for Soldered Connectors. J Prosthet Dent.1993;69(1):16-19.

Tiossi R, Falcão-Filho H, Aguiar Júnior FA, Rodrigues RC, Mattos MG, Ribeiro RF. Modified section method for laser-welding of ill fitting cp Ti and Ni–Cr alloy one-piece cast implant-supported frameworks. J Oral Rehabil. 2010 1;37(5):359-63.

Wang RR, Chang CT. Thermal modeling of laser welding for titanium dental restorations. J Prosthet Dent. 1998;79(3):335-41.

Wang RR, Fenton A. Titanium for prosthodontic applications: a review of the literature. Quintessence Int. 1996 Jun; 27(6):401-8.

Wang RR, Welsch GE. Joining titanium materials with tungsten inert gas welding,laser welding, and infrared brazing. J Prosthet Dent. 1995;74(5):521-30.

Wataha JC. Biocompatibility of dental casting alloys: a review. The jornal of prosthetic destistry, v. 83, n.2, p. 223-234, 2000.

Watanabe F, Uno I, Hata Y, Neuendorff G, Kirsch A. Analysis of stress distribution in a screw-retained implant prosthesis. Int J Oral Maxillofac Implants. 2000; 15(2):209-218.

Watanabe I, Topham DS. Laser welding of cast titanium and dental alloys using argon shielding. J Prosthodont. 2006;15:102–107.

Wee AG, Aquilino SA, Schneider RL. Strategies to achieve fit in implant prosthodontics: a review of the literature. Int J Prosthodont. 1999;12 (2): 167-178.

Williams DF. Titanium and titanium alloys. In: Williams DF, editor. Biocompatibility of clinical implant materials. Boca Raton: CRC Press, 1984:44-7.

Zavanelli RA, Guilherme AS, Pessanha-Henriques GE, Arruda Nobilo MA, Mesquita Mf. Corrosion-fatigue of laser-repaired commercially pure titanium and Ti-6Al-4V alloy under different test environments. J Oral Rehabil. 2004;31(10):1029.

Zupancic R, Legat A, Funduk N. Tensile strength and corrosion resistance of brazed and laser-welded cobalt-chromium alloy joints. J Prosthet Dent. 2006 96(4):273-82.

Chapter 17

Optimized Stud Arc Welding Process Control Factors by Taguchi Experimental Design Technique

Riyadh Mohammed Ali Hamza

Additional information is available at the end of the chapter

1. Introduction

Welding is an ancient craft that combines art, science and human skill. It can be traced back to around 3000 BC, with the Sumerians and the Egyptians. The Sumerians used swords with parts joined by hard soldering. The Egyptians found that, after heating iron, it was much easier to work with welding just by hammering the parts to join them. Several objects that have been found in tombs and excavations, etc., indicate the exploitation of several welding techniques, such as "pressure" (hammering) welding, applied with several metal materials (gold, iron, bronze and copper, etc.) during those ancient times.

In the sixteenth century, the basic welding techniques were well known but not used to any great extent. In 1540, the Italian Engineer Vannoccio Biringuccio (as cited in Smith and Gnudi, 1990) explains in his book "The Pirotechnia", published in Venice, that welding "seems to me an ingenious thing, little used, but of great usefulness". During these middle ages, the art of blacksmithing was further developed and it was possible to produce any items of iron welded by hammering. The welding, as we know it today, was not invented until the nineteenth century.

A number of different processes can be used for joining studs to sheets or structure: resistance, friction and arc welding (stud arc welding or manual metal arc welding). Manual metal Arc Welding is sometimes used, but often only fillet welds are possible, and it is very slow. Stud Arc Welding (SAW) was invented just prior to World War II at the New York Navy Yard and developed for necessity to attach wood planking to naval aircraft carriers, and later it was used in the shipbuilding and construction industries. To undertake a weld, the welder first cleans the workpiece to bright shining metal. A stud is fitted with its ferrule into the chuck. The gun is pressed against the workpiece in the correct position and the trigger squeezed. There are four steps of SAW process. First step the automatic solenoid of gun is energized, withdrawing the stud from the workpiece and starting the current to create, an arc. The arc melts the end of the stud and the workpiece. When the preset time is complete, the current cut off. The spring in the gun plunges the stud into the molten pool to complete the weld. Once the weld is done, the welder removes the gun, breaks off the ferrule and inspects the weld. Figure (1) illustrates a stud arc welding process steps (Miller Welds Electrical Mng. Co., 2005).

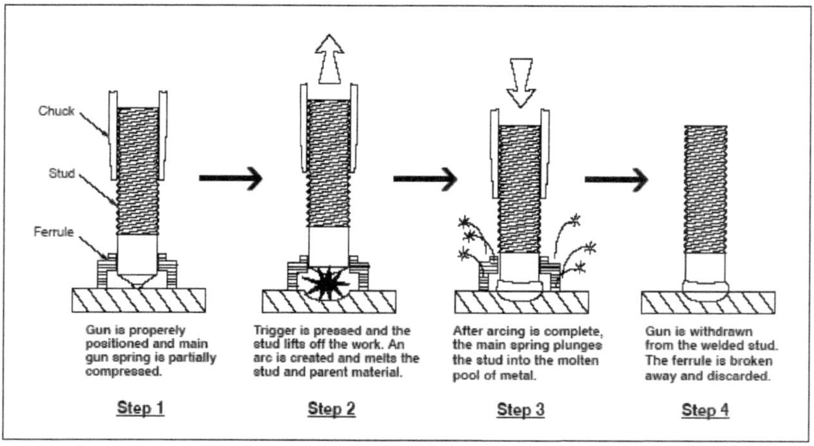

Figure 1. The stud welding process steps (Miller Welds Electrical Mng. Co., 2005).

The stud arc welding process includes the same electrical, mechanical and metallurgical principles as with other arc welding processes (Lee J. S. et al., 2009). The quality of the weld joint in the Drawn Arc Welding (DAW) process with a ceramic ferule depends upon a number of factors such as: the type of the base metal and the stud material, the welding position, the welding time and other factors; however, the proper selection of welding parameters has an important role. The literature survey shows that, due to the short time of the welding cycle, simplicity in the use of equipment, cost efficiency and the application of the stud arc welding process are all well known in various manufacturing fields. Reviewing the previous literature surveys shows that the researchers have been concerned with the search for this process in two directions: the first direction is the examination of the process factors affecting mechanical properties - such as tensile stress weld or strain pieces - influenced by multiple factors, such as welding current, welding time, stud plunge and lift, and other factors - see reference (Klarić et al., 2009). References (Bursi O. S. & Gramola G., 1999) and (Lee et al., 2009) describe the ability of studs to develop full strength welds and discuss the fact that, in some cases, some welds were less than full strength. Reference (Anderson N. S. & Meinheit D. F., 2000) documents the embedment shear and tension tests of deformed bar anchors where no weld failures occurred and summarizes the results of extensive testing and studies from many sources in relation to the performance of the stud welded anchor and other types of anchor devices. Reference (Hsu C. & Mumaw J., 2011) presents the findings of a weldability study of the drawn arc stud welding of various advanced high-strength steels (AHHS), including two grades of boron steel and one grade of dual-phase steel of various thicknesses, coatings from several automakers and benchmarked against mild steel. Researchers like (Strigel R. M., Pincheira J. A. & Oliva M. G., 2000) also consider examining failure in stud welding joints and they show that 19 % of samples examined fail - the weld's fail in the vicinity of the weld area. Eşme (2009) reports on an investigation of the effect and optimization of welding factors on the tensile shear strength in the Resistance Spot Welding (RSW) process. The experimental studies were conducted under varying electrode forces, welding currents, electrode diameters and welding times. The settings of the welding factors were determined by using the Taguchi experimental design method. The confirmation tests indicated that it is possible to increase tensile shear strength significantly by using the Taguchi method. The experimental results confirmed the validity of the Taguchi method for enhancing welding performance and optimizing the welding factors in the RSW process.

A second direction of research studies is the application of automated systems in the control procedure in relation to an interest in the research on the development of technology; the previous research indicates the evolution trend, especially since the procedure can be worked by automation, such as with robots - see (Samardžić I. & Klari Š., 2007), (Hsu et al., 2007) and (Hsu et al. 2008). In addition, the researchers studied the possibility of using neural network systems for optimization process parameters (Riyadh Mohammed Ali Hamza R. M. A., 2011).

In this chapter, an experimental study is conducted under varying welding times, sheet thicknesses, sheet coatings, welding currents, stud designs, stud materials, preheat sheets and surface conditions. The effectiveness of the welding factors levels on the joint and tensile strength is determined via Analysis of Variance (ANOVA). The optimum welding parameter combination is obtained using the analysis of the signal-to-noise (S/N) ratio and the quality loss function. The confirmation tests indicated that it is possible to increase the tensile strength significantly using the Taguchi method, by which 225 samples are tested.

Due to the mentioned importance of proper parameter selection, the main aim of this optimization technique is to ascertain the assumption that the specific selection of welding factors will influence weld tensile strength and that the proper selection of factors will give a weld joint the desired tensile strength.

2. Factors of the stud arc welding process

A process can be defined as a combination of inputs - such as materials, machines, manpower, measurements, environments and methods - that results at various outputs as the measurements of performance (Conti, Kondo & Watson, 2003). The inputs x_1, $x_2 \ldots x_p$ are controllable factors, such as temperature, pressure, feed rates and other process variables. The inputs z_1, $z_2 \ldots z_q$ are uncontrollable (or difficult to control) input factors, such as environmental factors or the properties of a raw materials provided by the supplier, as shown in figure (2). The manufacturing process transforms these inputs into an output that has several quality characteristics (Schmidt & Launsby, 1992).

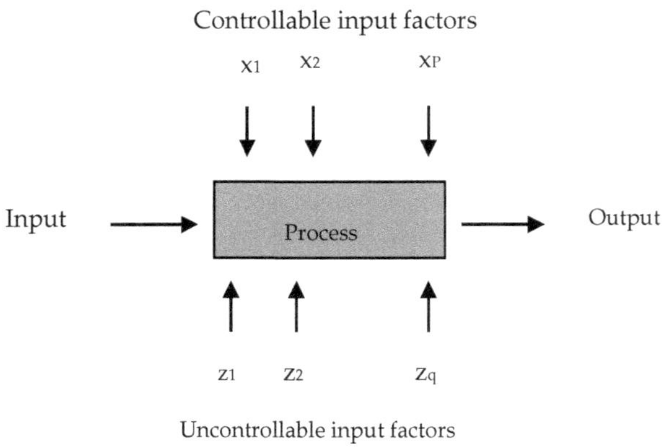

Figure 2. General model of a process (Schmidt & Launsby, 1992).

There are two types of arc-stud welding processes: Capacitor Discharge Welding and Arc Stud Welding.

2.1. Capacitor discharge welding

In this process, the Direct Current (DC) produced by the rapid discharge of stored electrical energy from a bank of capacitors is used to create an arc between a stud and the sheet or structure. Pressure is applied immediately following electrical discharge to form the weld and no flux or ferrule is required. The arc stud processes are quick and access to the other side of the joint is not required (as is necessary for bolted connections). Because of the short welding cycle, the HAZs are narrower than for other arc processes. (Samardžić I., 2007) explains that Capacitor Discharge Stud Welding (CD) can be accomplished by a specially-drawn arc stud welding process - known as the "short cycle" process - whereby stud welding to sheet metal is characterized by the use of a high current and a short time.

The stud is held in a gun. When the trigger is operated, the capacitor is discharged so as to fuse the end of the stud and the base material; then, the stud is plunged into the weld pool. Welds are produced using very high currents (6000A) for very short durations of about 3 to 15 milliseconds. Because of the percussive nature of the process, surface coatings are removed more effectively than with the arc stud process. Less similar combinations can be welded (e.g., brass to steel), than with the arc stud process because of the short duration. The process is also suitable for welding studs to thin sheets without damaging the surface coating on the opposite side.

The capacitor discharge method is limited to studs of 8 mm and less for economic reasons. It is less tolerant to rust and scale. Because of these limitations, this process is used less than with the arc stud welding process for heavy fabrication. The most common application of capacitor discharge welding is to join the thermocouple to the steel structure for monitoring preheat and post-weld heat treatment. The scar that remains after the removal of the thermocouples is insignificant (Taylor, 2001).

2.2. The arc stud welding process

During this process, an arc is established between the stud and the workpiece using a conventional welding power source. After a brief time, the stud is plugged into the weld pool and the current is shut off. The process is quick and there is little time for detrimental phases to form. The main limitation is that it is intolerant to contamination and the surface to be welded should be free of rust, scale, paint and other contaminants.

The welding factors (the current and arc time) depend upon the material type and the size of the stud base. The current used is between 250 and 600 Amperes and the cycle time is 0.13 seconds to 1 second for studs of a diameter of 3 mm to 22 mm. An average of around six studs can be welded per minute.

2.3. The required process equipment

The most basic equipment is a stud gun connected to a control unit that is connected to a source of DC power. Some modern stud welding equipment includes the controller and the power source as one unit, but it is possible to obtain a controller and a gun utilizing an existing DC welding power source. Figure (3) illustrates that the process equipment consists of a stud gun, a control unit for timing the weld, a DC power source and a suitable weld cable.

The stud gun consists of the following components (Taylor, 2001):

- A spring-load chuck for holding the stud.
- An adjustable spacer for holding the stud gun against the workpiece.
- A solenoid coil to lift the stud away from the workpiece by a preset distance of approximately 3 mm.
- A trigger for initiating the welding cycles.

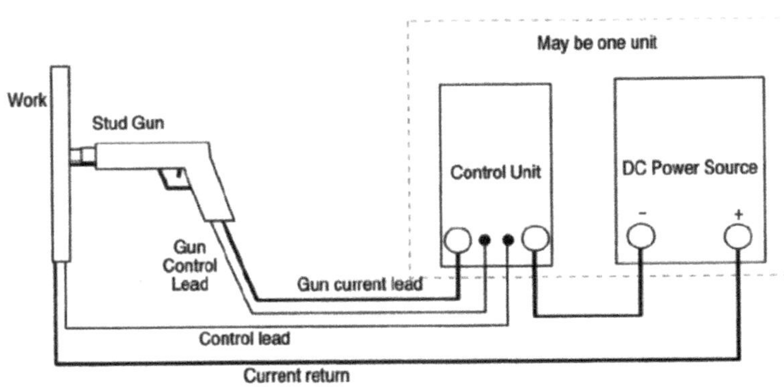

Figure 3. Arc Stud Welding Equipment (Taylor, 2001).

Most welding is undertaken using a hand-held gun. An automatic stud gun - which is fixed to robot arm or another fixture - can be used to automate the process. The controller has a solenoid switch to turn the current on and off rapidly as well as timers to control the automatic welding cycle and the adjustment of the current and the cycle time.

2.3.1. Studs and ferrules

Studs can have circular, square or rectangular bases. If the base is rectangular, the width should not be more than five times the thickness. It must have a shape that is capable of being held in the chuck; otherwise, the form of the stud is limitless. The most common stud types are screw fasteners and shear studs, but hooks, rings, rings, brackets and many other items can be made. Studs are available in a variety of materials. Carbon steel studs are semi-killed or fully-killed carbon steel of grads 1010 to 1020 in the cold drawn condition (Taylor, 2001).

The studs for must materials have a flux tip. They have to be supplied by a reputable stud-welding supplier, who is required by code to perform qualification tests. Those from other than reputable suppliers risk not producing satisfactory welds. Studs and ferrules should be from the same supplier.

Each stud is supplied with a matching ceramic ferrule so as to:

- Protect the arc by restricting air flow.
- Concentrate the arc heat to the weld area.
- Mould the weld flash.
- Prevent the charring of adjacent materials.

The ferrule is broken off when the weld is complete.

2.4. Application of the stud arc welding process

The stud arc welding process is applied in different production areas, such as boiler production, the motor vehicle industry, bridge construction and shipbuilding, due to the efficiency of the process.

The application of draw arc welding with a ceramic ferrule plays an important role in steam boiler production. This process is successfully used in ship building and the automobile industry, etc. The stud welding process is used for fixing in place the cryogenic insulation of membrane tanks in ship building (Lee et al., 2009). In addition, stud welding is widely used in the construction industries and in bridge construction in particular (composite steel/concrete structures). There are many different stud welded products that are commonly used in the manufacture of precast/pre-stress components, including threaded, headed and deformed bars (Bursi & Gramola, 1999).

2.5. Stud welding failures

The stud butt fully welds with the base material such that there is no unfused central area that is a feature of fillet welded attachments. Because the weld is a full penetration, the small amount of flash interference is much less with an attachment than with a fillet weld would. For the full strength of the stud, the base metal thickness should be at least 1/3 of the stud base diameter. Studs can be closer to a flange edge than with threaded connections. The basis for loading is the smallest cross section of the stud (Taylor, 2001).

When the proper operation of stud welding equipment is combined with good quality control and inspection procedures, full strength welds can be obtained consistently and can result in the optimal performance of the studs. However, improper stud welding process factors cause stud failures. The root causes for weld or stud failures can usually be attributed to one or more of the following factors (Chambers, 2001):

- Unacceptable base plate materials or plate surface conditions.
- Inappropriate weld settings.
- Malfunctioning or obsolete equipment.
- Little or no formal training for stud welding operators.
- A lack of quality control and inspection procedures.

3. The Taguchi experimental design methodology

Experimental design is a subject with a set of techniques and body of knowledge which assists investigators in conducting experiments by better analysing the results of experiments and finding the optimal factor combinations to achieve the intended objectives – see (Montgomery D.C., 2009) and (Antony J. & Kaye M., 1999). Stud arc welding technology has generally continued to grow vigorously because of new applications. Tensile strength quality is one of the key factors in achieving good welding process performance and so the purpose of this study is to improve the tensile strength of stud joints by using the Taguchi Experimental Design Technique. In the following sections, some of the most important concepts in the design of the experimental technique will be explained.

3.1. Measure of variation (measure of dispersion)

This describes how the data is spread out or scattered on each side of the central value (mean). The elements involved in the measurement of variation are explained in two sections below.

3.1.1. The range of data

For a series of numbers, the range is the difference between the largest and the smallest values of observation. The range equation is:

$$r = x_h - x_l \tag{1}$$

Where

- r= range
- x_h= highest observation in a data

- x_l= lowest observation in a data

3.1.2. Standard deviation

Which of a set of (n) numbers $x_1, x_2, \ldots, x_n$ denoted by (S) and defined by:

$$S = \sqrt{\frac{\sum_{i=1}^{N} (x_i - \bar{x})^2}{N-1}} \tag{2}$$

where (S) is the root mean square of the deviations of each number x_i from the mean $\bar{x}$.

3.2. Target value

In the data analysis, the target value - or an objective value - is a parametric quantity identified as the standard against which all measurements or calculations of the same response are to be evaluated. The target value is represented by T (Buyske S. & Trout R., 2003).

3.3. Sum of Squares (SS)

The sum of squares (SS) of a factor i at level k was calculated according to the equation (Buyske S. & Trout R., 2003):

$$SS_i = \sum_{K}^{L} \frac{\left(\sum_{J}^{N} Y_j\right)^2}{N_K} - \frac{\left(\sum_{J}^{N} Y_J\right)^2}{N} \tag{3}$$

where N is the total number of experiments, N_k is the number of levels and Y_j is the mean response. The total sum of squares (SS_T) is calculated using equation:

$$SS_T = \sum_{J}^{N} Y_J^2 - \frac{\left(\sum_{J}^{N} Y_J\right)^2}{N} \tag{4}$$

Experimental error (S_e) is calculated from:

$$S_e = SS_T - \sum SS_i \tag{5}$$

3.4. Degree of freedom

The degree of freedom, as an integer associated with a statistic, is the number of available independent squares of the associated statistic. If the independent sum of the squares is n, then the number of degrees of freedom denoted by f is equal to n-1.

3.5. Variance

The variance is defined as the sum of the squares of the deviations of the observation data from a specific value, divided by the degrees of freedom f. The variance - sometimes called the mean square - is denoted by V (Steiner S. H. & MacKay R. J., 2005).

$$V_i = {SS_i}/{f_i} \tag{6}$$

3.5.1. Analysis of variance

The relative magnitude of the effect of different factors can be obtained by the decomposition of the variance, namely ANOVA - this is given in table (1). The experimental design permits the effects of numerous factors to be investigated at the same time. When many different factors dynamically affect a given quality characteristic, ANOVA is a systematic and meaningful way of statistically evaluating experimental results (Montgomery D. C., 2009).

Sources of variation	Degrees of freedom *F*	Sum of squares SS	Mean square V	Pure sum of squares Ś	F-ratio	Percent contribution (%)
Factor(a)	1	Sa	Va	Śa	Fa	á
Error(e)	n-1	Se	Ve	Śe	1	é
Total(t)	N	St		Śt		100.0

Table 1. ANOVA table

Where:

1. Variance ratio

$$F_a = {v_a}/{v_e} \tag{7}$$

2. Sum of squares

$$s'_a = s_a - (f_a \times v_e) \tag{8}$$

$$s'_e = s_e + (f_a \times v_e) \tag{9}$$

3. %age contribution:

$$a' = \left(\frac{s'_a}{s_t}\right) \times 100 \tag{10}$$

$$e' = \left(\frac{s'_e}{s_t}\right) \times 100 \tag{11}$$

After n pieces of experimental data are collected and after the values of á and é are calculated, significant testing provides the criterion for making such decisions. The F-tests are used to statistically determine whether the constituents - the total sum of squares which are decomposed - are significant with respect to the components that remain in the error variance. The specific numerical confidence levels, depending upon which F-table is used, are called the level of significance. When the variance ratios Fa are larger than the F-table at the 5% level, then the effect is called significant at the 5% level (Montgomery D.C., 2009).

3.6. The Larger-the-better Signal to Noise (S/N) ratio

A signal-to-noise (S/N) ratio is a measure of performance which estimates the effect of the noise factors on the quality characteristic (Taguchi G., Chowdhury S., & Wu Y., 2005; Ross, P. J., 1986). The S/N is defined as:

$$S/N_{LTB} = -10\log\left(\frac{\sum \frac{1}{y_i^2}}{n}\right) \tag{12}$$

where y= response, n= run experiment number.

3.7. The Taguchi Losses Function

The Taguchi quality losses' function for the larger-the-better is (Taguchi G., Chowdhury S. & Wu Y., 2005; Ross, P. J., 1986):

$$L(y) = A_0 \Delta_0^2 \frac{1}{y^2} \tag{13}$$

A_o is the loss (stated in monetary or scaled monetary units) at a specified distance, Δ_o, from the target, T, and y is the performance measure.

3.8. The Orthogonal Array (OA)

Orthogonal Arrays (OA) are a special set of Latin squares, constructed by Taguchi in order to lay out the product design experiments. For each OA, a code is available in the form of $L_a b^c$, where (a) is the number of experiments, (b) is the number of levels for each factor and (c) is the number of columns in the array (Taguchi G., Chowdhury S. & Wu Y., 2005; Ross P. J., 1986).

4. Experimental work

The Taguchi experimental design is a statistical technique that allows the running of the minimum number of experiments to optimize the process.

4.1. (The DABOTEK stud welding) machine

The experimental work was executed using the DABOTEK stud welding device. The welding current can be set at five grades, such as (350, 540, 750, 900 and 1250 Amperes). The welding time can be set at grades of 0.05 seconds (from 0.05 seconds to 1 second). The machine that was used in the experiments is shown in figure (4).

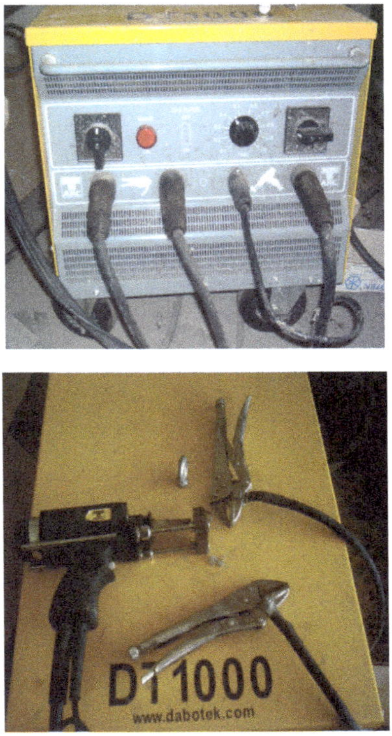

Figure 4. The DABOTEK stud welding machine

4.2. The identification of process factors

Problem identification is critical for any industrial experiment, since the experimental and analysis stages are based on this. One of the most frequently used methods for identifying the problem is brainstorming. Brainstorming is an activity that promotes team participation, encourages creative thinking and generates various ideas over a short period of time. For an investigation into the possible causes of undesirable variability in the stud welding process, the researcher modified a cause and effect diagram that lists several suspected causes of this variability. Figure (5) illustrates the cause and effect of the problem under study. The researcher used brainstorming in conjunction with Cause and Effect Analysis (CEA) to identify the control factors which are to be considered for the experiment.

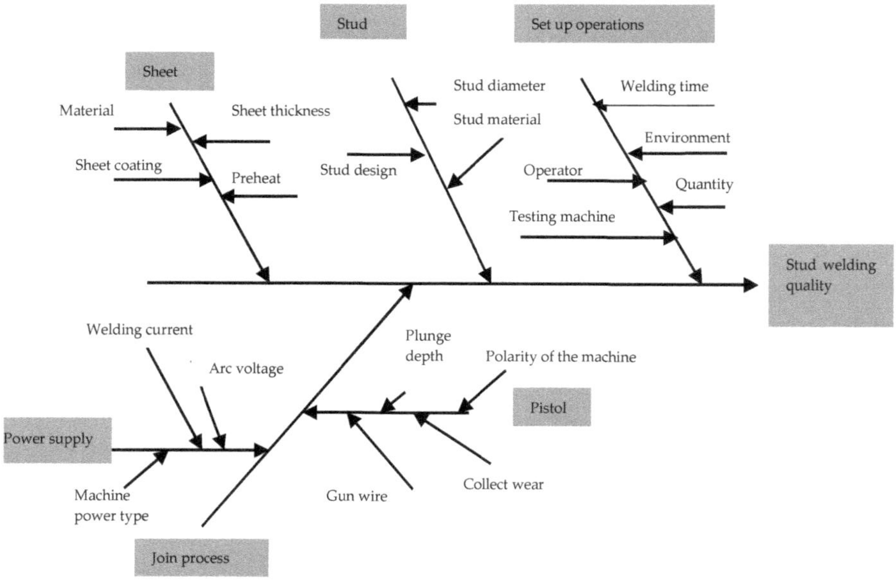

Figure 5. The suggested stud welding cause and effect diagram

Figure (5) shows that many factors play an important role in the stud welding process; they are separated into five main groups:

1. The sheet group

The factors that can be distinguished for these groups are:

- Sheet material.
- Sheet thickness.
- Sheet coating.
- Sheet preheating.

2. The stud group

The factors that can be distinguished for this group are:

- Stud design.
- Stud material.
- Stud diameters.

3. The welding machine group

The factors that can be distinguished for this group are:

- The power supply properties (voltage, current, machine power type (Continuous Electric Arc or Direct Capacitor Arc)).

- The pistol properties (gun wear (new or used), the polarity of machine and the gun wire length).

4. The setup welding operations group

The factors that can be distinguished for this group are:

- The welding time adjustment.
- The quantity of the studs to be welded.
- The operator performance.
- The environment.

5. The arc machine pistol group

The factors that can be distinguished for this group are:

- The polarity of the machine.
- The plunge depth.
- The gun wire.
- The collect wear.

To implement the experimental welds samples, eight independent control factors were chosen to improve the stud welding process. These factors are: welding time, sheet thickness, sheet material, welding current, stud design, stud material, preheat sheet and surface cleaning.

4.3. Selection of the factor levels and the range of factor settings

The selection of a number of levels depends upon how the outcome (tensile strength) is affected due to the different level settings. The levels for control factors are shown in table (2).

				Sheet material (C)															
				K52355								K14358							
Thickness mm (B)	symbol	Process factor	unit	Level 1	Level 2	Level 3	Level 4	Level 5	Level 6	Level 7	Level 8	Level 1	Level 2	Level 3	Level 4	Level 5	Level 6	Level 7	Level 8
1.6	(A)	Welding time	(second)	0.15	0.2	0.25	0.3	0.35	0.4	0.45	0.5	0.15	0.2	0.25	0.3	0.35	0.4	0.45	0.5
	(D)	Welding current	Ampere	350	540							350	540						

Thickness mm	symbol	Process factor	unit	Sheet material (C)															
				K52355								K14358							
				Level 1	Level 2	Level 3	Level 4	Level 5	Level 6	Level 7	Level 8	Level 1	Level 2	Level 3	Level 4	Level 5	Level 6	Level 7	Level 8
	(E)	Stud design	None	Small stud	Flange stud							Small stud	Flange stud						
	(F)	Stud material	None	54NiCrMoS6	40CrMnMoS8-6							54NiCrMoS6	40CrMnMoS8-6						
	(G)	Preheating	None	Preheating	No preheating							Preheating	No preheating						
	(H)	Surface cleaning	None	Oil sheet	Clean sheet							Oil sheet	Clean sheet						
3.175	(A)	Welding time	(second)	0.15	0.2	0.25	0.3	0.35	0.4	0.45	0.5	0.15	0.2	0.25	0.3	0.35	0.4	0.45	0.5
	(D)	Welding current	Ampere	350	540							350	540						
	(E)	Stud design	None	Small stud	Flange stud							Small stud	Flange stud						
	(F)	Stud material	None	54NiCrMoS6	40CrMnMoS8-6							54NiCrMoS6	40CrMnMoS8-6						
	(G)	Preheating	None	Preheating	No preheating							Preheating	No preheating						
	(H)	Surface cleaning	None	Oil sheet	Clean sheet							Oil sheet	Clean sheet						

Table 2. The Levels of the welding time control factors for the experiments.

4.4. Method of measurement

The researcher took a sample containing ten pieces for stud welding depending upon the value for the welding time and the current in order to define the variety of the tensile strength of the samples. The results are in table (3). The dot plot for the data is shown in figure (6). The mean is 330.53 N/mm^2, the standard division is 57.560 N/mm^2 and the range is 189.90 N/mm^2.

Piece Number	1	2	3	4	5	6	7	8	9	10
Tensile Strength (N/mm^2)	310.5	377.8	352.1	243.1	350.3	342.4	253.8	354.6	432.4	289.7

Table 3. The tensile strength of the samples before the experiments

Figure 6. The dotplot for the observed data.

4.5. The Orthogonal Array (OA) design

The number of degrees of freedom required for the experiment must be greater than 14 (7+7). A Taguchi $L_{16}2^{7}1^{8}$ orthogonal array (OA) design, with seven in two levels and one in eight levels is shown by table (4) for the code design matrix.

Run	Welding time	Sheet thickness	Sheet material	Welding current	Stud design	Stud material	Preheat	Surface cleaning
1	1	1	1	1	1	1	1	1
2	1	2	2	2	2	2	2	2
3	2	1	1	1	1	2	2	2
4	2	2	2	2	2	1	1	1
5	3	1	1	2	2	1	1	2
6	3	2	2	1	1	2	2	1
7	4	1	1	2	2	2	2	1
8	4	2	2	1	1	1	1	2
9	5	1	2	1	2	1	2	1
10	5	2	1	2	1	2	1	2
11	6	1	2	1	2	2	1	2
12	6	2	1	2	1	1	2	1
13	7	1	2	2	1	1	2	2
14	7	2	1	1	2	2	1	1
15	8	1	2	2	1	2	1	1
16	8	2	1	1	2	1	2	2

Table 4. Code design matrix orthogonal array $L_{16}2^{7}1^{8}$.

4.6. Experimental preparation and the process run

In this step, the main task was to construct the uncoded design matrix for the experiment. The uncoded design matrix is shown by table (5).

Run	Welding time	Sheet thickness	Sheet material	Welding current	Stud design	Stud material	Preheat	Surface cleaning
1	0.15	1.6	K14358	350	Small	54NiCrMoS6	Preheat	Clean sheet
2	0.15	3.175	K52355	540	Large	40CrMnMoS8-6	No Preheat	Oil sheet
3	0.2	1.6	K14358	350	Small	40CrMnMoS8-6	No Preheat	Oil sheet
4	0.2	3.175	K52355	540	Large	54NiCrMoS6	Preheat	Clean sheet
5	0.25	1.6	K14358	540	Large	54NiCrMoS6	Preheat	Oil sheet
6	0.25	3.175	K52355	350	Small	40CrMnMoS8-6	No Preheat	Clean sheet
7	0.3	1.6	K14358	540	Large	40CrMnMoS8-6	No Preheat	Clean sheet
8	0.3	3.175	K52355	350	Small	54NiCrMoS6	Preheat	Oil sheet
9	0.35	1.6	K52355	350	Large	54NiCrMoS6	No Preheat	Clean sheet
10	0.35	3.175	K14358	540	Small	40CrMnMoS8-6	Preheat	Oil sheet
11	0.4	1.6	K52355	350	Large	40CrMnMoS8-6	Preheat	Oil sheet
12	0.4	3.175	K14358	540	Small	54NiCrMoS6	No Preheat	Clean sheet
13	0.45	1.6	K52355	540	Small	54NiCrMoS6	No Preheat	Oil sheet
14	0.45	3.175	K14358	350	Large	40CrMnMoS8-6	Preheat	Clean sheet
15	0.5	1.6	K52355	540	Small	40CrMnMoS8-6	Preheat	Clean sheet
16	0.5	3.175	K14358	350	Large	54NiCrMoS6	No Preheat	Oil sheet

Table 5. Uncoded design matrix array $L_{16}2^71^8$

Run	Actual run order	Tensile strength (N/mm²)						Mean N/mm²	Standard deviation N/mm²
1	5	175.73	213.23	143.66	195.09	210.50	155.60	182.302	28.860
2	9	288.70	251.20	330.40	284.99	225.90	300.70	280.315	36.946
3	13	284.39	198.56	225.89	245.87	276.24	263.54	249.082	32.539
4	3	359.99	420.50	428.42	300.03	387.38	367.54	377.310	46.790
5	12	190.70	245.87	235.90	298.46	164.33	289.46	237.453	52.977
6	11	370.45	392.68	191.74	360.38	288.70	383.26	331.202	77.637
7	8	321.60	139.00	349.05	310.00	362.93	457.50	323.375	104.318
8	1	331.96	326.32	331.15	401.60	387.26	314.78	348.828	36.095
9	4	388.10	233.60	372.20	287.95	225.43	278.00	297.547	68.611
10	2	530.00	460.72	549.85	375.12	410.53	375.89	450.352	76.343
11	15	305.40	383.20	456.00	378.00	478.00	375.00	395.933	62.388
12	7	152.09	160.74	170.76	166.80	250.88	132.45	172.287	40.835
13	16	219.19	152.97	250.85	257.16	266.78	198.75	224.283	43.258
14	10	155.65	180.45	289.40	220.68	225.35	248.78	220.052	47.705
15	14	289.36	215.62	318.43	256.84	288.23	145.63	252.352	62.900
16	6	185.32	178.45	223.21	155.82	298.33	188.43	204.927	50.651

Table 6. Tensile strength of the samples.

5. Results, analysis and discussions

The results of the experiments conducted depend upon the $L_{16}2^{7}1^{8}$ OA with randomized order, as shown in table (6).

5.1. Determination of the optimum condition for the process

One objective is to reduce the variability in the tensile strength and to bring the mean as close as possible to the target. The target is 728.48 N/mm^2, which is the tensile strength of the stud. The optimization procedure by Taguchi for the study is:

Stage (1): Calculate the SNR for each experimental design point. The SNR for the larger-the-best quality characteristic is calculated by equation (12). Substitute the values into the above equation. The SNR values for the experimental trials are shown in table (7).

Trial no.	1	2	3	4	5	6	7	8	9	10	11	12	13	14	15	16
S/N (dB)	44.9	48.7	47.7	51.3	46.9	49.4	48.1	50.7	48.9	52.7	51.6	44.2	46.5	46.3	47.0	45.7

Table 7. The SNR values for the experimental trials.

After obtaining the SNR values, the next step was to obtain the average response values of a SNR at low and high levels of each factor and, hence, the effect of each factor on the SNR. The results are shown in tables (8) and (9).

Factor A	Average SNR at level 1	Average SNR at level 2	Average SNR at level 3	Average SNR at level 4	Average SNR at level 5	Average SNR at level 6	Average SNR at level 7	Average SNR at level 8	Effect of the factor	rank
Factor Effect dB	46.83	49.53	48.19	49.43	50.84	47.96	46.41	46.38	4.52	1

Table 8. Average SNR table for factor A.

Factors	Average SNR at level 1 dB	Average SNR at level 2 dB	Effect of the factor dB	Rank
B	47.73	48.69	0.96	6
C	47.10	49.31	2.21	2
D	48.18	48.23	0.05	8
E	48.23	48.46	0.23	7
F	47.41	49.00	1.69	3
G	48.98	47.43	-1.65	4
H	47.55	48.86	1.31	5

Table 9. Average SNR table for factors (B, C, D, E, F, G and H).

Tables (8) and (9) show that factors A and C have a dominant effect on the SNR, followed by factors F, G, H, B, E and D. The main effects plot for the SNR is shown in figure (7).

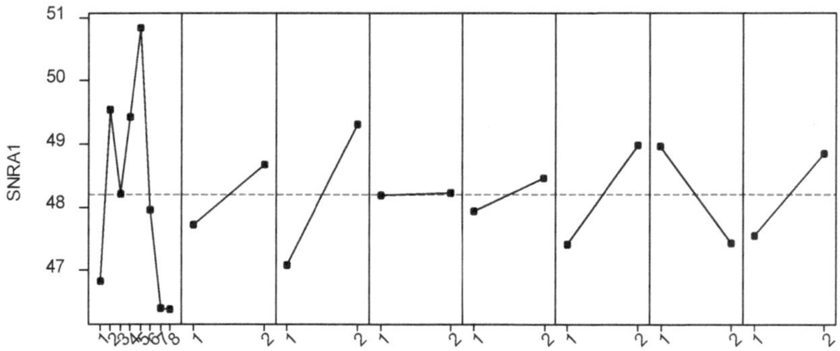

Figure 7. The main effects plot for the S/N ratio.

The calculations of ANOVA for the factors using the Minitab software package are shown in table (10):

Source of variation	Sum of Squares	df	Mean Square	F-ratio
A	37.384	7	5.341	0.88
B	3.529	1	3.529	0.58
C	19.769	1	19.769	3.26
D	0.004	1	0.004	0.00
E	1.129	1	1.129	0.19
F	9.899	1	9.899	1.63
G	9.402	1	9.402	1.55
H	6.679	1	6.679	1.10
error	6.070	1	6.070	1
Total	93.865	15	6.257	

Table 10. ANOVA for the SNR

The second column in Table (10) was calculated using equations 3, 4 and 5, the fourth column with equation 6 and the fifth column from equation 7. The ANOVA table has shown that the most dominant factor effects are D (welding current), E (stud design) and A (welding time). The optimal condition settings of the factors, which will maximize the SNR (i.e., the best control factor settings) based on the SNR are A_5, B_2, C_2, D_2, E_2, F_2, G_1 and H_2.

Trial no.	1	2	3	4	5	6	7	8	9	10	11	12	13	14	15	16
S N/mm^2	28.8	36.9	33.1	46.7	52.9	77.6	104.3	36.1	68.6	76.3	62.3	40.8	43.2	47.7	62.9	50

Table 11. The standard deviation values for the experimental trials.

The following step studies the effect of the factors on the standard deviation (S) of the process. The standard deviation for each experimental design trial is shown in table (11). The average response effect values of factor A on the standard deviation is shown in table (12). The low and high levels of the other factors are shown in table (13).

Factor A	Average St. at level 1	Average St. at level 2	Average St. at level 3	Average St. at level 4	Average St. at level 5	Average St. at level 6	Average St. at level 7	Average St. at level 8	Effect of the factor	Rank
Factor Effect N/mm²	32.9	39.6	65.3	70.2	72.4	51.6	45.4	56.7	39.5	1

Table 12. The average standard deviation for factor A.

Factors	Average St at level 1 N/mm²	Average St at level 2 N/mm²	Effect of the factor N/mm²	rank
B	56.98	51.62	-5.36	6
C	54.27	54.32	0.05	8
D	50.56	58.04	7.48	5
E	49.80	58.79	8.99	4
F	46.01	63.1759	16.58	2
G	51.75	56.84	5.09	7
H	59.70	48.9	-10.8	3

Table 13. The average standard deviation for the factors (B, C, D, E, F, G and H).

Tables (12) and (13) show that factors A and F have a dominant effect on the St, followed by factors H, E, D, B and C. The main effects plot for the St is shown in figure (8).

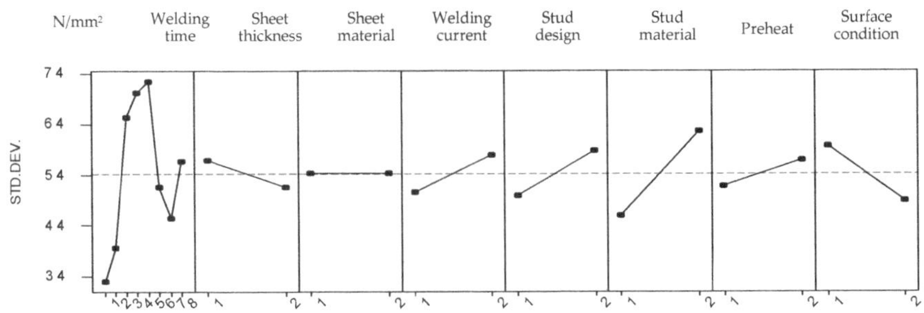

Figure 8. The main effects plot for the standard deviation.

In order to obtain the statistical significance of the effects, the ANOVA table for the standard deviation was performed, as shown in table (14).

Source of variation	Sum of Squares	df	Mean Square	F-ratio
A	2935.4	7	419.34	0.538
B	114.7	1	114.7	0.147
C	0.0	1	0.0	0.00
D	224.1	1	224.1	0.287
E	323.3	1	323.3	0.415
F	1100.6	1	1100.6	1.413
G	103.7	1	103.7	0.133
H	467.2	1	467.2	0.599

Source of variation	Sum of Squares	df	Mean Square	F-ratio
error	778.9	1	778.9	1
Total	6047.9	15	403.193	

Table 14. ANOVA for the standard deviation.

It can be seen from table (13) that C (sheet material) has a large affect on the tensile strength's standard deviation, while F (Stud material) has less of an effect. The next step was to determine the optimal settings for these factors that will minimize the standard deviation. The optimum conditions (i.e., the best control factor settings) based on the standard deviation are A_1, B_2, C_2, D_1, E_1, F_2, G_2 and H_2. Comparing this result with the result of the SNR setting, it was found that for factors B, C, F and H it was the same. Meanwhile, for factor A it was found that there was a big difference in the values between the two choices and that A_6 gets a balance between the two criteria. For factor D, the effect of this factor on the SNR was very small though it had more of an effect on the standard deviation - as such, the choice for the factor level is D_1. The same holds for factor E and so the choice for this factor level is E_1. For factor G, the effect of this factor on the SNR is less than on the standard deviation - thus, the level of this factor is G_2. After analysing SNR, the standard deviation tables for the best settings for the factor levels were:

$$A_6, B_2, C_2, D_1, E_1, F_2, G_2 \text{ and } H_2$$

Stage (2): Performing the SNR analysis and the standard deviation analysis, the next step was to identify the factor effects that have a significant impact on the mean response. The average response values at each level of factor A and the effects are present in table (15) while the average response values at low and high levels for the other factors and their effects are present in table (16).

Factor A	Average mean at level 1	Average mean at level 2	Average mean at level 3	Average mean at level 4	Average mean at level 5	Average mean at level 6	Average mean at level 7	Average mean at level 8	Effect of the factor	Rank
Factor Effect N/mm^2	231.3	313.1	284.3	336.1	382.3	284.1	222.1	228.6	160.6	1

Table 15. The average response of the welding time control factor.

Factors	**Mean response at level 1** N/mm^2	**Mean response at level 2** N/mm^2	**Effect** N/mm^2	**Rank**
B	270.29	298.16	29.96	6
C	257.07	313.47	56.4	3
D	278.73	291.81	13.08	8
E	278.43	292.11	13.68	7
F	255.61	314.93	59.32	2
G	310.17	260.37	-49.8	4
H	269.55	300.99	31.44	5

Table 16. The average response values at each level of the factors (B, C, D, E, F, G and H) and their effects are shown in figure (7).

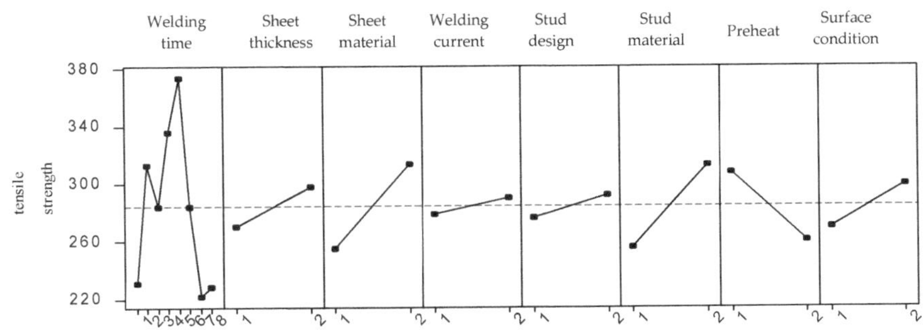

Figure 9. Main effects plot for the mean response.

Figure 7 shows that factors A, C, E and F have a significant impact on the mean response (i.e., the mean tensile strength).

Source of variation	Sum of Squares	Df	Mean Square	F-ratio	% contribution (ρ)
A	42644	7	6092	42.35	40.11
B	3107	1	3107	21.6	2.92
C	13686	1	13686	95.14	12.87
D	482	1	482	3.35	0.51
E	9099	1	9099	63.25	8.55
F	13095	1	13095	91.04	12.31
G	9099	1	9099	63.25	8.55
H	3444	1	3444	23.94	3.23
error	11651	81	143.84	1	10.95
total	106307	95	1119.02	-	100

Table 17. ANOVA for the response.

It can be seen from table (17) that factor A (welding time) has a large affect on the mean of the stud welding tensile strength (40.11% fraction of importance) - see equations 10 and 11. The factors C (sheet material) and F (stud material) have just (12.87%) and (12.31%) respectively. The added factors B, D, E, G and H can be pooled. A new table without these factors was constructed as table (19). The sum of the squares of the pooled factors was added to the error term. The new mean square of the error term was calculated using equation:

$$V_e = \frac{\sum_i s_i^p + S_e}{\sum_i f_i^p + f_e} \quad (14)$$

where the superscript p indicates the pooled factors.

Since the degree of freedom of factor A is 7 and that of the error term is 86, from F-table at a level of significance of (95% confidence) we obtain $F_{7,86} = 2.11$.

Source of variation	Sum of Squares	df	Mean Square	Variance ratio (F-ratio)	% contribution (ρ)
A	42644	7	6092	14.2	40.36
C	13686	1	13686	31.91	13.5
F	13095	1	13095	30.53	12.89
error	28779	86	428.86	1	33.25
total	98204	95	1033.72		100

Table 18. The pooled ANOVA for the response.

Because the computed values of the variance ratio in table (18) are bigger than the value from the F–table, there is a 95% degree of confidence that this factor (welding time) has an effect on the stud welding process. For factors C and F, the degree of freedom is 1; as such, $F_{1,86}$= 3.97, since the computed F-ratio is 31.91 and 30.53 respectively is higher than that from F-table, then these two factors also have an effect on the stud welding process. After identifying the significant factor effects, the next step was to determine the optimal settings for these factors that will bring the mean response as close as possible to the target. The optimum condition (i.e., the best control factor settings) based on the mean response figure was:

$$A_5, B_2, C_2, D_2, E_2, F_2, G_1 \text{ and } H_2$$

Here, factors B, C, F and H are the same as with the last setting. Meanwhile, for factor A there is significant difference when we choose A_5 or A_6, and when we choose A_5 (the welding time is 0.35 seconds) the tensile strength will be 382.341N/mm^2 and the standard deviation will be 72.47 N/mm^2. Furthermore, when choosing A_6 (the welding time is 0.4 seconds) the tensile strength will be 284.110 N/mm^2 and the standard deviation will be 51.61N/mm^2. Because the welding time is a continuous value, the researcher's choice of the new level for this factor will be intermediate between 0.35 and 0.4 seconds, namely $\hat{A}_6$=0.38 second. For factor D, the effect for the standard deviation of this factor is more and opposite to that for the mean. As such, the level for this factor is D1. The same applies for factor E_1. For factor G, the effect of this factor on the mean is more and opposite that for the standard deviation. Thus, the level of this factor is G_1. The factor levels are:

$$\hat{A}_6, B_2, C_2, D_1, E_1, F_2, G_1 \text{ and } H_2$$

In order to arrive at the optimal factor settings, the factor setting is the one which yields the minimum quality loss. The Taguchi quality losses function for the larger-the-better is shown in equation (13). The summarized calculation is shown in table (19).

Run	1	2	3	4	5	6	7	8	9	10	11	12	13	14	15	16
$(\hat{y})^2$	**299094.4**	202216.8	230924.9	125509.6	243914.1	163858.9	174992.3	145438.5	190410.7	83183.4	160655.2	290938.1	256371.3	260776.3	230653.3	276680.6
L(y)/K (money unit/piece)	**3.3*10^{-6}**	4.9*10^{-6}	4.33*10^{-6}	7.97 *10^{-6}	4.1*10^{-6}	6.1*10^{-6}	5.71*10^{-6}	6.89*10^{-6}	5.25*10^{-6}	1.2*10^{-5}	6.22*10^{-6}	3.43*10^{-6}	3.9*10^{-6}	3.83*10^{-6}	4.33*10^{-6}	3.61*10^{-6}

Table 19. The loss function calculation.

From table (19), run (1) (represented in bold) yields the minimum loss. The optimal factor settings based on the loss-function analysis was, therefore, obtained as:

A_1, F_1, C_1, G_1 and H_1

For factor A, level 1 will yield a very low tensile strength (182.302N/mm^2), so this level is not take. For the three factors F, C and G, the level is the same. For factor H in level 1, the tensile strength is (269.55N/mm^2), while in level 2 it is (300.99N/mm^2). The reduction is also high, so the final optimum stetting is:

$\hat{A}'_6$, B_2, C_2, D_1, E_1, F_2, G_1 and H_2.

These factors are summarized in table (20).

factor	$\hat{A}'_6$: welding time	B_2 :sheet thickness	C_2 :sheet material	D_1: welding current	E_1: stud design	F_2: stud material	G_1	H_2: Surface cleaning
level	0.38 second	3.175 mm	non-galvanized (K14358 steel)	350 Ampere	Small stud	40CrMnMoS8-6 steel	Preheating	Clean sheet

Table 20. The optimum stud welding condition based on Taguchi methodology optimization.

The predicted mean response at the optimal conditions is estimated only from the significant main and interaction effects. For the study, the main factor effects which have a significant impact on the mean response were A, F, C, G and H. The predicted mean response based on the optimal factor levels of A, F, C, G and H is given by:

$$R= T+ (\hat{A}'_6-T) + (C_2-T) + (F_2-T) + (G_1-T) + (H_1-T) \tag{15}$$

Where

R = predicted mean response at the optimal condition
T = overall mean of all observations in the data

Then: R = 284.225 + (310.5-284.225) + (313.47-284.225) + (314.93-284.225) + (310.17-284.225) + (300.99-284.225)

$$R=413.185 \text{ N/mm}^2$$

5.2. Experimental conclusions and the confidence interval for the predicted mean response

The confidence interval (CI) is the variation of the estimated result at the optimum condition, calculated as:

$$99 percentCI = R \pm \sqrt{\frac{F \times MSE}{Ne}} \tag{16}$$

MSE = error variance =143.84 N/mm^2, $F_{1,\,96}$ = 3.96, $N_e = \dfrac{96}{7+1+1+1+1+1} = 8$

Therefore, the 99 % confidence interval for the mean tensile strength is given by:

$$99 percent CI = 413.185 \pm \sqrt{\frac{3.96 \times 143.84}{8}}$$

= 413.185 ±8.43 N/mm^2

Accordingly, the result at the optimal condition is 413.185±8.43 N/mm^2 at the 99 % confidence level.

5.3. Confirmation run

A confirmatory run is necessary in order to verify the results from the statistical analysis. A confirmatory run should be carried out to confirm the optimal factor settings obtained from step 10. A sample taken contains ten pieces were produced under the optimal condition that is in table (21):

Sample	1	2	3	4	5	6	7	8	9	10
Tensile strength N/mm^2	443.52	421.32	410.63	390.48	472.40	422.67	398.93	431.88	408.33	524.55

Table 21. The sample tensile strength based on Taguchi methodology optimization

The mean tensile strength from the confirmation run was 432.47 N/mm^2; the standard deviation is 39.950 N/mm^2 and the range is 134.07 N/mm^2. The distribution of this data is explained in figure (10):

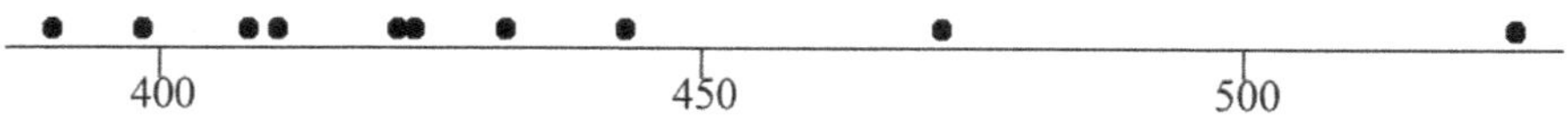

Figure 10. Dot plot for the sample at optimal condition

6. Conclusion

The reduction in the standard deviation was approximately (30.06 %) while for the range the reduction was approximately (29.39%). On the other hand, the increase in the tensile strength mean was approximately (30.84 %). The tensile strength of stud welding process is mostly affected by welding time factor, followed sheet coating factor and stud material factor. The specific conclusions from this study are as follows:

- Dominant factors in the performance of stud welds – the performance of stud welds in this (welding time), (sheet material) and (stud material) dominated the study. In this case, the attached sheet thickness was found to be the dominant variable, with the thicker material demonstrating nearly double the strength compared to using the thinner material. In such cases, thicker materials will have implied higher strengths.

This, in fact, appears to be the case with tensile strengths varying nearly in proportion to the attached sheet's thickness

- Effect of preheating the sheet —preheating has positive effects on increasing the tensile strength while reducing variability.
- Effect of stud design — increasing the stud area appeared to decrease the measures of mechanical performance. This was true even though the levels of internal porosity also increased with the larger studs.
- Effect of sheet thickness — increasing thickness led to increases in the mechanical measure (tensile strength) of the weld quality. The benefits appeared to come from the increased stiffness of the joint as well as the increased peel strengths associated with the thicker material.
- Effect of the sheet material — welding onto galvanized sheets appears to result in substantial porosity in the joint; as such, the non-galvanized sheets have better tensile strength.

7. Future work

There are two tracks to be followed for the use of the proposed Taguchi experimental design. First, to use the output of the experiment as an input for artificial intelligence techniques - like neural networks and fuzzy logic - to get a processes relationship between inputs and outputs. In particular, if this relationship between input and output cannot be represented by lower-order equations, then these techniques can result in accurate factor levels for optimization.

Second, to extend the work of this chapter in multi-objective optimization. This could be optimized with respect to torque testing and bending testing.

Author details

Riyadh Mohammed Ali Hamza
Gulf University, Kingdom of Bahrain

8. References

Anderson N. S. & Meinheit D. F. (2000) "Design criteria for headed stud groups in shear: part1-steel capacity and back edge effects." *PCI Journal*, pp. 48-75.

Antony J. & Kaye M. (1999) "Experimental Quality: A Strategic Approach to Achieve and Improve Quality." Kluwer Academic Publishers, London, ISBN: 9780412814402.

Behrens B., Grub D. & Jenicek A. (2011) "Stud welding within sheet metal working tools." *Prod. Eng. Res. Devel.*, Vol. 5, pp. 283–292.

Bursi O. S. & Gramola G. (1999) "Behaviour of headed stud shear connectors under low cycle high amplitude displacements." *Materials and structures*, Vol. 32, May, pp. 290-297.

Buyske S. & Trout R. (2003) "Robust Design and Taguchi Methods." *Journal of Quality Technology*, Vol. 22, No. 1, January pp. 15–22.

Chambers H. A. (2001) "Principles and Practices of stud welding." *PCI Journal*, September-October, pp. 46-58.

Conti T., Kondo Y. & Watson G. H. (2003) "Quality into the 21st Century: Perspectives on Quality and Competitive for Sustained Performance." *American Society for Quality*, William A. Tony, USA.

Eşme U. (2009) "Application of Taguchi method for the optimization of resistance spot welding process." *The Arabian Journal for Science and Engineering*, Vol. 34, No. 2B, pp. 519-528.

Hsu C., Mumaw J., Thomas J. & Maria P. (2008) "Robotic Stud Welding Process Optimization with Designed Experiment." *Welding Journal*, October, Vol. 87, pp 265-272

Hsu C. & Mumaw J. with Nelson Stud Welding, Inc. (2011) "Weldability of Advanced High-Strength Steel Drawn Arc Stud Welding." *Welding Journal*, Vol. 90, March, pp. 45-53

Hsu C., Phillips D., Champney C. & Mumaw J. (2007) "Portable and Intelligent Stud Welding Inverter for Automotive and Sheet Metal Fabrication." *Robotic Welding, Intelligence and Automation*, Vol. 362, pp. 367–374.

Klarić Š., Kladarić I., Kozak D., Stoić A., Ivandić Ž. & Samardžić I. (2009)" The influence of the stud arc welding process parameters on the weld penetration." Fascicle: *Mechanics, Tribology, Machine Manufacturing Technology*, Serie C, Volume XXIII.

Lee J. S., Ryu Y. S., Kim N. I., Kim B. J., Kim Y. K. & Kim M. H. (2009)" Stud welding for fixation of cryogenic insulation of membrane tanks in LNG ship building." *Trans. Nonferrous Met. China*, Vol. 19, pp. 271-275. Elsevier Press.

Miller Welds Electrical Mng. Co., July (2005), www.MillerWelds.com.

Hamza R. M. A. (2011) "Multi-objective neural network modeling for improving stud arc welding process joining." *Journal of Engineering, Science and Technology*, School of Engineering, Taylor's University Vol. 6, No. 3, 2011, pp. 382–391.

Ross P. J. (1986) "Taguchi Techniques for Quality Engineering." New York, McGraw-Hill.

Samardžić I. & Klari Š. (2007) "On Line Monitoring For Defects In Electric Arc Stud Welding." 11th International Research/Expert Conference *Trends in the Development of Machinery and Associated Technology*. TMT, Hammamet, Tunisia, 5-9 September 2007.

Schmidt S. R. & Launsby R. G. "Understanding Industrial Designed Experiments." Air Academy Press, Colorado Springs Co., 1992.

Smith C. S. & Gnudi M. T. (1990) "The PIROTECHNIA of Vannoccio Biringuccio, The classic Sixteenth-Century Treatise on Metals and Metallurgy." Dover Publications Inc., New York.

Steiner S. H. & MacKay R. J. (2005) "Statistical Engineering: an Algorithm for Reducing Variation in Manufacturing Processes." American Society for Quality ASQ.

Strigel R. M., Pincheira J. A. & Oliva M. G. (2000) "Reliability of 3/8 in. stud-welded deformed bar anchors subject to tensile loads." *PCI Journal*, November-Decembers 2000, pp. 72-82.

Taguchi G., Chowdhury S. & Wu Y. (2005) "Taguchi's Quality Engineering Handbook." John Wiley & Sons Inc. New Jersey.

Taylor J. (2001) "AISC-An Engineer's guide to fabricating steel structure." *Fabrication methods Australian institute of steel construction*, Vol. 1, John Taylor, pp. 107-109.

Permissions

The contributors of this book come from diverse backgrounds, making this book a truly international effort. This book will bring forth new frontiers with its revolutionizing research information and detailed analysis of the nascent developments around the world.

We would like to thank Radovan Kovacevic, Ph.D., for lending his expertise to make the book truly unique. He has played a crucial role in the development of this book. Without his invaluable contribution this book wouldn't have been possible. He has made vital efforts to compile up to date information on the varied aspects of this subject to make this book a valuable addition to the collection of many professionals and students.

This book was conceptualized with the vision of imparting up-to-date information and advanced data in this field. To ensure the same, a matchless editorial board was set up. Every individual on the board went through rigorous rounds of assessment to prove their worth. After which they invested a large part of their time researching and compiling the most relevant data for our readers. Conferences and sessions were held from time to time between the editorial board and the contributing authors to present the data in the most comprehensible form. The editorial team has worked tirelessly to provide valuable and valid information to help people across the globe.

Every chapter published in this book has been scrutinized by our experts. Their significance has been extensively debated. The topics covered herein carry significant findings which will fuel the growth of the discipline. They may even be implemented as practical applications or may be referred to as a beginning point for another development. Chapters in this book were first published by InTech; hereby published with permission under the Creative Commons Attribution License or equivalent.

The editorial board has been involved in producing this book since its inception. They have spent rigorous hours researching and exploring the diverse topics which have resulted in the successful publishing of this book. They have passed on their knowledge of decades through this book. To expedite this challenging task, the publisher supported the team at every step. A small team of assistant editors was also appointed to further simplify the editing procedure and attain best results for the readers.

Our editorial team has been hand-picked from every corner of the world. Their multi-ethnicity adds dynamic inputs to the discussions which result in innovative outcomes. These outcomes are then further discussed with the researchers and contributors who give their valuable feedback and opinion regarding the same. The feedback is then collaborated with the researches and they are edited in a comprehensive manner to aid the understanding of the subject.

Apart from the editorial board, the designing team has also invested a significant amount of their time in understanding the subject and creating the most relevant covers. They scrutinized every image to scout for the most suitable representation of the subject and create an appropriate cover for the book.

The publishing team has been involved in this book since its early stages. They were actively engaged in every process, be it collecting the data, connecting with the contributors or procuring relevant information. The team has been an ardent support to the editorial, designing and production team. Their endless efforts to recruit the best for this project, has resulted in the accomplishment of this book. They are a veteran in the field of academics and their pool of knowledge is as vast as their experience in printing. Their expertise and guidance has proved useful at every step. Their uncompromising quality standards have made this book an exceptional effort. Their encouragement from time to time has been an inspiration for everyone.

The publisher and the editorial board hope that this book will prove to be a valuable piece of knowledge for researchers, students, practitioners and scholars across the globe.

List of Contributors

J. Zhou
Department of Mechanical Engineering, Pennsylvania State University, The Behrend College, USA

H.L. Tsai
Department of Mechanical and Aerospace Engineering, Missouri University of Science and Technology, USA

Abdel-Monem El-Batahgy
Manufacturing Technology Department, Central Metallurgical R & D Institute, Cairo, Egypt

Hitoshi Ozaki
Graduate School of Engineering, Mie University, Japan

Muneharu Kutsuna
Advanced Laser Technology Research Center Co., Ltd., Japan

Junjie Ma, Fanrong Kong and Radovan Kovacevic
Center for Laser-aided Manufacturing, Lyle School of Engineering, Southern Methodist University,
Dallas, TX, USA

Blair Carlson
General Motors R&D Center, Warren, MI, USA

Kelvii Wei Guo
MBE, City University of Hong Kong, Hong Kong, China

Ikhwan Naim Md Nawi
Centre of Foundation Studies & Faculty of Applied Sciences, Universiti Teknologi MARA, Malaysia

Jalil Ali
Universiti Teknologi, Malaysia

Mohamed Fadhali
Physics Department, Faculty of Science, Ibb University, Yemen

Preecha P. Yupapin
King Mongkut's Institute of Technology Ladkrabang, Bangkok, Thailand

Zhiguo Gao
University of Manitoba, Canada

Fanrong Kong and Radovan Kovacevic
Research Center for Advanced Manufacturing, Lyle School of Engineering, Southern Methodist University, Dallas, TX, USA

Farag Soul
University of Elmergeb, Faculty of Engineering, Libya

Hamdy Nada
University of Menoufia, Faculty of Engineering, Egypt

Jauhari Tahir Khairuddin and Jamaluddin Abdullah
School of Mechanical Engineering, Universiti Sains Malaysia, Nibong Tebal, Penang, Malaysia

Zuhailawati Hussain
School of Materials and Mineral Resources Engineering, Universiti Sains Malaysia, Nibong Tebal,
Penang, Malaysia

Indra Putra Almanar
Mechanical Engineering Department, Universiti Teknologi Mara, Permatang Pauh, Penang,
Malaysia

Miroslav Mijajlović and Dragan Milčić
University of Niš, Faculty of Mechanical Engineering Niš, Serbia

Marek Stanisław Węglowski
Instytut Spawalnictwa (Institute of Welding), Poland

Wei Jie Zhang, Yu Kang Liu and Yu Ming Zhang
Department of Electrical and Computer Engineering Institute for Sustainable Manufacturing,
University of Kentucky, Lexington, USA

P. S. Wei
Department of Mechanical and Electro-Mechanical Engineering,
National Sun Yat-Sen University, Kaohsiung, Taiwan, China

Clébio Domingues da Silveira-Júnior, Morgana Guilherme de Castro, Letícia Resende Davi,
Flávio Domingues das Neves, Verediana Resende Novais and Paulo Cézar Simamoto-Júnior
Federal University of Uberlândia, Brazil

Riyadh Mohammed Ali Hamza
Gulf University, Kingdom of Bahrain